Widegap II–VI Compounds for Opto-electronic Applications

ELECTRONIC MATERIALS SERIES

Electronic Materials series:
This series, devoted to electronic materials subjects of active research interest, provides coverage of basic scientific concepts as well as relating the subjects to the electronic applications and providing details of the electronic systems, circuits or devices in which the materials are used. The series will be a useful reference source at senior undergraduate and graduate level as well as for research workers in industrial laboratories who wish to broaden their knowledge into a new field.

Series editors:

A.F.W. Willoughby,
Professor of Electronic Materials,
University of Southampton,
UK.

R. Hull,
AT & T Bell Laboratories,
Murray Hill,
USA.

Widegap II–VI Compounds for Opto-electronic Applications

Edited by

Harry E. Ruda

Assistant Professor, University of Toronto, Canada

CHAPMAN & HALL
London · New York · Tokyo · Melbourne · Madras

Published by Chapman & Hall, 2–6 Boundary Row, London SE1 8HN

Chapman & Hall, 2–6 Boundary Row, London SE1 8HN, UK

Chapman & Hall, 29 West 35th Street, New York NY10001, USA

Chapman & Hall Japan, Thomson Publishing Japan, Hirakawacho Nemoto Building, 7F, 1-7-11 Hirakawa-cho, Chiyoda-ku, Tokyo 102, Japan

Chapman & Hall Australia, Thomas Nelson Australia, 102 Dodds Street, South Melbourne, Victoria 3205, Australia

Chapman & Hall India, R. Seshadri, 32 Second Main Road, CIT East, Madras 600 035, India

First edition 1992

Typeset in 10/12 Times by Thomson Press (India) Ltd., New Delhi
Printed in Great Britain by St. Edmundsbury Press, Bury St Edmunds

ISBN 0 412 39100 7

A catalogue record for this book is available from the British Library

Library of Congress Cataloging-in-Publication data available.

Contents

Contributors

Sg. Fujita
Department of Electrical Engineering
Kyoto University
Kyoto
Japan

Sz. Fujita
Department of Electrical Engineering
Kyoto University
Kyoto
Japan

R.L. Gunshor
School of Electrical Engineering
Purdue University
Indiana
USA

M. Isshiki
Department of Materials Science
Tohoku University
Sendai
Japan

M. Kobayashi
School of Electrical Engineering
Purdue University
Indiana
USA

L.A. Kolodziejski
Department of Electrical Engineering and Computer Science
Massachusetts Institute of Technology
Massachusetts
USA

P. Lilley
The Electrical Engineering Department
Manchester University
Manchester
UK

H. Mar
Department of Electrical Engineering
University of Toronto
Ontario
Canada

A. Miller
Department of Physics and Electrical Engineering and
Center for Research in Electro Optics and Lasers
University of Central Florida
Florida
USA

G.F. Neumark
Henry Krumb School of Mines
Columbia University
New York
USA

J.-i. Nishizawa
Semiconductor Research Institute
Sendai
Japan

N. Otsuka
School of Materials Engineering
Purdue University
Indiana
USA

R.M. Park
Department of Materials Science and Engineering
University of Florida
Florida
USA

H.E. Ruda
Department of Metallurgy and Materials Science and
Department of Electrical Engineering
University of Toronto
Ontario
Canada

D. Shaw
Department of Applied Physics
Hull University
Hull
UK

K. Suto
Research Institute of Electrical Communication
Tohoku University
Sendai
Japan

J. Woods
Department of Applied Science and Engineering
University of Durham
Durham
UK

M. Yamaguchi
NTT Opto-Electronics Laboratories
Ibaraki-ken
Japan

A. Yoshikawa
Department of Electrical and Electronics Engineering
Chiba University
Chiba
Japan

Preface

This book is intended for readers desiring a comprehensive analysis of the latest developments in widegap II–VI materials research for opto-electronic applications and basic insight into the fundamental underlying principles. Therefore, it is hoped that this book will serve two purposes. Firstly, to educate newcomers to this exciting area of physics and technology and, secondly, to provide specialists with useful references and new insights in related areas of II–VI materials research. The motivation for preparing this book originated from the need for a current review of this fertile and important field. A primary goal of this book is therefore to present an eclectic synthesis of these sometimes diverse fields of investigation.

This book consists of three main sections, namely (1) *Growth and Properties*, (2) *Materials Characterization* and (3) *Devices*. Part One presents an overall perspective of the state of the art in the preparation of the widegap II–VI materials. Part Two concentrates on current topics pertinent to the characterization of these materials from the unique perspective of each of the authors. Part Three focuses on advances in the opto-electronic applications of these materials. The material in this section runs the gamut from addressing recent advances in device areas which date back to some of the earliest reported research in these materials, to tackling some quite new and exciting future directions.

Part One encompasses research on the preparation of the widegap II–VI materials by both equilibrium and non-equilibrium techniques. Advances in the preparation of both bulk material and epitaxial layers are highlighted. In particular, recent progress in preparing high quality material essential for device development is shown to hinge on two factors, namely close control of source purity, while at the same time placing ever-increasing emphasis on non-equilibrium approaches. Gradually, conventional approaches such as liquid phase epitaxy (LPE) and vapour phase epitaxy (VPE) are being supplanted by techniques such as photo-assisted metal–organic VPE (MOVPE), molecular beam epitaxy (MBE) and metal–organic MBE (MOMBE). In this book, authors report on how the latter techniques are finally reproducibly yielding controllable n- and p-doped epilayers which are stable (to diffusion) and suitable for junction device development. These developments are reported for probably the most widely studied and perhaps historically most unyielding of the widegap materials, ZnSe. Although we are now only at the dawn of a new era of research in these materials, there is finally intense optimism that the vast potential of these elusive materials may now finally begin to be tapped. This first section ends with a chapter which focuses on the most novel developments in the materials area, namely looking at the use of non-equilibrium techniques to micro-engineer structures incorporating the widegap II–VI materials. The fabrication of sophisticated heterojunctions is discussed. In particular, the rapid recent progress in multiple quantum well (QW) structures and strained-layer superlattices (SLSs) based on II–VI/II–IV and II–VI/III–V multilayers is presented. The chapter also touches on the area of dilute magnetic semiconductors (DMSs) and ideas pertaining to the tuning of physical structures and electronic states in the materials. These exciting concepts close this buoyant first section.

Part Two reviews topics on the current state of understanding of characterization of the widegap II–VI materials. Emphasis is placed on establishing the fundamental principles that underlie understanding the material characterization. A review of optical characterization techniques clearly illuminates the power and sensitivity of the techniques as tools for studying materials and helping to advance material quality. The chapter on transport characterization points to our present understanding of the mechanisms which determine transport characteristics. In particular, essential mechanistic factors prove to be common to a wide range of transport characteristics: these include low and high field direct current (d.c.) and alternating current (a.c.) transport characteristics of three-dimensional systems, as well as low-dimensional systems. Optical and transport techniques are highlighted as important tools suitable for helping to identify and quantify impurity-related influences. The chapter on transmission electron microscopy (TEM) shows how this technique now plays a vital role as a key structural characterization tool. Two particular areas of current interest in this field relate to interfacial problems, namely interfacial layer formation and lattice-mismatch-related defect generation. All of these factors can have profound implications on the materials' properties. As increasing attention is now being placed on controlled doping of the widegap II–VI materials, especially by non-equilibrium epitaxial techniques, the need to understand diffusional behaviour of dopants is becoming ever more acute. In this section, the chapters on diffusion, doping and conductivity control summarize current thinking on these problems and conclude that a careful choice of dopant is required. Dopant solubility, electronic state characterization and diffusivity are major current concerns. However, diffusion behaviour itself is shown to be extremely complex in these materials depending not only on concentration gradients but also on stoichiometric deviations. Furthermore, the lack of data pertinent to low-temperature epitaxial systems is anticipated to fuel much future work in this area. Another important future area of activity will probably be associated with looking at intermixing in multilayer II–VI-based heterostructures as research steadily moves in that direction.

Part Three is intended to provide a cross-section of current research progress in some of the most interesting areas of current device research on widegap II–VI materials. Historically, electroluminescent II–VI devices have always been important opto-electronic (OE) devices. They were also probably amongst the first widegap II–VI OE devices that were commercialized. However, as the first chapter in this section points out, present commercial blue electroluminescent light-emitting diodes (LEDs) are still not of sufficient brightness or longevity to satisfy current demands. Hope is pinned on commercializing the recent successes reported by a number of authors on fabricating doped homojunctions. Indeed the second chapter in this section focuses on this topic and presents results for such devices. These authors recognize that although the early thrust was towards blue emitters (as the last of the three primary colours), the present impetus is much broader encompassing injection LEDs and laser diodes suitable for spectral coverage from the green to the ultraviolet (UV). Another approach discussed in this section is that of ion implantation for advanced OE device development. However, it is clear that future work on the selection of appropriate ion species and implantation–annealing conditions will have to be addressed first. Promising work on electron-beam-pumped devices is also reported. Such devices carve out a unique niche, with application to scanning microscopy, photolithography and optical telemetry, for example. A most fascinating chapter in this section is concerned with optical non-linearities and bistability in the widegap II–VIs. Dramatic progress in this area is outlined and, in particular, the encouraging success of optically bistable II–VI devices operating via optothermal non-linearities. Such devices are playing an important role in the development of parallel architectures for optical computing, as well as other potential applications such as image processors and radar array processors, for example. In the future, as widegap II–VI materials technology continues to advance rapidly, we can expect to see hybrid optical switching devices coming to fruition.

Finally, I would like to express my appreciation to all of the contributors to this volume and to Mr Michael Dunn, publisher at Chapman & Hall, for his patience and understanding.

The contributors were selected for their long and continuing expertise in the field of widegap II–VI materials. Since the boundaries of present knowledge have been outlined by each author, we hope that this volume will serve as a stepping stone for future progress.

H.E. Ruda
Toronto
Canada

Part One: Growth and Properties

1

Bulk growth of widegap II–VI single crystals

M. Isshiki

1.1 INTRODUCTION

The most important factor in the application of semiconducting materials is the ability to prepare structurally and chemically pure crystals. Opto-electronic devices made of III–V compound semiconductors are a recent success mainly because they can be formed as large and nearly perfect single crystals. Two other reasons why a high-quality crystal is important are to characterize the compounds and to find a new function. Because the crystals of these compound semiconductors are not perfect but only nearly perfect, it is essential to control their impurities and native defects so the materials can be put to practical use.

Because of their direct gap, semiconducting II–VI compounds have been expected to be the materials applicable to opto-electronic devices. Wide-gap II–VI compounds are especially promising for light-emitting devices in the short-wavelength region of visible light. However, it has been difficult to grow crystals that are structurally and chemically pure because widegap II–VI compounds have high melting points and, at their melting points, the dissociation pressure is high, too.

Another problem with their application to devices is difficulty in controlling the conductive type. Recent improvements to low-temperature epitaxy may make it possible to overcome this problem. Growth of widegap II–VI compound single crystals will become more and more significant for characterizing them, fabricating active devices and supplying the substrates for epitaxial growth.

This chapter discusses the thermodynamics necessary for preparing a well-defined single crystal, techniques applicable to the growth of II–VI compounds and current progress on the single-crystal growth of widegap II–VI compounds with zinc blende structure such as cadmium telluride (CdTe), zinc chalcogenides and their solid solutions. For theories and studies of nucleation and growth, the reader is referred to the literature, because materials preparation and crystal growth will be approached from the practical side.

1.2 PHASE EQUILIBRIA

Since single crystals are grown mainly from vapour phase or liquid phase, equilibria between these phases and solid phase should be known in order to grow single crystals of which the compositions are controlled. For the melt growth, the importance of simultaneous three-phase equilibria (solid–liquid–vapour), which is indicated by *P–T–x* diagram, has been pointed out, but the amount of data available is much more limited than that pertaining to solid–vapour equilibria. Because of the high melting points of II–VI compounds, even the data of their melting points are scattered [1].

T–x diagrams of CdTe [2] and zinc chalcogenides [3] obtained by experiments and/or thermodynamic calculation are shown in Figs 1.1–1.4. ZnSe and ZnS have phase transformations from zinc blende structure to wurtzite structure at 1698 K [4] and 1293 K [5], respectively. (The zinc blende structure is stable at low temperatures.)

In the II–VI systems, phase equilibria of CdTe have been studied most extensively. Figure 1.5 shows the entire existence region of CdTe on an enlarged scale in order to demonstrate a very small existence region [6]. It should be noted that the composition of the compounds at their congruent melting points is generally not exactly stoichiometric. As shown in Figure 1.5, the composition of the liquid phase should be controlled in order to grow a stoichiometric compound.

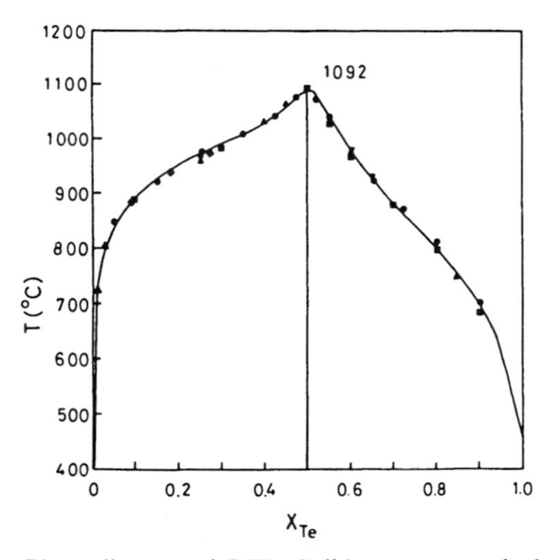

Fig. 1.1 Phase diagram of CdTe. Solid curves are calculated [2].

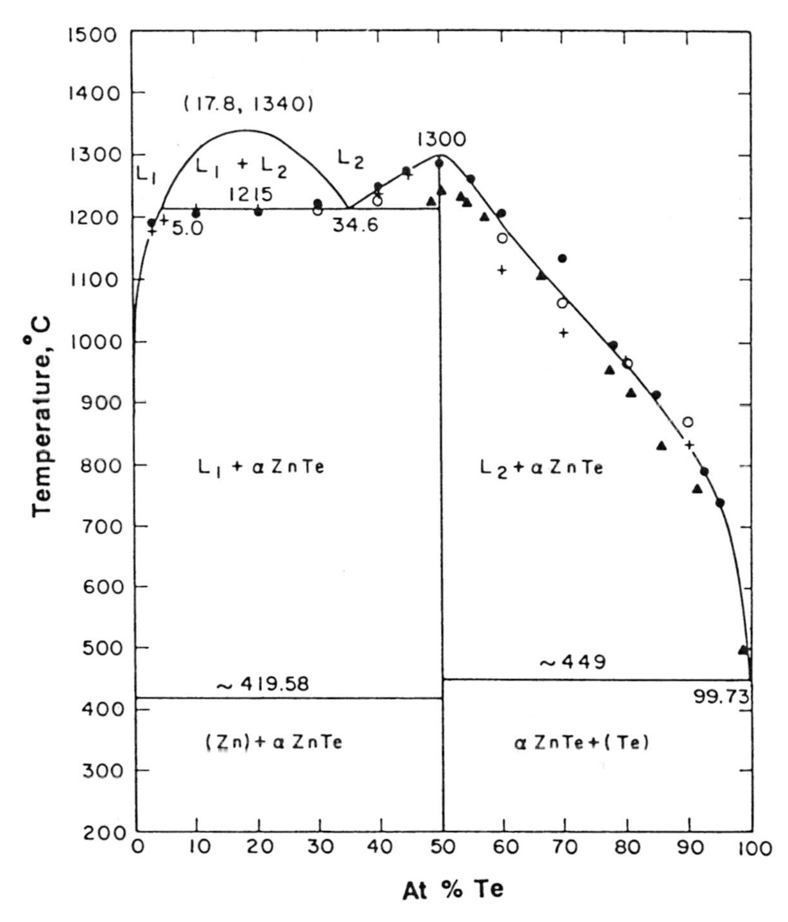

Fig. 1.2 Phase diagram of ZnTe. Solid lines are calculated [3].

Equilibria involving pressure and temperature are generally represented by $\log P$ versus $1/T$ plots. Figure 1.6 indicates the P–T diagram obtained on CdTe [6]. For other systems, similar P–T diagrams are reported by Sharma and Chang [3]. The upper straight line is simply the vapour pressure of pure Cd. Similarly, the lower boundary is established by pure tellurium in terms of P_{Cd}. P^{b}_{Cd} represents the three-phase line, in which the existence region is placed. In this region, the solid phase is stable and equilibrates with the vapour phase.

Because the chalcogen species in the vapour phase is the diatomic molecule, the equilibrium reaction between solid (AB compound) and vapour phases is written as

$$2AB = 2A(g) + B_2(g) \tag{1.1}$$

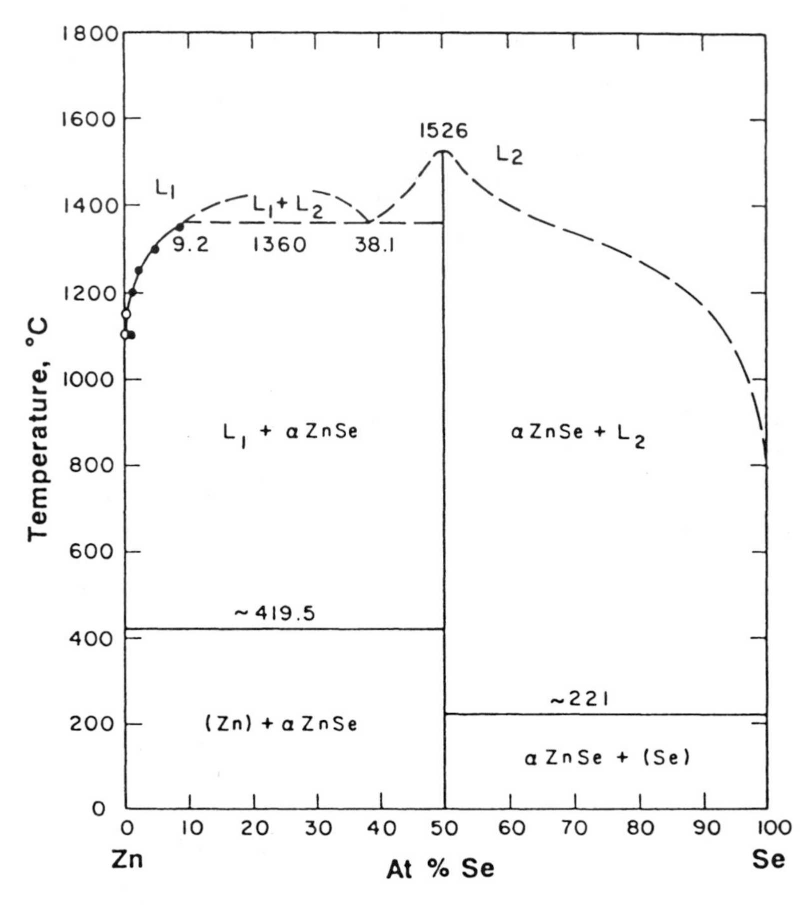

Fig. 1.3 Phase diagram of ZnSe. Solid and broken lines are calculated [3].

The total pressure (P) in the ampoule is given by

$$P = P_A + P_{B_2} = P_A + K \cdot P_A^{-2} \tag{1.2}$$

$$= (K/P_{B_2})^{1/2} + P_{B_2} \tag{1.3}$$

where P_A and P_{B_2} are the partial pressures of group II and VI elements respectively and K $(= P_A^2 \cdot P_{B_2})$ is the equilibrium constant of Equation 1.1.

At any temperature, there is a minimum total pressure (P_{min}), which corresponds to the condition,

$$P_A = 2P_{B_2} = 2^{1/3} \cdot K^{1/3}. \tag{1.4}$$

Under this condition, the vapour phase composition is stoichiometric and the composition of the solid phase equilibrated is defined in the existence region. The values of K are plotted in Fig. 1.7 [7].

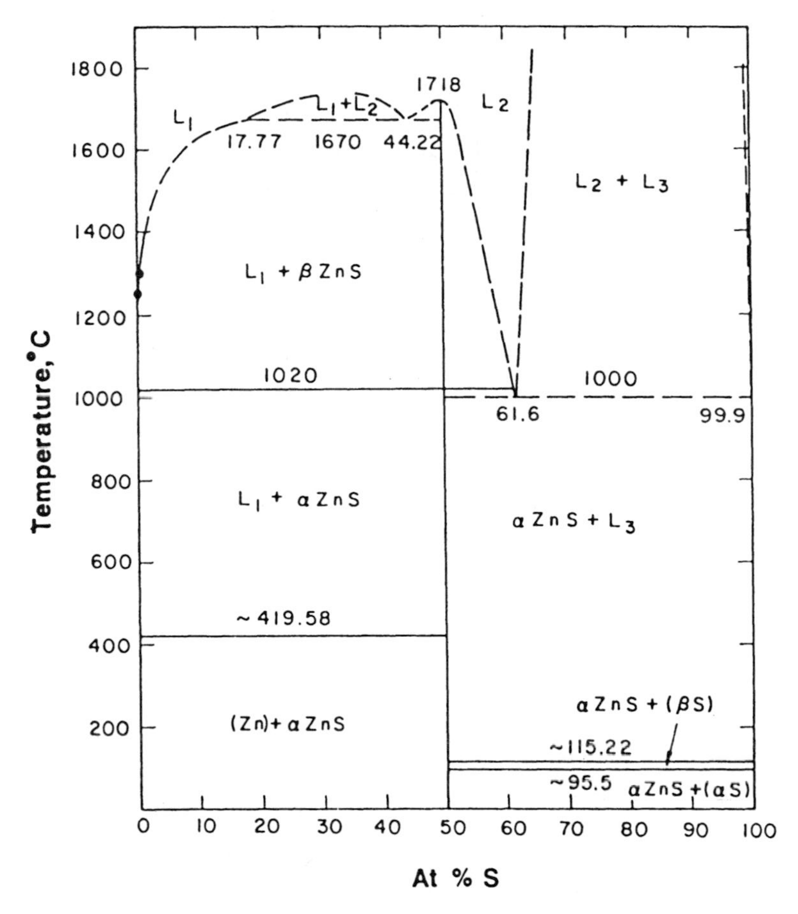

Fig. 1.4 Calculated phase diagram of ZnS [3].

1.3 CRYSTAL GROWTH TECHNIQUE

1.3.1 Growth from liquid phase

(a) Bridgman

Growth by the Bridgman method possesses an advantage that large single crystals are easily obtained due to its high growth rate. Although most of the widegap II–VI compounds have high melting points and high vapour pressures at melting point, all of the zinc and cadmium chalcogenides have been grown by the Bridgman method. Because of the relatively low melting point, CdTe and ZnTe can be melted using quartz ampoule.

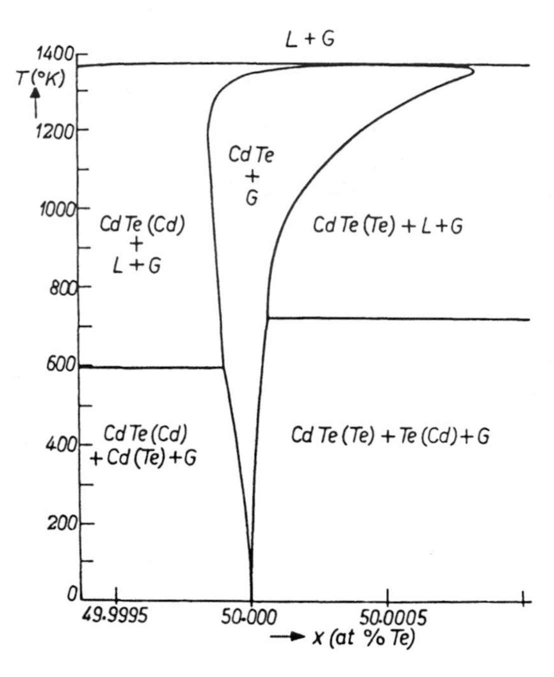

Fig. 1.5 Part of the T–x diagram of Figure 1, showing the entire existence region of solid CdTe [6].

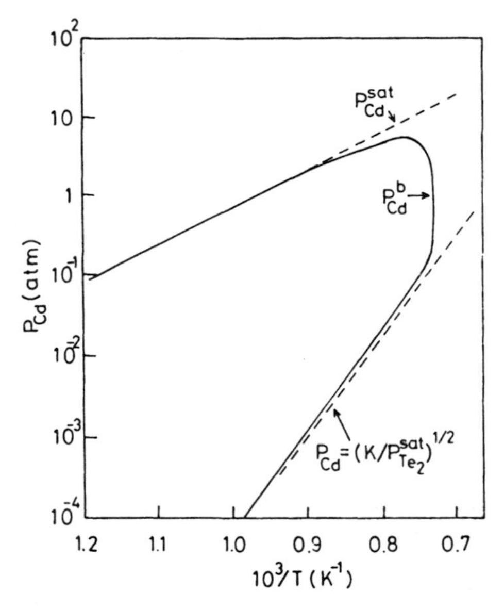

Fig. 1.6 P_{Cd}–T diagram of CdTe [6]. P_{Te}^{sat} is converted to P_{Cd}.

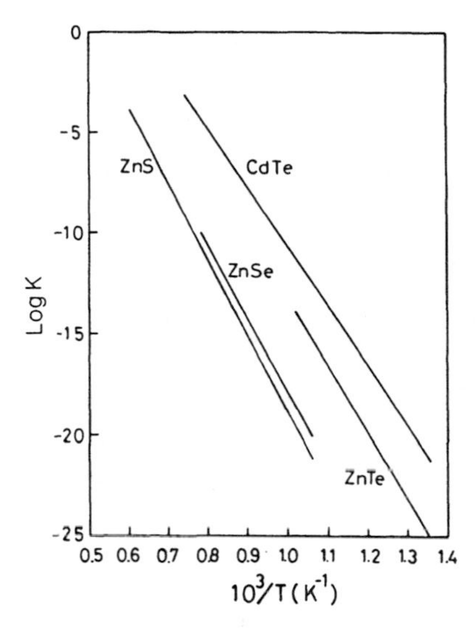

Fig. 1.7 The equilibrium constants for II–VI compounds as a function of temperature [7].

The most straightforward melt growth technique is normal freezing. When freezing is achieved by the use of a two zone furnace, it is called the Bridgman (or Bridgman–Stockbarger) method. Usual configuration is vertical and the ampoule is lowered from the hot zone to the cooler zone which is below the melting point. This is called the Vertical Bridgman method (VB). On the other hand, a horizontal configuration using the boat has been applied for growth of CdTe. This is called the Horizontal Bridgman (HB) method.

There are three ways of moving the solid–liquid interface:

1. The ampoule moves through the temperature gradient.
2. The furnace moves and the ampoule is stationary.
3. The ampoule and furnace are stationary and the temperature is gradually reduced by keeping the temperature gradient at the interface constant.

The third one is called the Gradient Freezing (GF) method.

Nucleation is controlled in various ways as depicted in Fig. 1.8. Seed crystal can be used, but this has rarely been attempted. Twinning and deviation from stoichiometric composition are the main problems against obtaining the crystals with good crystallinity. Growth can be controlled by the growth rate, the shape of the solid–liquid interface, the temperature gradient in the growth zone and the atmosphere in the ampoule. High pressure is applied to prevent loss of the constituent elements by dissociation, especially in the case of compounds with a high melting point.

In the case of growing solid solution or impurity doping, a distribution effect between liquid and solid should be taken into account. If the distribution

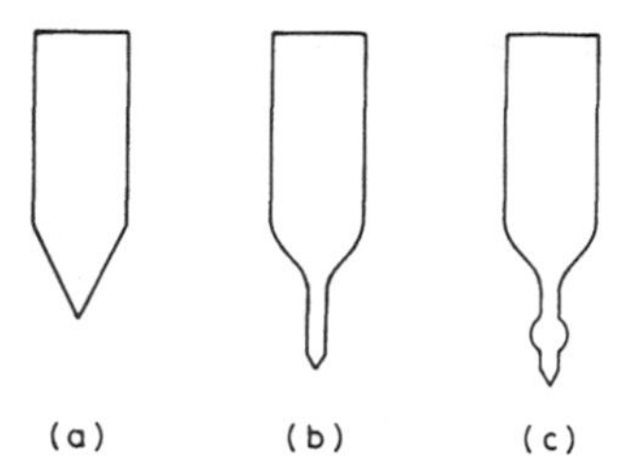

Fig. 1.8 Techniques for nucleating single crystals from the melt: (a) conical bottom, (b) capillary, (c) necking.

coefficient k is not close to unity, the doping element segregates through the crystal and a uniform concentration cannot be obtained.

(*b*) *Travelling heater*

A schematic illustration of the Travelling Heater Method (THM) is shown in Figure 1.9. A molten solvent zone is moved slowly through a solid source material. In this process, the dissolution of feed material occurs at the receding liquid–solid interface (T_2) and the crystallization of dissolved feed material occurs continuously at the advancing interface (T_1).

The major advantage of this method is a lower growth temperature as compared with the growth from the near-stoichiometric melt. Low growth temperature reduces the possible contamination and the defect density in the crystals. In addition to that, THM has the purification effect by the solvent. The drawbacks of this method are low growth rate (5 mm per day) and difficulty to obtain near-stoichiometric compounds. The selection of the solvent material is important. Te solvent has been used mainly for Cd and Zn tellurides.

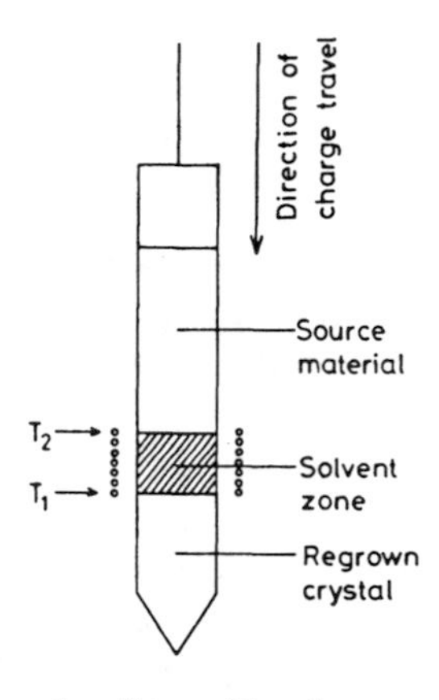

Fig. 1.9 Schematic of travelling heater method (THM).

(*c*) *Zone melting*

Zone melting has been applied to CdTe [8,9] because of its low melting point. The configuration is similar to THM except for the solvent zone. Only a small portion of the charge is melted and the molten zone is moved slowly through the charge. Compared to the Bridgman method, this reduces contamination from the ampoule and evaporation of the melt. Zone melting can grow ternary alloys of relatively uniform composition [8], another advantage.

(*d*) *Liquid encapsulation Czochralski* (*LEC*)

This method, which is extensively used for growing III–V compounds, has been applied only for CdTe and its ternary alloys [10, 11] using B_2O_3 as an encapsulant. But they have a low thermal conductivity so single crystals have not been grown. B_2O_3 reacts with other II–VI compounds, so until a suitable encapsulant is found, there is little hope of making them by LEC. The search continues.

(*e*) *Growth from solvent*

All II–VI compounds have been grown by this method. Supersaturation of the compounds is achieved by decrease in temperature or evaporation of solvent.

1.3.2 Growth from vapour phase

Growth from the vapour phase has the advantage that crystals can be grown at low temperatures – below the phase transition temperature – especially for ZnSe and ZnS. This reduces the magnitude of contamination and the defect density. High quality crystals have been grown from the vapour phase, the growth rate, however, is low. Chemical vapour transport and sublimation methods have been used for growing sizeable crystals. The latter methods consist of physical vapour transport (PVT), Piper–Polich and Prior methods.

(*a*) *Chemical transport*

This is based on chemical transport reactions which take place in a sealed ampoule having two different temperature zones. As an example, transport of ZnS with I_2 is represented by the following reaction:

$$2ZnS(s) + 2I_2 \rightleftharpoons 2ZnI_2(g) + S_2(g). \qquad (1.5)$$

In the higher temperature zone, where the source material is located, this reaction proceeds from left to right. The gaseous products of the reaction

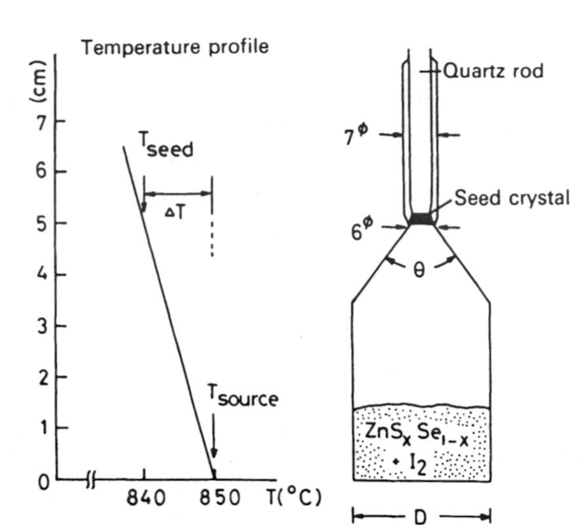

Fig. 1.10 Ampoule and temperature profile used for the growth of ZnS_xSe_{1-x} single crystals by iodine transport method [12].

diffuse to the lower temperature region where the reverse reaction takes place. The whole process continues by back diffusion of the I_2 generated in the lower temperature region. When iodine is used as a transport agent, the method is usually called iodine transport. Its disadvantage is that it incorporates the transport agent. The schematic of the ampoule and the temperature profile are shown in Fig. 1.10, used by Fujita *et al.* to grow ZnS_xSe_{1-x} single crystals [12]. The details of their experiments are described later.

(*b*) *Piper–Polich*

This method is developed by Piper and Polich [13]. The experimental arrangement is shown in Fig. 1.11. The charge is the sintered material. The crucible with a blunt conical tip, into which a closed inner quartz tube fits snugly, is placed in a mullite tube. The mullite tube is evacuated and heated to remove volatile impurities. After baking, argon gas at 1 atm is introduced and maintained throughout the growth run. The initial position of the crucible is such that the tip is near the maximum temperature. The entire mullite tube, containing the crucible, is mechanically pushed so that the tip moves into a cooler region at the rate of 0.3–1.5 mm h^{-1}. The crucible velocity is constant during a particular run. The optimum velocity depends on the temperature profile. As the tube moves, the supersaturation at the tip increases so that nucleation and growth occur. By this method, crystals of centimetre size have been grown.

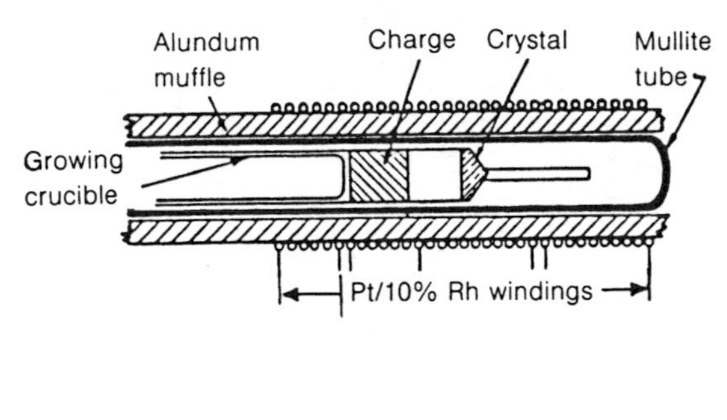

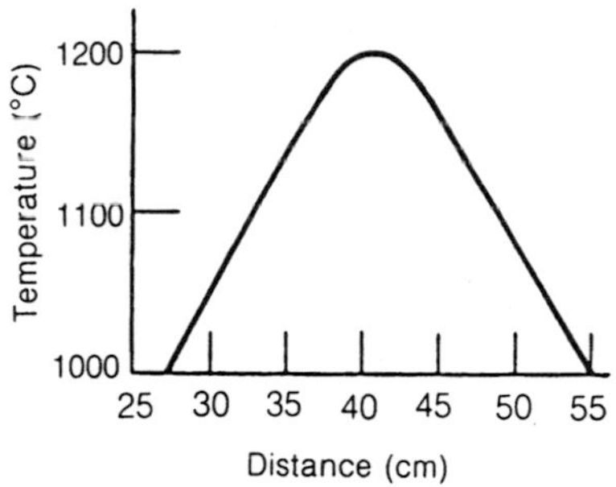

Fig. 1.11 Furnace cross-section and temperature profile used by Piper and Polich [13].

(*c*) *Prior*

Prior [14] established a vapour growth method which uses a reservoir of the constituent element to control deviation from stoichiometry. After modifying it to use a closed ampoule, it was applied to grow high quality crystals of II–VI compounds. The configuration of the growth ampoule is shown in Fig. 1.12 [15]. The reservoir temperature for the growth of crystals with near-stoichiometric composition is derived by taking into account the solid–vapour equilibrium represented by Equation 1. Single crystals are grown under a P_{min} condition assessed from Equation 4. Under this condition, the maximum growth rate is obtained and near-stoichiometric compounds can be grown [15].

1.4 CRYSTAL GROWTH OF CUBIC II–VI COMPOUNDS

1.4.1 Zinc sulphide

(*a*) *Growth from liquid phase*

Zinc sulphide has a phase transition from hexagonal to cubic at 1293 K [5]. Because of small difference of free energy between the two phases, the kinetics

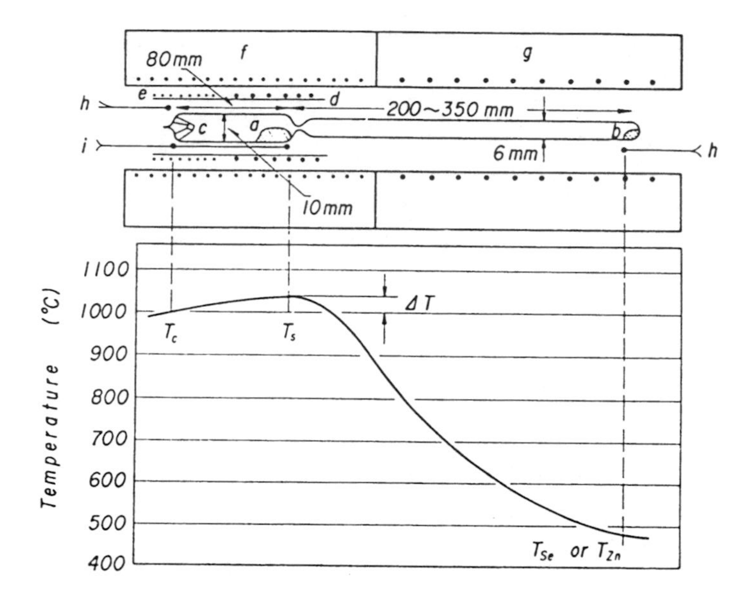

Fig. 1.12 Schematic construction of the electric furnace and temperature profile for the modified Prior method [15]: (a) source chamber; (b) Zn or Se reservoir; (c) growth chamber, (d), (e) auxiliary heater, (f), (g) main heater; (h) thermocouple; (i) differential thermocouple.

of the phase transition are so slow that the conversion is not observed [16]. For this reason, the cubic ZnS single crystal is hardly grown from the near-stoichiometric melt.

Cubic single crystals of ZnS have been grown from solvents such as Te [17] and some halides [18, 19] at temperatures significantly lower than its melting point. Washiyama *et al.* [17] grew ZnS single crystals using Te solvent. The Te solvent, in which ZnS was dissolved at 1373–1473 K, was slowly cooled to 773–823 K at a rate of 15–40 Kh^{-1}. They grew plate-like transparent cubic crystals with the dimensions of 3 mm × 3 mm × 1 mm. They found that the grown crystal contained 0.1–0.8 at% Te.

Linares [18] grew optically isotropic ZnS crystals from $PbCl_2$ solvent at temperatures from 773–1073 K by slow cooling or by a temperature gradient method. Colourless crystals free from birefringence or flux inclusions can be grown at a slow growth rate. When Parker and Pinnell [19] used Halide solvents, they investigated the optimum composition of (KI or KCl)–MCl_2(M = Zn and Cd) at temperatures 723–1173 K. They grew crystals up to 10 mm on an edge from a solvent of KI (20 wt%)–$ZnCl_2$. They found the change of crystal habit from dendritic platelets – thin platelets became thicker platelets with increasing growth temperature. The thick crystals had a low dislocation density, but the thin-platelet and well-faceted crystals frequently showed twinning.

(b) Growth from vapour phase

Cubic ZnS single crystals have also been grown by chemical transport methods using HCl [20, 21], NH_4Cl [22, 23] and I_2[12, 22, 24–27] as transport agents. NH_4Cl dissociates and acts as HCl. Although crystals grown below 1273 K usually have a cubic structure and those grown above 1273 K have a hexagonal structure [22], Dangel and Wuensch [24] reported that, even in the temperature range where the cubic structure is stable, the hexagonal structure predominates at iodine concentrations below $5 \times 10^{-4}\,g\,cm^{-3}$, where surface reactions are rate controlling.

Fujita *et al.* [12] grew large ZnS, ZnSe and ZnS_xSe_{1-x} single crystals with dimensions as large as 24 mm × 14 mm × 14 mm using the seed crystal by iodine transport method. The ampoule and the temperature profile of the furnace are shown in Figure 1.10. They reported that the optimal conditions to obtain large single crystals with good reproducibility were $\Delta T = 7\,K$, $20 < D < 25\,mm$ and $45 < \Theta < 70°$. Their average linear growth rate was about $1.0 \times 10^{-6}\,cm\,s^{-1}$. The grown ZnS single crystal with cubic structure was free from twins. The size of the single crystal depended strongly on the ampoule geometry and the ΔT value. Electrical properties of these crystals were investigated after annealing in molten Zn at 1223 K for more than 48 h. Low resistivity *n*-type crystals were obtained by this treatment. Typical values of the electrical properties were: $\rho = 1.2\,\Omega\,cm$, $n = 4.8 \times 10^{16}\,cm^{-3}$ and $\mu = 107\,cm^2\,V^{-1}\,s^{-1}$.

Ohno *et al.* [27] grew cubic ZnS single crystals as large as $1.5\,cm^3$ by the iodine transport method without seed. By means of Zn-dip treatment, they prepared a low resistivity *n*-type crystal and used it as a substrate for homoepitaxial MOCVD (Metal-Organic Chemical Vapour Deposition) growth. They fabricated a MIS (Metal-Insulator-Semiconductor)-structured blue light-emitting diode which yielded an external quantum efficiency as high as 0.05%. They found that the quality of the crystal grown was significantly improved by prebaking the ZnS powder in H_2S gas prior to growth. After prebaking in H_2S, the growth rate became about three times larger than with a charge pretreated in vacuum.

Although the modified Piper–Polich method [28] and PVT [26] have been applied to grow ZnS single crystals, large high quality cubic single crystals have not been grown. Hartmann [26] grew a single crystal with dimensions 5 mm × 3 mm × 3 mm by PVT after optimizing the growth conditions. The grown crystals have a hexagonal structure due to the high optimum growth temperature, 1623 K.

1.4.2 Zinc selenide

(a) Growth from liquid phase

Zinc selenide has a high melting point, 1793 K. Its vapour pressure at melting is about 4 atm for the stoichiometric composition. But judging from the *P–T* diagram, the pressure rises to about 65 atm for a Zn-rich melt. For a selenium-rich melt it can rise above 200 atm [29]. Growth techniques that employ either high inert gas pressure or a closed system have been used to prevent mass transport from the melt during the growth. Unlike the case of ZnS, cubic ZnSe crystals can be grown from the melt. By means of a high temperature X-ray oscillation technique, Kikuma and Furukoshi [30] directly observed the transformation from cubic zinc blende to hexagonal wurtzite structure on heating and the reverse transformation on cooling.

Kikuma and Furukoshi [31, 32] have grown zinc selenide single crystals by the Bridgman method using a graphite crucible under an argon pressure of about 100 atm. The grown crystals had a cubic structure. Part of them had twins, voids and rod-like low angle grain boundaries. The twins extended completely across the grown crystal on one set of (1 1 1) planes. The spacing between twins ranged from 0.05–1.5 mm. The void density was dependent on the lowering rate of the crucible. They suggested that the origin of the voids was the argon entrapped in the melt. The formation of rod-like low angle grain boundaries depended on the angle between the growth axis and the (1 1 1) twin plane normal. At angles less than 75° the boundaries did not occur. Many crystals grown at rates in the range 0.8–2.5 mm h^{-1} had neither boundaries nor voids. The problem with this method is the high vapour pressure of the ZnSe melt. Loss of ZnSe during growth was 2–10% of the charge weight. This makes the melt composition Se-rich [31]. The excess Se in the ZnSe melt leads to growth of polycrystals.

To avoid the loss of the charged material and to maintain a nearly stoichiometric melt composition, Kikuma *et al.* [33] grew ZnSe from the melt under Zn partial pressure in a high pressure chamber. The Zn partial pressure was varied from 0.5–10.6 atm. A crucible of 10 mm i.d. and a Zn reservoir were contained in a growth vessel placed in a high pressure container. The total pressure was 55 atm. They found the evaporation loss decreased with increasing temperature of the Zn reservoir. The composition of the melt depended on the temperature of the Zn reservoir. They grew low resistivity *n*-type ZnSe crystals from a Zn-rich melt. This indicates a possibility that the Zn reservoir could control deviation from the stoichiometric composition of the grown crystals. But voids and twins were still observed on the crystals and this may be due to the lowering rate of 3 mm h^{-1}, slightly higher than in other cases.

Other attempts have been made to prevent the loss of charged materials during growth. Debska *et al.* [34] grew $Zn_{1-x}Mn_xSe$ single crystals in self-sealing graphite crucibles by an RF-heated Bridgman method. The crucible assembly consisted of a graphite tube with one solid end and an open threaded end as shown in Fig. 1.13. After loading the powdered charge, a threaded graphite plug was screwed into place and the crucible assembly was inverted with the plug at the bottom. This was placed on a fused silica pedestal in the centre of an RF induction coil so that the bottom

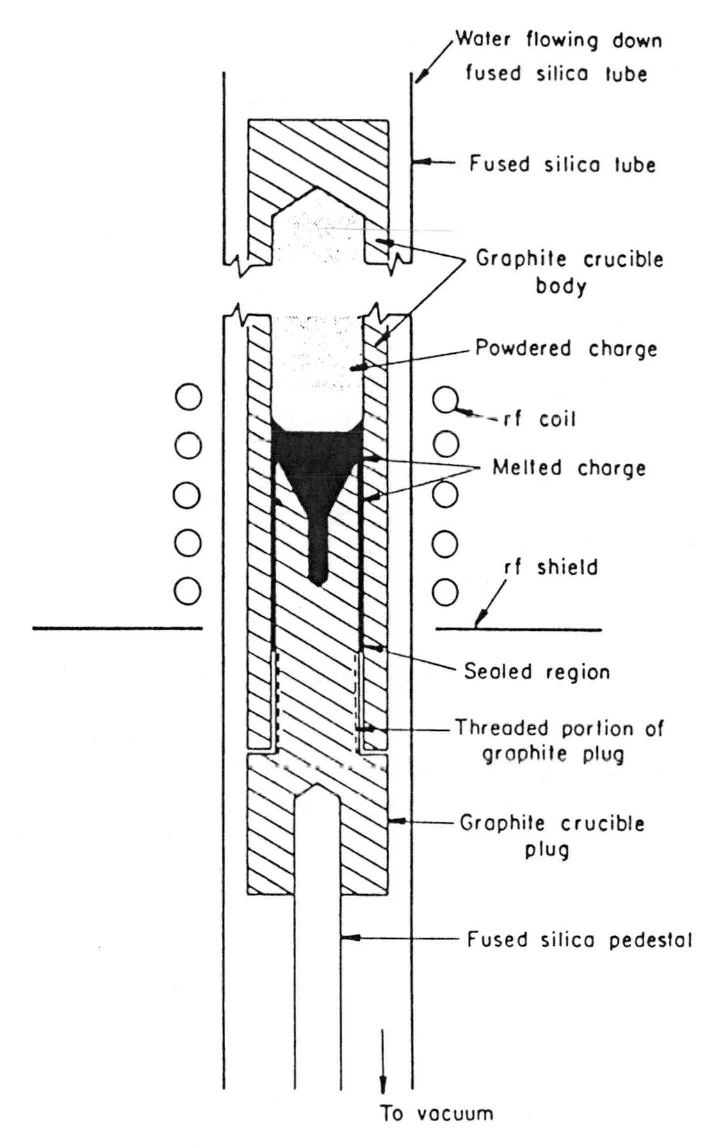

Fig. 1.13 Self-sealing graphite crucible [34].

half of the threaded plug extended below the coil. A fused silica tube was slipped over the inverted closed unsealed crucible assembly and the system was evacuated to about 10^{-6} Torr. The temperature at the centre of the RF coil was slowly raised to beyond the melting point. Some of the molten material flowed down between the crucible wall and the plug then solidified in the cooler region of the crucible below the RF coil, thereby sealing the crucible. This method has the benefit of needing no high pressures. Debska *et al.* grew twin-free cubic crystals, x range from 0–0.28. The single crystal dimensions were 9 mm diameter, 5–10 mm length.

Fitzpatrick *et al.* [35] developed a self-sealing self-releasing (SSSR) technique for II–VI compounds. They used a relatively narrow RF coil to melt a small amount of material at a time as shown in Fig. 1.14. Some of the material volatilized from the molten zone and condensed on the threads of the screw cap, forming a seal. The crucible was lowered through the coil so that the hot zone traversed the charge. Eventually, the coil was opposite the sealing threads, and volatilized the seal material, allowing easy removal of the screw cap. The growth was carried out at the lowering rate of the crucible, 1.6–3.2 mm h^{-1}, under the N_2 pressure 5–7 atm. They grew ZnSe boules with twins weighing about 352 g. The bottom of the boule usually contained some voids, but no voids were seen above this region.

Shone *et al.* [36] reported the influence of doping on the occurrence of twinning. They found that doping with Mn (5×10^{18}–1×10^{19} cm^{-3}) effectively prevented twin formation. They obtained a result similar to that of Debska *et al.* [34], however, their concentration range was different. Shone *et al.* used a concentration range that allowed the ZnSe to preserve some of its room temperature physical properties, such as lattice constant and bandgap.

ZnSe single crystals have been grown from solvent such as Te [17], Zn [37], Se [37], $Se_{1-x}As_x$ [38], $Se_{1-x}Sb_x$ [38], Ga [39], In [40] and Sn [41]. Unlike ZnS, an attempt to grow ZnSe from $PbCl_2$ was unsuccessful [18].

The most important result was obtained by Nishizawa *et al.* [37], who succeeded in growing low resistivity *p*-type ZnSe single crystals by the temperature difference solution growth method under controlled vapour pressure. They obtained pure blue light emission from a ZnSe *pn* junction. The growth temperature from the Se solvent was 1323 K and Li was added

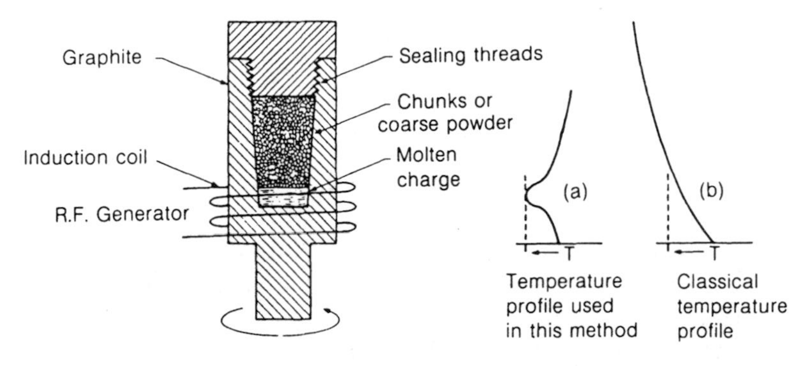

Fig. 1.14 Cross-section of the crucible used for SSSR technique, with schematic representations of the temperature profiles used in (a) the SSSR method and (b) in the usual method [35].

to the Se solvent in order to obtain a *p*-type crystal. The grown crystals were about 10 mm long. The best value of the FWHM of the X-ray double crystal rocking curve was 15 arc sec in spite of the existence of twins. The carrier concentration and Hall mobility of the As-grown *p*-type crystals were 6.4×10^{12} cm^{-3} and 78 cm^2 V^{-1}, respectively. They reported that the electrical resistivity and the emission spectrum were dependent upon the applied Zn pressure.

(*b*) *Growth from the vapour phase*

ZnSe single crystals have often been grown by the iodine transport method [12, 42–45]. Results are similar to those of ZnS. Common growth temperature and ΔT value are 1023–1223 K and 10–20 K, respectively. Koyama *et al.* [45] obtained ZnSe single crystals with values of FWHM and etch pit density (EPD) of 6.2 arc sec and 8×10^2 cm^{-2}, respectively. This FWHM value is close to the ideal value of 4.5 arc sec, estimated by a dynamic theory calculation [46]. Robinson and Kun [42] found that incorporation of Ga, In and Tl by diffusion into ZnSe, grown by the iodine transport method, converted the *n*-type ZnSe to *p*-type. The reason has not been clarified.

The Piper–Polich method has been used to grow single crystals of ZnSe [47–50]. Nakau *et al.* [50] prepared (1 1 0) ZnSe substrates of about 4 mm × 3 mm × 1 mm from single crystals grown by this method. They incorporated Tl into the ZnSe substrates and obtained a *p*-type layer, as reported by Robinson and Kun [42]. When they used a charge of powdered material, they found that pretreatment of the charge was important to obtain a high growth rate as reported for iodine transport of ZnS by Ohno *et al.* [27].

The PVT method has been applied to grow ZnSe [51, 52] and $ZnSe_xTe_{1-x}$ [53]. Cheng *et al.* [51] grew ZnSe single crystals as large as 10 mm × 3 mm × 3 mm using an ampoule of 10 mm i.d. with a quartz cold finger, in which a needle-shaped cavity was made to provide a channel for the development of a seed crystal. The growth temperature and the value of ΔT are 1155 K and 60 K, respectively. They suggested that the transport rate, which ranged from 1–12 mg h^{-1}, was strongly dependent on the purity of the starting material. The lower purity resulted in a lower transport rate which hampered crystal growth.

Koyama *et al.* [52] applied PVT to grow ZnSe single crystals using seeds. The growth temperature was 1343 K and ΔT was about 10 K. On the $(1\,1\,1)_B$ seeds there grew well-faceted single crystals of centimetre size. On $(1\,1\,1)_A$ seeds there grew only polycrystals. The transport rate was about 9 mg h^{-1} for their experimental conditions. The morphology of the grown crystal was a hexagonal prism constructed by $\{1\,1\,0\}$ faces.

The modified Prior method has been applied to grow ZnSe single crystals [15, 54, 55, 57–59]. Kiyosawa *et al.* [15] grew ZnSe single crystals as large as about 7 mm × 5 mm × 5 mm with the shape of a hexagonal prism constructed by {1 1 0} planes. Growth direction was [1 1 1]. They measured the transport rate as a function of Zn or Se partial pressure and compared it with the theoretical calculation. Experimental data obtained using a Zn reservoir are shown in Fig. 1.15. The transport rate is constant in the low P_{Zn} region and starts to decrease with increasing P_{Zn} at the value of P_{Zn} corresponding to the minimum total pressure (P_{min}).

Assuming that the rate-determining process is a diffusion process caused by the difference in the vapour phase composition between the source chamber and the growth chamber, they obtained the following equation to express the transport rate J for the Zn reservoir.

$$J(\text{g per day}) = A \cdot P_{Zn}\{K(T_s) - K(T_c)\}[\{(P_{Zn}/2) - P_{Se_2}\}\{P_{Zn}^3 - 2K(T_c)\}]^{-1} \quad (1.6)$$

Where T_c and T_s represent the temperatures at the growth chamber and at the source chamber, respectively. K is the equilibrium constant of the Equation 1.1. A is given by

$$A = 227S \cdot W\{T'^3(M_{Zn} + M_{Se_2})/2M_{Zn} \cdot M_{Se_2}\}^{1/2} \cdot [N \cdot k \cdot T \cdot \Delta X \{(\sigma_{Zn} + \sigma_{Se_2})/2\}]^{-1} \quad (1.7)$$

where S is a cross-sectional area of the ampoule, W the molar weight of ZnSe, M the molar weight of gas molecules, N the Avogadro number, k the Boltzman constant, σ the molecular diameter and ΔX the distance between source chamber and growth chamber. T' is $(T_s + T_c)/2$.

In the case that P_{Zn} is high and predominant in total pressure, the measured

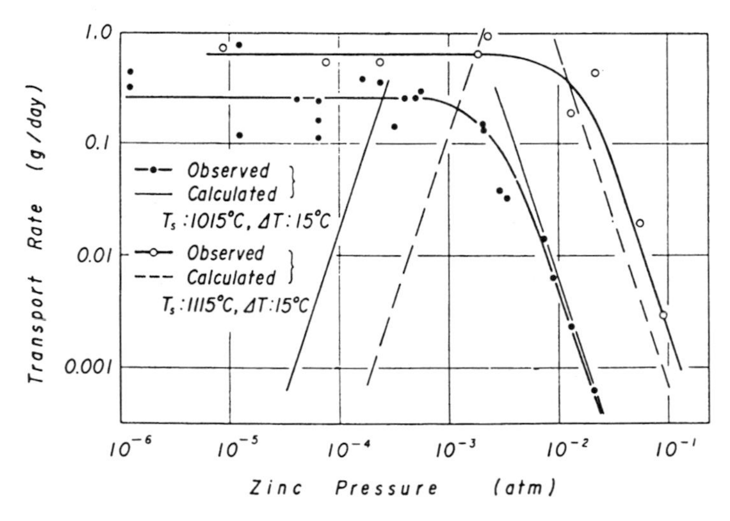

Fig. 1.15 Relation between transport rate and partial pressure of Zn [15].

transport rate is proportional to P_{Zn}^{-3}. On the other hand, the transport rate is constant and independent of P_{Zn} in a low P_{Zn} region. Calculated results are also shown in Fig. 1.15. Good agreement between experimental and calculated results in a higher P_{Zn} region indicates that the diffusion process is the rate-determining process and that the P_{Zn} can be controlled by the Zn reservoir in this region. On the other hand, this deduction cannot explain the experimental results in the region of P_{Zn} lower than that corresponding to P_{min}. In this region, another process should be the rate-determining process and the P_{Zn} cannot be controlled by the Zn reservoir. The P_{min} condition has been selected to grow the single crystals, because this condition gives maximum growth rate in the region where the deviation from stoichiometric composition can be controlled.

Huang and Igaki [55] grew high purity ZnSe single crystals, starting from the purification of the selenium. Before the synthesis, selenium was purified by a distillation method. ZnSe polycrystals were synthesized at about 1273 K by chemical reaction of commercial Zn (6N grade) with refined Se. Before growth, ZnSe polycrystals were purified by sublimation under a controlled Zn partial pressure corresponding to P_{min}. Growth temperature and ΔT were 1273 K and 3–5 K, respectively. The necking they designed into the growth ampoule gave a considerable increase in the reproducibility of obtaining high quality single crystals. The photoluminescence (PL) spectra measured on their crystals showed a strong emission due to the radiative recombination of free excitons and very sharp donor-bound exciton lines.

Isshiki *et al.* [56] purified zinc by a process consisting of vacuum distillation and overlap zone melting in pure argon. Using refined zinc and commercial high purity selenium, high purity ZnSe single crystals were grown [57] by the same method as reported by Huang and Igaki [55]. The PL spectra in the near band edge region [58] is shown in Fig. 1.16. The emission intensities of donor-bound excitons (I_2) are remarkably small. The emission intensities of the radiative recombinations of free excitons (Ex) are very strong. These intensities indicate the crystal had a very high purity and a very low donor concentration, and they suggest that purity of the grown crystal strongly depends on purity of the starting materials. This method is suitable for preparing high purity crystals, since a purification effect is expected during growth. Impurities with higher vapour pressure will condense at the reservoir portion and those with lower vapour pressure will remain in the source crystal. This effect was confirmed by the PL measurements [59]. On these crystals, photoexcited cyclotron resonance measurements have been attempted and cyclotron resonance signals due to electron [60, 61] and heavy holes [62, 63] have been detected for the first time. The cyclotron mobility of electrons under $B = 7\,T$ is $2.3 \times 10^5\,cm^2\,V^{-1}\,s^{-1}$. This indicates that the quality of the grown crystals is very high. The donor concentration in the crystal is estimated to be $4 \times 10^{14}\,cm^{-3}$ by analysing the temperature dependence of the cyclotron mobility [59].

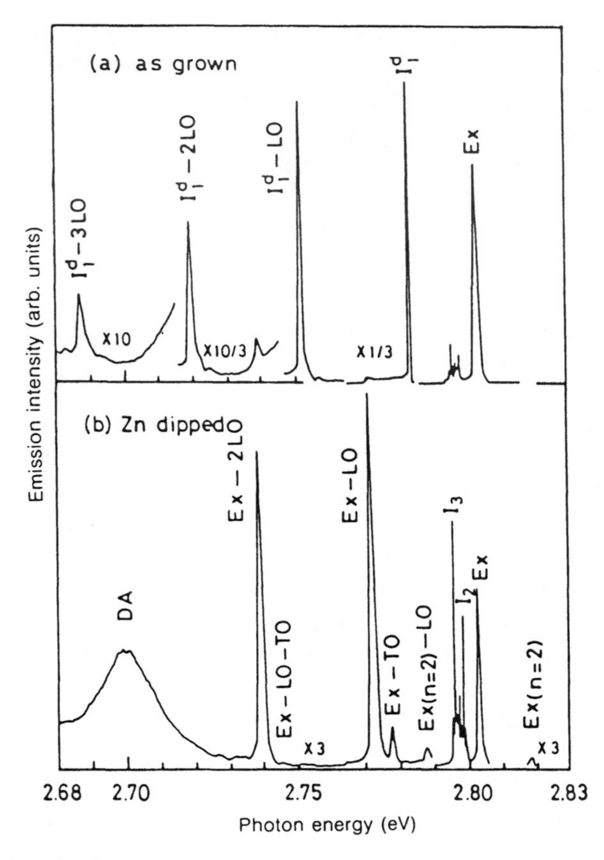

Fig. 1.16 Photoluminescence spectra measured on high purity ZnSe grown by the modified Prior method [58].

(*c*) *Growth from solid phase*

Zinc selenide single crystals were grown by recrystallization THM. Yamada *et al.* [64] developed this method and grew very pure ZnSe single crystals. They recrystallized polycrystalline ZnSe rod from one end by using a travelling heater, consequently the purity of the grown crystal depended upon the purity of the starting material.

1.4.3 Zinc telluride

(*a*) *Growth from liquid phase*

Crystal growth from the melt of nearly stoichiometric composition was reported by Fischer *et al.* [65]. The melt of slightly excess Zn was passed through a temperature gradient from about 1553 K–1373 K in 5 h under 30–50 atm of argon pressure. Translucent red ingot was obtained; this consists of crystals with average dimensions 5–8 mm. Using the gradient freeze technique, Lynch [66] grew single crystals from near-stoichiometric melts in a graphite crucible under controlled Zn pressures. The crystal ingots were about 12 mm diameter by 25 mm long. There were several runs where 80–90% of the ingots were in the form of a single crystal.

Title *et al.* [67] reported crystal growth from solutions containing up to 50% excess tellurium. The excess tellurium lowers the melting point to around 1470 K so that unsupported quartz ampoules can be used without softening. Steininger and England [68] grew single crystals with dimensions up to 35 mm × 20 mm × 10 mm from the Te excess solution using a sealed quartz ampoule. THM growth of ZnTe was reported by Taguchi *et al.* [69].

(*b*) *Growth from vapour phase*

Growth by chemical vapour transport in the ZnTe–I_2 and ZnTe–Ge–I_2 systems was experimentally and thermodynamically studied by Kitamura *et al.* [70]. In the former system, they observed the simultaneous growth of Te crystals as well as ZnTe crystals and the decrease in transport rate with growth time at temperatures lower than 1173 K. In the case of this system, the convection flow largely affected the transport process and the deposition of Te was attributed to the convection effect of the vapour phase. In the latter system, neither the growth of Te nor the Ge inclusion was observed in the crystals grown. The transport rate of ZnTe was large enough and almost independent of the growth time. In this system, a large ZnTe crystal was grown at a relatively low temperature, 973 K.

Taguchi *et al.* [71] developed a travelling heater method to grow high purity ZnTe single crystals from the vapour phase. The temperature of the sublimation interface was fixed at 1088 K and that of the growth interface was varied from 1058–1073 K. The growth rate was 3 mm per day. The grown crystals showed the strong free exciton line at 4.2 K in the photoluminescence spectra. Te precipitates, which were often observed in ZnTe crystals grown by other methods, were hardly detected in the crystals grown by the sublimation THM.

1.4.4 Cadmium telluride

(*a*) *Growth from liquid phase*

Cadmium telluride can be melted under relatively mild conditions compared to other II–VI compounds. Growth of single crystals by the Bridgman method has been investigated [72–80] most extensively, because CdTe is in great demand as the substrate for epitaxial growth of $Hg_{1-x}Cd_xTe$(MCT) films and this demand becomes more and more severe for large-area uniformity. The ampoule is commonly lowered at 1–5 mm h^{-1}. There is usually a pointed end on the quartz ampoule; this is to seed a single crystal. To avoid possible contamination from the quartz ampoule due to the high congruent melting point of CdTe, growth from the Te-rich solution has been also pursued [75, 78].

The problem with this method is the high dissociation pressure of Cd at its melting point. This makes the grown crystals Te-rich, consequently Te precipitates appear in the crystals. To control deviation from the stoichiometric composition, growth by the Bridgman method under controlled Cd partial pressure was performed by Wen-Bin *et al.* [80]. They found that the lattice parameter and the electrical properties were dependent on the applied pressure of cadmium. Under a high Cd applied pressure, 3.8 atm, they grew *n*-type CdTe crystal, whereas under lower Cd pressures they grew *p*-type crystals. This makes it possible to control deviation from the stoichiometric composition. Kennedy *et al.* [81] reported that post-annealing at 1248 K effectively annihilated the Te precipitates.

The shape of the solid–liquid interface is important for growing single crystals by the Bridgman method. A slightly convex liquid–solid interface (towards the melt) is favourable for grain selection and prevention of spontaneous nucleation at the ampoule wall. Mathematical modellers have attempted to determine the thermal configuration [82, 83]. However, the lack of thermophysical data means that some calculations are based on assumptions. Using an ampoule of 28 mm diameter, Route *et al.* [77] investigated the influence of the temperature gradient in the growth zone upon the solid–liquid interface. By autoradiography they found the interface shape was almost flat and the thermal modification generated a slightly convex shape, but there is no significant improvement in grain structure.

The quality of the grown crystals is influenced by the furnace design and the thermal properties of the charge materials, especially when the diameter of the ampoule becomes larger. The most important thermal property is thermal conductivity. Thermal conductivity affects thermal coupling between the furnace and the material, so a more precise temperature control system is required for growing a larger single crystal. Sen *et al.* [83] has used gradient freezing in a multizone furnace controlled by computer. The furnace consisted of three main zones. The upper zone maintained a constant Cd overpressure over the melt. The middle zone was the hot zone, the temperature of which was held above the melting point of CdTe. This part consisted of 10 segments with a separate power supply for each segment. The lower zone was the cold zone held between 1023 and 1273 K.

Like other cubic II–VI compounds, CdTe suffers from twinning. Recent development of crystal growth techniques makes it possible to grow large twin-free CdTe single crystals. Fig. 1.17 shows a twin-free CdTe single crystal of 75 mm diameter grown by gradient freezing.

Fig. 1.17 Photograph of a twin-free large single crystal CdTe grown by the GF method. (By courtesy of Nippon Mining Co. Ltd)

Demands for the substrate have been changed from CdTe to $Cd_{1-x}Zn_xTe$(CZT) which can be lattice matched to MCT. Addition of Zn is also effective in strengthening the lattice due to a solution hardening effect. Hardening the lattice prevents the generation of dislocations and consequently the formation of a cellular structure. In fact, X-ray double crystal rocking curves indicate that the quality of CZT is better than that of CdTe. The values of FWHM measured on CdTe and CZT are about 14 and 8 arc sec, respectively [83].

Since the segregation coefficient of Zn in CdTe is larger than unity, CZT with uniform concentration of Zn throughout the crystal cannot be obtained. Tanaka *et al.* [84] attempted codoping of Zn and Se for the following reasons.

1. Both elements reduce the lattice constant of CdTe.
2. Because the segregation coefficients of Zn or Se in CdTe are larger or smaller than unity, respectively, it can be expected that the reduction of the Zn concentration due to the segregation is compensated by the increase in the Se concentration.
3. Furthermore, these two elements are isoelectronic with Cd and Te.

The lattice constant of a Zn- and/or Se- doped crystal is shown in Figure 1.18. Good uniformity of the lattice constant was obtained within the fraction

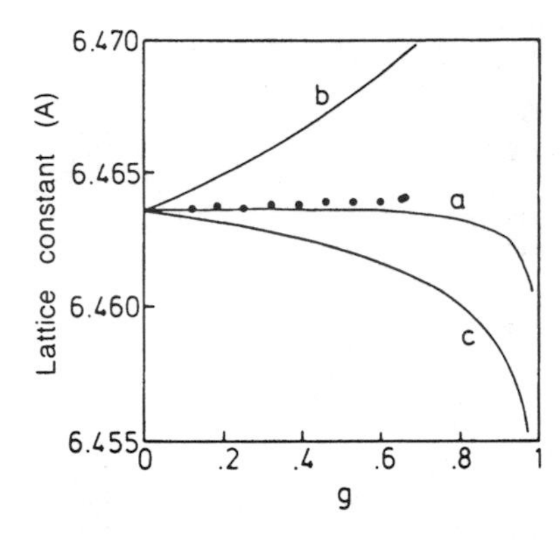

Fig. 1.18 Lattice constant of a Zn- and/or Se- doped CdTe crystal as a function of fraction solidified (g). Solid lines show the calculated results for different dopant concentrations: (a) 0.96 at% Zn and 3.6 at% Se; (b) 3.7 at% Zn; (c) 4.9 at% Se.

solidified range 0–0.7. Furthermore, experimental and theoretical results show good agreements in this region. It was reported that $Zn_{0.0096}Cd_{0.9904}$-$Se_{0.06}Te_{0.964}$ was the best composition to grow the crystal with uniform lattice constant [84].

CdTe crystals have been grown also by the horizontal Bridgman method [85, 86], which has the potential to reduce the magnitude of thermomechanical stress on grown crystals. Crystal growth under low stress conditions can significantly reduce the dislocation density. Kahn *et al.* [85] grew CdTe crystal with FWHM of 9.5 arc sec by the HB method. Lay *et al.* [86] reported that the low FWHM value, 11 arc sec, was obtained on CdTe crystal at the region close to the top free surface, while 33 arc sec at the region close to the bottom of the ingot where the crystal was in contact with the boat. This means that the HB method can grow a higher quality crystal than the VB method, however, a crystal of uniform high quality has not been grown.

Zone melting [8, 9], THM [87, 88] and LEC [10, 11, 89] have been used to grow CdTe crystals as described in Section 1.3.

(*b*) *Growth from vapour phase*

Chemical vapour transport has been used to grow CdTe single crystals using NH_4Cl [90] and $H_2 + I_2$ [91] as transport agents. However, large single crystals of high quality have not been grown by this method. Paorici *et al.* [91] proposed a thermodynamic model and compared it with their experimental results obtained by using $H_2 + I_2$ as transport agents. They suggested that the closed-tube chemical transport mechanisms were mainly governed by: (1) diffusive transport of the sublimation products of CdTe at low pressures; (2) thermal-convection-assisted diffusion and laminar flow of the gaseous species such as I_2(g), I(g) and HI(g) at intermediate pressures;

and (3) SLV processes at still higher pressures. SLV is transported from a vapour to a solid through a liquid phase. In this case, formation of Te(l) occurs and results in dendritic growth. The iodine content of the grown crystals was measured as (3–40) × 10^{18} cm^3 by neutron activation analysis. In spite of the high iodine concentration, crystals showed high resistivity.

PVT techniques have been applied to grow large CdTe single crystals using seeds [92–94]. Unseeded PVT growth has also been applied [95–97]. Kuwamoto [94] grew large single crystals up to 18 cm^3 using a semi-open vertical ampoule. The growth temperature and ΔT were 953–1123 K and 5–10 K, respectively. Seed crystals were Bridgman grown {1 1 1} CdTe plates.

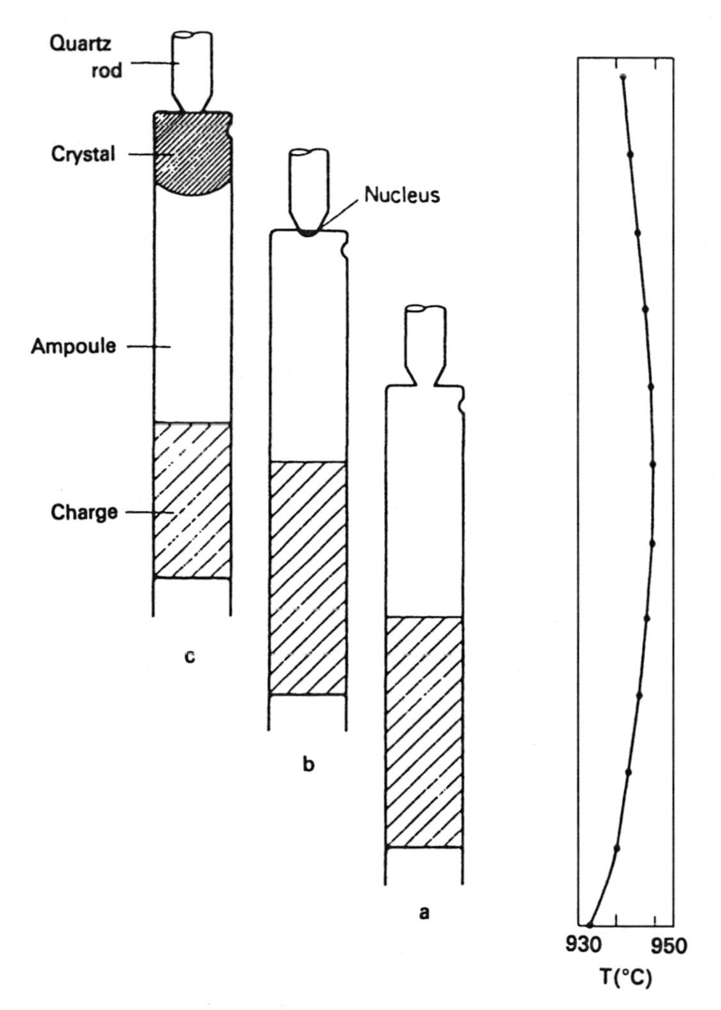

Fig. 1.19 Temperature profile and growth ampoule positions in the furnace during VUVG of CdTe [95]: (a) reverse transport; (b) nucleation; (c) final position.

He reported that epitaxial growth was observed on the $(1\,1\,1)_A$, but not on the $(\bar{1}\,\bar{1}\,\bar{1})_B$surfaces. The presence of thin voids oriented parallel to the growth direction may be the result of growth that was too rapid, although the details of the growth rate are not given.

Yellin *et al.* [95] used the vertical unseeded vapour growth (VUVG) technique to grow high quality CdTe single crystals, 23 mm diameter and 5 mm long. Figure 1.19 shows the temperature profile of the furnace and the relative positions of the growth ampoule during the VUVG process. After reverse transport, the ampoule was raised to the nucleation position. The growth temperature was about 1223 K. A cold finger made of quartz rod was attached to the centre of the upper bottom to cool the point by about 2 K. The cold finger is effective for single nucleation. The EPD of the grown crystals was as low as $2 \times 10^3\,cm^{-2}$. Emission spectrography analysis indicated that purification of the source materials was also achieved during the vapour transport process. The maximum growth rate was about 2 g per day and the growth rate was dependent upon the amount of excess Te in the charge material.

The dependence of the transport rate upon the partial pressure was examined in detail by Mochizuki [98] in a way similar to that reported by Kiyosawa *et al.* [53] in their study of ZnSe. In addition to the two regions shown in Fig. 1.15, he found another region which became dominant with decreasing transport temperature, where the transport rate was proportional to $P_{Te_2}^{-1/2}$. In this region, a sublimation process is considered to determine the rate [98]. According to Mochizuki, the transport rate also depended strongly on how much the charge material deviates from stoichiometry in the range where P_{Cd} is lower than P_{min}. The transport rate did not depend on the controlled partial pressure in this region.

1.5 CONCLUSION

The main problems which should be solved for growing large high quality single crystals of cubic II–VI compounds are twinning, deviation from the stoichiometric composition and purity. There are clues to solving these problems in an extensive literature. Although frequent twinning is due to the low stacking fault energy of II–VI compounds [99], twin-free CdTe (see Figure 1.17), ZnSe:Mn [36] and ZnSe [100] were already grown by gradient freezing, Bridgman and modified Prior methods, respectively.

Crystals of very high quality have been grown by vapour methods that allow their deviation from the stoichiometric composition to be controlled during growth. The growth rate of the vapour method is low, however, an improvement in growth rate has been made by Akhekyan *et al.* [101], in which crystals of 30–50 mm diameter by 20 mm length were grown in 80 h. To obtain high purity crystals, it is necessary to refine the starting materials and to pretreat the charged materials. In addition, EPD should be decreased.

REFERENCES

1. Narita, K., Watanabe, H. and Wada, M. (1970) *Japan. J. Appl. Phys.*, **9**, 1278.
2. Brebrick, R.F. (1988) *J. Cryst. Growth*, **86**, 39.
3. Sharma, R.C. and Chang, Y.A. (1988) *J. Cryst. Growth*, **88** 193.
4. Kulakov, M.P., Kulakovskii, V.D., Savchenko, I.B. and Fadeev, A.V. (1976) *Soviet Phys., Solid State*, **18**, 526.
5. Allen, E.T. and Crenshaw, J.L. (1912) *Am. J. Sci.*, **34**, 310.
6. Albers, W. (1967) in *Physics and Chemistry of II–VI Compounds*, (eds M. Aven and J.S. Prener), North-Holland, Amsterdam p. 210–211.
7. Lorenz, M.R. (1967) in *Physics and Chemistry of II–VI Compounds*, (eds M. Aven and J.S. Prener), North-Holland, Amsterdam, p. 80.
8. Woodbury, H.H. and Lewandowski, R.S. (1971) *J. Cryst. Growth*, **10**, 6.
9. Fitzpatrick, B.J., McGee, T.F., III and Harnack, P.M. (1986) *J. Cryst. Growth*, **78**, 242.
10. Klausutis, N., Adamski, J.A., Collins, C.V. and Hunt, M. (1975) *J. Electron. Mater.*, **4**, 625.
11. Hobgood, H.M., Swanson, B.W. and Thomas, R.N. (1987) *J. Cryst. Growth*, **85**, 510.
12. Fujita, S., Mimoto, H., Takebe, H. and Noguchi, T. (1979) *J. Cryst. Growth*, **47**, 326.
13. Piper, W.W. and Polich, S.J. (1961) *J. Appl. Phys.*, **32**, 1278.
14. Prior, A.C. (1961) *J. Electrochem. Soc.*, **108**, 82.
15. Kiyosawa, T., Igaki, K. and Ohashi, N. (1972) *Trans. Japan Inst. Metals*, **13**, 248.
16. Fitzpatrick, B.J. (1988) *J. Cryst. Growth*, **86**, 106.
17. Washiyama, M., Sato, K. and Aoki, M. (1979) *Japan. J. Appl. Phys.*, **18**, 869.
18. Linares, R.C. (1968) *Trans. Metall. Soc. AIME*, **242**, 441.
19. Parker, S.G. and Pinnell, J.E. (1968) *J. Cryst. Growth*, **3/4**, 490.
20. Samelson, H. (1962) *J. Appl. Phys.*, **33**, 1779.
21. Ujiie, S. and Kotera, Y. (1971) *J. Cryst. Growth*, **19**, 320.
22. Lendvay, E. (1971) *J. Cryst. Growth*, **10**, 77.
23. Glazunova, V.K. and Gorbunova, K.M. (1971) *J. Cryst. Growth*, **10**, 721.
24. Dangel, P.N. and Wuensch, B.J. (1973) *J. Cryst. Growth*, **19**, 1.
25. Pałosz, W., Kozielski, M.J. and Pałosz, B. (1982) *J. Cryst. Growth*, **58**, 185.
26. Hartmann, H. (1977) *J. Cryst. Growth*, **42**, 144.
27. Ohno, T., Kurisu, K. and Taguchi, T. (1990) *J. Cryst. Growth*, **99**, 737.
28. Russell, G.J. and Woods, J. (1979) *J. Cryst. Growth*, **47**, 647.
29. Holton, W.C., Watts, R.K. and Stinedurf, R.D. (1969) *J. Cryst. Growth*, **6**, 97.
30. Kikuma, I. and Furukoshi, M. (1985) *J. Cryst. Growth*, **71**, 136.
31. Kikuma, I. and Furukoshi, M. (1977) *J. Cryst. Growth*, **41**, 103.
32. Kikuma, I. and Furukoshi, M. (1978) *J. Cryst. Growth*, **44**, 467.
33. Kikuma, I., Kikuchi, A., Yageta, M., Sekine, M. and Furukoshi, M. (1989) *J. Cryst. Growth*, **98**, 302.
34. Debska, U., Giriat, W., Harrison, H.R. and Yoder-Short, D.R. (1984) *J. Cryst. Growth*, **70**, 399.
35. Fitzpatrick, B.J., McGee, T.F., III and Harnack, P.M. (1986) *J. Cryst. Growth*, **78**, 242.
36. Shone, M., Greenberg, B. and Kaczenski, M. (1988) *J. Cryst. Growth*, **86**, 132.
37. Nishizawa, J., Itoh, K., Okuno, Y. and Sakurai, F. (1985) *J. Appl. Phys.*, **57**, 2210.
38. Aoki, M., Washiyama, M., Nakamura, H. and Sakamoto, K. (1982) *Japan. J. Appl. Phys.*, Suppl. **21**, 11.
39. Wagner, P. and Lorentz, M.R. (1966) *J. Phys. Chem. Solids*, **27**, 1749.
40. Shirakawa, Y. and Kukimoto, H. (1980) *J. Appl. Phys.*, **51**, 2014.

41. Rubenstein, M. (1966) *J. Electrochem. Soc.*, **113**, 623.
42. Robinson, R.J. and Kun, Z.K. (1975) *Appl. Phys. Lett.*, **27**, 74.
43. Catano, A. and Kun, Z.K. (1976) *J. Cryst. Growth*, **33**, 324.
44. Poindessault, R. (1979) *J. Electron. Mater.*, **8**, 619.
45. Koyama, T., Yamashita, K. and Kumata, K. (1989) *J. Cryst. Growth*, **96**, 217.
46. O'Hara, S., Halliwell, M.A.G. and Childs, J.B. (1972) *J. Appl. Cryst.*, **5**, 401.
47. Clark, L. and Wood, J. (1966) *Brit. J. Appl. Phys.*, **17**, 319.
48. Burr, K.F. and Woods J. (1971) *J. Cryst. Growth*, **9**, 183.
49. Toyama, M. and Sekiwa, T. (1969) *Japan. J. Appl. Phys.*, **8**, 855.
50. Nakau, T., Fujiwara, T., Yoshitake, S., Takenoshita, H., Itoh, N. and Okuda, M. (1982) *J. Cryst. Growth*, **59**, 196.
51. Cheng, H.Y. and Anderson, E.E. (1989) *J. Cryst. Growth*, **96**, 756.
52. Koyama, T., Yodo, T. and Yamashita, K. (1989) *J. Cryst. Growth*, **94**, 1.
53. Yamamoto, M., Ebina, A. and Takahashi, T. (1973) *Japan. J. Appl. Phys.*, **12**, 232.
54. Gutter, J.R. and Woods, J. (1979) *J. Cryst. Growth*, **47**, 405.
55. Huang, X.M. and Igaki, K. (1986) *J. Cryst. Growth*, **78**, 24.
56. Isshiki, M., Tomizono, T., Yoshida, T., Ohkawa, T. and Igaki, K. (1984) *J. Japan Inst. Metals*, **48**, 1176.
57. Isshiki, M., Yoshida, T., Tomizono, T., Satoh, S. and Igaki, K. (1985) *J. Cryst. Growth*, **73**, 221.
58. Isshiki, M., Yoshida, T., Igaki, K., Uchida, W. and Suto, S. (1985) *J. Cryst. Growth*, **72**, 162.
59. Isshiki, M. (1988) *J. Cryst. Growth*, **86**, 615.
60. Ohyama, T., Otsuka, E., Yoshida, T., Isshiki, M. and Igaki, K. (1984) *Japan. J. Appl. Phys.*, **23**, L382.
61. Ohyama, T., Otsuka, E., Yoshida, T., Isshiki, M. and Igaki, K. (1985) in *Proc. of 17th Int. Conf. Physics of Semiconductors, San Francisco, 1984* (eds D.J. Chadi and W.A. Harrison), Springer, Berlin, p. 1313.
62. Ohyama, T., Sakakibara, K., Otsuka, E., Isshiki, M. and Masumoto, K. (1987) *Japan. J. Appl. Phys.*, **26**, L136.
63. Ohyama, T., Sakakibara, K., Otsuka, E., Isshiki, M. and Igaki, K. (1988) *Phys. Rev.*, **B37**, 6153.
64. Yamada, Y., Kidoguchi, I., Taguchi, T. and Hirata, A. (1989) *Japan. J. Appl. Phys.*, **28**, L837.
65. Fischer, A.G., Carides, J.N. and Dresner, J. (1964) *Solid State Comm.*, **2**, 157.
66. Lynch, R.T. (1968) *J. Cryst. Growth*, **2**, 106.
67. Title, R.S., Mandel, G. and Morehead, F.F. (1964) *Phys. Rev.*, **136**, A300.
68. Steininger, J. and England, R.E. (1968) *Trans. Metallurgical Soc. AIME*, **242**, 444.
69. Taguchi, T., Shirafuji, J. and Inuishi, Y. (1977) *Rev. Phys. Appl.*, **12**, 117.
70. Kitamura, N., Kakehi, M. and Wada, T. (1977) *Japan. J. Appl. Phys.*, **16**, 1541.
71. Taguchi, T., Fujita, S. and Inuishi, Y. (1978) *J. Cryst. Growth*, **45**, 204.
72. Kyle, N.R. (1971) *J. Electrochem. Soc.*, **118**, 1790.
73. Kimura, H. and Komiya, H. (1973) *J. Cryst. Growth*, 20, 283.
74. Popova, M. and Polivka P. (1973) *Czech. J. Phys.*, **B23**, 110.
75. Zanio, K. (1974) *J. Electron. Mater.*, **3**, 327.
76. Muranevich, A., Roitberg, M. and Finkman, E. (1983) *J. Cryst. Growth*, **64**, 285.
77. Route, R.K., Wolf, M. and Feigelson, R.S. (1984) *J. Cryst. Growth*, **70**, 379.
78. Mochizuki, K., Yoshida, T., Igaki, K., Shoji, T. and Hiratate, Y. (1985) *J. Cryst. Growth*, **73**, 123.
79. Oda, O., Hirata, K., Matsumoto, K. and Tsuboya, I. (1985) *J. Cryst. Growth*, **71**, 273.

80. Wen-Bin, S., Mei-Yun, Y. and Wen-Hai, W. (1988) *J. Cryst. Growth*, **86**, 127.
81. Kennedy, J.J., Amirtharaj, P.M. and Boyd, P.R. (1988) *J. Cryst. Growth*, **86**, 93.
82. Jasinski, T. and Witt, A.F. (1985) *J. Cryst. Growth*, **71**, 295.
83. Sen, S., Konkel, W.H., Tighe, S.J., Bland, L.G., Sharma, S.R., and Taylor, R.E. (1988) *J. Cryst. Growth*, **86**, 111.
84. Tanaka, A., Masa, Y., Seto, S. and Kawasaki, T. (1989) *J. Cryst. Growth*, **94**, 166.
85. Khan, A.A., Allred, W.P., Dean, B., Hooper, S., Hawkey, J.E. and Johnson, C.J. (1986) *J. Electron. Mater.*, **15**, 181.
86. Lay, K.Y., Nichols, D., Mcdevitt, S., Dean, B.E. and Johnson, C.J. (1988) *J. Cryst. Growth*, **86**, 118.
87. Triboulet, R. (1977) *Rev. Phys. Appl.*, **12**, 123.
88. Triboulet, R. and Marfaing, Y. (1981) *J. Cryst. Growth*, **51**, 89.
89. Blackmore, G.W., Courtney, S.J., Royle, A., Shaw, N. and Vere, A.W. (1987) *J. Cryst. Growth*, **85**, 335.
90. Paorici, C., Attolini, G., Pelosi, C. and Zuccalli, G. (1973) *J. Cryst. Growth*, **18**, 289.
91. Paorici, C., Attolini, G., Pelosi, C. and Zuccalli, G. (1974) *J. Cryst. Growth*, **21**, 227.
92. Gołacki, Z., Górska, M., Makowski, J. and Szczerbakow, A. (1982) *J. Cryst. Growth*, **56**, 213.
93. Zhao, S.N., Yang, C.Y., Huang, C. and Yue, A.S. (1983) *J. Cryst. Growth*, **65**, 370.
94. Kuwamoto, H. (1984) *J. Cryst. Growth*, **69**, 204.
95. Yellin, N., Eger, D. and Shachna, A. (1982) *J. Cryst. Growth*, **60**, 343.
96. Yellin, N. and Szapiro, S. (1984) *J. Cryst. Growth.*, **69**, 555.
97. Gołacki, Z., Majewski, J. and Makowski, J. (1989) *J. Cryst. Growth*, **94**, 559.
98. Mochizuki, K. (1981) *J. Cryst. Growth*, **51**, 453.
99. Takeuchi, S., Suzuki, K., Maeda, K. and Iwanaga, H. (1984) *Phil. Mag.*, **A50**, 171.
100. Huang, X.M. (1983) Thesis, Tohoku University.
101. Akhekyan, A.M., Kozlovskii, V.I., Korostelin, Yu.V., Nasibov, A.S., Popov, Yu.M. and Shapkin, P.V. (1985) *Sov. J. Quantum Electron.*, **15**, 737.

2

Liquid phase epitaxy and vapour phase epitaxy of the widegap zinc chalcogenides

P. Lilley

2.1 VAPOUR PHASE EPITAXY

2.1.1 Introduction

This review is largely concerned with developments since about 1980. However, for completeness it is instructive to mention the beginnings briefly. The first published work on vapour phase epitaxy (VPE) in vapour flow was in 1970 [1], in which Lilley *et al.* described the growth of single-crystal layers of ZnS on silicon substrates in hydrogen flow. This followed several years of experimentation with quite complex growth systems; eventually the authors succeeded in producing films epitaxial with the substrates using a comparatively simple apparatus (Fig. 2.1). The quartz system contained a vaporization zone (1050 °C) for the ZnS powder source, a deposition zone (400–600 °C) for growth, and a remote silicon etching zone, in which the oxide surface of the Si wafer could be removed (H_2–HCl flow at 1250 °C, r.f. heated). The difficulties had been associated with the isolation of the source material from the substrate etching gas. As most of the vaporized source material condensed on the reactor walls, and low substrate temperatures were essential to avoid chemical reaction with the substrates, growth rate at best was 130 nm/h^{-1}; the higher substrate temperatures could not be utilized.

Interest turned to the production of thick insulating layers (~ 100 μm) for electro-optic modulators; consequently growth at higher temperatures was necessary, and so other substrates were investigated [2, 3]. The feature of this work was the problem of layer cracking as the films cooled from the growth temperature (550–700 °C); GaAs was found to be more suitable than the lattice-matched GaP, probably because of the better thermal coefficient match. To achieve mirror-smooth surfaces it was necessary to use HCl–H_2 as the transport flow. Further work followed, including a qualitative

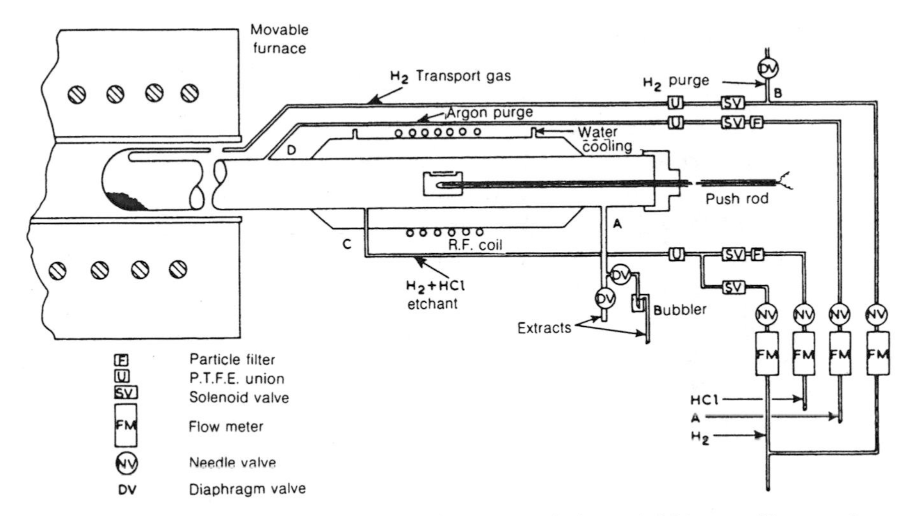

Fig. 2.1 Apparatus for the growth of single-crystal zinc sulphide on silicon wafers. (From ref. 1.)

assessment of transport kinetics [4], with the objective of optimizing growth rate. The growth apparatus had become much simpler in that the III–V compound substrates did not require r.f. heating for oxide removal.

The layers were unintentionally doped with the donor chlorine from the transport gas stream; the layers tended to be photoconductive, making them unsuitable for Pockels cell light modulators. However, interest in the materials was revived in 1975 by the publication [5] of p–n junction electroluminescence in bulk crystals of ZnSe: VPE was clearly a more commercial proposition, and publications in this area began to appear [6–8]. The growth systems were simplified further by the lower source temperature (~ 1000 °C) needed for this compound. In addition, another route to layer growth was investigated: the use of pure metallic zinc and selenium sources, rather than a powder of the II–VI material.

A feature of the electrical properties of widegap II–VI films at this time was that they were always n-type (except for ZnTe which is naturally p-type). As-grown materials were heavily compensated even when intentionally doped with donors.

This review will describe in some detail a selection of papers published in the last decade (1980–1990). The number of publications is quite impressive, considering the lack of success in the device area. The most popular of the II–VI materials is, by a wide margin, zinc selenide.

2.1.2 The period 1980–1985

This period provided a more thorough assessment and characterization of epitaxial films; this was particularly important, bearing in mind that growth was on foreign substrates. It had become apparent that although the VPE growth technique provided a reasonably straightforward means for producing thin epitaxial films, the characterization of the opto-electronic properties of layers had not been a feature of the published work. In addition to the conventional electrical assessment techniques such as van der Pauw and Hall measurements, it was natural to make full use of the techniques and expertise developed specifically for the III–V compounds, particularly low temperature photoluminescence (generally the word 'photoluminescence' will be abbreviated to PL in this text, and excitation is from an Hg lamp unless otherwise stated; the word 'epitaxy' will be used for all films, including heteroepitaxial layers).

The purity of the starting materials had long been associated with the electronic properties of other semiconducting materials; it is therefore not surprising to find reports of layers grown from pure elemental sources (Zn, Zn, S). In addition to the prospect of improved purity compared with crystalline or powder sources, the technique provides means for controlling the stoichiometry of layers; at that time vacancies were thought to be the main cause of dopant compensation. Unfortunately, with regard to purity, the reactor temperatures are not significantly reduced and so contamination with impurities from the growth systems is similar to that in systems using a solid source.

Scott *et al.* [9] described the growth of ZnSe on GaAs by this technique (1981). In their four-zone growth apparatus (Fig. 2.2), elemental sources were

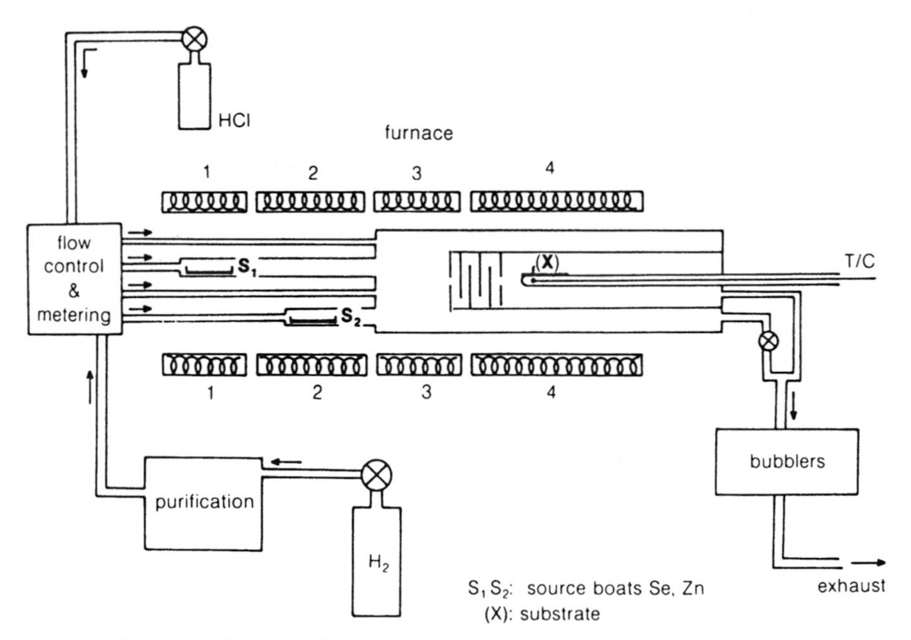

Fig. 2.2 Schematic diagram of the growth apparatus. (From ref. 9.)

vaporized in H_2 flow and further diluted in H_2 before passing through a high-temperature mixing zone; deposition occurred in zone 4 of the system, with unreacted material condensing out at the cold end beyond the substrate. Before growth, the substrates were degreased and etched in the standard 3:1:1 solution of H_2SO_4–H_2O_2–H_2O; additionally, an *in situ* HCl etch was carried out to remove an oxide surface layer left from the liquid etch (as detected by photoelectron spectroscopy) immediately prior to deposition. Deposition on the reactor walls took place in the range 900–500 °C, but the substrates were attacked by the vapour species at temperatures above 500 °C; for this reason it was necessary to grow a buffer layer at 500 °C before growth at higher temperatures. Optimum growth was established at 760 °C with source depletions at about 0.7 μg min^{-1}, resulting in a growth rate of 0.7 μg min^{-1}. The layers were assessed by 77 K PL using an argon laser and by X-ray diffraction; typical PL results are shown in Fig. 2.3. The layers

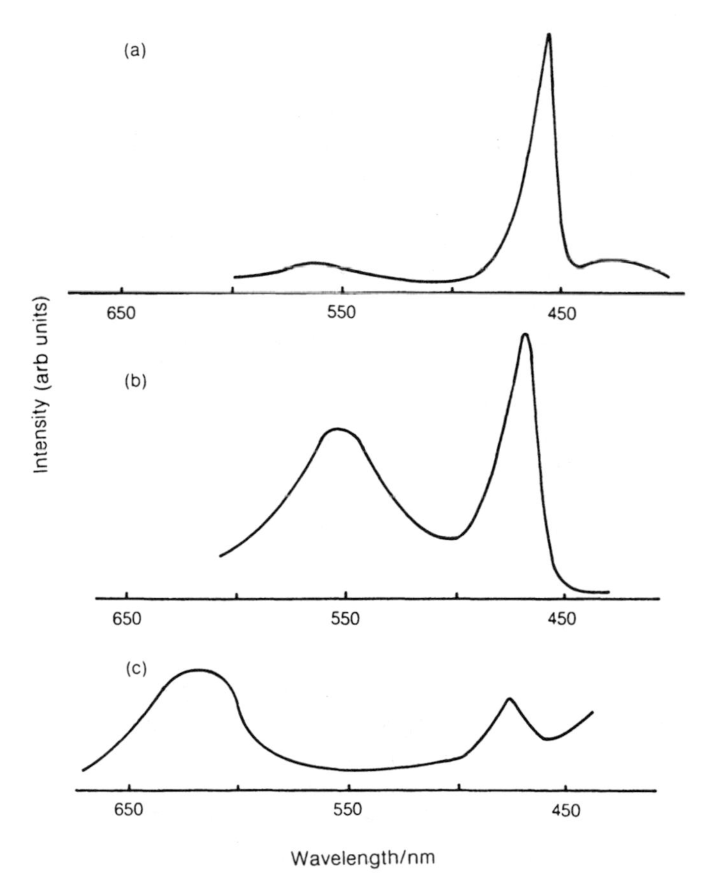

Fig. 2.3 PL spectra at 77 K of ZnSe epitaxial layer: (a) 1:1 Zn:Se stoichiometry; (b) containing Zn excess; (c) containing Se excess. (From ref. 9.)

were polycrystalline at low substrate temperatures, and single crystal with growth rate increasing with higher temperatures up to 760 °C (similar to the pattern of growth for ZnS reported in ref. 3 above). The PL results highlighted the significance of controlling the Zn:Se ratio to unity for optimum optical properties, combining minimum deep level and strong near-band-edge (NBE) emissions. This paper does not report on room-temperature 300 K PL (suggesting perhaps an absence of blue emission).

A particularly detailed investigation of the growth of ZnS on GaP using this technique was reported by Matsumoto *et al.* [10]. However, the growth temperatures in their system are rather high (663–780 °C), suggesting significant autodoping from the substrates. They investigated most of the system variables in relation to surface and crystal quality; growth on the (111)P substrates resulted in featureless and smooth surfaces but with twinned structure (a common feature of ZnS). Cracks were a feature of all layers, caused by the thermal mismatch between layer and substrate (lattice match is quite close in this system).

A more detailed analysis of the crystal structure of the layers grown by this technique was given by Matsumoto and Ishida [11] in 1984. The authors appreciated the significance of polymorphism in these epilayers, in relation to opto-electronic properties. However, they did not make use of buffer layers in their investigation. The experiments had been extended to all the Zn chalcogenides, and to growth on (111)Ga, (111)As, and (100) surfaces of GaAs substrates. They were able to relate crystallinity, as assessed by electron diffraction, to the electronegativity differences between the constituent elements (Table 2.1): the tendency for crystallization in the wurtzite form, or to twin, increases with ionicity. For instance, the ionicity of ZnTe is low enough for the stable zinc blende structure, whereas ZnSe and ZnS have near-critical ionicity and can therefore form a mixed structure.

One of the earliest reports of VPE growth at lower temperatures was in 1983 by Muranoi and Furukoshi [12] in which they describe the growth of ZnSe from Zn and Se sources in H_2 flow, on GaAs substrates in the temperature range 180–400 °C; the (100)GaAs substrates were not only chemically etched in a standard 3:1:1 solution but thermally treated in H_2 flow prior to deposition. Growth rate was necessarily low to avoid powder deposits. Epitaxial deposits were obtained on substrates above 230 °C; layers grown at lower temperatures could be converted to epitaxial films by annealing in H_2–Ar for about 1.5 h at 400 °C (through the process known as solid phase epitaxy). Electrical properties of the films were established; the material was n-type with low carrier concentration and quite high mobility ($\sim 250\,cm^2\,s^{-1}$); resistivity was about 1 Ω cm for layers grown at 400 °C with a Zn:Se ratio ~ 1, whereas the layers were insulating for ratios departing appreciably from this value. Clearly these films are of high quality; however, there is an absence of PL spectra, and therefore no discussion on impurities.

There were several reports in this period describing advances in the growth

Table 2.1 Effect of substrate materials and orientations on structural properties of vapour phase epitaxial layers of ZnS, ZnSe and ZnTe grown at temperatures between 650 and 850 °C (from ref. 11)

Epitaxial layer	Substrate	Crystal structure
ZnS	GaP(111)Ga[a]	W[b]
$\Delta X = 0.9$[d]	GaP$(\bar{1}\bar{1}\bar{1})$P[a]	Twinned ZB
	GaP(100)[a]	ZB[c]
	GaAs(111)Ga	W
	GaAs$(\bar{1}\bar{1}\bar{1})$As	Twinned ZB
	GaAs(100)	ZB
	Ge(111)[e]	W, twinned ZB
	Ge(100)[e]	Twinned ZB
ZnSe	GaAs(111)Ga	Twinned ZB
$\Delta X = 0.8$	GaAs$(\bar{1}\bar{1}\bar{1})$As	ZB, twinned ZB
	GaAs(100)	ZB
ZnTe	GaAs$(\bar{1}\bar{1}\bar{1})$As	ZB
$\Delta X = 0.5$	GaAs(100)	ZB
	GaP(111)Ga	ZB
	GaP$(\bar{1}\bar{1}\bar{1})$P	ZB
	GaP(100)	ZB

[a] Reference 11.
[b] Wurtzite structure.
[c] Pauling's electronegativity difference between the constituent elements.
[d] Zinc blende structure.
[e] Reference 14.

from powder or crystalline sources of these compounds. Heime *et al.* [13] reported on a rather ambitious project, involving a direct comparison between epitaxial ZnS_xSe_{1-x} layers of various compositions on GaP substrates, and bulk crystals of the same compositions grown in reasonably similar conditions, including iodine transport, and the high temperatures needed for bulk growth. The 1.6 K PL spectra of the layers were dominated by deep-level emissions, which is not surprising considering the high growth temperatures and the consequent large amounts of Ga and P in the layers (detected at the 1–2% level).

Besomi and Wessels [14] reported on the growth of ZnSe on GaAs from a powder source, using growth temperatures as high as 800 °C. The substrates were 2° off (100), etched in a 5:1:5 solution at 40 °C for 30 s; both substrates and source were outgassed in H_2 before growth. The authors monitored source depletion rates and growth rate, and found general agreement with theory as modelled by Lilley [4]. A feature of this report is the C–V measurement data for the heterojunction; these results suggested abrupt interfaces, which seems surprising for deposition at such high temperatures. The net donor

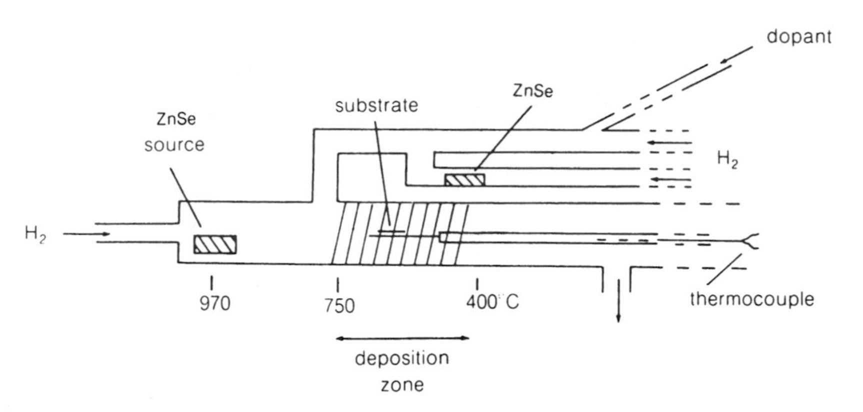

Fig. 2.4 The quartz growth system. 'Standard' growth conditions (resulting in a 1.2 μm ZnSe deposit) for the VPE growth system are shown in the following table. (From ref. 16.)

Source temperature (°C)	970
Substrate temperature (°C)	550
Transport flow (H_2) ($cm^3\,min^{-1}$)	250
Secondary flow (H_2 ($cm^3\,min^{-1}$)	1300
Zn source temperature (°C)	530
Zn transport flow (H_2) ($cm^3\,min^{-1}$)	250
Substrate material	(100) GaAs
Substrate etchant	1:5:1 H_2SO_4:H_2O_2:H_2O
Growth time (min)	90

density was established as $2 \times 10^{16}\,cm^{-3}$; the compensation ratio is unknown (but is likely to be high at these high growth temperatures).

A thorough investigation of the VPE growth of ZnSe was carried out by the Manchester University group, in collaboration with a group at Hull University who were able to record and analyse 4 K PL data (the results of this work are recorded in detail in a thesis by Czerniak [15]). The first of three publications on this work [16] describes the initial objective, namely the growth of epilayers on GaAs with quality good enough to allow the use of 4 K PL for assessment, and hence to establish the role of certain growth variables in the incorporation of trace impurities. To this end it was found informative to monitor 300 K PL using a low excitation mercury lamp in addition to the 4 K PL spectra (the former avoids the saturation of deep levels). Figure 2.4 shows the system and standard growth conditions; the lowest practical growth temperature was ~500 °C. The initial approach was to assess the importance of source material on the purity of layers. The following materials were used:

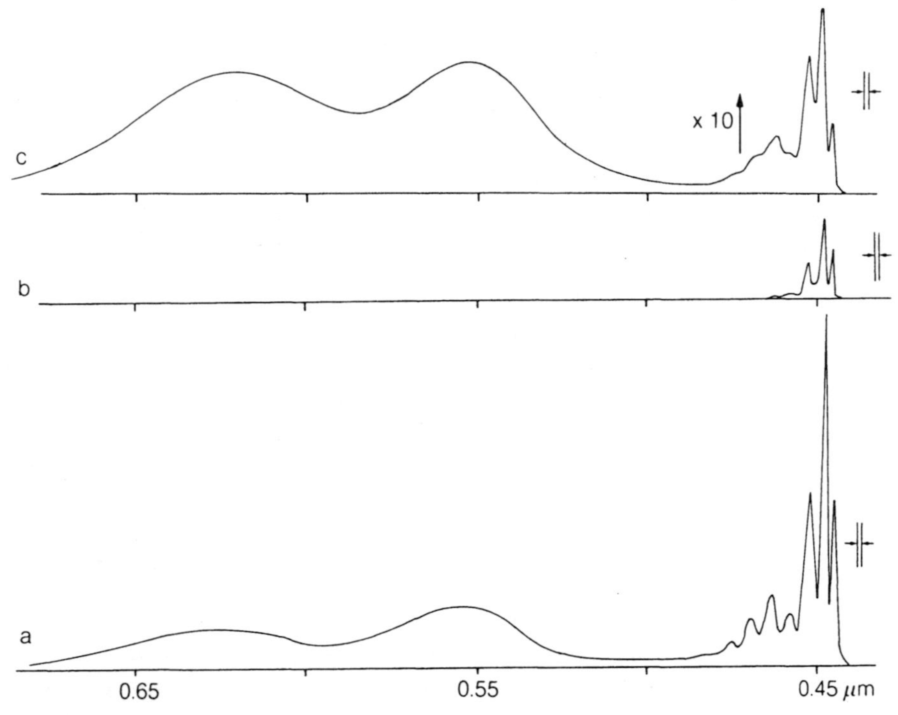

Fig. 2.5 2 K PL spectra for layers grown from different source materials: (a) laser window polycrystalline ZnSe; (b) Merck powder ZnSe; (c) high-purity zinc and selenium. (From ref. 16.)

1. polycrystalline 'laser window' ZnSe;
2. high-purity Merck Powder ZnSe;
3. separate elemental Zn and Se sources.

The powder source gave the cleaner layers, as indicated by the 4 K PL spectra (Fig. 2.5), whereas the elemental sources seemed to yield layers of inferior quality. Detailed work showed that if polycrystalline source material was re-used then the layer quality was greatly improved: deep-level emission at 300 K was reduced by an order of magnitude, and resistivity reduced by 4 orders to $\sim 0.1\,\Omega\,\text{cm}$. It was concluded that an impurity such as copper was readily leached from grain boundaries near the surface of the source material, and consequently first-run layers were contaminated; significantly, the solid source is luminescent in a surface region, but only after use in the growth system, indicating changes in impurity–native defect state. A low substrate temperature ($< 600\,^{\circ}\text{C}$) was found to be essential for achieving good quality layers, and thicker layers (longer growth times) resulted in further improvements in the PL spectra. The 300 K spectra remained dominated by deep-levels, although NBE emission was in evidence.

A detailed account of the above work was published in 1982 [8], describing 300 K PL and van der Pauw–Hall measurements; at this stage, the work concentrated on the effect of system variables on 300 K opto-electronic properties, although 4 K PL was used to monitor layer quality. Layers grown under standard conditions give 4 K PL spectra with well-defined NBE peaks: an I_2 line associated with the shallow donor Ga, and a strong I_{1d}(Cu) line; at longer wavelengths are weaker donor–acceptor (D–A) lines corresponding to Al–Li and Al–Na complexes on zinc sites, which are shown to originate mainly from the source materials, and broad deep-level emissions at 620 nm and 570 nm, associated with Ga–As–Zn vacancy complexes. The 300 K PL BE emission is greatly increased if source material is re-used many times; the 4 K PL spectrum of the 'used' source material was found to be comparatively clean, compared with As-received ZnSe. Pre-growth substrate etching was found to be significant; layers grown on substrates etched in a 1:5:1 solution provided layers with weak deep-level 300 K emission, compared with 1:1:5 or 5:1:1 etchant. The BE emission also increased with layer thickness, but saturated for layers thicker than 5 μm. High substrate temperature (700 °C) produced weak BE and strong deep-level emissions, mainly associated with compensation kinetics rather than increased contamination from the GaAs substrate; resistivity was low at 0.2 Ω cm with the usual anomalous low donor activation energy of 9 meV (now known to be characteristic of heavily or even moderately doped material). Zn overpressure was a necessary factor in achieving BE emission. The use of buffer layers, grown at 500 °C under high zinc overpressure, as substrates for growth at 640 °C, resulted in good quality high resistivity layers, probably because of less outdiffusion from the substrates. The deep-level spectra were found to be sensitive to zinc and selenium overpressure; in addition the latter resulted in insulating layers, which suggests that defect–impurity complexes dominate the opto-electronic properties. Deposition on GaP and Al_2O_3 substrates under otherwise standard conditions failed to produce high quality layers, and BE emission was not observed; Ge substrates, however, produced improved PL properties. The authors conclude that trace acceptors dominate the properties of the layers and suggest the band diagram of Fig. 2.6. An analysis of the impurity state of the layers, as assessed from the 4 K PL spectra, is given by Lilley *et al.* [17].

Umar-Syed and Lilley grew layers of ZnSe on GaAs at substantially lower substrate temperatures [18] from a powder source. The novel growth apparatus (Fig. 2.7) provides a source vaporization chamber separated from the deposition zone by a transport tube used in conjunction with a high temperature gradient between them; in this way the transporting flux was not totally lost to the reactor walls at higher temperatures. Figure 2.7 also shows the standard conditions used for growth in this reactor. Layers grown at temperatures above 500 °C resulted in the usual strong deep-level 300 K PL emissions; depositions below 480 °C resulted in dramatic changes in the PL spectra: the 300 K luminescence was negligible throughout the visible region.

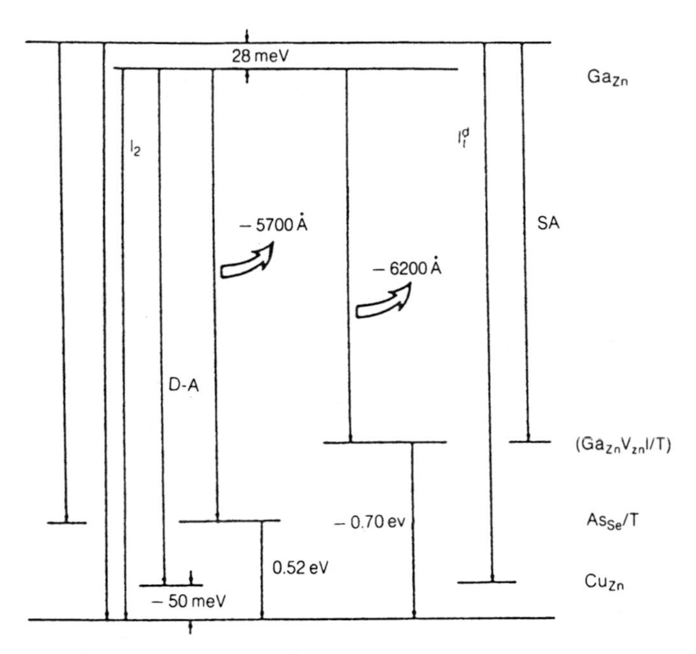

Fig. 2.6 Schematic band diagram for the recombination levels in the visible region of the spectrum: VPE ZnSe grown on GaAs. (From ref. 8.)

Even at 77 K the PL was weak, except for layers grown below 300 °C. A typical 4 K PL NBE spectrum is shown in Fig. 2.8, and is dominated by a rather broad NBE line, thought to be due to several donors even though the emission is at a wavelength associated with acceptors in ZnSe, (stress has caused a shift in what actually is a donor emission); significantly, there is an absence of shallow compensating acceptors. The quality of these layers was clearly due to an advancement in the VPE system technology. Attempts were made at Li doping from an LiS_2 source (the available material was rather impure); this resulted in blue emission at room temperature (Fig. 2.9), probably as a result of donor impurities. Doping with In also gave strong blue, but not without deep-level emission.

Nishio *et al.* [19] reported the growth of ZnTe homoepitaxial layers at 862 °C. This work describes growth rate kinetics for the ZnTe–H_2–I, ZnTe–H_2, and ZnTe–Ar systems, along with electrical properties of the layers. Another interesting ZnTe paper was published by Radautsan *et al.* [20], describing the electroluminescent properties of p ZnTe–n InP heterojunctions: interface states played an important role in enhancing emission.

The works described above indicate the complexity of the subject, in relation to the production of high-quality layers with controlled opto-electronic properties, and the different routes followed in attempting to achieve

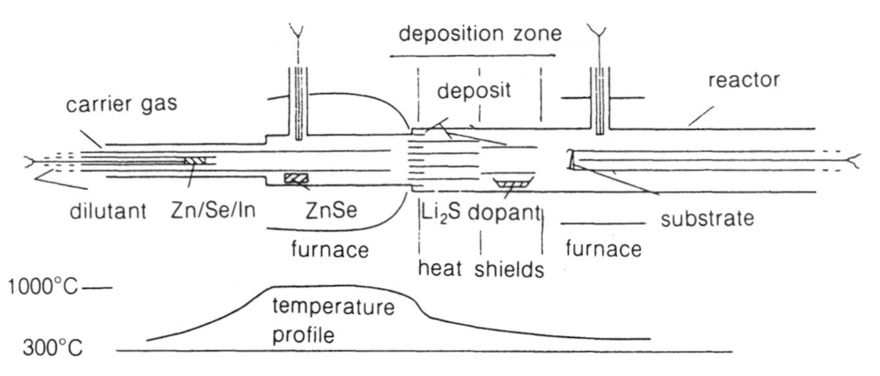

Fig. 2.7 A system for the low-temperature growth of ZnSe by VPE. (From ref. 18).

Source temperature (°C)	970
Substrate temperature (°C)	250–700
Carrier H_2 flow rate ($cm^3\,min^{-1}$)	600
Dilutant H_2 flow rate ($cm^3\,min^{-1}$)	600
Substrate material	(100)GaAs
Pre-growth etch	1:5:1 H_2SO_4:H_2O_2:H_2O (60 s at 50°C)
Source material	Merck ZnSe powder or AWRE laser window ZnSe

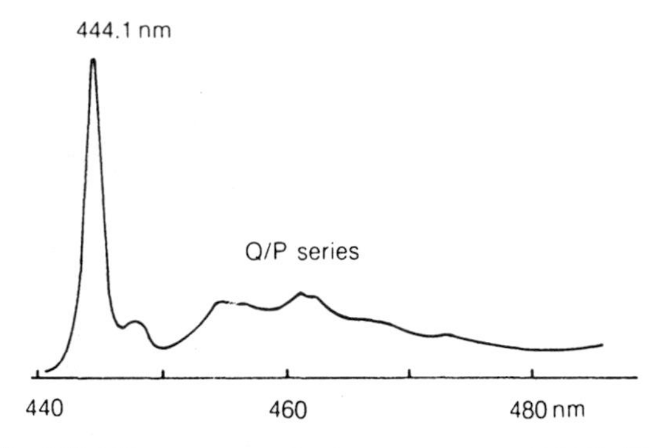

Fig. 2.8 The BE 2 K spectrum of a layer grown at 300 °C. (From ref. 18.)

this goal. This period provided a mass of detailed information on several aspects of the subject; the interpretation of this data as a unified whole appears to be impossible, and different views could be expected. In addition, the particular skills associated with particular groups are not likely to be abandoned in favour of alternate techniques for which new skills would be necessary. It is therefore not surprising that the next half of the decade

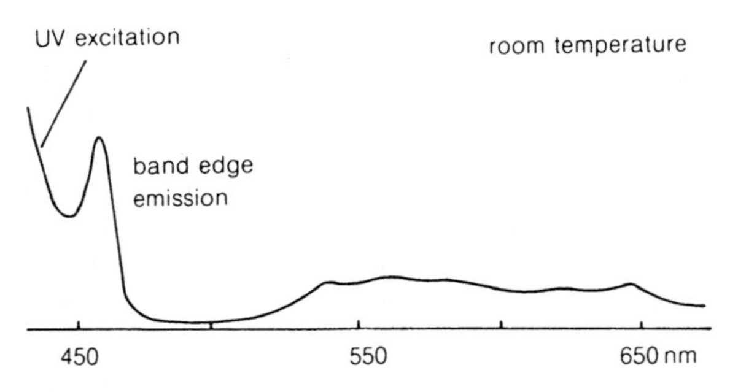

Fig. 2.9 Room-temperature PL from a ZnSe layer grown at 300 °C doped from an Li_2S source. (From ref. 18.)

continued to produce advances on many fronts. For example, we will find that, despite the current trends at that time, low substrate temperature work is by far the least favoured approach.

2.1.3 The period 1986–1990

Matsumoto *et al.* [21] reported the growth of layers of ZnSe on GaAs substrates from elemental sources, using an improved VPE apparatus involving a hydrogen plasma; low temperatures (270–320 °C) and reduced pressure (H_2 gas at 10 kPa) are features of the growth conditions. The advantage of the plasma is that the substrates are more thoroughly cleaned within the apparatus prior to deposition. The layers, grown in the plasma, were assessed by 4.2 K and 300 K PL. The former indicated a dominant shallow donor line (I_2), and, for (100) orientations, an absence of deep-level emissions; at 300 K the emission is mainly blue, with a weaker but broad deep-level emission over the remaining part of the visible spectrum. This method provides a new advance in the pre-growth treatment of substrates in this field. In a subsequent paper from this group [22] very-high-quality layers were grown without plasma during growth, but again at a low pressure (21 kPa). In this case, 4.2 K PL indicated that free-exciton emission was some 7 times stronger than that of the residual bound-donor exciton. These layers were grown at 420 °C.

Indium vapour introduced into the H_2 transport flow ahead of the source has been shown by Goto *et al.* [23] to improve the crystal quality of ZnS layers on GaP. The growth rate on the (111)B surfaces at 700 °C was independent of source temperature in the range 800–900 °C; this was in contrast to (100) deposition, suggesting differences in growth kinetics. Surface quality was greatly improved, particularly for the (100) orientation, with an absence of facets and microcracks normally observed in ZnS grown on GaP. Furthermore, similar but homoepitaxial layers deposited on

(100)ZnS substrates were not smooth. The authors conclude that indium provides a stress-relieving mechanism at the initial interface (the stress is associated with the lattice and thermal mismatch). The donor concentration in the layers was estimated as 10^{18}–10^{19} cm^{-3}, as measured by secondary ion mass spectrometry (SIMS). A feature of this paper is the range of different PL measurements, at various temperatures and using different optical sources; the results indicated a substantial increase in BE emission, and a reduction in Ga and P autodoping from the substrates, compared with the heteroepitaxial layers.

Mirror-like featureless layers of ZnSe grown on (100) surfaces of GaAs under low pressure and indium doping conditions have been reported by Matsumoto *et al.* [24]; in this case low substrate temperatures were used (~280 °C), with the usual elemental source growth system associated with the Yamanashi group. Resistivity could be controlled over the range 10 000 to 0.01 Ω cm by varying the In source temperature in the range 400–550 °C; mobility was high (> 200 cm^2 V^{-1} s^{-1}), and carrier concentration varied from 10^{16} to 10^{18} cm^{-3} almost linearly with indium source temperature. The low-doped materials appeared to have a donor level energy of 6.3 meV and $N_a = N_d$ (heavily compensated); Fig. 2.10. summarizes the electrical properties. 4.2 K PL spectra are dominated by a shallow donor emission line (as were undoped layers); the free-exciton lines were also resolved. At 300 K, the dominant emission was BE blue.

Goto *et al.* [25] have reported a thorough numerical analysis of the mass transport kinetics associated with the growth of ZnS_xSe_{1-x} from solid sources of ZnS and ZnSe in hydrogen flow. The authors were particularly interested in compositions with high sulphur fraction, to give layers lattice-matched to the GaP substrates. The solution of some 15 sets of equations produced interesting growth rate–substrate temperature characteristics: growth rate rises to a maximum at the highest temperatures for deposition and decreases to a minimum at lower temperatures, before rising to a secondary maximum at even lower temperatures. However, it is not apparent which variables are responsible for the secondary peak; this inability to 'see' the physics of the processes is of course a feature of all numerical analysis. The modelling indicates that by simply varying the substrate and/or source temperatures, the sulphur fraction can be controlled over the range 0.44–0.92: this was observed in practice. 300 K PL spectra indicate strong BE emissions, with an absence of deep-level emission for sulphur fraction above 0.60; however, a pulsed nitrogen laser was used to excite the luminescence, and this may have saturated the deep-level luminescence. A second paper [26] analyses the control of alloy composition by using a bypass flow as part of the gas flow system. The authors conclude that this method is far superior; in addition to allowing the fine control of layer thickness, layer composition can be changed more abruptly.

Another report of a computer analysis of mass transport and growth rate

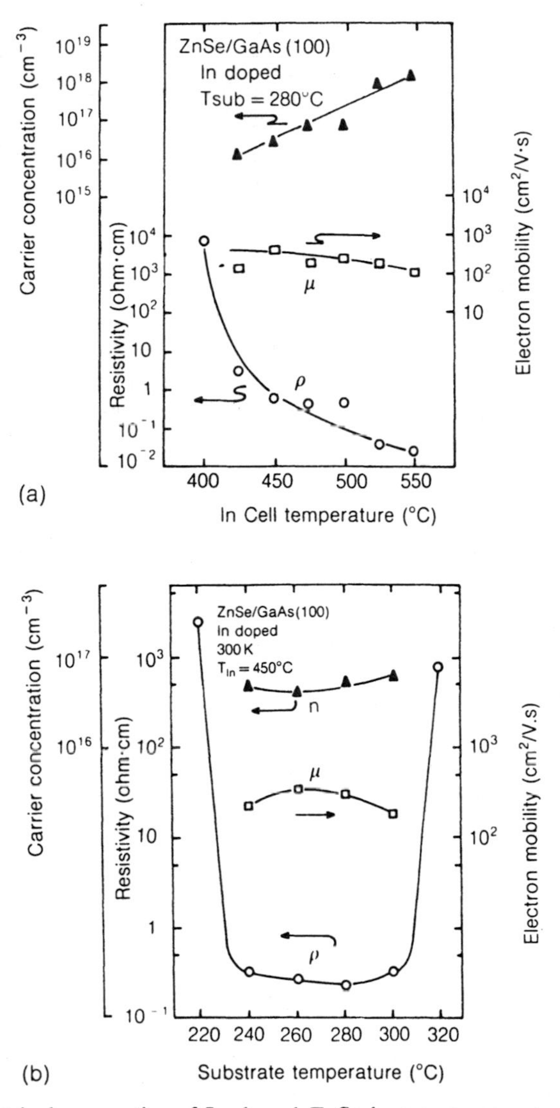

Fig. 2.10 Electrical properties of In-doped ZnSe layers at room temperature as a function of (a) In cell temperature and (b) substrate temperature. (From ref. 24.)

is given by Hartmann *et al.* [27]. They concentrate efforts on maximizing mass transport from solid sources in hydrogen with iodine as a means for growth at lower substrate temperatures (450–600 °C). The specific details of the model are not given. Growth of ZnSe on GaAs in hydrogen flow at high substrate temperatures ($\sim$ 700 °C) is reported. Significantly, the 300 K PL spectra (N_2 laser) are dominated by BE emissions; the layers are of course highly conducting (iodine being a shallow donor).

From 1988 onwards, the publications move towards reporting experiments directed at device properties; the impression is that the earlier literature has given the base knowledge for the deposition of high quality ZnS and ZnSe layers on both GaAs and GaP substrates, and the impurity–defect state of the layers is fully understood although not totally controllable. It is therefore reasonable to expect device-specific work to begin to dominate the literature. An example of this is the work of Zhang *et al.* [28], describing experiments aimed at doping ZnS with the acceptors nitrogen and phosphorus. The growth conditions, which include high substrate temperatures, normally yield highly conducting n-type layers of ZnS on GaAs substrates (0.2–0.02 Ω cm); the addition of acceptors in significant amounts would be expected to compensate this conductive state and to result in insulating material. As acceptor states are deep compared with donors, insulating material would indeed indicate heavy acceptor doping, as long as the donor density was not significantly reduced by the changed experimental procedure and the material consequently intrinsic. The authors provide a particularly convincing report showing that the particular objective of incorporating acceptor dopant had been achieved. The result is presumably significant to a metal–insulator–semiconductor (MIS) blue emitting device, but not to producing p–n junctions. Nitrogen doping was achieved during growth, by including a few per cent of ammonia in the H_2 transporting gas stream in conjunction with a powder source of ZnS, whereas phosphorus doping was by the thermal diffusion technique (after growth) in a sealed quartz ampoule at 500 °C. The incorporation of phosphorus was detected by an electron probe microanalyser; Auger spectroscopy was used for nitrogen; the largest concentration was estimated at $\sim 10^{20}\,cm^{-3}$. A SIMS analysis established that there was no significant change in the Ga and As autodoping profile, as compared with undoped layers; the same applied to copper ($\sim 10^{18}\,cm^{-3}$). Resistivity was estimated from I–V characteristics of the ZnS–GaAs heterojunction sandwiched between indium ohmic contacts; N-doped layers were $\sim 10^{6}\,\Omega\,cm$, whereas P doping gave 10^{3}–$10^{4}\,\Omega\,cm$ material. Figure 2.11 shows features of 77 K PL spectra as found in the various doped and undoped layers, and using different excitation wavelengths. The undoped samples and P-doped layers produced the A, B, C, and D peaks; the B peak appeared for all samples and is assigned to edge emission. The A peak is identified as due to free excitons, and the E emission appears only for N-doped material and with excitation energy just below the band gap. Trap levels were monitored by transient thermoluminescence; it was found that doping did not produce additional trap levels.

A paper with particular significance with respect to p-doping of ZnSe layers on GaAs was published by Umar-Syed and Lilley [29]. The authors have modified the system reported earlier [18] to have a steeper temperature gradient between source and substrate. This had a significant effect on the impurity state of layers grown at low temperatures (350–550 °C): the 10 K

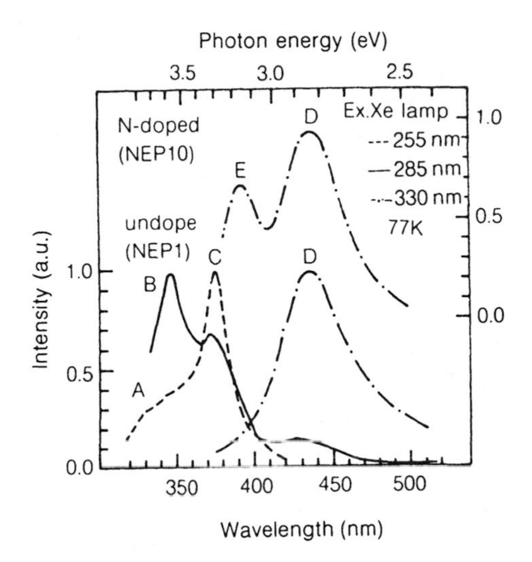

Fig. 2.11 Ultraviolet PL spectra at 77 K for N-doped and undoped ZnS/GaAs samples under various excitations with different photon energies from the Xe lamp. (From ref. 28.)

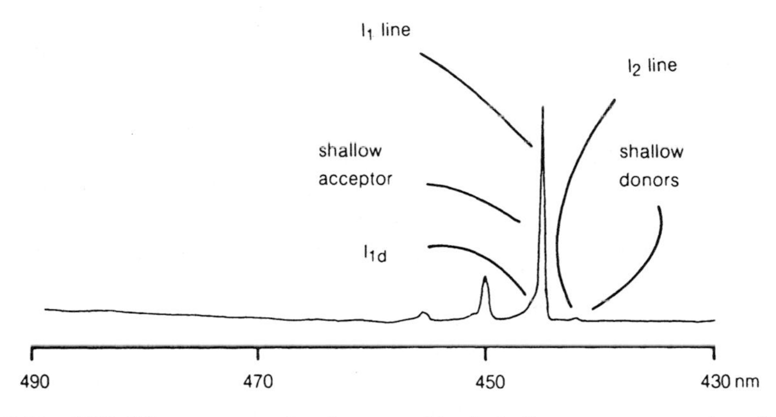

Fig. 2.12 10 K PL spectrum showing a residual shallow acceptor and negligible concentrations of shallow donors. (From ref. 29.)

PL spectra are dominated by a single shallow acceptor emission, with an extremely weak donor emission, as shown in Fig. 2.12. The electrical properties are reported as insulating n-type, but the likelihood of erroneous measurement is very high; indium contacts were used, and these would be highly rectifying on p-type material. The growth conditions for these films appear to be ideal for p-doping and the production of light-emitting diodes (LEDs) (unfortunately, the ever-increasing difficulties in attracting support for this work have temporarily halted this research).

Matsumoto *et al.* [30] have reported on an investigation into the relation between the quality of heteroepitaxial layers and the quality of the GaAs substrates. It was hitherto assumed that the substrate would not be a limiting factor with regard to layer quality; improvements in deposition technology have been seen as the way ahead towards device-quality material. This type of investigation required good-quality material as a starting point; the authors' low-pressure and low-temperature (280 °C) growth conditions, along with plasma substrate cleaning, ensure layers of high quality. Particularly thin layers were deposited on the substrates; this was to ensure that substrate-related growth features were not relaxed by misfit dislocations, and that these features were stable, even after the sample was cooled. A liquid encapsulation Czochralski (LEC) grown (100)GaAs wafer was used for the depositions; ten samples were taken from a wafer along a diameter parallel to the [100] axis, etched in 5:1:1 solution, and then plasma etched within the system. The grown layers were characterized by monitoring two factors: the free-exciton peak intensity in the 4.2 K PL spectra, and X-ray diffraction peak width and intensity. The results indicated a definite correlation between these factors and the etch-pit density profile as measured across the wafer surface; the authors conclude that epitaxial layer quality is currently limited by the substrate quality.

A more basic paper on the growth of ZnSe films on GaAs, using elemental sources, was published by Kyotani *et al.* [31] in 1989. The general pattern of results observed by others for growth from a powder source was found to apply to this system, at least in the range of substrate temperatures investigated (500–650 °C); there are, however, additional experimental results that are worth noting. The authors have experimental evidence for the long-appreciated need (a) for source material of the highest purity, and (b) to restrict the maximum system temperature to minimize reactions with the quartz reactor; the latter avoids contamination with silicon, as reported in much earlier works [3, 32]. Transport rate was controlled accurately by the temperature (to ± 0.1 °C) of the elemental sources; a Zn:Se ratio of unity was found by experiment to give both highest growth rate and BE photoluminescence. Figure 2.13 shows SIMS profiles for various elements including Li and Na; these particular impurities are thought to originate from the substrate surface. The improvement in PL emission characteristics is a consequence of the lower diffusivity of the impurities trapped at the mismatched ZnSe–GaAs interface in addition to the reduction in autodoping. One possibility is that the substrate etchant is the source of impurities; a feature of this work is the length of the thermal treatment (30 min) given to the substrate before deposition of ZnSe. The substrates are heated in flowing H_2 at 523 °C for 30 min as a means for achieving low concentrations of D–A pairs. Further experiments involving homoepitaxial growth indicated that impurities originating in the substrates have a strong influence on the purity of these layers. Comparisons with the layers grown on GaAs suggest that

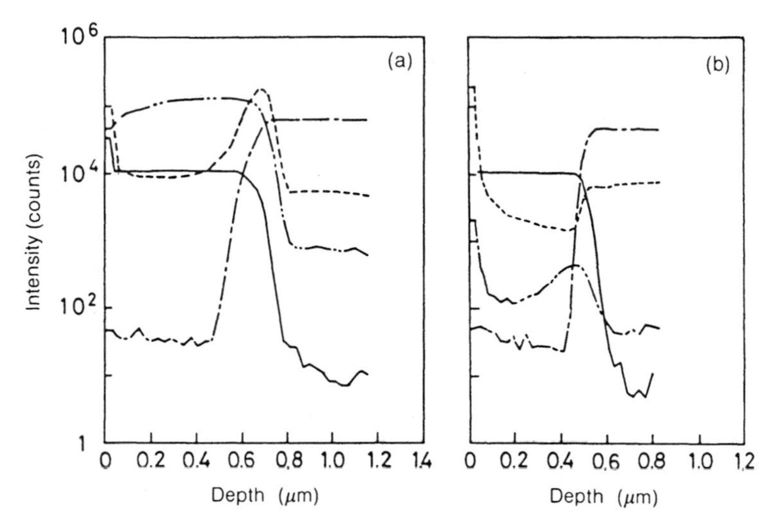

Fig. 2.13 Profiles of the impurities in the epitaxial films grown on the (100)GaAs substrate at substrate temperatures of (a) 873 K and (b) 723 K: —— - ——, ^{75}As; ————, ^{74}Se; - - - - -, ^{23}Na; —— - ——, ^{7}Li. (From ref. 31.)

the defective GaAs–ZnSe interface prevents much of the interdiffusion associated with autodoping. This suggestion of what is basically a gettering effect is particularly interesting, and could well help to explain certain results in other publications.

In 1989, there were two reports of the growth of p-type heteroepitaxial layers, with low resistivity; significantly, these depositions were at high substrate temperature and from powder sources of the compound. Iida *et al.* [33] report p-type conduction in ZnS grown on GaAs, and give the results of several different experiments in favour of accepting the main conclusion, namely that p-type layers have been produced (rather than the possibility of a p-type conversion layer in the GaAs at the interface). Layers were grown at 670 °C, with the source at ~1000 °C, with a transport gas of $H_2 + NH_3$; layers without NH_3 are conducting n-type. P-type layers were grown on n- and p-type GaAs, in addition to n-ZnSe–n-GaAs substrates. The results are summarized as follows.

1. Thermopower measurements indicate that a Zn overpressure was necessary to achieve p-type deposits; otherwise the layers are insulating n-type (this agrees with unpublished work in our laboratories for which Zn overpressure was not used).
2. The above is observed for deposition on both n-type GaAs and n-ZnS–n-GaAs, giving apparently strong evidence for the absence of a p-type conversion of the GaAs.

3. Hot probes were also employed on layers after the substrate had been removed by etching, again indicating p-type material.
4. *I–V* characteristics of the various heterojunctions show clearly that layers previously identified as p-type on n-GaAs form highly rectifying junctions with the substrates.
5. Electroluminescence from the forward-biased diodes is at infrared wavelengths, corresponding to the bandgap of GaAs; the authors assume this is caused by hole injection from the p-type ZnS (it can be expected that the barrier to holes is smaller than that for electrons for this p–n heterojunction, owing to the differences in the band offsets; the injection of electrons into the ZnS and the resulting UV-blue luminescence is therefore highly unlikely).
6. Hall effect measurements at room temperature gave the hole concentration as $\sim 6 \times 10^{18}\,\mathrm{cm}^{-3}$, and mobility $40\,\mathrm{cm}^2/\mathrm{V}^{-1}\,\mathrm{s}^{-1}$.
7. X-ray diffraction spectra were the same for both n- and p-doped ZnS layers, indicating that the two layers were the same basic material, eliminating the possibility of a III–V:II–VI compound.

The authors conclude that the zinc overpressure reduces the autodoping by Ga from the substrates and increases the probability of nitrogen doping on sulphur sites (the real explanation may be more complex than this).

Finally, there has been another report on p-type heteroepitaxial layers, in this case ZnSe on GaAs substrates; in contrast to the work of Iida *et al.*, the forward-biased junction emits in the blue region of the spectrum. The paper, by Stucheli and Bucher [34] of Universitat Konstanz, describes the growth of ZnSe films by iodine transport in H_2 from a powder at low source temperature ($\sim 800\,°\mathrm{C}$), on GaAs substrates in the temperature range 550–650°C. The use of iodine to achieve p-type conductivity in this material presents the II–VI community with a re-run of an old story that first appeared in the works of Robinson and Kun over ten years ago; these authors grew bulk crystals by iodine transport, and type-converted them after annealing in zinc with a group III element such as indium. It was generally considered that the whole complex process allowed trace impurities (Li and Na) to be fully activated as acceptors. With this in mind we must look at the current report in some detail, in the hope of understanding the nature of the acceptor; the report does not comment at all on what the acceptor may or may not be. It must be said that the Manchester group and others have found layers (not grown in iodine) with apparently high p-type conductivity; these were assumed to be erroneous results caused by an interface layer in the GaAs substrate.

Hall measurements, using gold contacts, on layers grown in the temperature range 550–650 °C, indicated that p-type resistivity decreased with growth temperature by 6 orders of magnitude from 6×10^5 to $0.1\,\Omega\,\mathrm{cm}$. However, this was for the case of the lowest iodine source temperature ($T_i = -10\,°\mathrm{C}$); for higher T_i (0, 10, 20 °C) the changes in resistivity were smaller, ranging between 1 and 10 Ω cm. The resistivity, mobility, and carrier concentration were measured as a function of sample temperature. Resistivity was constant, indicating degenerate doping levels; mobility appears to follow a normal pattern of increasing with temperature to a maximum ($\sim 150\,cm^2\,V^{-1}\,s^{-1}$) at about 150 K, and decreasing to about $90\,cm^2\,V^{-1}\,s^{-1}$ at room temperature. Carrier concentration is constant except for a dip at low temperatures. These measurements were repeated for homoepitaxial layers grown in otherwise similar conditions: the authors feel this is strong evidence for accepting the results of p-type highly conducting epilayers. Photoluminescence at 1.8 K is particularly interesting; the BE spectra are quite well resolved, and indicate several shallow acceptor and donor emissions (Fig. 2.14). Bearing in mind that current knowledge suggests that donors are substantially more shallow than acceptors, the p-type conductivity is surprising. However, there are unidentified acceptor emissions that have not been observed elsewhere; significantly the strength of one of these (I_c) increases with growth temperature, leading the authors to suggest that the acceptor associated with I_c causes the p-type conductivity.

The $I-V$ characteristics of the metal–heterojunction–metal system are also reported in the above publication. Reasonable rectifying behaviour was observed for gold contacts to the ZnSe films, and not-so-ideal characteristics for silver and aluminium contacts. The reverse currents are the same for all three metals. The authors suggest the results are in agreement with the Anderson model for this p–n heterojunction. It is noticeable that the Au metal $V-I$ curve has a turn-on voltage of just over 1V: this does not appear to fit with the model, if it is assumed that the forward current is carried by electron injection into the ZnSe (as the authors suggest); the barrier to electrons appears to be about 2.5 eV. A possible answer to this feature is that holes may tunnel through the spike in the valence band at the ZnSe–GaAs interface, causing turn- on at a lower voltage. Another possibility is that the metal contacts are in fact rectifying as well as the junction; the 'reverse' characteristic controlled by the interface, and 'forward' by the contact; the Au contact may be ohmic owing to a low breakdown voltage. There is further evidence for this in the paper: the diodes emit in the blue region of the spectrum in the 'forward' direction; the authors suggest electron injection from the GaAs, but this is unlikely, and hole injection into the GaAs would result in IR emission, as in the Iida report (above). So, we are left with the metal contacts and a further statement by the authors that the emission is from under the metal and in the form of spots—this is probably a field-defect phenomenon. To sum up, this paper has new but unidentified acceptor data in the PL spectra, strong but not total evidence of p-type ZnSe (congratulations to the authors if they really have such control of the p-type doping concentration), and blue electroluminescence from a contact region; the latter is not unusual in this field, but blue is of course unusual, and therefore the results are particularly interesting.

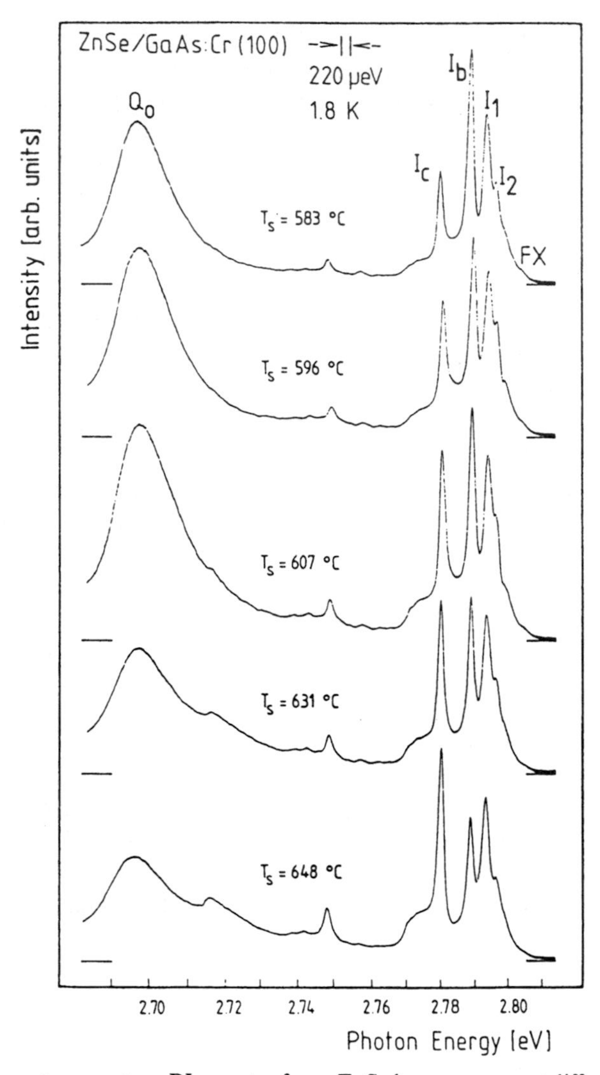

Fig. 2.14 Low-temperature PL spectra for p-ZnSe layers grown at different substrate temperatures T_s. (From ref. 34.)

The possibility of low-cost blue LEDs still dominates the research in VPE of the wideband II–VI compounds. P-type material is the major objective, and reports of achievements in this area are not only from VPE work, but also the more exotic (and expensive) systems such as molecular beam epitaxy (MBE). In all cases there remains the problem of convincing assessment of the electronic properties of the materials, and beyond that we still need to identify the nature of the acceptors. Meanwhile, the best route is to continue to produce emitting structures under a variety of experimental conditions, and then to analyse the results using as many different techniques as possible. We are probably about 5 years away from a full understanding of these materials, and controllable opto-electronic properties.

It may be that a novel VPE process is needed to finally achieve the objectives; atomic layer epitaxy (ALE) [35] could for instance be applied to VPE, as a means of providing additional control over the growth and doping processes.

2.2 LIQUID PHASE EPITAXY

Liquid phase epitaxy (LPE) is seen by many as a suitable method of producing high quality epitaxial layers of, for instance, ZnSe; growth occurs at low temperature and at near-equilibrium conditions. The successful use of this technique for the growth of device-quality epitaxial structures of III–V alloys is the main stimulus for investigating LPE of II–VI compounds. The early part of the decade 1980–1990 saw an assortment of activities in this area, mainly by three groups; these were at the University of Tokyo and Chubu Institute of Technology, Nagoya-Shigai in Japan, and the Philips Laboratories, USA. Significantly, recent work has concentrated on the narrowgap materials (not within the bounds of this text), rather than ZnSe. The following papers provide much information about the basic science such as solubilities and segregation coefficients; in addition there are important contributions to the knowledge of the wideband II–VI compounds. Most of the publications are on ZnSe. To set the scene, a slightly earlier report will be described.

Zn–Ga melt was used by Fujita *et al.* [36], with the main objective of minimizing Zn vacancy concentration and enhancing room-temperature blue PL (Ga was known to increase this emission). A closed system was necessary to cater for the high Zn vapour pressure. A tilting system with graphite boat was used to grow smooth ZnSe films on ZnS_xSe_{1-x} substrates prepared by the iodine transport technique. The ZnSe solute was pre-heated at 950 °C for 48 h as a means of purification, before an etch in boiling NaOH solution. Figure 2.15 shows the growth system. Prior to sealing the ampoule, Zn and Ga are baked in a quartz tube in hydrogen for 2 h at 500 °C. Finally the system is loaded and sealed at a vacuum of 10^{-6} Torr. The growth variables were as follows: start temperature 850–1050 °C, cooling rate 0.27–1 °C min^{-1}, pre-growth hold period 3 h, cooling range 200–350 °C, and Ga mole fraction 0–76%; this represents a thorough experimental approach to the investigation. Smooth mirror-like layers were grown on (100) and (1 1 1)A surfaces, but only at low cool rates. Ion microprobe analysis revealed Na, Si, and Al contamination in the layers. The layers were all n-type, moderately doped, but with quite low mobility; the electrical parameters were independent of growth conditions. Ga in the melt produced a marked effect

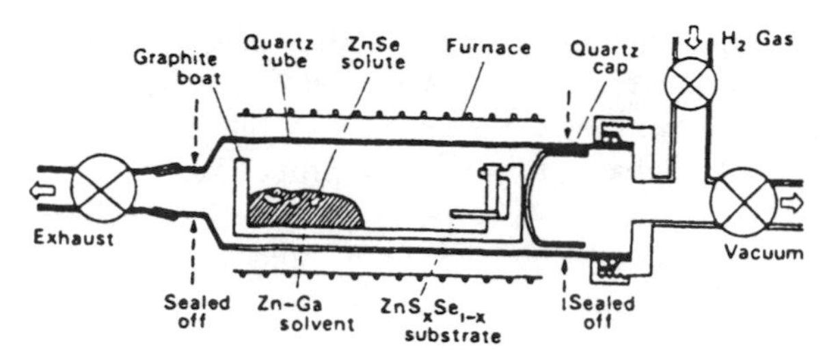

Fig. 2.15 Experimental arrangement for the preparation of the growth ampoule in LPE. (From ref. 36.)

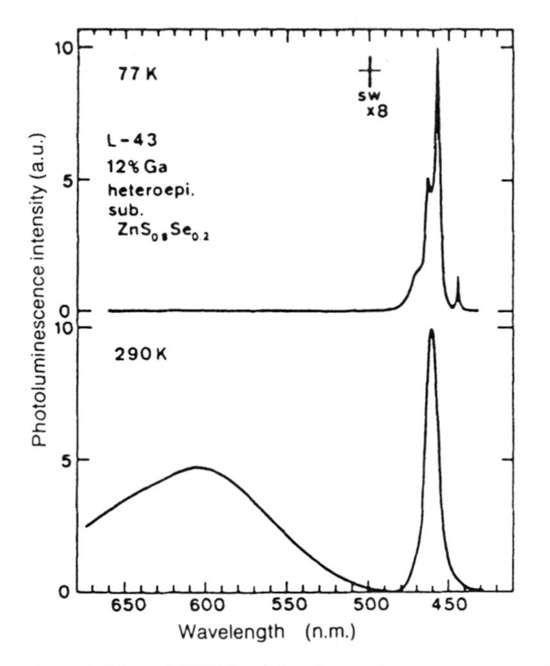

Fig. 2.16 PL spectra at 77 and 290 K of the heterolayer grown on the $ZnS_{0.8}Se_{0.2}$ substrate from Zn–12% Ga alloy solution. (From ref. 36.)

on 300 K PL; a strong blue emission was observed. However, the deep-level emission, although less intense at the broad peak, dominated the spectrum (Fig. 2.16); presumably the initial high melt temperatures are not compatible with low impurity concentrations.

A group at the University of Tokyo has published several papers on the growth of both ZnS and ZnSe layers from various melts. Its initial work [37] was aimed at finding more suitable melts; the zinc melt used by Fujita requires a closed system to cope with the vapour pressure of this element at the growth temperatures needed for reasonable solubility and growth rate.

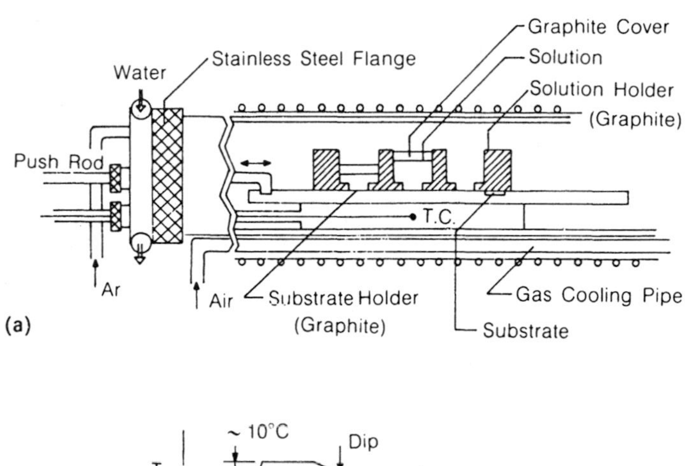

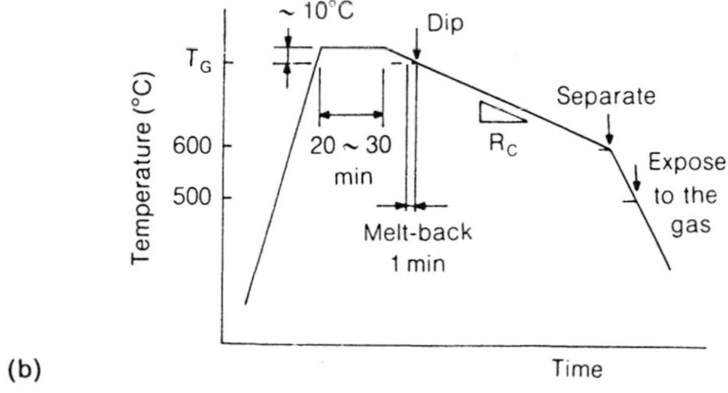

Fig. 2.17 (a) Main part of the growth apparatus for LPE of ZnS, ZnSe and $ZnS_{1-x}Se_x$. (b) Typical temperature program employed for LPE. (From ref. 37.)

It had earlier experimented with the solution growth of ZnS from Te melt, producing crystals with low Te content. Consequently, an open-tube sliding-boat LPE Te melt system was used to deposit epitaxial layers on various orientations of ZnS substrates (Bridgman). The ZnS solute was a sintered powder ingot whereas, for ZnSe, powder was used, both prepared from 5N-grade materials. Figure 2.17 shows the apparatus and temperature cycle. As T_g was restricted to be less than 900 °C to avoid excessive vaporization of the melt, only thin ($\sim 1\ \mu$m) layers of ZnS could be grown; the surfaces were smooth for growth on both $\{1\,1\,1\}$ surfaces. The Te content was found to be 0.1%, and the layers produced a white–blue photoluminescence at 300 K. Lower temperatures were used for ZnSe ($T_g \sim 750$ °C); the $\{1\,1\,1\}$B surfaces gave the best surfaces, but at a lower growth rate compared with $\{1\,1\,1\}$A deposits. In the case of ZnSe, the concentration of Te was excessive (3–5%), resulting in a ternary compound. 77 K PL indicated strong NBE emissions, in addition to weaker deep-level Cu red, although the latter was less apparent in the case of $\{1\,1\,1\}$A deposits, suggesting that the segregation coefficient of Cu is lower in this case; BE emission was not observed at 300 K.

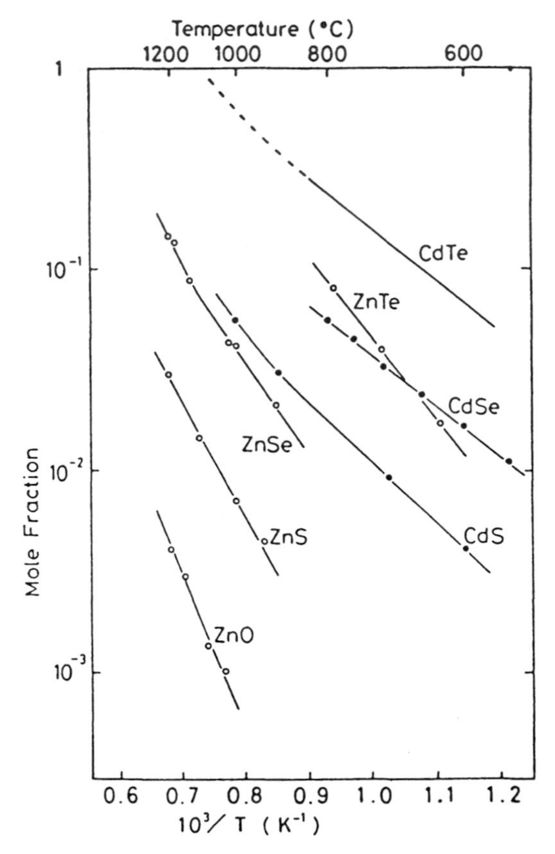

Fig. 2.18 Solubilities of various II–VI compounds in Te solvent. (From ref. 38.)

In 1983 the group published its work [38] on the growth of ZnSe on ZnTe substrates for solar cell applications. The problem of using Te melts in this case is apparent from Fig. 2.18, namely melt-back of the substrates; this was overcome by the use of a Te–Se mixture as solvent. To further this work, a phase diagram was established (Fig. 2.19). LPE was performed by a closed-tube tipping process on solution-grown ZnTe substrates, using a solvent composition of 97% Te + 3% Se; initial conditions correspond to point B in the diagram whereas, without Se in the melt, point A would apply, causing significant melt-back. Again, the {1 1 1}B surfaces were the smoothest, but the layers contained microcracks due to thermal mismatch. 4.2 K PL shows strong but broad BE emission, along with deep levels. The photovoltaic properties are reported and appear to be encouraging.

Nakamura *et al.* [39] have also reported an investigation of the growth of particularly thick layers of ZnSe from Sb–Se melt; Fig. 2.20 shows that ZnSe has largest solubility in this melt (Te is perhaps an exception, but tends

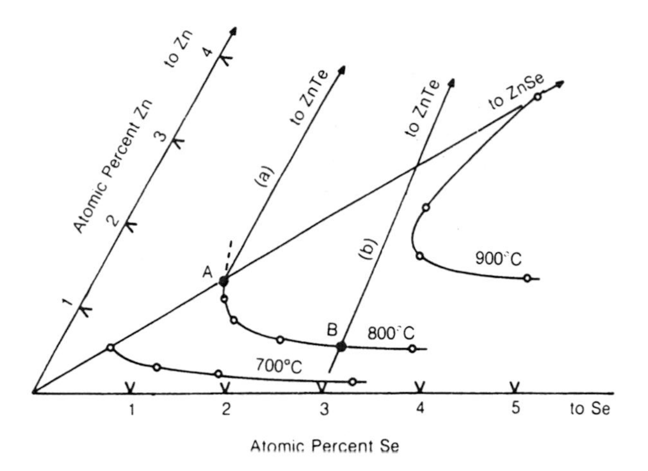

Fig. 2.19 Diagram explaining the melt-back of the ZnTe substrate into the solution. (From ref. 38.)

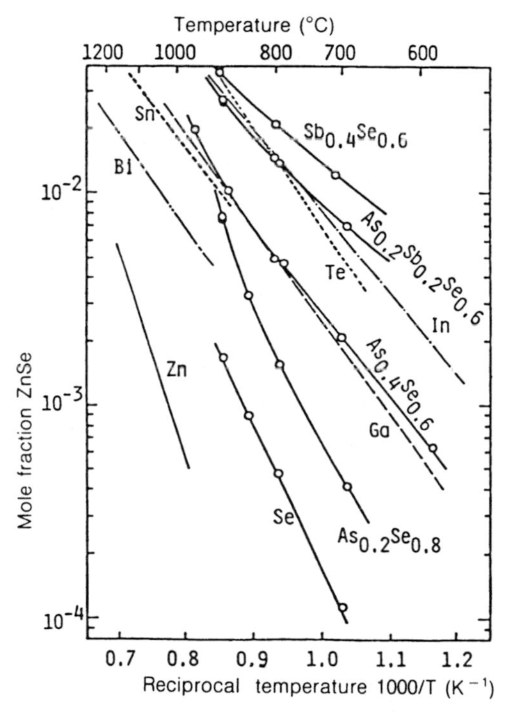

Fig. 2.20 Molar solubilities of ZnSe in various kinds of solvents. The solubilities in Zn, Bi, In, Ga, Te and Se are from the literature. (From ref. 39.)

to be incorporated in the lattice). Another important advantage of the Sb–Se melt is its low vapour pressure, enabling a sliding-boat open-tube system to be used. The solution was pre-synthesised from the pure elements. Typical growth conditions were as follows: starting temperature 800 °C, cooling rate 0.7 °C min^{-1}, 800 °C hold time 30 min, pre-growth melt-back without Zn 2 min. Surface morphologies were smooth for layers less than 20 μm thick on {1 1 1}B ZnSe substrates; {1 1 1}A surfaces gave a rough overgrowth with inclusions and holes. PL at 4.2 K from the layers is somewhat disappointing in that the dominant emissions are $I_{1(deep)}$, the Q_0 series, and Cu red deep level. The authors conclude that much of the associated impurities originate in the ZnSe substrates; this would appear to be an area for further improvement.

Researchers at Chubu Institute of Technology were also active in the earlier part of the decade. Thick homoepitaxial layers (> 40 μm) of ZnSe were grown from a $ZnCl_2$ solvent at lower temperatures, in a closed tube system [40]. First, they obtained the solubility data (Fig. 2.21). The growth cycle included sealing the ZnSe substrate and $ZnCl_2$ into separate regions of a quartz ampoule, and keeping at 700 °C for 5 h; epitaxial growth was commenced by rotating the ampoule and cooling to 550 °C at a rate of 1–5 °C min^{-1}. For some of the experiments a zinc pellet was added to the charge to provide Zn overpressure. High conductivity was achieved by post-growth annealing in molten zinc saturated with ZnSe. The resistivity of As-grown layers is normally high (> 10^4 Ω cm); even the Zn overpressure during growth failed to produced As-grown layers with high conductivity, and in both cases, mobility was low (80 cm^2 V^{-1} s^{-1}). Following the Zn extraction process, 1 Ω cm 100 cm^2 V^{-1} s^{-1} material was achieved. 300 K PL spectra (nitrogen laser excitation) include BE and deep emissions; the former increases significantly for the case of Zn overpressure and dominates for Zn-extracted specimens.

A slightly later paper, by Ido and Miyasato [41], reports on the growth of ZnSe layers from a Ga melt at high temperatures; again Zn pellets were included to prevent attack on the substrates and the formation of GaSe. Two

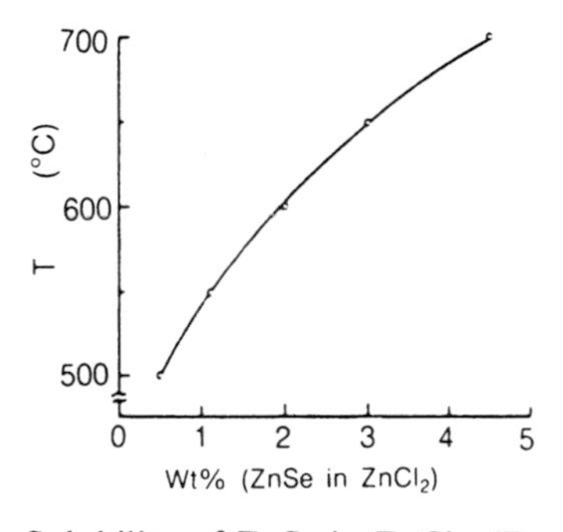

Fig. 2.21 Solubility of ZnSe in $ZnCl_2$. (From ref. 40.)

Table 2.2 Depth E_T, concentration N_T, and capture cross-section σ of traps observed in undoped, P-doped and Ga-doped samples, and the donor concentration N_D (from ref. 42)

Sample	E_T (eV)	N_T (cm^{-3})	σ (cm^2)	N_D (cm^{-3})
Undoped	0.30	1.8×10^{15}	2.2×10^{-15}	1.8×10^{16}
P-doped	0.32	1.0×10^{15}	2.1×10^{-13}	1.2×10^{16}
Ga-doped	0.027[a]	8.5×10^{15}	1.9×10^{-21}	2.5×10^{16}
	0.15	4.7×10^{15}	3.6×10^{-18}	2.5×10^{16}
	0.19[a]	3.4×10^{15}	6.4×10^{-15}	2.5×10^{16}
	0.29	1.5×10^{15}	3.1×10^{-15}	2.5×10^{16}
	0.52	7.3×10^{14}	2.1×10^{-13}	2.5×10^{16}

[a] Under illumination with a steady, 460 nm, light source.

growth techniques were employed: (a) a temperature difference was maintained between source (ZnSe crystals) and the ZnSe substrates and (b) a gradual cooling of the melt; both provided means for deposition. The 300 K PL spectra (nitrogen laser) revealed strong BE and weak deep-level emissions for the particular case of growth by the former technique, with the substrate at the highest of the temperatures investigated (1000 °C) and a temperature difference of 10 °C.

In 1985, Ido and Okada [42] reported on deep-level transient spectroscopy (DLTS) measurements of trap levels in ZnSe layers grown from Sn solution, and doped with Ga and P. The results are summarized in Table 2.2. The 0.30 eV level is associated with the Se vacancy.

A series of papers by the group at the Philips Laboratories, USA, appeared in the early part of the decade. The main thrust of this work was to use modern LPE techniques to produce high-quality material with impurity concentrations low enough to yield high resolution in the BE PL spectra. This objective was achieved, and a deeper understanding of both the compensating processes and the nature of shallow donors emerged. Fitzpatrick *et al.* [43] describe spectroscopic studies on homoepitaxial ZnSe grown in various new and modified III–V LPE systems. Extreme measures were taken to minimize impurity levels; these included the use of platinum–rhodium wire for furnace windings for some of the layers. Bismuth and tin were used as solvents for deposition on {1 1 1}A substrates, under a cooling range of 900–825 °C. The best surfaces were featureless (1600 ×). The layers were pure enough (impurity concentration $< 10^{16}\ cm^{-3}$) for the exciton emissions from the different impurities to be clearly resolved; a buffer layer and 'good' substrates were found to be essential to achieve this level of purity; Fig. 2.22 shows an example of the BE spectra of the layers, with lines assigned to impurities by careful experimentation (the I_{x_0} line is particularly difficult;

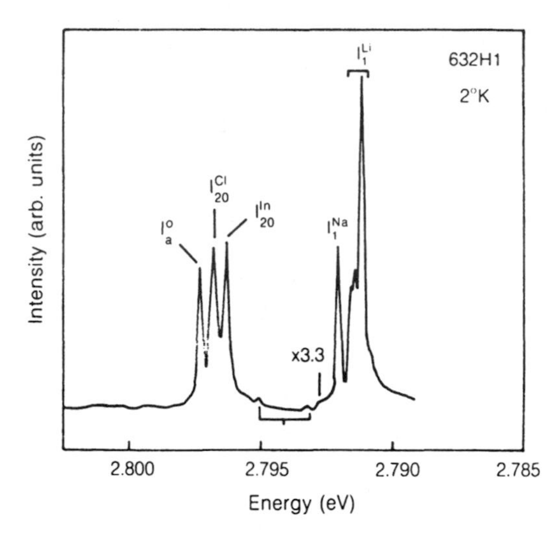

Fig. 2.22 The BE region of a nominally undoped sample showing the distinct resolution of donor BEs and the splitting of the Li BE. (From ref. 43.)

it may be an Li_i–Li_{Zn} or a V_{Zn}–Zn_i complex). The main conclusion was that all the shallow donors (I_2 lines) are due to impurities and not native defects. The shallow acceptors Li and Na are always seen; mass spectrometry suggests concentrations $< 10^{14}\,cm^{-3}$ for Li. The other alkali elements do not seem to give shallow levels. The snag with the smaller acceptor elements, such as lithium, is that they may not be stable; in particular, they can form interstitial donors. The group V elements may be useful acceptors, although the specific nature of the resulting centres is in doubt.

At about the same time as the above publication, Werkhoven *et al.* [44] published further details of the results achieved with an Sn solvent. Figure 2.23 shows clearly the influence of the substrate quality on the impurity content of the layers. Hence, a buffer layer was employed in conjunction with fast growth cycles. Mass spectrometry indicated a concentration of Al at 0.25 ppm, all other impurities being below the detection limit, particularly Si which is associated with n-type conductivity in this material.

Kosai [45] has reported on the trap levels in ZnSe grown from Bi solution onto Al-doped (n-type) substrates using a vertical dipping technique. Schottky barriers were formed by evaporating Au dots onto freshly etched surfaces. Table 2.3 shows a summary of the results from various measurement techniques (different techniques were necessary for the range of samples investigated). Basically, four levels have been established: 0.17, 0.3, 0.64, and 1.4 eV below the conduction band.

McGee and Werkhoven [46] produced data on segregation coefficients of selected impurities in ZnSe; in this way the importance of controlling particular impurities became apparent. The solvent used for the experiments

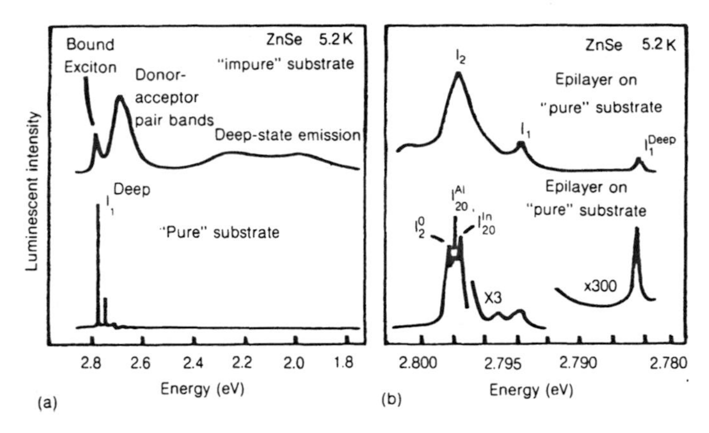

Fig. 2.23 (a) PL spectra of two substrates taken before growth. The presence of intense D–A pair bands and deep-state emission classifies a wafer as 'impure', (b) PL spectra in the bound-exciton region of epilayers grown on 'pure' and 'impure' substrates. (From ref. 44.)

was Sn, and a sliding graphite susceptor allowed the substrates to contact the melt held in a graphite chamber. A transparent gold-coated reflector furnace with Pt–10%Rh windings provided a suitable furnace. The resulting segregation coefficients are shown in Table 2.4. The coefficients are generally moderate; the value for copper is particularly small; it is suggested that this may be due to diffusion from the layer into the substrate during cooling. Clearly, Al and In must be carefully controlled, as they are shallow donors. Mn is regarded as isoelectronic, and should not be the problem it appears to be from these data.

A theoretical analysis of the Zn–Sn–Se system, and experimental verification, was presented by Heurtel *et al.* [47]. Layers were grown in a closed system, cooling over the temperature range 948–900 °C; lower temperatures encouraged the precipitation of $SiSnO_3$. The main impurity found was aluminium; as the concentration appeared to peak near the layer–substrate interface, it was concluded that the Al originated from the substrates.

Significantly, there has been little activity in the LPE area for the last five or six years. However, an abstract of a report (Japanese) by Koyama [48] of the Nippon Sheet Glass Co. Ltd indicates that the LPE of ZnSe and ZnSSe by the dipping technique, using GaAs substrates, has been achieved. The melt temperature was 500 °C, but the nature of the solvent is not given.

Earlier (1986), Skobeeva *et al.* [49] published an interesting paper on their observations on the growth of ZnTe on ZnSe substrates, in both Sn and Bi melts. The authors appreciated one of the current problems associated with the growth technology of heterojunctions for blue LEDs: the ZnSe substrate crystals should exhibit a strong PL edge emission, and the subsequent layer

Table 2.3 Summary of samples, their characteristics, and the results of various deep-state measurements (from ref. 45).

Sample	Dopants	$N_D - N_A$ (10^{16} cm^{-3})	Trap label	DLTS $E_c - A_A$ (eV)	DLTS N_T (10^{14} cm^{-3})	DLTS σ_α (10^{-15} cm^2)	TSCAP $E_c - E_A$ (eV)	TSCAP N_T (10^{14} cm^{-3})	PHCAP $E_c - E_T$ (eV)	PHCAP N_T (10^{14} cm^{-3})	Transient PHCAP $E_c - E_T$ (eV)
565-V	Si	1	E2	0.38[a]		30	0.34	2	0.42	2	0.40
			E3				0.62	2	0.90	4	0.53
			E4								1.4
708-V-B	N	7	E1	0.17	10	200					
			E2	0.29	2	3					
			E3	0.63	8	300	0.61	10			0.67
			E4								1.3
709-V-B	N, Na	4	E2	0.33	5	8	0.33	4			
			E3	0.65	60	700	0.60	60			0.68
			E4								1.4

The quantity σ_α is the $T = \infty$ value of the electron capture cross-section as calculated from the intercepts of the e/T^2 data shown in Fig. 4.
[a] Measured by current transient DLTS.

Table 2.4 Analysis results and segregation coefficient (from ref. 46)

	C_L[a]	C_S[b]	k
Al	0.02	0.16	8
In	0.005	0.05	10
Fe	0.01	0.004	0.4
Si	0.023	0.003	0.1
Mn	0.013	5.3	400
Ga	0.007	0.001	0.1
Cr	0.02	0.03	1.5
Cu	0.01	0.0004	0.04

[a] C_L represents the mole per-cent of the impurity in the liquid.
[b] C_S represents the mole per-cent of the impurity in the solid.

processing should not adversely affect it. The report describes the changes in luminescence as seen beneath the ZnTe layer and in adjacent regions masked from the melt by the graphite susceptor, at various stages of the LPE process; in addition, a dummy substrate was annealed in the LPE temperature cycle without melt. 77 K PL (HeCd laser) was monitored, in addition to electroluminescence. First, the substrates were assessed; these gave strong NBE along with free-excitonic emissions. Following deposition of ZnTe, in either melt system, the PL spectra from regions under the overgrowth were unchanged, whereas the emission from unexposed regions of the substrate was now dominated by deep levels (this was also the case for the dummy samples). The electroluminescence spectra for the heterojunction were also identical to those of the original substrate PL spectra (Fig. 2.24). The authors conclude that the growing layer prevents changes in

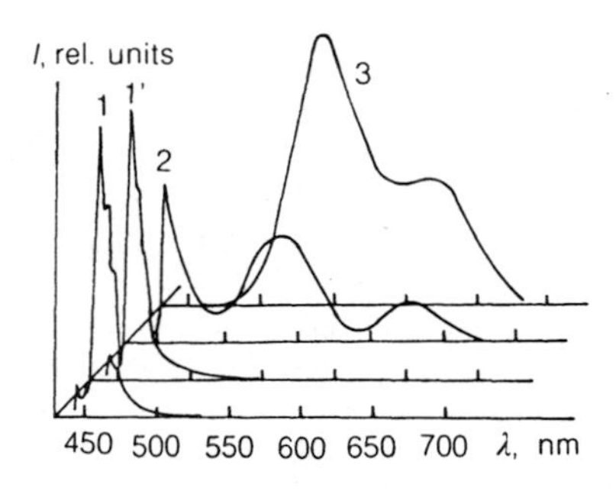

Fig. 2.24 PL spectra of ZnSe substrates (curves 1–3) and electroluminescence spectrum of ZnTe–ZnSe heterojunctions (curve 1′) at 77 K. $T_{ep} = 950$ K (curves 1 and 2) and 1070 K (curve 3). (From ref. 49.)

stoichiometry in the substrates that would otherwise occur. Sadly, there is no mention of 300 K PL or electroluminescence.

At the moment (1990), we must conclude that LPE is not a favourable route to device-quality structures in the widegap II–VI devices: the opto-electronic properties cannot be controlled.

REFERENCES

1. Lilley, P., Jones, P.L. and Litting, C.N.W. (1970) The epitaxial growth of zinc sulphide on silicon by forced vapour transport in hydrogen flow, *J. Mater. Sci.*, **5**, 891–897.
2. Yim, W.M. and Stofko, E.F. (1972) Vapor-phase epitaxial growth and some properties of ZnSe, ZnS, and CdS, *J. Electrochem. Soc.*, **119**, 381–388.
3. Kay, P.M.R. and Lilley, P. (1975) The epitaxial growth of thick smooth films of ZnS on GaAs, *J. Cryst. Growth*, **31**, 339–344.
4. Lilley, P. (1978) Transport kinetics in horizontal ZnS epitaxial growth systems, *J. Cryst. Growth*, **44**, 446–452.
5. Robinson, R.J. and Kun, Z.K. (1975) p–n junction zinc sulpho-selenide and zinc sulphide light-emitting doides, *Appl. Phys. Lett.*, **27**, 74–76.
6. Etienne, D. and Bougneot, G. (1976) Photoluminescence des couches minces ZnSe/GaAs obtenues par transport avec H_2, *Thin Solid Films*, **35**, 363–371.
7. Besomi, P. and Wessels, B.W. (1980) High-conductivity heteroepitaxial ZnSe films, *Appl. Phys. Lett.*, **37**, 955–957.
8. Czerniak, M.R. and Lilley, P. (1982) VPE ZnSe on GaAs: photoluminescence and conductivity, *J. Cryst. Growth*, **59**, 455–467.
9. Scott, M.D., Williams, J.O. and Goodfellow, R.C. (1981) Epitaxial growth of ZnSe on (100) GaAs by open-tube transport of elemental vapours in H_2 flows, *J. Cryst. Growth*, **51**, 267–272.
10. Matsumoto, T., Morita, T. and Ishida, T. (1987) Epitaxial growth of ZnS on GaP by Zn–S–H_2 CVD method, *J. Cryst. Growth*, **53**, 225–233.
11. Matsumoto, T. and Ishida, T. (1984) Chemical vapour deposition of zinc chalcogenides using elemental source materials, *J. Cryst. Growth*, **67**, 135–140.
12. Muranoi, T. and Furukoshi, M. (1983) Vapor and solid phase epitaxies of ZnSe films on (100) GaAs using metallic Zn and Se, *Jpn. J. Appl. Phys.*, **22**, L517–L519.
13. Heime, A., Senske, W., Tews, H. and Matthes, H. (1982) Photoluminescence of ZnS_xSe_{1-x} epilayers and single crystals, *IEEE Trans. Electron Devices*, **28** (4), 436–439.
14. Besomi, P. and Wessels, B.W. (1981) Growth and characterisation of heteroepitaxial zinc selenide, *J. Cryst. Growth*, **55**, 477–484.
15. Czerniak, M.R. (1982) *Ph.D. Thesis*, University of Manchester.
16. Lilley, P., Czerniak, M.R., Nicholls, J.E. and Davies, J.J. (1982) Control of optoelectronic properties of ZnSe films grown on GaAs by VPE, *J. Cryst. Growth*, **59**, 161–166.
17. Lilley, P., Czerniak, M.R. and Nicholls, J.E. (1984) Photoluminescence and electrical properties of vapour phase epitaxial ZnSe grown on GaAs, *Phys. Status Solids*, **85**, 235–242.
18. Umar-Syed, M. and Lilley, P. (1984) The growth and photoluminescence of ZnSe on GaAs by VPE in the temperature range 300–500 °C, *J. Cryst. Growth*, **66**, 21–25.
19. Nishio, M., Nakamura, Y. and Ogawa, H. (1983) Homoepitaxial growth of ZnTe by horizontal open-tube methods, *Jpn. J. Appl. Phys.*, **22**, (7), 1101–1105.

20. Radautsan, S.I., Rebrov, S.A. and Tsurkan, A.E. (1984) Some properties of ZnTe–InP heterojunctions, *Phys. Status Solidi A*, **84**, K169–K171.
21. Matsumoto, T., Yoshida, S. and Ishida, T. (1986) Growth of high quality layers in hydrogen plasma *Jpn. J. Appl. Phys.*, **25** (5), L413–L415.
22. Matsumoto, T., Kobayashi, N. and Ishida, T. (1987) Low pressure vapor phase epitaxy of high purity ZnSe using zinc and selenium as source materials, *Jpn. J. Appl. Phys.*, **26** (3), L209–L211.
23. Goto, H., Zhou, J., Sawaki, N. and Akasaki, I. (1986) VPE growth of ZnS incorporating indium on GaP, *Jpn. J. Appl. Phys.*, **25** (7), 1036–1039.
24. Matsumoto, T., Iijima, T., Katsumata, Y. and Ishida, T. (1987) Electrical and luminescent properties of In-doped ZnSe grown by low-pressure vapor-phase epitaxy, *Jpn. J. Appl. Phys.*, **26** (10), L1736–L1739.
25. Goto, H., Zhou, J., Sawaki, N. and Akasaki, I. (1987) Thermodynamic analyses and luminescence properties of vapor grown ZnS_xSe_{1-x}, *Jpn. J. Appl. Phys.*, **26** (8), 1300–1304.
26. Goto, H., Zhou, J., Sawaki, N. and Akasaki, I. (1989) Thermodynamic analysis and luminescence properties of vapor-grown ZnS_xSe_{1-x} II—bypass flow effect, *Jpn. J. Appl. Phys.*, **28** (7), 1154–1159.
27. Hartmann, H., Mach, R. and Testova, N. (1987) Vapour phase epitaxy of wide gap II–VI compounds, *J. Cryst. Growth*, **84**, 199–206.
28. Zhang, S., Kinto, H., Yatabe, T. and Iida, S. (1988) Nitrogen and phosphorus doping in ZnS layers grown by vapor phase epitaxy on GaAs substrates, *J. Cryst. Growth*, **86**, 372–376.
29. Umar-Syed, M. and Lilley, P. (1988) Heteroepitaxial ZnSe on GaAs: high growth rates at low temperatures by conventional VPE, *J. Cryst. Growth*, **88**, 415–418.
30. Matsumoto, T., Iijima, T. and Ishida, T. (1988) Quality variation of ZnSe heteroepitaxial layers correlated with nonuniformity in the GaAs substrate wafer, *Jpn. J. Appl. Phys.*, **27** (10), L1942–L1944.
31. Kyotani, T., Isshiki, M. and Masumoto, K. (1989) VPE growth of ZnSe thin films on GaAs (100) and ZnSe (110) substrates, *J. Electrochem. Soc.*, **136** (8), 2376–2381.
32. DiLorenzo, J.V. (1972) Vapor growth of epitaxial GaAs: a summary of parameters which influence the purity and morphology of epitaxial layers, *J. Cryst. Growth*, **17**, 189–206.
33. Iida, S., Yatabe, T. and Kinto, H. (1989) p-type conduction in ZnS grown by vapor phase epitaxy, *Jpn. J. Appl. Phys.*, **28** (4), L535–L537.
34. Stucheli, N. and Bucher, E. (1989) Low resistivity p-type ZnSe: a key for an efficient blue electroluminescent device, *J. Electron. Mater.*, **18** (2), 105–109.
35. Goodman, C.H.L. and Pessa, M.V. (1986) Atomic layer epitaxy, *J. Appl. Phys.*, **60** (3), R65–R81.
36. Fujita, S., Mimoto, H. and Noguchi, T. (1978) Liquid-phase epitaxy of ZnSe from Zn–Ga solution, *J. Cryst. Growth*, **45**, 281–286.
37. Nakamura, H. and Aoki, M. (1981) Liquid-phase epitaxial growth of ZnS, ZnSe and their mixed compounds using Te solvent, *Jpn. J. Appl. Phys.*, **20** (1), 11–16.
38. Nakamura, H., Sun, L.Y., Asano, A., Nakamura, Y., Washiyama, M. and Aoki, M. (1983) Liquid-phase epitaxial growth of ZnSe on ZnTe substrate, *Jpn. J. Appl. Phys.*, **22** (3), 499–503.
39. Nakamura, H., Kojima, S., Washiyama, M. and Aoki, M. (1984) Liquid-phase epitaxial growth of ZnSe using $Sb_{0.4}Se_{0.6}$ as solvent, *Jpn. J. Appl. Phys.*, **23** (8), L617–L619.
40. Ido, T. (1981) Liquid-phase epitaxy of ZnSe at low temperature, *J. Cryst. Growth*, **51**, 304–308.
41. Ido, T. and Miyasato, K. (1982) Liquid-phase epitaxy of ZnSe by Temperature difference method, *J. Cryst. Growth*, **59**, 178–182.

42. Ido, T. and Okada, M. (1985) Detection of traps in ZnSe grown by liquid phase epitaxy, *J. Cryst. Growth*, **72**, 170–173.
43. Fitzpatrick, B.J., Werkhoven, C.J., McGee, T.F., III, Harnack, P.M., Herko, S.P., Bhargava, R.N. and Dean, P.J. (1981) Spectroscopic studies of ZnSe grown by liquid phase epitaxy, *IEEE Trans. Electron Devices*, **28** (4), 440–444.
44. Werkhoven, C., Fitzpatrick, B.J., Herko, S.P. and Bhargava, R.N. (1981) High-purity ZnSe grown by liquid phase epitaxy, *Appl. Phys. Lett.*, **38** (7), 540–542.
45. Kosai, K. (1982) Electron traps in ZnSe grown by liquid-phase epitaxy, *J. Appl. Phys.*, **53** (2), 1018–1022.
46. McGee, T.F., III and Werkhoven, C. (1982) Segregation coefficients of selected impurities in ZnSe grown by LPE, *J. Cryst. Growth*, **59**, 649–650.
47. Heurtel, A., Marbeuf, A., Tews, H. and Marfaing, Y. (1982) Liquid phase epitaxy of ZnSe in Sn: calculation of the ternary phase diagram and electrical properties, *J. Cryst. Growth*, **59**, 167–171.
48. Koyama, T. (1988) Preparation of heteroepitaxial films, *Jpn. Kokai Tokkyo Koho.*
49. Skobeeva, V.M., Serdyuk, V.V., Semenyuk, L.N. and Malushin, N.V. (1986) Influence of technological conditions upon the luminescence properties of ZnTe–ZnSe heterostructures grown by liquid-phase epitaxy, *J. Appl. Spectrosc.*, **44** (2), 164–167.

3

Photo-assisted metal–organic vapour phase epitaxy of zinc chalcogenides

Sz. Fujita and Sg. Fujita

3.1 INTRODUCTION

Low-temperature growth is one of the key factors in obtaining high-quality epilayers of II–VI semiconductors. Metal–organic vapour phase epitaxy (MOVPE) and molecular beam epitaxy (MBE) have successfully performed low temperature growth, and the crystalline quality has gradually been improved by optimizing the growth conditions. In the MOVPE growth of zinc chalcogenides, the growth temperature was below 300 °C when using alkylzinc and hydrochalcogens as group II and group VI precursors, respectively [1–8]. However, these precursors cause room-temperature pre-reactions, resulting in poor surface morphology and crystallinity unless special care is taken in the gas-supplying techniques and low-pressure growth. Further, hydrochalcogens have very high toxity (H_2Se has a similar LD_{50} value to AsH_3). In order to overcome these problems, alkyl or heterocyclic compounds of chalcogens have been developed as novel precursors, but the growth temperature must be as high as 500 °C [9–12]. The photo-assisted technique, therefore, is a promising tool for reduction in growth temperature and for obtaining high-quality epilayers when using these source combinations. By this technique, selective growth or selective doping can also be expected.

In the course of previous studies on photo-assisted growth of various materials, direct photodecomposition or photoexcitation of precursors in the gas phase has been found to be the most fundamental and essential process for promoting the material growth (wavelength dependence of the growth rate in photo-assisted processes appears in some papers for photochemical vapour deposition (photo-CVD) of amorphous-silicon-related thin films, and they reported little enhancement of growth rate unless direct photodecomposition or photoexcitation of reactant sources was used [13]). Also, for the

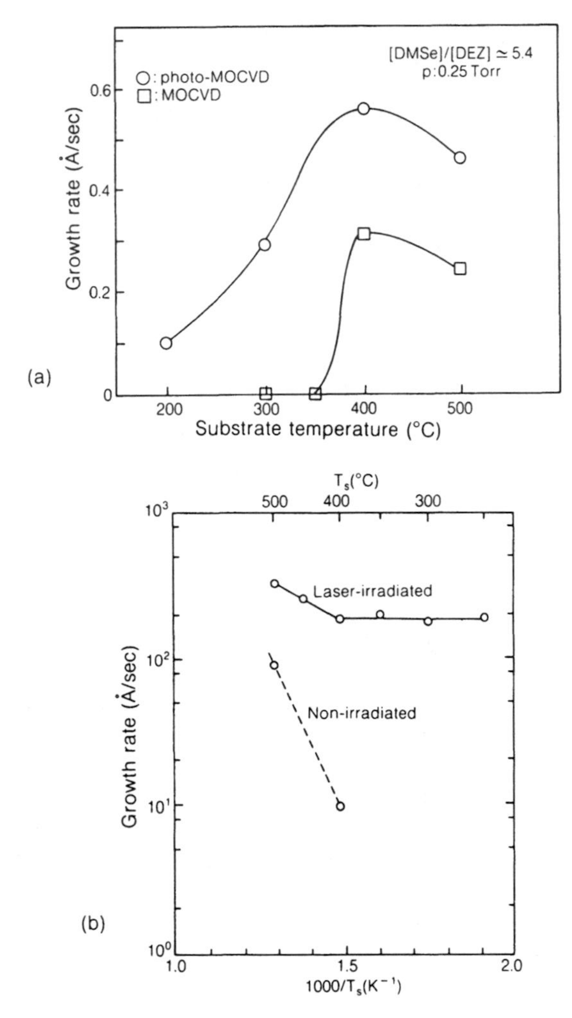

Fig. 3.1 Growth rate of ZnSe as a function of growth temperature with and without irradiation of (a) a low-pressure mercury lamp [19] and (b) ArF excimer laser [20].

photo-assisted MOVPE of II–VI semiconductors, the ultraviolet light which can be absorbed by the source precursors in the gas phase, shorter than 200–280 nm [14–18] in wavelength, was considered to be necessary for the growth rate enhancement. Therefore, 253.7 and 184.9 nm lines from a low-pressure mercury lamp [19] or 193 nm ArF excimer laser [20] were used, so that the source precursors could effectively absorb the irradiating light. The growth rate in these processes is shown in Fig. 3.1, which clearly indicates that the growth temperature was successfully reduced to as low as 300 °C. In these papers, reduction in intrinsic defects and residual impurities has also been reported.

On the other hand, we have for the first time observed growth rate enhancement in ZnSe [21–25] and ZnS [22, 26] by light irradiation, by which direct photodecomposition or photoexcitation of the source precursors in gas phase is hardly expected. We have suggested that the mechanism for the growth rate enhancement is the promotion of surface reactions with the assistance of electrons and/or holes generated at the growth surface under the irradiation [22]. Such a process has never been recognized for the promotion of epitaxial growth and, from an applicational standpoint, it is important that no special light sources with high intensity are necessary. In this chapter, we review the previous and recent experimental results on this new type of photo-assisted MOVPE growth and discuss the photoassociation mechanism.

3.2 EPITAXIAL GROWTH

3.2.1 Source precursors and growth systems

As the source precursors, we used dimethylzinc (DMZn) or diethylzinc (DEZn) as a Zn source, dimethylselenide (DMSe) or diethylselenide (DESe) as an Se source, and diethylsulphide (DES) or methylmercaptan (CH_3SH) [27] as an S source. These precursors were supplied with H_2 carrier gas. The (100)-oriented GaAs wafers were used as the substrates. A 500 W xenon lamp with maximum irradiation intensity of $100\,mW\,cm^{-2}$ was used as the light source.

We have observed the growth rate enhancement caused by the irradiation for growth at both normal pressure (760 Torr) growth and low pressure (100–200 Torr). From now on, we mainly focus on the results obtained with low-pressure growth, where gas phase reactions can be suppressed compared with the normal-pressure growth. In this experiment, we used a horizontal furnace in which the light can be irradiated in a direction either vertical or horizontal to the substrate surface. Both the gas phase and the growth surface are irradiated by the former configuration, while the latter can irradiate only the gas phase. In other words, horizontal irradiation eliminates the contribution of surface reactions, and provides information on how gas phase reactions are influenced by the irradiation.

3.2.2 Temperature dependence of the growth rate

Figure 3.2 shows the growth rate of ZnSe as a function of the growth temperature [23]. Without the irradiation, shown by the squares, the growth rate is very low ($< 0.1\,\mu m\,h^{-1}$) without the irradiation. On the other hand, when the light was irradiated vertically onto the substrate, the growth rate, shown by the circles, is as high as $1\,\mu m\,h^{-1}$ even at 330 °C. The drastic increase of the growth rate compared with that without the irradiation is

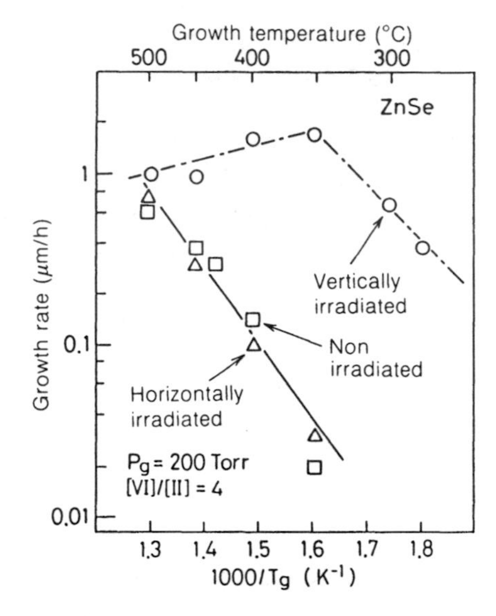

Fig. 3.2 Growth rates at low-pressure ($P_g = 200$ Torr) growth of ZnSe epilayer from DEZn–DESe source combination as a function of growth temperature [23]. Flow rates of DEZn and DESe are 10 and 40 μmol min^{-1} respectively. The xenon lamp irradiates vertically or horizontally with respect to the substrate with irradiation intensity of 47 mW cm^{-2}.

confirmed not only for ZnSe but also for ZnS. However, when the light was irradiated horizontally, the growth rates, shown by triangles, were as low as those without the irradiation. We consider, therefore, that gas phase reactions such as gas phase decomposition or adduct formation are barely enhanced under the light irradiation. It seems essential to irradiate the light onto the growth surface.

3.2.3 Irradiation wavelength dependence of growth rate

For the investigation of the irradiated wavelength dependence of the growth rate, a cut-off filter for shorter-wavelength light was inserted between the lamp and the reactor. Figure 3.3 shows the growth rates of ZnSe, ZnS, and ZnSSe as a function of irradiated light wavelength, which is changed by using the filters [25]. The horizontal axis of this figure shows the photon energy corresponding to the wavelength where the transmittance of the filter inserted becomes 50%. In other words, the energy value shown in the horizontal axis nearly corresponds to the highest value in the incident photons.

The growth rate of ZnSe increases abruptly at around 2.5 eV (500 nm) and that of ZnS around 3.5–3.7 eV (350–335 nm). The threshold energy values, which were not very different for different source combinations, are close to the bandgaps of ZnSe and ZnS at the growth temperature. Furthermore, for

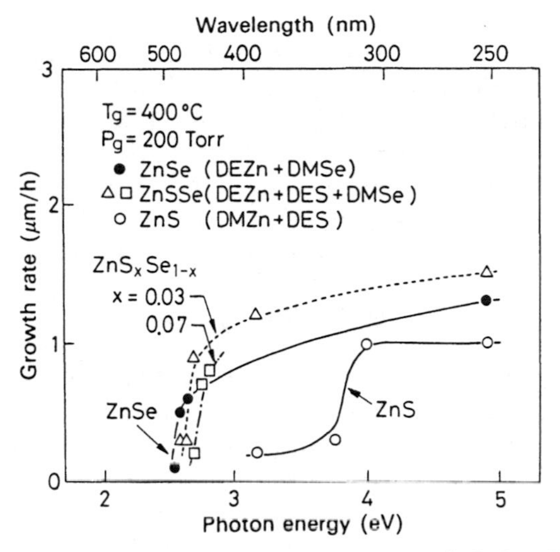

Fig. 3.3 Wavelength dependence of growth rates of ZnSe, ZnS, and ZnSSe, investigated by using xenon lamp + filter [25]. Here, the growth temperature T_g is 400 °C.

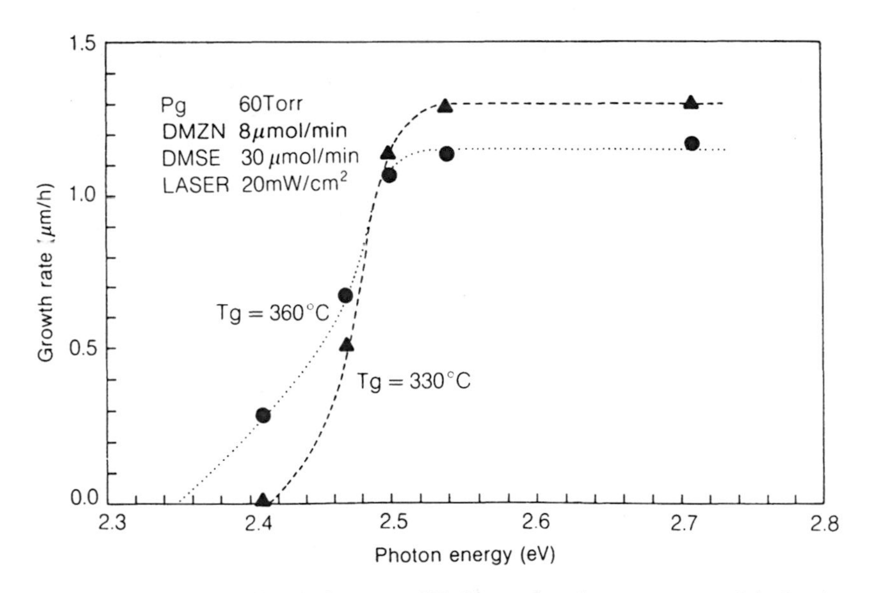

Fig. 3.4 Dependence of growth rates of ZnSe on the photon energy of Ar ion laser lines [28].

ZnSSe alloys, the threshold energies shifted corresponding to the alloy composition and were nearly equal to their bandgaps.

More recent results by Yoshikawa *et al.* [28], where they used monochromatic light from an Ar ion laser as the irradiation source, demonstrate that the threshold energy values shift slightly to the lower-energy side with increasing growth temperature (Fig. 3.4). They considered that this phenomenon corresponds to the decrease in ZnSe bandgap with temperature.

3.2.4 Irradiation intensity dependence of growth rate

The growth rate is shown as a function of the light intensity in Fig. 3.5 [23]. It increases almost linearly with increasing total irradiation intensity of the xenon lamp from 0 to 50 mW cm^{-2} under vertical irradiation, and beyond 50 mW cm^{-2} it saturates. The saturation is attributed to the total amount of supplied sources. From the growth rate and the light intensity, we calculated the number of adhered molecules and the number of incident photons which were responsible for the growth rate enhancement, i.e. whose energy was higher than the bandgap of ZnSe at the growth temperature, 2.5 eV. The correlation is shown in Fig. 3.6. Both values have the same order of 10^{15} cm^{-2} s^{-1} and they give a quantum efficiency, i.e. the number of adsorbed molecules divided by the number of incident photons, as high as few tens of per cents.

3.2.5 Hydrogen partial pressure dependence of growth rate

The growth rate drastically decreases if we dilute the supplying H_2 gas with He. Figure 3.7 shows the growth rate as a function of the partial pressure of

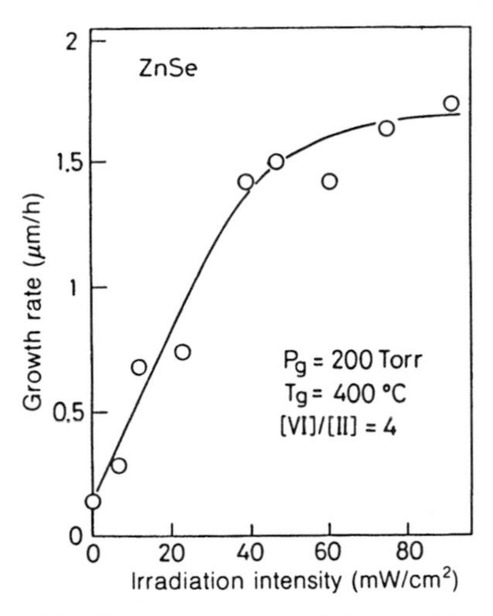

Fig. 3.5 Growth rate of ZnSe as a function of the total irradiation intensity of the xenon lamp [23]. Here, the growth temperature T_g is 400 °C.

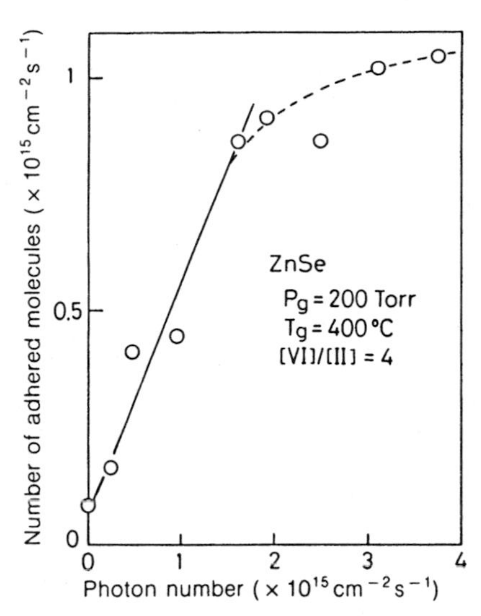

Fig. 3.6 Correlation between number of adhered molecules and that of incident photons which are responsible for the growth rate enhancement, i.e. whose energy is higher than 2.5 eV [23].

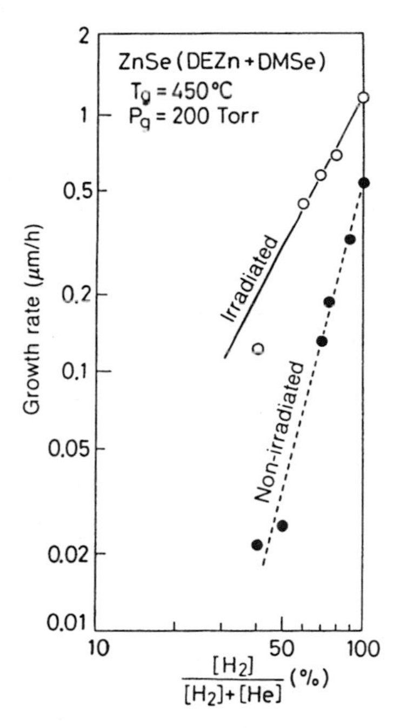

Fig. 3.7 Growth rate of ZnSe as a function of H_2 concentration in a carrier gas which was diluted by He [25].

H_2 in the carrier gas [25]. The growth rate r_g is proportional to $[H_2]^4$ without irradiation but to $[H_2]^2$ with irradiation.

3.3 DISCUSSIONS ON THE PHOTOASSOCIATION MECHANISM

3.3.1 Gas phase or surface reaction?

Here, we discuss what mechanism is responsible for the growth rate enhancement under light irradiation. The experimental results shown in section 3.2.2 (Fig. 3.2) strongly suggest the association of surface reactions rather than gas phase reactions. Association of surface reactions is also supported by the results shown in section 3.2.3 (Fig. 3.3). Since the growth rate is increased even by the light which cannot be absorbed by the source precursors, the direct dissociation of source precursors in the gas phase is not expected. If adducts or intermediate species formed in gas phase under irradiation enhance the growth rate, the photon energy where the growth rate increases will be different for different reactant sources. However, the threshold energy was nearly the same if we changed the source combinations. These results suggest that the photo-assisted processes in the gas phase do not play an important role in the growth rate enhancement.

The result shown in section 3.2.4 (Fig. 3.6) also suggests the responsibility of surface reactions. This result indicates that most of the incident photons are absorbed and become responsible for the epitaxial growth. Considering the low density of molecules in the gas phase, we can hardly expect such a high absorption coefficient there. On the other hand, the growth surface can absorb photons very effectively with high quantum efficiency. It is, therefore, reasonable to consider that surface reactions rather than gas phase reactions are enhanced by the irradiation.

Suppose that incident photons excite some precursors on the surface and that the excited precursors are responsible for the growth of ZnSe. It is reasonable to assume that one photon excites one precursor, and one excited precursor grows one ZnSe molecule. Then, one photon is responsible for the growth of one ZnSe molecule. This one-to-one correlation between an incident photon and an adhered ZnSe molecule is also expected even if we suppose that an excited precursor forms an adduct which contributes to the epitaxial growth.

3.3.2 Surface pyrolysis?

We shall discuss effects of substrate heating by irradiation. The substrate temperature was raised by 10–20 °C when irradiated by ultraviolet light at room temperature. However, from the relationship between the substrate temperature and the growth rate shown in Fig. 3.1, a great increase of the growth rate under the irradiation, by more than 10 times, cannot be explained by only taking into account the increase in the substrate temperature of 10–20 °C. Further, infrared light whose photon energy is lower than the bandgap may raise the substrate temperature, but does not enhance the growth rate. From these results, the growth rate enhancement seems not to be due to surface pyrolysis.

3.3.3 Surface photolysis or photocatalysis?

The photon energy where the growth rate increases almost coincides with the bandgap of the growing material, as shown in Fig. 3.3. Therefore, it seems that electrons and/or holes generated by light absorption promote some surface reactions, e.g. elimination of alkyl groups from molecules which are adhered on the growth surface. These chemical processes remind us of photocatalysis where the growing material itself works as catalyst. In this sense, we may well refer to the processes as autophotocatalysis.

However, we must also discuss the possibility of photolysis of precursors adhered to the growth surface. Ritz-Froidevaux *et al.* [29] reported that the photoabsorption edge of organometallics extended to longer wavelengths when adhered to the surface, compared with the gas phase. If the absorption edge extends to the visible light region, adhered precursors can absorb the light of this spectral region and be decomposed or excited by photolysis. In our experiments shown in section 3.2.3 (Fig. 3.3), however, the threshold wavelength was different for different compositions of ZnSSe alloys. For the growth of the alloys, existing precursors in the growth atmosphere are not different for different compositions, and thus we cannot explain the variation of the threshold wavelength with the alloy composition if we suppose surface photolysis by the red shift of the absorption band of precursors.

On the other hand, for the Ar-laser-assisted MOVPE of GaAs, GaP and GaAsP, Sugiura *et al.* [30] reported that the growth rates of the irradiated areas exhibited the same value whether the bandgap energies of the grown films were greater or less than the photon energy. They considered, therefore, that the growth rate enhancement due to the irradiation at 400–500 °C is caused by photolytic decomposition of TEGa adsorbed on a heated substrate. The difference in the conclusions on the photoassociation mechanism between II–VI and III–V semiconductor growth will be disclosed through a detailed understanding of the chemical processes on the surface under irradiation.

3.3.4 Plausible model of photocatalytic reactions

One of the most fundamental and important chemical processes in MOVPE has been believed to be an alkyl group elimination mechanism from precursors. The low growth rates at low temperatures are due to insufficient thermal energy to eliminate the alkyl groups. We have shown in Fig. 3.2 that the growth rate enhancement by the irradiation occurred at the temperature range of reaction-limited growth; hence we may consider that the photoirradiation enhances the chemical reactions which limit the growth rate at low temperatures, i.e. alkyl group elimination.

In the alkyl group elimination reactions, H_2 gas plays an important role. One of the possible mechanisms is the formation of a stable hydrocarbon gas by the reaction between alkyl groups in precursors and H_2. The results shown in section 3.2.5 (Fig. 3.7) seem to suggest an enhancement in alkyl group elimination under light irradiation.

During the growth, the growth surface is suspected to be covered by physisorbed or chemisorbed precursors, then they release alkyl groups and form a new molecular layer. Fig. 3.8 demonstrates a model where alkylzinc and alkylselenium are chemisorbed to the surface and form Se–Zn–C and Zn–Se–C bonds. By the irradiation, electrons and holes are generated at the surface. If the anti-bonding state or reduction level of Zn–C or Se–C bonds lie at the lower energy than the conduction band of ZnSe, these bonds can accept electrons in the conduction band of ZnSe, as shown in Fig. 3.9. This is a reduction reaction. If the bonding states or oxidation level of Zn–C or Se–C bonds lie at the higher energy than the valence band of ZnSe, these bonds can release electrons to the holes in the valence band, as shown in the same figure. This is an oxidation reaction. In both cases, Zn–C and/or Se–C bonds can break very easily. These are plausible primary chemical reactions which can contribute to the alkyl group elimination.

```
             CH3   CH3
              |     |
              Zn    Se
   |          |     |     |
 — Zn —— Se —— Zn —— Se --- Surface
   |      |     |     |
 — Se —— Zn —— Se —— Zn —
   |      |     |     |
```

Fig. 3.8 A model of the growth surface covered with chemisorbed precursors.

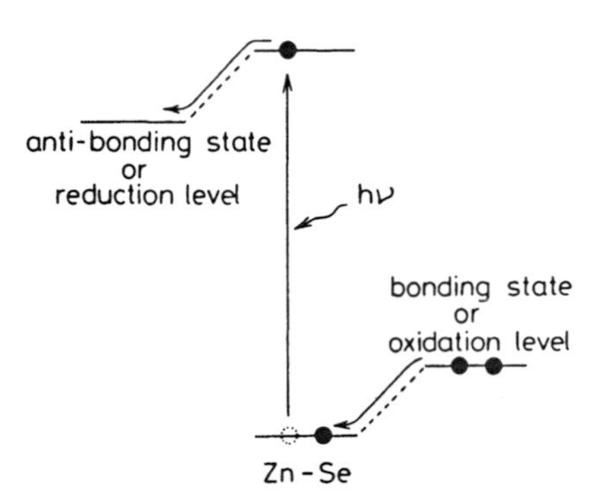

Fig. 3.9 Energy scheme for charge transfer between ZnSe and chemisorbed bonds, which results in reduction and oxidation reactions respectively.

We are not sure at the present stage whether the oxidation or the reduction takes place and whether Zn–C or Se–C bonds are preferrably decomposed. Considering that the band of ZnSe is bending upward at the ZnSe surface because ZnSe is generally n-type, the oxidation reaction might occur dominantly [37]. In-situ analysis during the growth is strongly expected to disclose the detailed chemical processes. Very recently, under mass analysis, we observed enhancement of alkyl group elimination from Zn–C bonds under irradiation [38].

3.4 CHARACTERISTICS OF EPILAYERS

3.4.1 Surface morphology

In our experiments, ZnSe epilayers grown with irradiation generally showed poorer surface morphology than those grown without irradiation [21]. Yasuda *et al.* [31] investigated the optimum growth conditions with respect to irradiation intensity and DMZn flow rate for obtaining a smooth surface morphology at an acceptable growth rate of about 1 μm h. A lower irradiation intensity and a lower flow rate of DMZn tend to improve the surface morphology, as shown in Fig. 3.10.

3.4.2 X-ray rocking curve

Fig. 3.11 shows an example of double-crystal X-ray rocking curves of ZnSe epilayers [21]. Full widths at half-maximum (FWHMs) for various samples were not significantly different from those of samples grown under no irradiation. It seems that the FWHMs are limited by a lattice mismatch between the epilayers and the substrates [32, 33].

3.4.3 Photoluminescence

Fig. 3.12 shows examples of photoluminescence spectra measured at 4.2 K for ZnSe epilayers with a thickness of 3–5 μm [21], and thus the lattice distortion due to lattice mismatch was already relaxed [11]. Comparing the spectra for samples grown at 500 °C without and with irradiation, shown in Figs 3.12(a) and 3.12(b) respectively, we can recognize that emissions Y at 2.60 eV and I_v at 2.77 eV clearly appear for the former, while not for the latter. The emission Y is suggested to arise from a recombination of excitons bound at extended defects such as dislocations [34], and thus we may expect a decrease of the dislocation density in the irradiated samples. By lowering the growth temperature to 350 °C under irradiation, the intensity of deep-level emissions has been successfully reduced, as shown in Fig. 3.12(c), showing further reduction in defects in the epilayers.

However, the irradiated samples exhibit a strong emission I_x at 2.79 eV,

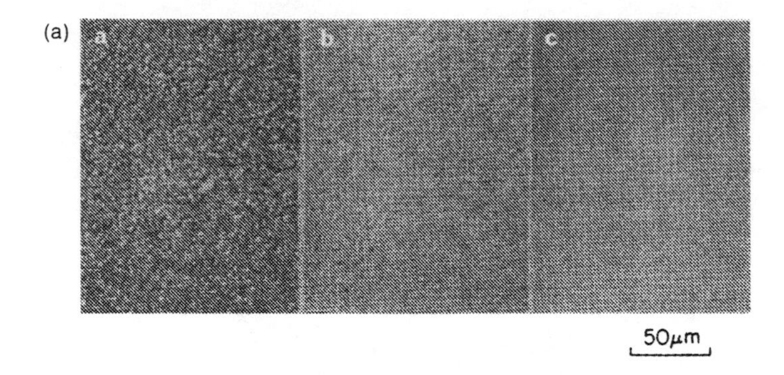

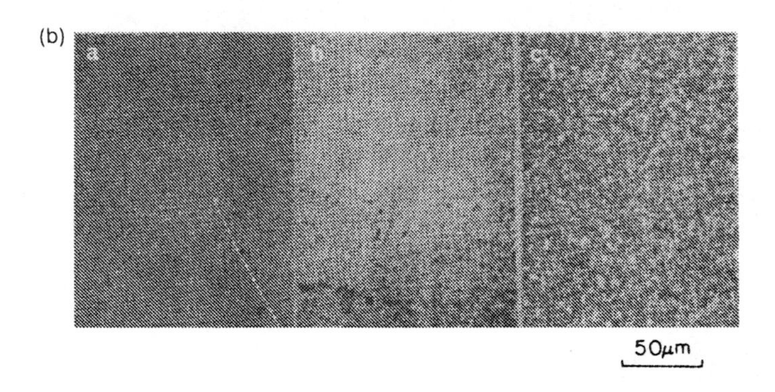

Fig. 3.10 Surface morphology observed under a Nomarski microscope for ZnSe layers grown under (a) different irradiation intensities (a, $1.6\,mW\,cm^{-2}$; b, $4.0\,mW\,cm^{-2}$; c, $8.0\,mW\,cm^{-2}$ under DMZn flow rate of $10\,\mu mol\,min^{-1}$ and VI:II ratio of 10) and (b) different flow rates of DMZn (irradiation intensity, $8.0\,mW\,cm^{-2}$; VI:II ratio of 10; a, $4.4\,\mu mol\,min^{-1}$; b, $2.5\,\mu mol\,min^{-1}$; c, $1.3\,\mu mol\,min^{-1}$) [31].

which has been suggested to originate from excitons bound to neutral donors [35]. According to the Hall effect measurements, samples grown without irradiation showed high resistivity, whereas those with irradiation at 350–500 °C showed n-type conductivity with an electron concentration of 5×10^{15}–$2 \times 10^{16}\,cm^{-3}$.

The low resistivity is attributed to either (i) higher doping efficiency and hence higher incorporation of residual impurities or (ii) donor-type defects generated in the epilayer. In order to identify which is more reasonable, we have investigated the doping efficiency of intentionally introduced impurities under irradiation. Fig. 3.13 shows the electron concentration of the iodine-doped epilayers grown at 350 °C under irradiation as a function of the flow rate of doping source, CH_3I [24]. Since we cannot grow ZnSe at 350 °C without irradiation at a reasonable growth rate, we cannot directly

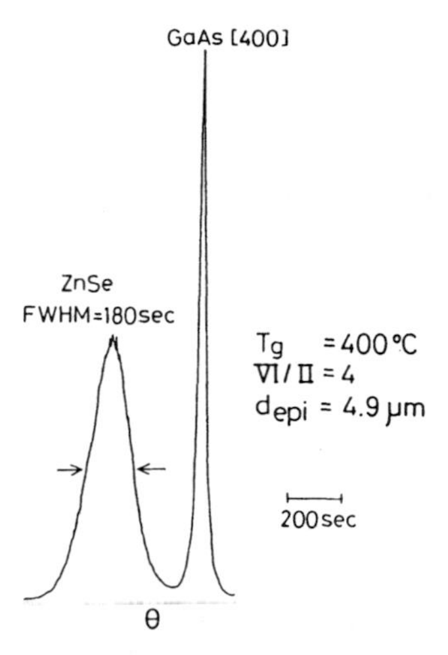

Fig. 3.11 Double-crystal X-ray rocking curve of ZnSe grown under irradiation [21]. Here, the thickness of ZnSe layer is 4.9 μm.

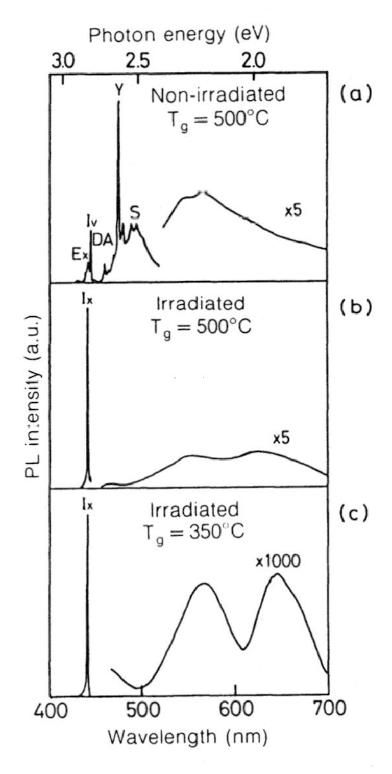

Fig. 3.12 Photoluminescence spectra at 4.2 K of ZnSe grown at different temperatures T_g with and without irradiation [21].

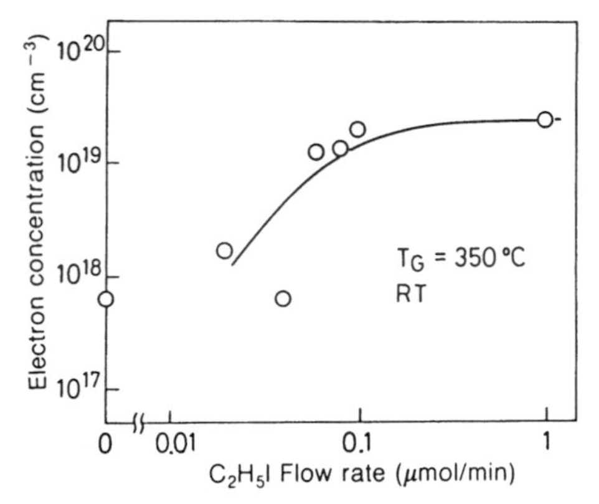

Fig. 3.13 Carrier (electron) concentration in ZnSe epilayer as a function of flow rate of Ch_3I in intentional doping under irradiation [24]. Here, epilayer thicknesses are about 1 μm.

discuss the effect of light irradiation on doping efficiency only from Fig. 3.13. However, the doping efficiency of CH_3I shown in Fig. 3.13 is more than 100 times higher than that reported for the growth using H_2Se as the Se source. Therefore, we may consider that the doping efficiency is remarkably enhanced by the irradiation. The low resistivity of non-doped epilayers grown under

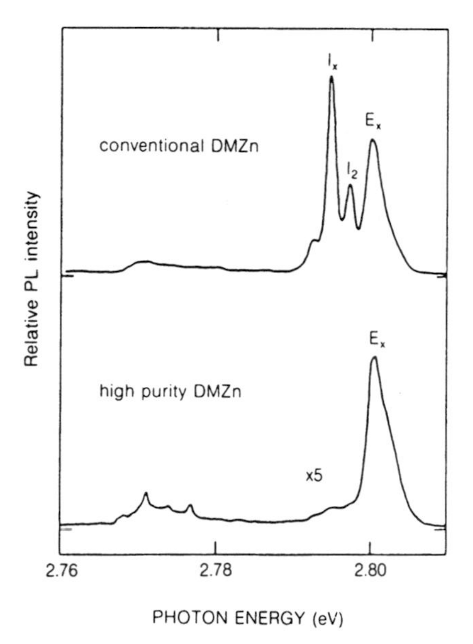

Fig. 3.14 Photoluminescence spectra at 10 K of ZnSe grown by using a conventional DMZn source in which about 15 ppm of Cl impurities are involved (upper trace) and by using a purified source for which the Cl content is below the detection limit of 5 ppm [36].

irradiation seems to be attributed to the incorporation of residual impurities. This phenomenon seriously obstructs high-quality epilayers and conductivity control, e.g. p-type doping.

Yasuda *et al.* [31] suggested that the origin of donor impurities in non-doped epilayers is probably Cl, coming from source DMZn. These impurities are easily incorporated in the ZnSe epilayers under irradiation and contribute to low-resistive n-type conductivity. They pointed out that the lower irradiation intensity and the lower flow rate of DMZn brought a reduction in I_x emission intensity and carrier concentration together with improvement of surface morphology. Recent work by Kukimoto [36] suggested that commercially available DMZn and DEZn typically contain Cl impurities with a concentration of 10–100 ppm at present. Special purification of DMZn resulted in a remarkable reduction in I_x emission intensity in photoluminescence, as shown in Fig. 3.14. Therefore, if the high purity source precursors are available, we expect photo-assisted MOVPE to become a promising technique for high-quality epilayers and excellent control of conductivity.

3.5 SUMMARY

In this chapter, we described the photo-assisted MOVPE growth of zinc chalcogenide II–VI semiconductors ZnSe, ZnS, and ZnSSe. A remarkable increase in the growth rate caused by irradiation of light whose photon energy is higher than the bandgap of the growing material allows a significant decrease in the growth temperature, e.g. from 500 °C without the irradiation to 350 °C with the irradiation. As a possible mechanism for the growth rate enhancement, we suggested autophotocatalysis at the growth surface, which has not been recognized so far in the growth of semiconductors. The low-temperature growth successfully reduced intrinsic defects, and the photo-MOVPE can become a promising technique for high-quality epilayers and excellent conductivity control if the high purity source precursors are available.

REFERENCES

1. Manasevit, H.M. and Simpson, W.I. (1971) *J. Electrochem. Soc.*, **118**, 644.
2. Manasevit, H.M. (1972) *J. Cryst. Growth*, **13–14**, 306.
3. Stutius, W. (1978) *Appl. Phys. Lett.*, **33**, 656.
4. Blanconnier, P., Cerclet, M., Henoc, P. and Jeanlouis, A.M. (1978) *Thin Solid Films*, **55**, 375.
5. Wright, P.J. and Cockayne, B. (1982) *J. Cryst. Growth*, **59**, 148.
6. Fujita, Sg., Tomomura, Y. and Sasaki, A. (1983) *Jpn. J. Appl. Phys.*, **22**, L583.
7. Yoshikawa, A., Tanaka, K., Yamaga, S. and Kasai, H. (1984) *Jpn. J. Appl. Phys.*, **23**, L424.
8. Fujita, Sg., Yodo, T., Matsuda, Y. and Sasaki, A. (1985) *J. Cryst. Growth*, **71**, 169.

9. Sritharan, S. and Jones, K.A. (1984) *J. Cryst. Growth*, **66**, 231.
10. Sritharan, S., Jones, K.A. and Mothl, K.M. (1984) *J. Cryst. Growth*, **68**, 656.
11. Mitsuhashi, H., Mitsuishi, I., Mizuta, M. and Kukimoto, H. (1985) *Jpn. J. Appl. Phys.*, **24**, L578.
12. Cockayne, B., Wright, P.J., Skolnick, M.S., Pitt, A.D., Williams, J.O. and Ng, T.L. (1985) *J. Cryst. Growth*, **72**, 17.
13. Tarui, Y., Hidaka, J. and Aota, K. (1984) *Jpn. J. Appl. Phys.*, **23**, L827.
14. Thompson, H.W. (1934) *J. Chem. Soc.*, **2**, 790
15. Chen, C.J. and Osgood, R.M. (1984) *J. Chem. Phys.*, **81**, 327.
16. Johnson, W.E. and Schlie, L.A. (1972) *Appl. Phys. Lett.*, **40**, 798.
17. Scott, J.D., Causley, G.C. and Rassel, B.R. *J. Chem. Phys.*, **59**, 6577.
18. Calvert, J.G. and Pitts, J.N., Jr. (1966) *Photochemistry*, Wiley, New York, p. 489.
19. Ando, H., Inuzuka, H., Konagai, M. and Takahashi, K. (1985) *J. Appl. Phys.*, **58**, 8021.
20. Kawakyu, Y., Sasaki, S., Hirose, M. and Beppu, T. (1986) *Extended Abstracts of the 18th International Conference on Solid State Devices and materials, Tokyo, 1986*, p. 643.
21. Fujita, Sg., Tanabe, A., Sakamoto, T., Isemura, M. and Fujita, Sz. (1987) *Jpn. J. Appl. Phys.*, **26**, L2000.
22. Fujita, Sz., Tanabe, A., Sakamoto, T., Isemura, M. and Fujita, Sg. (1988) *J. Cryst. Growth*, **93**, 259.
23. Fujita, Sz., Takeuchi, F.Y. and Fujita Sg. (1988) *Jpn. J. Appl. Phys.*, **27**, L2019.
24. Fujita, Sz., Tanabe, A., Kinoshita, T. and Fujita, Sg. (1990) *J. Cryst. Growth*, **101**, 48.
25. Fujita, Sz., Maruo, S., Ishio, H., Murawala, P.A. and Fujita, Sg. (1991) *5th International Conference on MOVPE, Aachen, FRG, 1990, J. Cryst. Growth*, **107**, 644.
26. Fujita, Sg., Tomomura, Y. and Sasaki, A. (1983) *25th Electronic Materials Conference, Burlington, 1983*, Paper E-7.
27. Fujita, Sg., Isemura, M., Sakamoto, T. and Yoshimura, N. (1988) *J. Cryst. Growth*, **86**, 263.
28. Yoshikawa, A., Okamoto, T., Fujimoto, T., Onoue, K., Yamaga, S. and Kasai, H. (1990) *Jpn. J. Appl. Phys.*, **29**, L225.
29. Ritz-Froidevaux, Y., Salathe', R.P., Gilgen, H.H. and Weber, H.P. (1982) *Appl. Phys. A*, **27**, 133.
30. Sugiura, H., Yamada, T. and Iga, R. (1990) *Jpn. J. Appl. Phys.*, **29**, L1.
31. Yasuda, T., Koyama, Y., Wakitani, J., Yoshino, J. and Kukimoto, H. (1989) *Jpn. J. Appl. Phys.*, **28**, L1628.
32. Matsumura, N., Ishikawa, K., Saraie, J. and Yokogawa, Y. (1985) *J. Cryst. Growth*, **72**, 41.
33. Mitsuhashi, H.,Mitsuishi, I. and Kukimoto, H. (1985) *Jpn. J. Appl. Phys.*, **24**, L864.
34. Dean, P.J. (1981) *Phys. Status Solidi A*, **81**, 6895.
35. Taguchi, T. and Yao, T. (1984) *J. Appl. Phys.*, **56**, 3002.
36. Kukimoto, H. (1990) *J. Cryst. Growth*, **101**, 953.
37. Okamoto, T. and Yoshikawa, A. (1991) *Jpn. J. Appl. Phys.*, **30**, L156.
38. Fujita, Sz., Hirata, S. and Fujita, Sg. (1991) *Jpn. J. Appl. Phys.*, **30**, L507.

4

ZnSe growth by conventional molecular beam epitaxy: a review of recent progress

R.M. Park

4.1 INTRODUCTION

The following chapter highlights recent results in the area of ZnSe growth by conventional molecular beam epitaxy (MBE) which, in the author's opinion, have significantly advanced both our understanding of the material itself as well as the MBE technology as it pertains to ZnSe epitaxial growth. The term, conventional MBE, is taken to mean epitaxial growth in which the constituent elements, in this case Zn and Se, are derived from Knudsen-style effusion sources containing solid-element material. The review is concerned solely with recent advances in the field of conventional MBE growth of ZnSe and does not include discussion on technologically related approaches to ZnSe growth such as metal–organic molecular beam epitaxy (MOMBE), for example.

Early work in the ZnSe growth by MBE field has previously been reviewed by Yao [1] while Gunshor and Kolodziejski [2] more recently reviewed the field with particular emphasis on ZnSe-based superlattice and quantum well structures. Research efforts concerning ZnSe growth by MBE on non-polar substrates (i.e. Ge and Si) have been reviewed by Park [3].

The present review article focuses on significant developments in the ZnSe growth by MBE field which have taken place primarily over the last two or three years. Studies are emphasized which concern the growth of device-quality n- and p-type ZnSe material and also the development of ZnSe p–n homojunction based blue light-emitting devices.

4.2 MOLECULAR BEAM EPITAXY GROWTH OF UNINTENTIONALLY DOPED HETERO- AND HOMOEPITAXIAL ZnSe

One of the most significant discoveries in the ZnSe growth by MBE field was made by Yoneda *et al.* [4] at the Sanyo Electric Co. in Japan and reported in late 1984. These authors determined that the principal donor impurity incorporated in unintentionally doped ZnSe originated from the Se source material. Yoneda *et al.* demonstrated that the free-electron concentration in unintentionally doped ZnSe material could be dramatically reduced from a typical level around $10^{17}\,cm^{-3}$ to a point where the material became highly resistive ($\rho > 10^4\,\Omega\,cm$) by employing a Se source material purification scheme. The Se purification scheme involved repeated distillation cycles [4]. This result was particularly significant in that it opened the door for meaningful n- and p-type doping studies of MBE-grown ZnSe material, since the influence of intentionally incorporated impurities on material properties could be investigated. Fortunately, not long after the report by Yoneda *et al.*, high-purity Se source material became commercially available from the Osaka Asahi Metal Manufacturing Co. Ltd. of Japan, which has allowed investigators around the world to grow ultrahigh purity, unintentionally doped ZnSe by MBE.

For instance, Shahzad *et al.* [5] recently reported a photoluminescence (PL) study of ultrahigh purity ZnSe grown on GaAs using high-purity Zn and Se source material obtained from the Osaka Asahi Metal manufacturing Co. Ltd. Like these authors, the vast majority of investigators working in this field currently employ GaAs as substrate material because of close lattice match to ZnSe and the fact that GaAs is readily available in the form of high-quality wafers. In their paper, Shahzad *et al.* [5] suggest that ultrahigh purity, unintentionally doped ZnSe material is characterized in terms of a low-temperature PL spectrum, by the presence of dominant free-exciton-related emission peaks together with a significantly strong unassigned peak, termed I_v, at 2.7738 eV. As also discussed by Shahzad *et al.* [5, 6], a thermal expansion coefficient mismatch exists between ZnSe and GaAs which results in ZnSe/GaAs heteroepitaxial layers thicker than $\sim 0.8\,\mu m$ experiencing in-plane biaxial tension. Such tensile strain is evidenced by the observation of exciton (both free and bound) peak splitting in low-temperature PL spectra [5, 6]. Splitting of exciton transitions results from light- and heavy-hole valence band splitting at $k = 0$ under conditions of biaxial tension. Consequently, although high-purity ZnSe can be grown by MBE on GaAs, a potential problem exists in this case because the ZnSe material is elastically stressed at room temperature. Whether or not the tensile strain will present a problem as far as device operation is concerned remains to be seen.

An obvious solution to the strain problem would be to grow homoepitaxial material. However, attempts to grow high-purity, homoepitaxial ZnSe layers by MBE have been hampered by the unavailability commercially of high-purity, high-quality ZnSe bulk wafers. Park *et al.* [7] in 1985 and later Menda *et al.* [8] in 1988 reported homoepitaxial growth of ZnSe by MBE using ZnSe substrate material ((100) oriented) obtained from Eagle-Picher Industries, Inc. In both cases, however, low-temperature PL analysis indicated the layers to contain a significant concentration of impurities, as evidenced by the presence of strong donor-bound exciton peaks as well as donor-to-acceptor pair transition peaks. Park *et al.* [7] showed the ZnSe substrate material itself to contain a high concentration of donor and acceptor impurities. In addition, unintentionally doped ZnSe layers grown by MBE on (1 1 0)-oriented Eagle-Picher ZnSe substrates were found to contain a significant concentration of donor impurities [9].

A recent investigation, however, into the homoepitaxial growth of unintentionally doped ZnSe by the author's group at the University of Florida has resulted, for the first time, in the production of high purity homoepitaxial ZnSe [10]. The ZnSe substrate material employed in this work was grown by Dr. Koyama at the Nippon Sheet Glass Co. Ltd. in Japan and consisted of (100)-oriented wafers cut from ingots prepared by the iodine vapour transport method. Double-crystal, X-ray rocking curve linewidths on the order of 12 seconds of arc were recorded from this substrate material [11] and, in addition, donor-to-acceptor pair transitions were not detectable in PL spectra recorded from the bulk crystals. Unintentionally doped ZnSe layers were grown in this work in a custom-designed (non-commercial) MBE system using 6N Zn and Se source materials obtained from Osaka Asahi Metal Manufacturing Co. Ltd. The homoepitaxial ZnSe layers were characterized by 10 K PL analysis and compared with heteroepitaxial (ZnSe/(100)GaAs) layers grown under the same conditions. Low-temperature PL spectra recorded from homo- and heteroepitaxial ZnSe layers can be compared and contrasted in Figs 4.1(a) and 4.1(b). As can be seen from these figures, both PL spectra contain features which are characteristic of high-purity ZnSe, namely comparably intense free exciton (E_x) and donor-bound exciton (I_2) peaks as well as a significantly intense I_v peak [5]. However, an important difference between the spectra is that in the case of the homoepitaxial layer (Fig. 4.1(b)), the excitonic transitions are unsplit in contrast to the peak splitting (both free and donor-bound excitons) evident in the spectrum recorded from the heteroepitaxial layer (Fig. 4.1(a)). Consequently, the ZnSe homoepitaxial material appears to be strain-free relative to the heteroepitaxial material as well as being of comparable purity. Furthermore, since the homoepitaxial material is strain-free, peak energies associated with the major transitions in relaxed high-purity ZnSe can be quoted as follows: free exciton, E_x, at 2.8027 eV, neutral donor-bound exciton, I_2, at 2.7970 eV, and I_v at 2.7768 eV. Shahzad *et al.* [5] detected I_v at 2.7738 eV in their material, which is extremely close in energy to the I_v peak shown in Fig. 4.1(a); however, a 3 meV peak shift to lower energy is evident relative to the peak position in unstrained material. It is thought that the peak shifting could be accounted for on the basis of the biaxial in-plane tension experienced by the heteroepitaxial material.

Assuming, therefore, that high-purity substrate material is employed, it appears that high-purity unintentionally doped ZnSe homoepitaxial material can be grown by MBE. However, it is highly probable that GaAs will remain

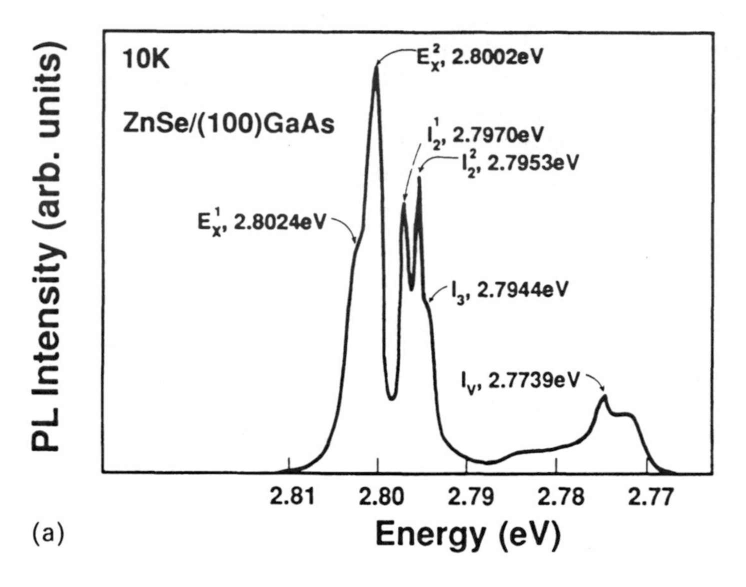

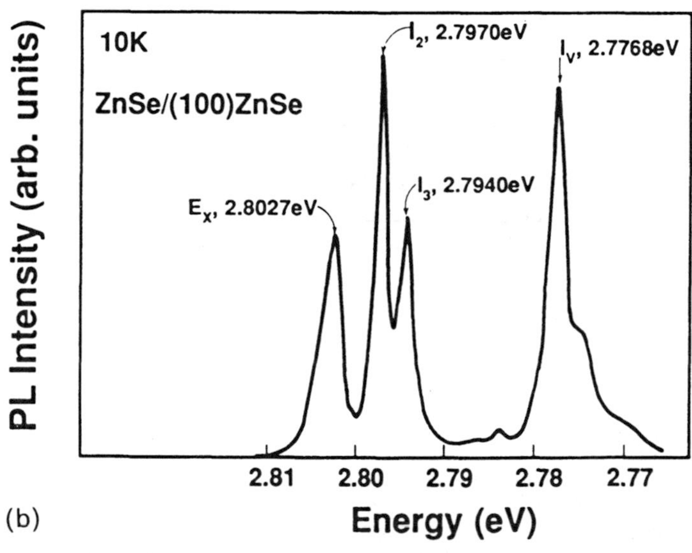

Fig. 4.1 Typical 10 K PL spectra recorded from 2 μm thick unintentionally doped ZnSe layers grown by MBE on (a) (100)GaAs and (b) (100)ZnSe. The ZnSe substrate material was prepared by the iodine vapour transport method.

the substrate of choice for MBE growth of ZnSe until high-purity, high-quality, large-area ZnSe bulk crystals become commercially available.

4.3 n-TYPE ZnSe GROWN BY MOLECULAR BEAM EPITAXY

Since the production of high free-electron density n-type ZnSe material is critical, not only for the provision of efficient p–n homojunctions, but also so that low-resistance ohmic contact can be made to devices, a considerable amount of effort has gone into studying n-type doping of MBE-grown ZnSe material during the last decade.

Prior to a report by Ohkawa *et al.* in 1987 concerning Cl doping [12], Ga appeared to be the most promising n-type dopant element for MBE-grown ZnSe. For instance, Niina *et al.* [13] reported in 1982 measuring free electron densities in Ga-doped material up to an apparent maximum of $\sim 5 \times 10^{17}\,cm^{-3}$, electron mobilities at these densities being around $300\,cm^2\,V^{-1}\,s^{-1}$. This upper limit for the free-electron density in Ga-doped material appears to be imposed as a consequence of the onset of strong compensation in the material at high doping levels [13]. In 1988, de Miguel *et al.* [14] reported a doubling of the maximum free-electron density achievable using Ga as the dopant element. These authors employed a planar (or delta) doping process which produced a maximum free-electron density around $1 \times 10^{18}\,cm^{-3}$. However, electron mobilities measured in the planar-doped material appear low. For example, an electron mobility of $\sim 150\,cm^2\,V^{-1}\,s^{-1}$ was measured for a carrier density of $\sim 1 \times 10^{18}\,cm^{-3}$ which only increased to $\sim 220\,cm^2\,V^{-1}\,s^{-1}$ at $\sim 2 \times 10^{17}\,cm^{-3}$ [14]. These rather low mobility values indicate the presence of a significant amount of compensation in the planar-doped material [15].

In the author's opinion, Cl is a more suitable n-type dopant element than Ga for the production of high-quality, high free-electron density ZnSe material. Cl-doped ZnSe grown by MBE was first reported by Ohkawa *et al.* [12]. These authors derived the Cl impurities from an effusion cell containing compound $ZnCl_2$ material. Ohkawa *et al.* [12] reported measuring free-electron densities up to $\sim 1 \times 10^{19}\,cm^{-3}$ in their Cl-doped ZnSe material, which represents an order-of-magnitude improvement over Ga-doping. The electron mobility values reported by Ohkawa *et al.* are also impressive; for instance, at carrier densities of $\sim 1 \times 10^{18}\,cm^{-3}$ and $\sim 1 \times 10^{17}\,cm^{-3}$, the electron mobilities reported for Cl-doped ZnSe were $\sim 300\,cm^2\,V^{-1}\,s^{-1}$ and $\sim 400\,cm^2\,V^{-1}\,s^{-1}$ respectively. These mobility values are approximately a factor of 2 higher than those reported for Ga-doped ZnSe using the planar doping technique [14]. The Cl doping results of Ohkawa *et al.* have been reproduced in the US, for example, by the 3M Co. (St. Paul, MN) group [16] and the author's group at the University of Florida, again employing $ZnCl_2$ source material. Hall effect data obtained from Cl-doped ZnSe grown at the University of Florida are illustrated in Fig. 4.2. These data are highly

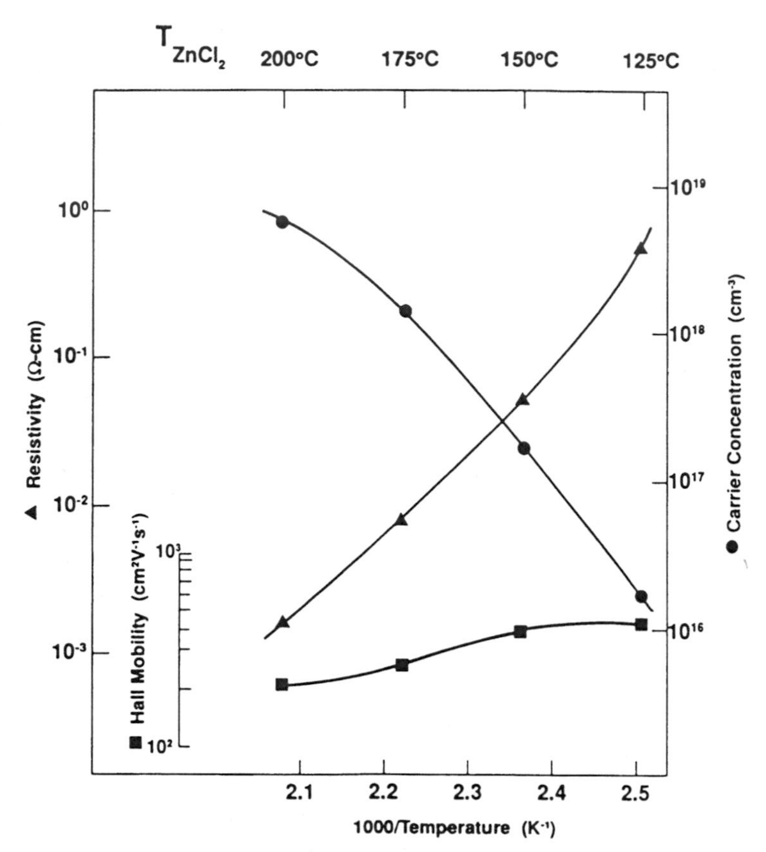

Fig. 4.2 Free-electron concentration, resistivity and electron mobility data recorded from Cl-doped ZnSe/GaAs layers grown by molecular beam epitaxy. Cl dopant impurities were derived from an effusion source containing compound $ZnCl_2$ material and the electrical data are plotted as a function of the $ZnCl_2$ source temperature.

reproducible and, furthermore, there is no apparent memory effect associated with the use of the $ZnCl_2$ source. Ultrahigh-purity unintentionally doped ZnSe material can be grown, as determined by low-temperature PL analysis, immediately following the growth of a Cl-doped ZnSe layer. In addition, Cl doping appears to be eminently suitable for the production of n-type and n^+-type ZnSe material with a view to device application, since diffusion of Cl atoms is negligible at typical growth temperatures [12].

4.4 p-TYPE ZnSe GROWN BY MOLECULAR BEAM EPITAXY

The major focus over the last few years regarding research on the MBE growth of ZnSe has been on producing low-resistivity p-type material. Development of a technology which can provide for the *in situ* production of epitaxial structures comprising ZnSe p–n junctions is considered of paramount importance for the fabrication of efficient light-emitting devices, such as light emitting diodes (LEDs) and diode lasers, operating in the blue region of the visible spectrum.

In terms of efforts to incorporate substitutional acceptor impurities in ZnSe epitaxial layers during crystal growth, the highest degree of reported success, until very recently, concerned Li-doping during MBE growth [17]. Two major problems, however, appear to hamper the employment of Li as a practical impurity in ZnSe. Firstly, a net acceptor density of approximately $8 \times 10^{16}\,cm^{-3}$ seems to represent the upper limit for Li doping. At higher Li concentrations, strong compensation occurs which renders the ZnSe material highly resistive [17]. Secondly, Li impurities are unstable in ZnSe at temperatures above approximately 300 °C. The latter problem would manifest itself should device processing procedures necessitate heating the material beyond 300 °C. P-type behaviour has also been reported employing the isoelectronic impurity, oxygen, as a dopant in ZnSe layers grown by MBE [18]. However, net acceptor concentrations in ZnSe:O layers appear to be low, the largest net acceptor density reported so far being $1.2 \times 10^{16}\,cm^{-3}$ [18]. Nitrogen has also received attention as a candidate p-type dopant element in MBE-grown ZnSe. For instance, Park *et al.* [19] in 1985 reported the incorporation of nitrogen acceptor impurities in ZnSe using the MBE technique. However, in their work, which involved the use of neutral N_2 and NH_3 molecules, the ZnSe layers remained highly resistive since only small concentrations of nitrogen impurities were incorporated during crystal growth. Later on (1986), Mitsuyu *et al.* [20] reported achieving a higher level of nitrogen incorporation in MBE-grown ZnSe by employing a partially ionized beam of NH_3. However, crystallographic damage at the ion current levels necessary for doping rendered the ZnSe material highly resistive [20]. Consequently, although N was thought to be potentially the most suitable p-type dopant element for ZnSe, an appropriate means of incorporating large concentrations of the impurity element into the material during MBE growth was not available. However, the author's group at the University of Florida has recently developed a novel technique to incorporate large concentrations of nitrogen acceptor impurities in ZnSe/GaAs epitaxial layers which involves nitrogen atom beam doping during MBE growth [21]. Net acceptor densities as large as $3.4 \times 10^{17}\,cm^{-3}$ have been measured to date in this material which represents the highest net acceptor density reported thus far for MBE-grown ZnSe layers.

The p-type (nitrogen-doped) ZnSe layers were grown on GaAs substrates in the custom-designed (non-commercial) MBE system previously used to grow the ultrahigh purity ZnSe hetero- and homoepitaxial layers discussed in section 4.2. Figure 4.3 illustrates the growth chamber configuration. As can be seen from the schematic diagram, p-type doping of the ZnSe layers was achieved by employing a specially designed free-radical source (FRS)

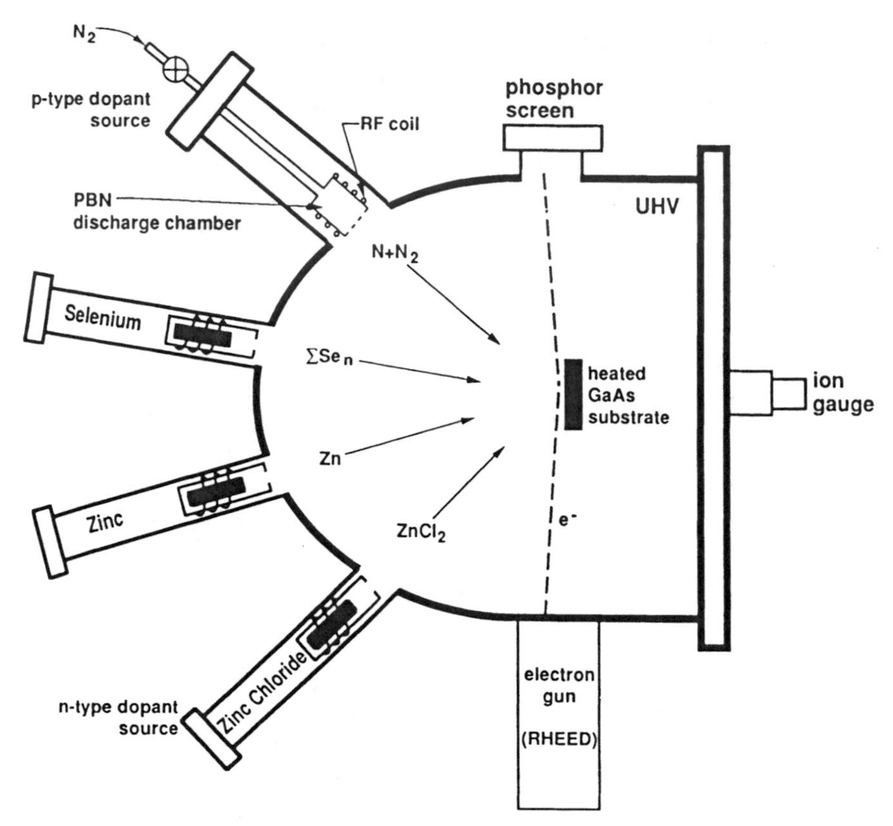

Fig. 4.3 MBE growth chamber configuration used to grow ZnSe:N/ZnSe:Cl, p–n homojunction structures on GaAs. The p-type dopant is atomic nitrogen which is created in the high frequency (13.5 MHz) plasma discharge source shown in the schematic diagram.

which is incorporated on the MBE system in place of a conventional effusion source. In the FRS, which was built by Oxford Applied Research, Oxfordshire, UK (model MPD21), an electrodeless high-frequency (13.5 MHz) discharge under conditions of molecular flow provides a flux of atomic nitrogen (together with a much larger flux of non-dissociated N_2), the source of nitrogen being a cylinder of ultrahigh-purity N_2. The atomic nitrogen flux level is controlled by suitably adjusting the intensity of the high-frequency plasma discharge. Such an approach to nitrogen incorporation in ZnSe was investigated since the reactivity of nitrogen atomic species at the growing ZnSe surface was expected to be much greater than that of molecular nitrogen which should promote enhanced substitutional incorporation.

The above hypothesis was, in fact, verified as evidenced by comparing 10 K PL spectra recorded from ZnSe layers grown with a flux of N_2 only and with a flux of $N + N_2$. As can be seen from Fig. 4.4(a), the 10 K PL

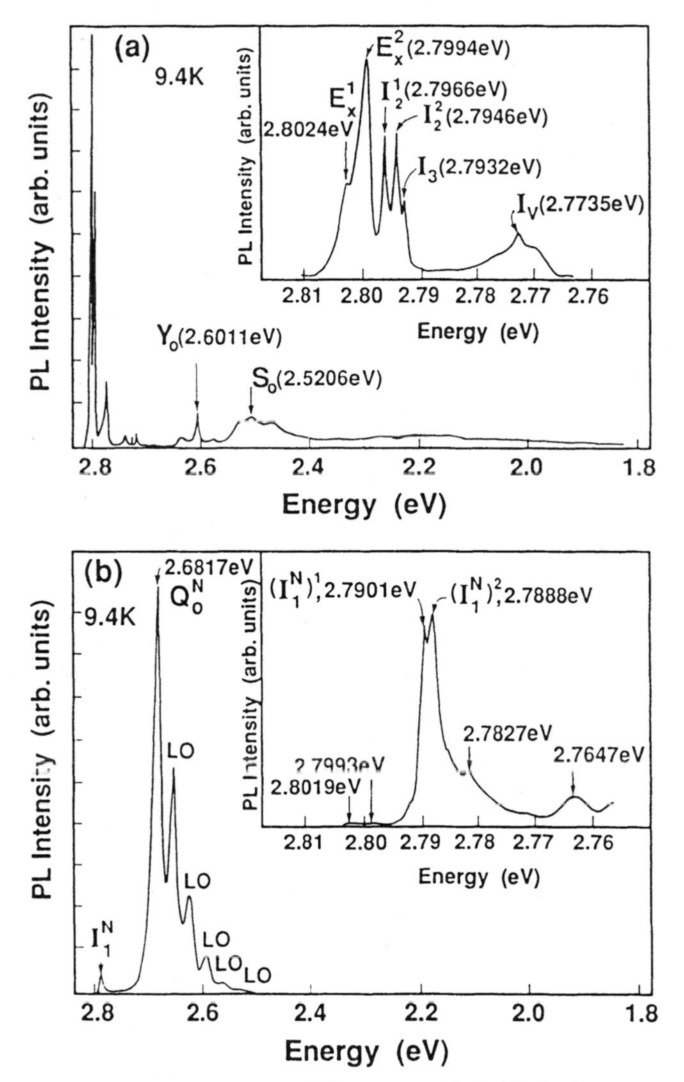

Fig. 4.4 PL spectra recorded at 10 K from (a) a ZnSe/GaAs layer grown with a coincident flux of N_2 only, and (b) a ZnSe/GaAs layer grown with an $N + N_2$ coincident flux. The net acceptor concentration in the ZnSe layer from which spectrum (b) was obtained is $1 \times 10^{17}\,cm^{-3}$.

spectrum recorded from a ZnSe layer grown using a flux of N_2 only (in this case an equilibrium background pressure of N_2 in the MBE chamber of 5×10^{-7} Torr was maintained) appears to be identical to that recorded from unintentionally doped ZnSe heteroepitaxial layers [10]. The dominant peaks in the excitonic regime are the split free exciton (E_x) and donor-bound exciton (I_2) transitions, the splitting being due to the thermal expansion coefficient mismatch between ZnSe and GaAs which renders the ZnSe layers under in-plane biaxial tension [6]. Consequently, at such low background N_2 partial pressures, molecular nitrogen is completely unreactive at the ZnSe surface. The situation changes dramatically, however, when a plasma discharge is created in the free-radical source, as can be seen from the 10 K spectrum shown in Fig. 4.4(b). Again the background N_2 partial pressure in the MBE chamber during growth was 5×10^{-7} Torr with the plasma discharge on in this case. The excitonic regime is now dominated by split acceptor-bound exciton (I_1^N) transitions owing to the incorporation of nitrogen acceptor impurities [22]. In addition, the complete PL spectrum is dominated by donor-to-acceptor (D–A) transitions (Q_0^N represents the no-phonon transition, with several longitudinal optical (LO) phonon replicas of Q_0^N also indicated). Thus, it can be concluded that the rate of substitutional incorporation of atomic nitrogen is much greater than that of molecular nitrogen at the growing ZnSe surface. The sample from which the PL spectrum shown in Fig. 4.4(b) was obtained was found to have a net acceptor concentration of $1 \times 10^{17}\,cm^{-3}$.

Net acceptor concentrations, N_A-N_D, in the nitrogen-doped ZnSe/GaAs layers were determined using capacitance–voltage (C–V) profiling. This work was performed by Drs James DePuydt and Michael Haase at the 3M Co. in St. Paul, MN. Since the ZnSe epitaxial layers were grown on semi-insulating GaAs, planar profiling between two Schottky contacts on the ZnSe surface was carried out. The surface contact pattern consisted of a series of 762 μm diameter Cr/Au dots physically isolated from a large Cr/Au surrounding electrode. The separation between the inner (dot) electrodes and the outer electrode was 25 μm, a small separation being necessary in order to maintain a low series resistance. The contact pattern was created by thermally evaporating 75 A of Cr followed by 1000 A of Au and performing photolithographic and lift-off processes. In all of these measurements the outer electrode was held at ground potential and bias was applied to the inner Schottky contact. With this sign convention the majority carrier type is given by the sign of the slope of the $1/C^2$ versus V plot; a positive slope would indicate the material to be p-type. The net acceptor (N_A-N_D) concentration is proportional to the slope of $1/C^2$ versus V. The $1/C^2$ versus V plot and the N_A-N_D versus depletion width profile obtained from the most heavily doped ZnSe layer grown to date are illustrated in Figs. 4.5(a) and 4.5(b) respectively. As can be seen from this figure, the material is p-type with a net acceptor concentration around $3.4 \times 10^{17}\,cm^{-3}$. As shown in Figure 4.5(b), the doping profile is rather flat from zero bias (0.068 μm) out to where reverse bias breakdown occurs (0.126 μm). Breakdown occurred at 3.8 V which is consistent with avalanche breakdown in ZnSe material doped at this level, i.e. $3.4 \times 10^{17}\,cm^{-3}$ p-type.

Again with a view to device application, it is the author's contention that of the impurity elements which have reportedly provided for measurable

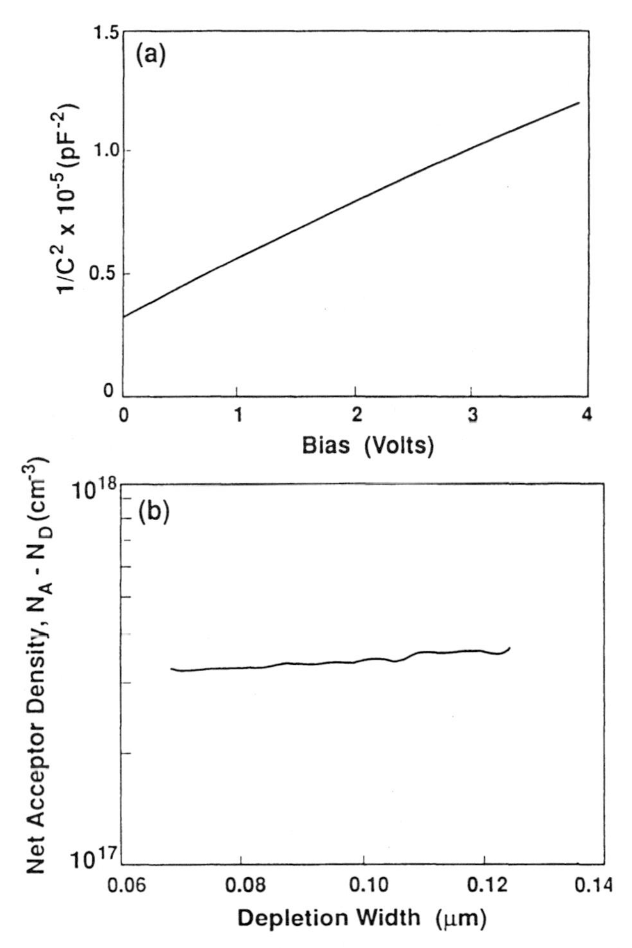

Fig. 4.5 Graphical representation of the capacitance–voltage (C–V) data obtained by profiling a nitrogen atom beam-doped ZnSe/GaAs layer. The positive sign of the slope in the $1/C^2$ versus V plot (graph (a)) indicates the material to be p-type while graph (b) illustrates the uniformity of the doping over the measurement range with an average value for the net acceptor concentration, N_A-N_D, of $3.4 \times 10^{17}\,cm^{-3}$.

p-type conduction in MBE-grown ZnSe material, i.e. Li, O and now N, the most suitable p-type dopant is N. In addition to the large net acceptor densities achievable with N, i.e. in the $10^{17}\,cm^{-3}$ range, N is a stable substitutional impurity in ZnSe to temperatures around 400 °C.

4.5 LIGHT-EMITTING DIODES BASED ON MOLECULAR BEAM EPITAXY-GROWN ZnSe p–n HOMOJUNCTIONS

Successful p-type doping of MBE-grown ZnSe material by means of Li incorporation led to a recent demonstration by the 3M Co. (St. Paul, MN) group of a room-temperature blue LED device based on an MBE-grown ZnSe p–n homojunction structure [17]. The 3M Co. device consisted of a ZnSe p–n junction structure grown on p-type GaAs substrate material and employed Li and Cl as the p- and n-type dopant elements respectively. Around the same time, Akimoto *et al.* [18] of the Sony Corporation in Japan reported room-temperature blue LED operation from an MBE-grown ZnSe homojunction device which consisted of an n-type, Ga-doped layer and a p-type, O-doped layer grown on an n-type GaAs substrate. However, although these blue LED device reports are encouraging, it is the author's opinion, as discussed in section 4.4, that neither Li nor O is an ideal p-type dopant element for ZnSe, particularly with regard to developing ZnSe diode lasers.

More recently, the novel nitrogen-doping technique developed in the author's laboratory at the University of Florida has resulted in the first demonstration of room-temperature blue LED device operation based on a ZnSe:N/ZnSe:Cl, p–n homojunction structure [21]. LED device fabrication work was performed by Drs James DePuydt and Michael Haase of the 3 M Co. (St. Paul, MN). A typical LED device structure employed in this work is shown schematically in Fig. 4.6. As can be seen from this schematic, the p-type ZnSe layer was grown first on a p^+-type GaAs substrate. This type

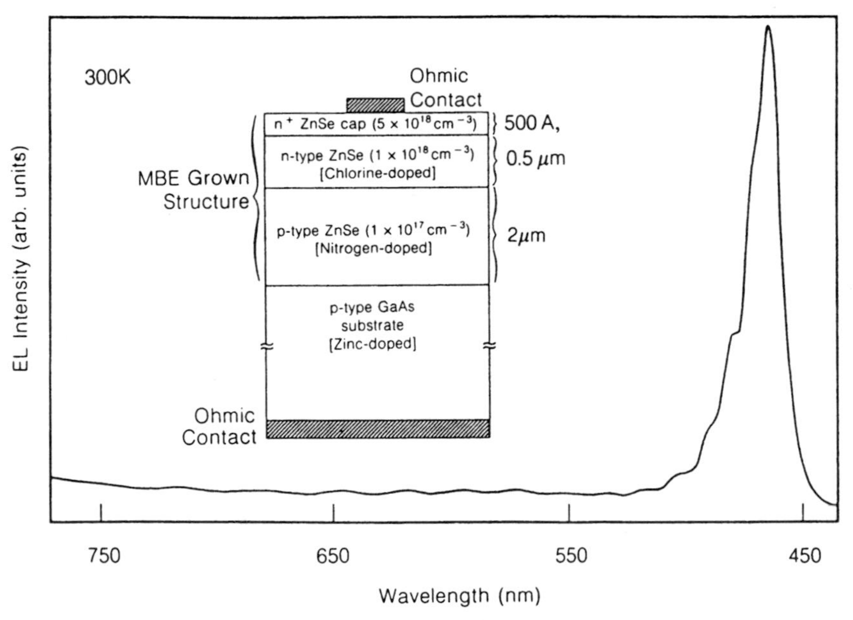

Fig. 4.6 Room-temperature electroluminescence spectrum (visible region only) recorded from the device structure shown schematically in the inset. The device structure is of the 'buried p-type layer' design.

of 'buried p-type layer' structure was chosen in order to avoid the serious problems at present associated with ohmic contact formation to p-type ZnSe [17]. However, a disadvantage with this device design is that a large hole barrier exists at the p^+-GaAs/p-ZnSe heterointerface [23]. In this type of device, hole injection across the p^+-GaAs/p-ZnSe heterointerface is only realized at avalanche breakdown. Consequently, large turn-on voltages are required in order to observe electroluminescence associated with the ZnSe p–n homojunction. LED fabrication was accomplished using conventional photolithographic techniques with device isolation being achieved by wet chemical etching to form 400 μm diameter mesas. The top electrode metallization was ring shaped and was patterned by vacuum evaporation and lift-off. Ultrasonic gold ball bonding was used to make contact to the devices for electroluminescence characterization. An electroluminescence spectrum recorded at room temperature from the device (visible region only) is shown in Fig. 4.6. As can be seen from the figure, dominant emission in the blue region of the visible spectrum is observed, peaking in intensity at a wavelength of 465 nm. For the particular spectrum shown, the voltage applied and current drawn were 22 V and 20 mA respectively. Infrared emission was also recorded from these devices (simultaneously with the blue emission) which is believed to be the result of electron injection into the p^+-GaAs material under avalanche breakdown conditions at the heterojunction (not shown in the figure). Further improvement in the performance of ZnSe:N/ZnSe:Cl, p–n junction light-emitting devices is anticipated, in particular a reduction in turn-on voltages (to less than 3 V), when a reproducible ohmic contact technology is developed for p-type ZnSe material. This would allow ZnSe p–n junction structures to be grown on n-type GaAs rather than p-type GaAs substrates which would eliminate the hole barrier problem discussed above.

4.6 CONCLUSIONS

The recent developments discussed above with respect, in particular, to the control now afforded over n- and p-type doping of ZnSe using conventional MBE indicate that this technology is eminently suitable for the production of ZnSe-based blue light-emitting device structures. It is the author's contention that Cl is currently the most suitable n-type dopant element in MBE-grown ZnSe and that N-doping is superior to either Li- or O-doping for the production of large net acceptor density, stable p-type ZnSe by the MBE technique.

As previously concluded by Haase *et al.* [17], it is clear that the problem of forming reliable ohmic contacts to p-type ZnSe has to be solved before high-performance ZnSe-based, light-emitting devices, in particular diode lasers, can be realized. In this regard, since elevated-temperature processing may be required, nitrogen again appears to be the most promising p-type dopant element due to the fact that N impurities are stable in ZnSe to temperatures as high as ~400 °C.

ACKNOWLEDGEMENTS

Research at the University of Florida has been supported by the Defense Advanced Research Projects Agency under ONR Grant number MDA-972-B5-J-1006 and the National Science Foundation under Grant number MSS-8909281.

High-calibre assistance in the MBE ZnSe project area has been provided by Mary Jane Troffer and Christopher Rouleau, both currently Graduate Research Assistants.

Finally, the author wishes to acknowledge a highly fruitful collaboration with Drs James DePuydt and Michael Haase of the 3M Co. (St. Paul, MN) particularly with regard to LED device development.

REFERENCES

1. Yao, T. (1985) *The Technology and Physics of Molecular Beam Epitaxy* (ed. E.H.C. Parker), Plenum, New York.
2. Gunshor, R.L. and Kolodziejski, L.A. (1988) *IEEE J. Quantum Electron.*, **24**, 1744.
3. Park, R.M. (1989) *Growth and Optical Properties of Wide-gap II–VI Low-dimensional Semiconductors* (eds T.C. McGill, C.M. Sotomayer Torress and W. Gebhardt), Plenum, New York.
4. Yoneda, K., Hishida, Y., Toda, T., Ishii, H. and Niina, T. (1984) *Appl. Phys. Lett.*, **45**, 1300.
5. Shahzad, K., Olega, D.J. and Cammack, D.A. (1989) *Phys. Rev. B*, **39**, 13016.
6. Shahzad, K. (1988) *Phys. Rev.* B, **38**, 8309.
7. Park, R.M., Mar, H.A. and Salansky, N.M. (1985) *J. Vac. Sci. Technol. B*, **3**, 1637.
8. Menda, K., Takayasu, I., Minato, T. and Kawashima, M. (1988) *J. Cryst. Growth*, **86**, 342.
9. Ohkawa, K., Karasawa, T., Yoshida, A., Hirao, T. and Mitsuyu, T. (1989) *Appl. Phys. Lett.*, **54**, 2553.
10. Park, R.M., Rouleau, C.M., Troffer, M.B., Koyama, T. and Yodo, T. (1990) *J. Mater. Res.*, **5**, 475.
11. Koyama, T., Yodo, T., Oka, H., Yamashita, K. and Yamasaki, T. (1988) *J. Cryst. Growth*, **91**,639.
12. Ohkawa, K., Mitsuyu, T. and Yamazaki, O. (1987) *J. Appl. Phys.*, **62**, 3216.
13. Niina, T., Minato, T. and Yoneda, K. (1982) *Jpn. J. Appl. Phys.*, **21**, L387.
14. de Miguel, J.L., Shibli, S.M., Tamargo, M.C. and Skromme, B.J. (1988) *Appl. Phys. Lett.*, **53**, 2065.
15. Ruda, H.E. (1986) *J. Appl. Phys.*, **59**, 1220.
16. Cheng, H., DePudyt, J.M., Potts, J.E. and Haase, M.A. (1989) *J. Cryst. Growth*, **95**, 512.
17. Haase, M.A., Cheng, H., DePuydt, J.M. and Potts, J.E. (1990) *J. Appl. Phys.*, **67**, 448.
18. Akimoto, K., Miyajima, T. and Mori, Y. (1989) *Jpn. J. Appl. Phys.*, **28**, L531.
19. Park, R.M., Mar, H.A. and Salansky, N.M. (1985)*J. Appl. Phys.*, **58**, 1047.
20. Mitsuya, T., Ohkawa, K. and Yamazaki, O. (1986) *Appl. Phys. Lett.*, **49**, 1348.

21. Park, R.M., Troffer, M.B., Rouleau, C.M., DePuydt, J.M. and Haase, M.A. (1990) *Appl. Phys, Lett.*, **57**, 2127.
22. Dean, P.J., Stutius, W., Neumark, G.F., Fitzpatrick, B.J., and Bhargava, R.N. (1983) *Phys. Rev. B*, **27**, 2419.
23. Kassel, L., Abad, H., Garland, J.W., Raccah, P.M., Potts, J.E., Haase, M.A. and Cheng, H. (1990) *Appl. Phys. Lett.*, **56**, 42.

5

Metal–organic molecular beam epitaxy growth and properties of widegap II–VI compounds

A. Yoshikawa

5.1 INTRODUCTION

In order to achieve conductivity control of widegap II–VI compounds, such as ZnS, ZnSe and ZnTe, many attempts to grow and dope epilayers with several kinds of dopants have been undertaken by low-temperature epitaxy. As discussed in other chapters in this book, molecular beam epitaxy (MBE) and metal–organic vapour phase epitaxy (MOVPE) are typical and promising low-temperature epitaxy methods even for II–VI compounds, as well as for III–V compounds.

In addition to these two methods, a new and interesting low-temperature epitaxy technique called metal–organic molecular beam epitaxy (MOMBE) has recently been introduced in this area. In MOMBE, it is expected to be able to grow very high-quality films at low temperature in a clean atmosphere (one of the merits of MBE) by using gas sources (one of the merits of MOVPE). Thus, MOMBE can utilize the merits of both MOVPE and MBE.

Basically, MOMBE is one of the variations of the MBE system, and metal–organic gaseous sources are used as at least one of the source materials in MOMBE [1]. For the gaseous source materials, metal–organic compounds are normally used, and hydride gases are also used [2]. Therefore, MOMBE is sometimes called gas source MBE (GSMBE) [3, 4]. Furthermore, in the field of III–V compounds, MOMBE is often called chemical beam epitaxy (CBE). Tsang [5] has classified them into three categories, i.e. CBE, MOMBE and GSMBE, depending on the combination of three types of sources, i.e. solid, MO and hydride sources. In this chapter, however, these 'gas source MBEs' are simply called MOMBE, because, for the growth of II–VI compounds, MO compounds are mainly used as gaseous sources except in a few reports.

First attempts to introduce gaseous source materials into MBE were made for the growth of III–V compounds, such as GaAs, GaP, AlN [1–4]. Accordingly, MOMBE has been mainly applied to the growth of III–V compounds until now and many papers have already been published in that area, but only a limited amount of work has been performed on the growth of widegap II–VI compounds. It should be noted, however, that MOMBE has potential advantages compared with MBE and MOVPE even for the growth of II–VI compounds.

Among various widegap II–VI compounds, MOMBE has been applied mainly to the growth of ZnSe epilayers [6–14], which are the most promising materials at present for p–n junction blue-light-emitting diodes. MOMBE growth of ZnS [7, 10, 15, 16] and ZnTe [17] epilayers has also been examined. Furthermore, examinations of atomic layer epitaxy (ALE) [18, 19] and growth of strained layer superlattices (SLSs) [17, 20–22] have also been performed by using MOMBE growth systems. Table 5.1 summarizes the MOMBE work on the growth of widegap II–VI compound epilayers.

In this chapter, the present status of MOMBE growth and properties of widegap II–VI compounds are discussed. First, features of MOMBE for the growth of widegap II–VI compounds are briefly shown, and the growth mechanism especially for ZnSe epilayers is discussed. Also, electrical and optical properties of MOMBE-grown ZnSe layers are discussed. Furthermore, growth and properties of novel structure II–VI compound layers, such as SLSs, are also discussed.

Table 5.1 Summary of the growth of widegap II–VI compound epilayers by MOMBE

Material	Group II reactant	Group VI reactant	Remarks	Year	References
ZnSe	DEZn	DESe		1986	[6]
ZnSe	DEZn	DESe		1987	[7]
ZnSe	DMZn, DEZn	DESc		1988	[11, 12]
ZnSe	DMZn	H_2Se		1988	[8]
ZnSe	DMZn	H_2Se		1989	[9, 10]
ZnSe	DMZn	H_2Se		1990	[13, 14]
ZnSe	DMZn	H_2Se	ALE	1990	[18]
ZnS	Zn	H_2S		1985	[15]
ZnS	Zn	H_2S		1986	[16]
ZnS	DEZn	DES, H_2S		1987	[7]
ZnS	DMZn	H_2S		1989	[10]
ZnS	DMZn	H_2S	ALE	1990	[19]
ZnSe–ZnS	DEZn	DESe, DES	SLS	1988	[17, 20]
ZnSe–ZnS	DMZn	H_2Se, H_2S	SLS	1989	[21]
ZnSe–ZnS	DMZn	H_2Se, H_2S	SLS	1990	[22]
ZnS–ZnTe	DEZn	DES, DETe	SLS	1988	[17]

5.2 FEATURES OF METAL–ORGANIC MOLECULAR BEAM EPITAXY FOR THE GROWTH OF WIDEGAP II–VI COMPOUNDS

As stated in the previous section, MOMBE can utilize the merits of MOVPE and MBE and further it can avoid some demerits of MOVPE and MBE. It should be noted, however, that the advantages expected of MOMBE for II–VI compounds are somewhat different from those for III–V compounds. We now consider the advantages of MOMBE from the following two viewpoints: (1) the advantages of MOMBE itself, i.e. the advantages utilizing the merits of MOVPE and MBE, and (2) the advantages arising from improving the demerits of MOVPE and MBE.

The advantages of MOMBE itself are summarized as follows.

1. Growth in a clean atmosphere (or in high vacuum). It becomes possible to make and keep the substrate surface extremely clean. Also, this results in a growth of high-purity and high-quality epilayers. Furthermore, the beam nature and separation of the beams in MOMBE simplifies, as in MBE, the gas flow patterns and this minimizes any chemical reactions between the source gases in the reactor.
2. Low-temperature epitaxy process. Low-temperature and/or a non-equilibrium growth process is very important in the conductivity control of widegap II–VI compounds.
3. Possibility of *in situ* process monitoring. Several process-monitoring instruments such as reflection high-energy electron diffraction (RHEED) and quadrupole mass spectrometry (QMS) etc., can be used during growth.
4. Precise thickness control at the atomic layer level. This is suitable for fabricating novel structure materials and devices such as SLSs.
5. Capability of using gaseous source materials as well as solid elemental sources. This enables us to use any kind of source material. In particular, the use of gaseous sources gives possibilities of (a) doping epilayers with various dopants, (b) selective area growth, (c) lowering the epitaxy temperature owing to the effect of the reaction of chemically reactive species at the substrate surface, and (d) easy change of source materials unlike in MBE.

Next, the features of MOMBE arising from improving the demerits of MBE and MOVPE are discussed. A problem in the MBE of widegap II–VI compounds is that the vapour pressures of group VI solid elemental sources, especially sulphur, are too high for their evaporation rate to be controlled. This is simply improved by the use of room-temperature vapour sources such as dialkyls and/or hydrides of chalcogens. In this case, precise, stable and reproducible control of their flow rates becomes possible with mass flow controllers. However, since these chalcogen gases are fairly chemically stable, their pyrolysis on the heated substrate surface cannot be expected. Therefore, the decomposition of such stable gases using a high-temperature cracker is often necessary for obtaining a certain degree of growth rate.

Compared with the MBE case, there are many problems in the MOVPE of widegap II–VI compounds. For the source materials in MOVPE of Zn chalcogenide films, dialkyl zincs (dimethyl or diethyl zincs; DMZn or DEZn) are used for a Zn source, and both dialkyl and hydrogen chalcogenides (e.g. DMSe, DMS, DESe, DES and H_2Se, H_2S) are used for chalcogen sources. Of two types of chalcogen sources, the hydrides are very reactive compared with the dialkyls. Therefore, when the hydrogen chalcogenides are used, the epitaxy temperature can be very low and this is desirable from the viewpoint of conductivity control (typical growth temperatures are 250–350 °C). However, because of high reactivity of source materials, a premature reaction takes place before they reach the substrate in a reactor, resulting in unwanted deposition on the reactor wall. This sometimes results in a poor surface morphology of the epilayers. In order to avoid the unwanted premature reaction, the reactor pressure is normally lowered to about 1 Torr (i.e. low-pressure MOVPE). On the contrary, when dialkyl chalcogenides are used, the unwanted premature reaction can be eliminated, resulting in a growth of epilayers with quite smooth surface morphology. However, in this case, the epitaxy temperature becomes fairly high (typically higher than 500 °C) owing to the higher decomposition temperature of the dialkyls. As stated above, higher growth temperatures are not desirable for the conductivity control of the widegap II–VI epilayers.

The problems in MOVPE stated above can be eliminated by the use of the MOMBE technique. In MOMBE, the beam of source gases directly impinges onto the substrate surface without any collision between the gases in the gas phase. This is attributed to the fact that, if the reactor pressures are below about 10^{-4} Torr, the mean-free path of the source gases is normally longer than the distance between the gas inlet and the substrate. Therefore, even when hydrogen chalcogenides are used as source materials, the unwanted premature reaction in the reactor can be minimized in MOMBE. Furthermore, with the high-temperature cracker for source gases, the epitaxy temperature can be lowered to about 250 °C even when the dialkyl chalcogens are used as source materials.

A disadvantage of MOMBE over MBE is the use of toxic gases as in MOVPE. Although the dialkyls are safer than hydrogen chalcogenides, MBE is superior to MOMBE and MOVPE from the viewpoint of safety.

5.3 GROWTH MECHANISM OF ZnSe AND ZnS FILMS BY METAL–ORGANIC MOLECULAR BEAM EPITAXY

For group VI source materials, both dialkyls and hydrides are used as stated in the previous section. It has been found that the growth mechanisms of ZnSe and ZnS films in MOMBE are quite different for the different combinations of source materials, and especially for the group VI source materials. When dialkyl chalcogenides are used, the growth mechanism is fairly simple, because the film growth does not proceed without precracking of the dialkyls by high-temperature crackers. So, the results for the case using group VI dialkyls are shown first.

Figure 5.1 shows the dependence of the growth rate of ZnSe grown on (100)GaAs on the growth (substrate) temperature when using DMSe for the Se source and DMZn and DEZn for the Zn source [11]. In this case, DMSe was cracked at 950 °C before reaching the substrate, and without DMSe cracking no film growth proceeds, because the bonding energy for M–Se–M bonds is as high as about 60 kcal mol^{-1}. The flux intensities for source materials are shown in the figure. For growth using DMZn, the growth rate is very sensitive to the growth temperature. The growth rate begins to increase steeply at about 400 °C and it soon reaches a maximum at 540 °C. Because of the thermal stability of DMZn, no film growth is observed at temperatures below 400 °C. The steep decrease of the growth rate above 540 °C is interpreted in terms of the congruent evaporation of ZnSe. However, when DEZn is used, film growth proceeds at temperatures above 200 °C as shown in the figure. The difference in the temperature at which the film begins to grow is attributed to the difference in the thermal stabilities of DMZn and DEZn. The mean bond dissociation energies of DMZn and DEZn are 42.0 and 34.6 kcal mol^{-1} respectively. Further, when using DEZn, the growth rate

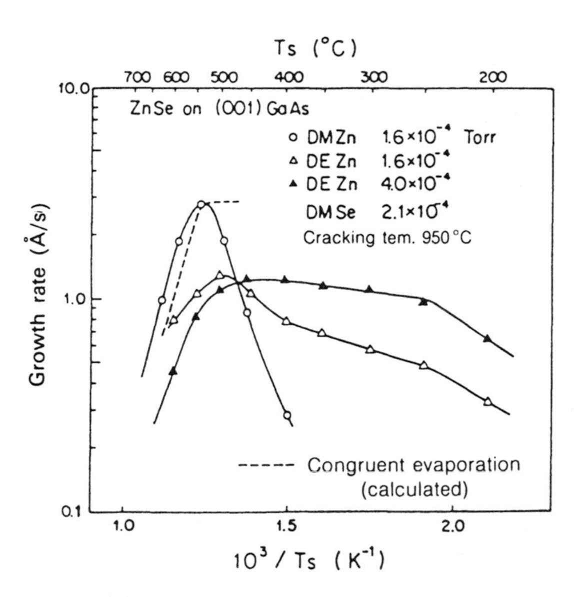

Fig. 5.1 Dependence of the growth rate of ZnSe grown on (100)GaAs on the growth temperature when using DMSe for the Se source and DMZn and DEZn for the Zn source.

changes gently compared with the case using DMZn. However, the observed values of the apparent activation energies of the growth rate when using DMZn (22.5 kcal mol^{-1}) and DEZn (4.5 kcal mol^{-1}) are much smaller than those expected from the bonding energies, and this is not understood well.

For growth of ZnS epilayers on (100)GaAs, DEZn and DES were used. Figure 5.2 shows the dependence of the growth rate of ZnS layer on the growth temperature [7]. Results for the case using H_2S are also shown for comparison. The hollow squares are the results for the use of DEZn and cracked DES and a fairly gentle temperature dependence is observed from room temperature (RT) to 500 °C, as in the growth of ZnSe using DEZn shown in Fig. 5.1. Since the bonding energy for DES is as high as 70 kcal mol^{-1}, of course it is necessary to crack DES before it reaches the substrate. Figure 5.3 shows the dependence of the ZnS growth rate on the

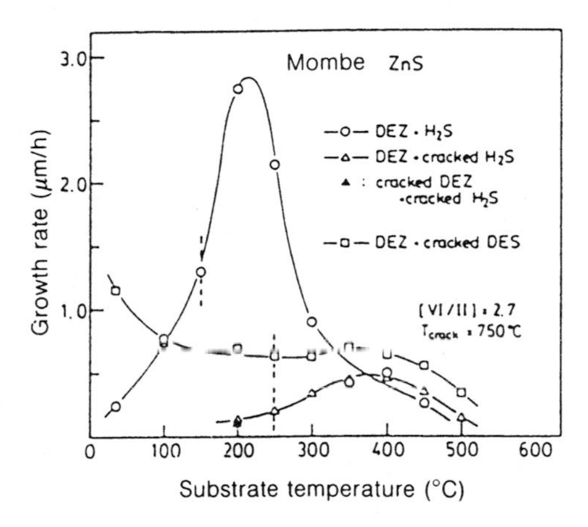

Fig. 5.2 Dependence of the growth rate of ZnS on th growth temperature when using DEZn for the Zn source and H_2S and DES for the S source.

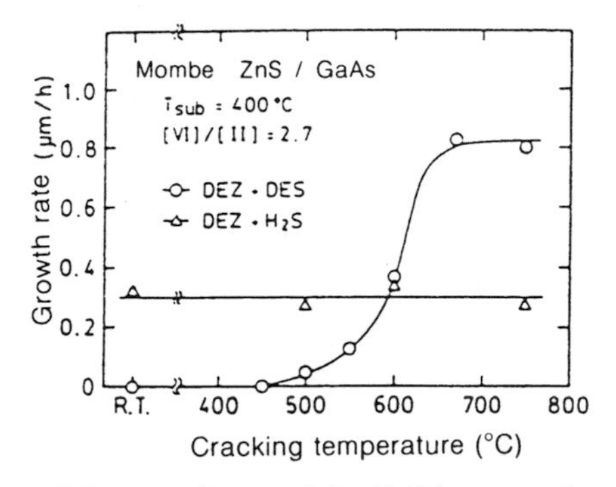

Fig. 5.3 Dependence of the growth rate of the ZnS layer on the cracking temperature of the sulphur sources, i.e. DES and H_2S.

cracking temperature of DES and H_2S [7]. For growth using H_2S, the growth rate is kept almost unchanged with respect to that for RT. Since the bonding energy for H_2S is as high as 87.4 kcal mol^{-1}, complete cracking by the cracker is very difficult. Therefore, at the cracking temperatures shown in Fig. 5.2, almost no cracking is achieved for H_2S. In this case, film growth proceeds as a result of the effect of the high chemical reactivity of H_2S. When using DES, however, the film growth starts at about 500 °C owing to the effect of cracking of DES and it increases until the temperature reaches about 650 °C.

For dialkyl chalcogenides, DESe and DETe were also used for the growth of ZnSe and ZnTe layers respectively.

For group VI hydrides, H_2Se and H_2S were used to grow ZnSe and ZnS layers [7–10, 13–16, 21, 22]. As already stated, dialkylzincs can react with H_2Se and H_2S to form ZnSe and ZnS layers even at RT and then without cracking of the group VI materials the film growth can proceed on the substrate surface even in high vacuum as in MOMBE. Therefore, the growth mechanism of the layers is greatly influenced by the cracking of the source gases [8–10]. For growth of ZnSe (ZnS) using DMZn and H_2Se (H_2S), there are four combinations of source–supply method, in which (1) both of, (2) and (3) one of, and (4) none of the source gases is cracked. The four cases are as follows: case I, cracked DMZn and cracked H_2Se (H_2S); case II, DMZn and cracked H_2Se (H_2S); case III, cracked DMZn and H_2Se (H_2S); case IV, DMZn and H_2Se (H_2S). On the basis of the cracking study using QMS, dominant precursors for Zn and chalcogens impinging upon the substrate surface can be assumed to be hot elemental Zn and Se_2 (S_2) molecules when the source gases are cracked. In contrast, when the source gases are not cracked, cold (about RT) source gases themselves impinge upon the substrate.

Figures 5.4 and 5.5 show how the growth rates of ZnSe and ZnS layers

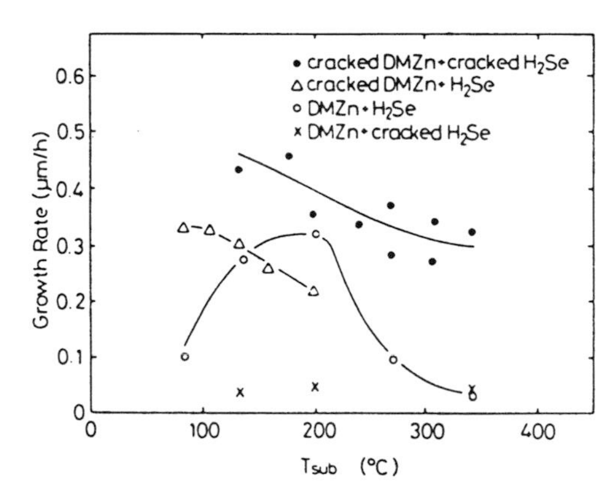

Fig. 5.4 Substrate temperature dependence of the growth rate of ZnSe layers. It is shown that the cracking of source gases (cases I–IV) greatly affects the growth rate and/or growth mechanism.

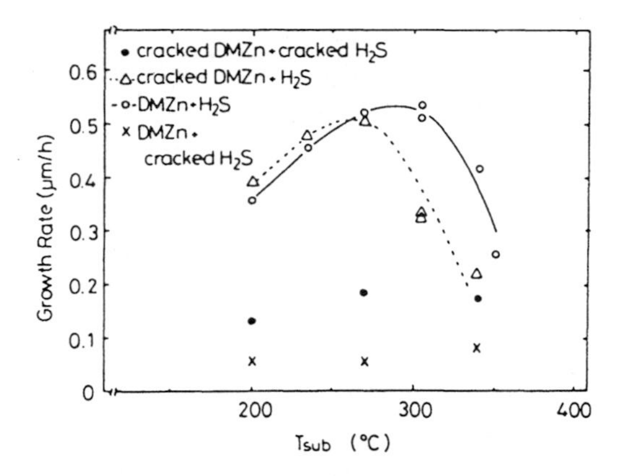

Fig. 5.5 Substrate temperature dependence of the growth rate of ZnS films for four different source supplies (cases I–IV).

vary with growth temperature in each source–supply case [8–10]. The cracking temperatures for DMZn, H_2Se and H_2S were 860, 680 and 1000 °C respectively. The most notable feature in these figures is that almost no film deposition has been observed when only hydride gases are cracked (case II), although significant film growth has been observed in the other three cases. In order to investigate how the cracking of the hydride gases affects the growth rate in case II, the dependence of the growth rates on their cracking temperatures has been examined, and the results are shown in Fig. 5.6. It is shown that the growth rates of both layers drastically decrease with cracking temperature, and practically no film growth is observed when the cracking temperatures reach those used in the experiments shown in

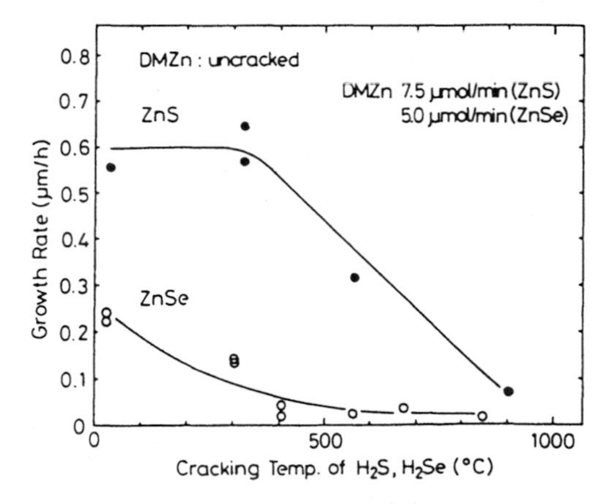

Fig. 5.6 Dependence of the growth rate of ZnSe and ZnS films on the cracking temperature of H_2Se and H_2S: DMZn is not cracked (case II).

Figs 5.4 and 5.5. From the experimental results for cases II and IV, it has been concluded that RT DMZn and hot Se (S) molecules cannot contribute to the film growth (case II), but RT DMZn and RT hydrogen chalcogenides can (case IV).

When both source gases are cracked (case I), hot Zn and hot Se molecules arrive at the substrate as dominant precursors contributing to the film growth. Therefore, the source–supply method in this case is essentially the same as that of conventional MBE using elemental Zn and Se as source materials. It is shown in Fig. 5.6 that the growth rate in case I monotonically decreases with temperature. This is indicative that no chemical reaction is involved in deriving constituent atoms at the substrate surface, because the source gases are already cracked into the constituent atoms in the cracking cells. Further, it has been found that the growth process in case I is mainly governed by the minority beam flux, as in MBE. Thus, it has been concluded that the growth kinetics for case I is similar to that for conventional MBE.

It should be noted that the difference in growth method between cases I and II is only in the supply method for Zn source, i.e. hot Zn atoms impinge upon the substrate in case I, while RT DMZn molecules do in case II. However, a quite notable difference in the growth rate has been observed between the two cases, i.e. almost no film growth occurs in case II, even though remarkable film growth does occur in case I. This indicates that, in case II, RT DMZn cannot dissociate into Zn on the substrate surface within the temperature range examined. This is because their surface lifetime is too short owing to the high vacuum in MOMBE, resulting in insufficient thermal energy for DMZn to dissociate. Compared with the result for the growth of ZnSe using DMZn and cracked DMSe shown in Fig. 5.1, it has been found that the substrate temperature must be higher than about 450 °C in order to dissociate DMZn on the substrate.

When only DMZn is cracked (case III), a significant growth rate has been observed even in a fairly low temperature range. This is in marked contrast to case II where only the hydrides are cracked. It is possible and likely that hot Zn catalyses the dissociation of hydride gases on the substrate and then they interact to form ZnSe (ZnS). This is similar to the substantial enhancement of the dissociation of hydride gases (AsH_3) by the catalytic effect of Ga metal particles [23].

Lastly, in case IV in which none of the source gases is cracked, RT DMZn and RT hydride gases directly impinge upon the substrate. It has been found that, although DMZn and the hydrides cannot dissociate independently on the substrate surface within the temperature range examined, they first interact to form adducts (DMZn–H_2Se (H_2S)) on the substrate surface and then the decomposition of the adducts can lead to the formation of the layers at fairly low temperatures [10, 24].

Furthermore, it has been found that the surface morphology of the layers reflects the growth mechanism stated above [8–10]. When both source gases are cracked (case I), a smooth and featureless surface can be achieved. However, when none of source gases is cracked (case IV), the surface becomes rather rough and hazy, as well as in case III.

5.4 PROPERTIES OF METAL–ORGANIC MOLECULAR BEAM EPITAXY-GROWN ZnSe AND ZnS FILMS

Crystallographic, optical and electrical properties of MOMBE-grown ZnSe and ZnS layers are briefly discussed here. Figure 5.7 shows the growth temperature dependence of the full width at half-maximum (FWHM) of the (400) diffraction peak for 1 μm thick ZnSe layers grown on (100)GaAs [8]. It is shown that the FWHM decreases with substrate temperature, and it becomes as small as 5.5 minutes of arc at temperatures above 250 °C, which is near the resolution of the equipment used. It has been found that epitaxial layers can be grown at as low a temperature as 135 °C, although the FWHM tends to increase with decreasing substrate temperature. As for the surface morphology, quite smooth and featureless surfaces can be obtained at temperatures from 200 to 300 °C [8–10]. Thus, MOMBE is a truly promising low-temperature epitaxy method for widegap II–VI compounds. Furthermore, even for ZnS layers grown on GaAs, the lowest epitaxy temperature is also as low as about 150 °C in MOMBE [7], though it is desirable to raise the growth temperature above 300 °C to obtain high quality layers as shown in Fig. 5.8 [25].

Figure 5.9 [8] shows the dependence on growth temperature of the photoluminescence spectra measured at 15 K of MOMBE-grown 1 μm thick

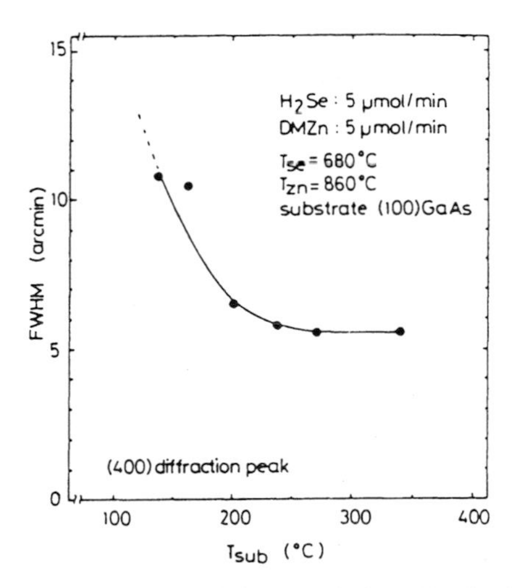

Fig. 5.7 Substrate temperature dependence of the FWHM of the (400) X-ray diffraction peak of the ZnSe layers grown on (100)GaAs.

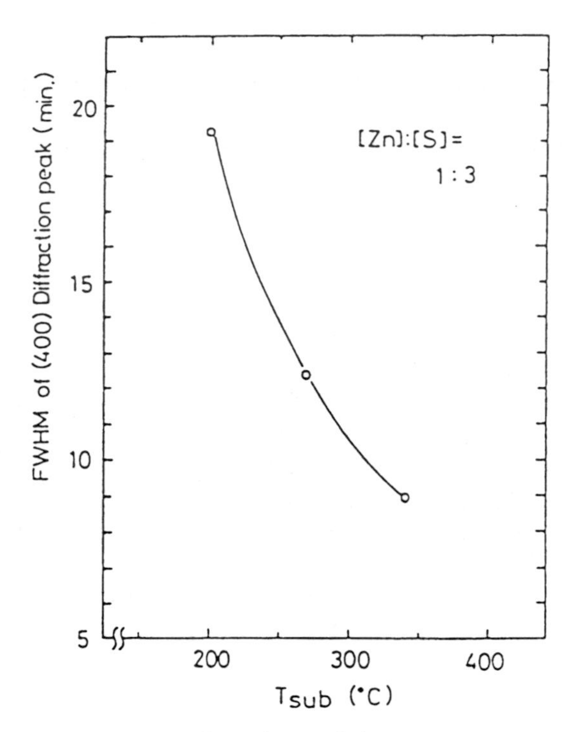

Fig. 5.8 Substrate temperature dependence of the FWHM of the (400) diffraction peak of the ZnS layers grown on (100)GaAs.

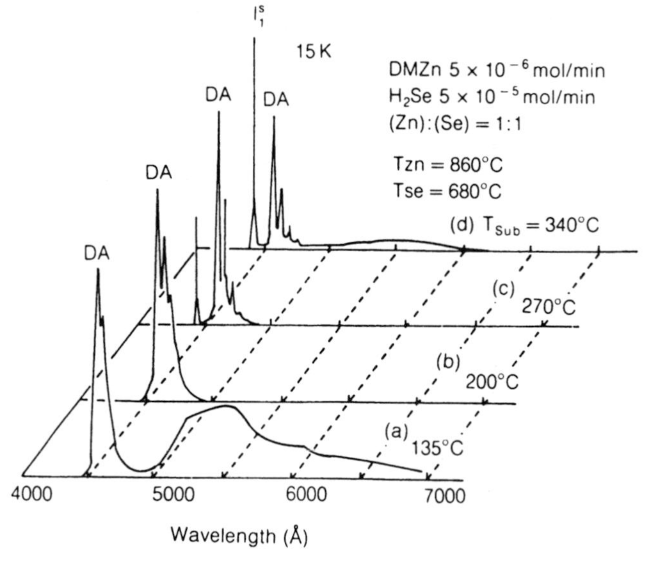

Fig. 5.9 Dependence on the growth temperature of the photoluminescence spectra of 1 μm thick ZnSe films grown at different temperatures.

ZnSe layers. As for the deep-level emissions, a fairly broad emission centred at 5500 Å is observed in the film grown at 135 °C. These broad emissions originate from so-called Cu green (~ 5500 Å) and self-activated (SA) (~ 6200 Å) centres. The intensity of these emissions decreases remarkably with increasing growth temperature, as shown in Fig. 5.8, and in the film grown at 270 °C it is smaller in magnitude by a factor of the order of 1000 than other dominant peaks of excitonic (I_1^s) and/or donor–acceptor (DA) pair emissions. It is shown, however, that further increase in growth temperature degrades the crystallinity, resulting in the appearance of deep-level emissions. Figure 5.10 [8] shows the excitonic emission region of the photoluminescence spectrum for the film grown at 270 °C. It is shown that the I_1^s emission originating from excitons bound to neutral shallow acceptors is dominant. Although the observed dominant emission peak originates from impurity centres in this case, it is very sharp and well defined. This indicates that the crystallinity of the MOMBE-grown ZnSe layer is quite good.

Figure 5.11 shows the photoluminescence spectra of MOMBE-grown ZnS layers [25]. It is shown that the intensity of SA emission decreases with increasing growth temperature, although the intensity of deep-level emission (labelled as DE) still remains high. Furthermore, it is shown that fairly strong sharp excitonic emissions originating from neutral donors and free-to-acceptor emissions are observed in the films grown at higher temperatures. Considering the results together with those shown in Fig. 5.9, higher growth temperatures are probably required for obtaining higher-quality ZnS layers.

Conductivity control, especially in p-type conduction, of widegap II–VI compounds has been a major interest for 'II–VI researchers'. Because of the recent progress in low-temperature epitaxy techniques such as MBE, MOVPE and also MOMBE, n-type carrier concentration control of ZnSe epilayers has been successfully achieved by doping with group III and VII

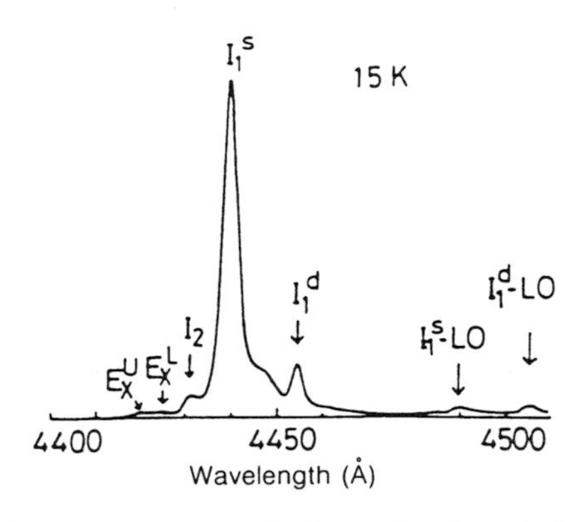

Fig. 5.10 Photoluminescence spectrum in the excitonic emission region of the ZnSe film grown at 270 °C.

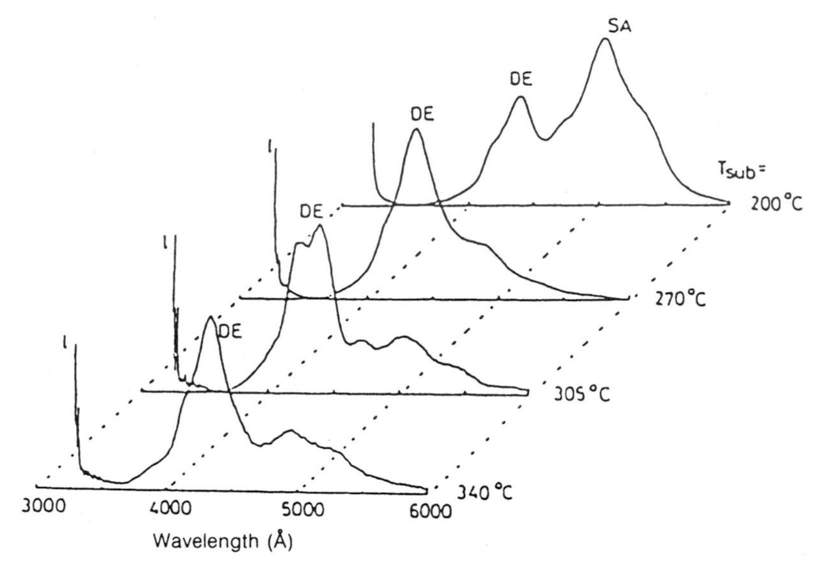

Fig. 5.11 Dependence on the growth temperature of the photoluminescence spectra of 1 μm thick ZnS films grown at different temperatures.

impurities. Also, it has been shown that p-type conductivity control of ZnSe epilayers is possible by lithium and/or nitrogen doping, as discussed in other chapters in this book. However, p-type control is still a major problem for us from the viewpoints of its reproducibility and achieving much higher carrier concentrations.

Recently, the MOMBE growth of highly conductive p-type ZnSe layers doped with nitrogen has been reported [13, 14]. DMZn and H_2Se were used as source materials and ammonia (NH_3) was used as a dopant source. All the gas cells were heated at 350 °C. This means that all source gases were not cracked but only heated in this work. Figure 5.12 shows electrical properties of the MOMBE-grown nitrogen-doped ZnSe films grown at 250–450 °C [13]. It is shown that the carrier (hole) concentration increases with increasing growth temperature and a resistivity of 0.57 Ω cm is reached while the hole concentration is 5.6×10^{17} cm^{-3} for the layer grown at 450 °C. It has been confirmed by using secondary ion mass spectroscopy (SIMS) that the nitrogen concentration in ZnSe tends to increase with increasing growth temperature. The concentration in the layer grown at 250 °C is as low as 10^{17} cm^{-3} (near the detection limit), while those for the layers grown at temperatures above 350 °C are about 10^{19} cm^{-3}. Figure 5.13 shows the low-temperature photoluminescence spectra of nitrogen-doped ZnSe films for different growth temperatures [13]. Details of the excitonic emission region for both nitrogen-doped and undoped films are shown in Fig. 5.14 [13]. It is shown that the emission originating from excitons bound to neutral

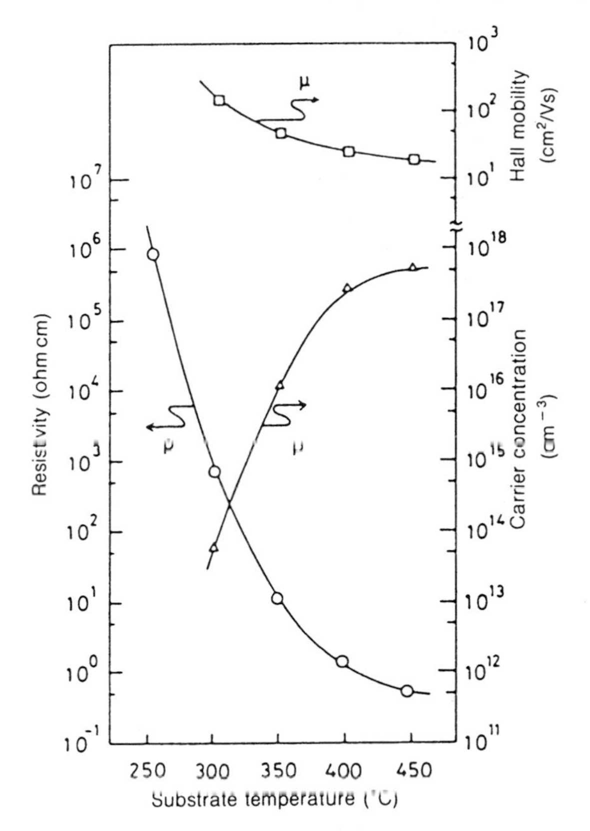

Fig. 5.12 Dependence of the electrical properties of nitrogen-doped p-type ZnSe films on the substrate temperature.

nitrogen acceptors (I_1^N) is observed only in the nitrogen-doped layers. It should be noted, however, that its intensity is much lower than that for neutral-donor-bound excitons I_2, although these layers exhibit p-type conduction. Similar results have also been reported in p-type ZnSe layers [26], and the reason why the p-type layers are not dominated by acceptor-associated emission lines is not understood at present. This may be one of the key points for achieving reproducible high-concentration p-type doping.

Preliminary results of experiments to make ZnSe p–n junction diodes on the basis of the conductivity control described above have been reported [13]. The diode structure is p-ZnSe/n-ZnSe/n-GaAs, whose carrier concentrations are about 10^{16}, 5×10^{16} and $10^{17}\,cm^{-3}$ respectively. A weak whitish-blue emission peaking at 460 nm was observed from this diode under a bias condition of 5 V, 25 mA, at 77 K.

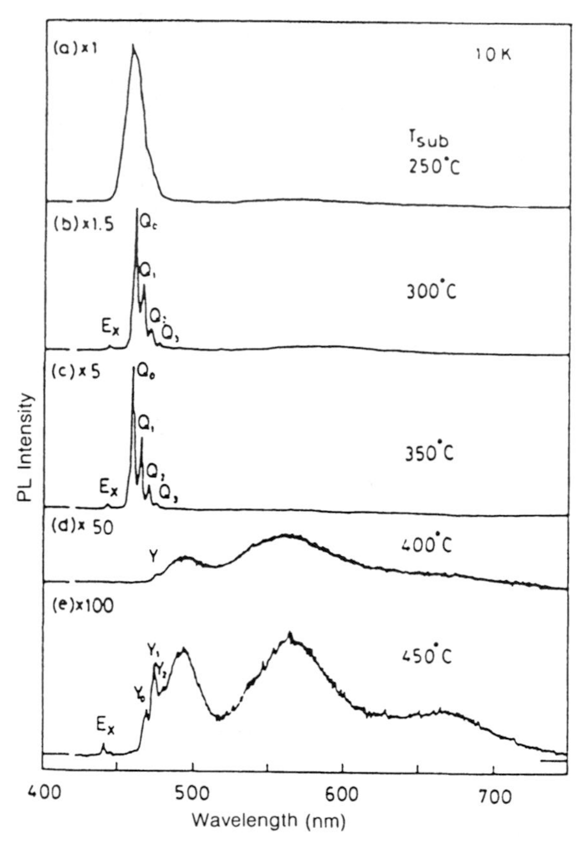

Fig. 5.13 Dependence of the photoluminescence spectra of the nitrogen-doped ZnSe films on the growth temperature.

5.5 GROWTH OF NOVEL STRUCTURE II–VI COMPOUND FILMS BY METAL–ORGANIC MOLECULAR BEAM EPITAXY

SLSs consisting of widegap II–VI compounds, such as ZnS, ZnSe, ZnTe, CdS and MnSe, are promising materials for various opto-electronic devices operating in the visible wavelength region, such as blue-to-green light-emitting diodes and non-linear optical devices. Also, the utilization of SLSs may be a useful idea even from the viewpoint of conductivity control because the technique of modulation doping can be used. MOMBE is one of the most promising methods for fabricating SLSs with widegap II–VI compounds. Furthermore, in order to fabricate high-quality SLSs, it is preferable to control the growth rate to an atomic-layer-level accuracy. In this section, first the results on the

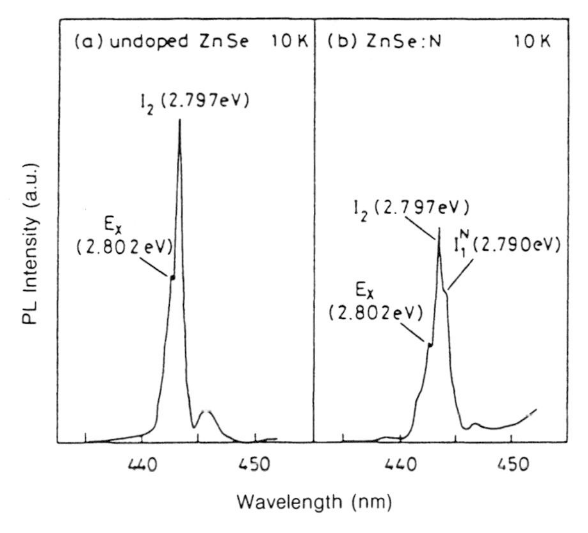

Fig. 5.14 Photoluminescence spectra in the band-edge region for undoped and nitrogen-doped ZnSe films grown at 350 °C.

MOMBE-based ALE of widegap II–VI compounds are shown, and then the results on growth and properties of SLSs consisting of them are given.

5.5.1 Atomic layer epitaxy of widegap II–VI compounds

The application of the ALE technique to the growth of widegap II–VI compounds is essentially fairly easy compared with the case of III–V compounds, since at the growth temperatures the vapour pressures of the constituent elements for II–VI compounds are much larger than those of the compound films formed from them. As is well known, the first attempt to grow thin films by ALE was performed in chemical vapour deposition (CVD) of ZnS by using $ZnCl_2$ and H_2S as source materials [27]. In this case, 'chemical reactions' between the source materials are essential for ALE growth, and we call this 'CVD-like' ALE. However, most attempts at epitaxial growth of Zn chalcogenides by the ALE technique were performed by a conventional MBE system [28–31]. In this case, no chemical reactions to dissociate source materials are necessary for ALE growth, because constituent elements are supplied as sources in the MBE system. We call this 'MBE-like' ALE. Thus, until now, two ALE modes have been applied for growing Zn chalcogenide films.

In the MOMBE of ZnSe (ZnS) using DMZn and H_2Se (H_2S) as source materials, both CVD-like and MBE-like film growth can be achieved depending on whether the source materials are cracked before reaching the substrate or not [18]. Figure 5.15 shows the dependence of the growth rate of ZnSe layers per source-supply cycle on the growth temperature in MBE-like mode (i.e. both source gases were cracked) [18]. In this experiment, in order to achieve ALE growth, pulsed molecular beams of both source gases were sequentially supplied for 3 s with an interval of 5 s. During one complete source-gas supply, the substrate was exposed to the molecular beam to produce a coverage of about five monolayers. It is clearly shown in Fig. 5.15 that ALE growth can be achieved in the temperature range 200–275 °C, i.e. self-limiting of the growth rate by 'one monolayer per cycle' has been achieved in this temperature range. At temperatures below 200 °C, the growth rate increases remarkably owing to the effect of the increased surface lifetime of the constituent elements, while, at temperatures above 275 °C, the growth rate gradually decreases owing to the effect of the reduced sticking coefficient of Zn which is a minority constituent element in this case. It has been found that the 'temperature window' for ALE growth conditions broadens to the higher temperature side when the supply rate of source gases increases.

Figure 5.16 shows the photoluminescence spectra measured at 20 K of ZnSe films grown at 300 °C for 1000 source-supply cycles (i.e. the layer thickness is about 0.28 μm) [18]. The two upper curves are for ALE films, but the lower curve is not. In all samples, the fairly sharp and intense excitonic emission called I_x is dominant. However, there is a difference among the samples in the deep-level emission region, i.e. when ALE growth is not achieved, the intensity of deep-level emission is remarkably

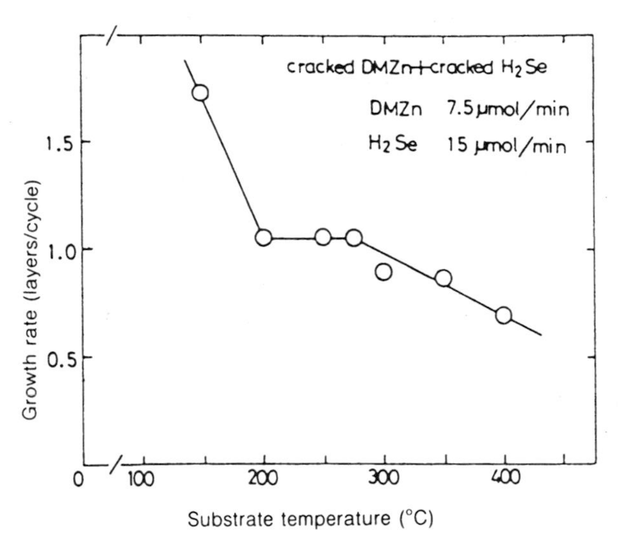

Fig. 5.15 Dependence of the growth rate of ZnSe films on the growth temperature in the 'MBE-like' growth mode, i.e both source gases were cracked in this case.

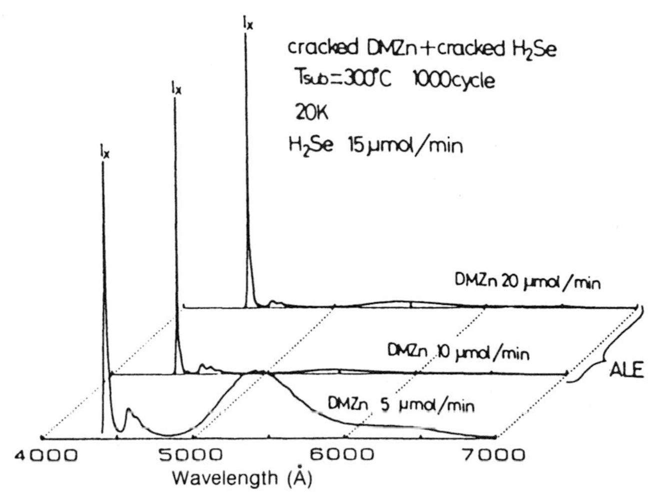

Fig. 5.16 Photoluminescence spectra measured at 20 K of ZnSe films grown at 300 °C for 1000 source-supply cycles. The two upper curves are for ALE films, but the lower curve is not.

high compared with those for ALE films. This indicates that the quality of the ALE films is quite good in terms of photoluminescence properties. Further, it has been shown that the surface morphology of ALE ZnSe films is excellent and superior to that of conventional MOMBE films [18].

Furthermore, even in the case of the CVD-like growth mode of ZnSe in a MOMBE system using DMZn and H_2Se, it has also been shown that ALE growth is possible [18]. That is, the self-limiting of the growth rate by 'one monolayer per cycle' has been confirmed. It should be noticed, however, that the sticking coefficient of H_2Se is very low and thus its flow rate must be several times higher to achieve ALE growth than that in MBE-like mode. As for the properties of the ALE films in CVD-like mode, their surface morphology and quality were not so good. To our surprise, although the apparent growth rate was similar to that for ALE, the surface morphology was very poor compared with the case for MBE-like ALE. This has been ascribed to the fact that the surface migration of the constituent elements in the CVD-like ALE mode is inadequate compared with the MBE-like ALE mode. Reflecting this, low-temperature photoluminescence properties were also poor, as shown in Fig. 5.17 [18]. It is shown that the luminescence is dominated by deep-level emissions. Thus, in CVD-like ALE mode, it is hard to enhance the surface migration of the constituent elements, although apparent self-limiting of the growth rate per cycle is achieved. Considering these results, the MBE-like ALE mode is superior to the CVD-like ALE mode for growing high quality films [18].

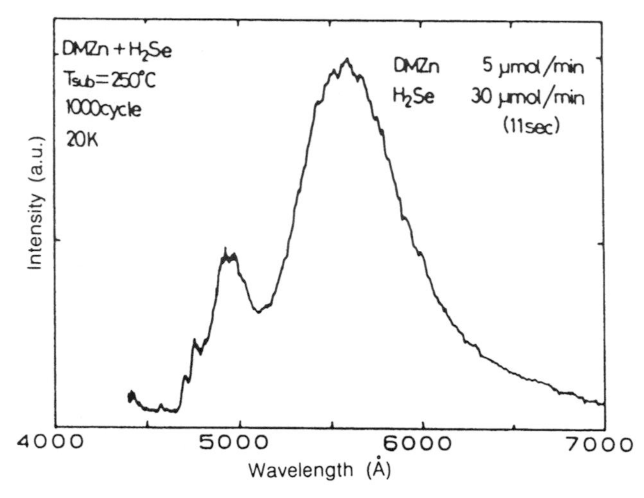

Fig. 5.17 Typical photoluminescence spectrum of a 'CVD-like' ALE ZnSe film grown at 250 °C for 1000 source-supply cycles.

As for the growth of the ZnS layer, ALE growth was achieved in MBE-like mode using cracked DMZn and cracked H_2S [19]. The ALE temperature window, in which one monolayer growth per source-supply cycle was achieved, was 250–310 °C. ALE ZnS layers grown on GaAs exhibited good surface morphology and strong near-band-edge photoluminescence [19]. It was found, however, that ALE growth by the CVD-like mode using RT DMZn and RT H_2S is practically impossible in the MOMBE system, because the sticking coefficient of H_2S is very small owing to the effect of the high vacuum in MOMBE and also because H_2S is too thermally stable to be decomposed on the substrate surface [32].

5.5.2 Growth and properties of widegap II–VI compounds strained layer superlattices

Until now, ZnSe–ZnS and ZnS–ZnTe SLSs have been grown by MOMBE. The former is basically a type I superlattice and the latter is a type II superlattice. In ZnSe–ZnS SLSs, ZnSe layers are the wells and ZnS layers are the barriers, and their effective energy bandgap can be varied between those of ZnSe and ZnS depending on the well layer's thickness, while, in ZnS–ZnTe SLSs, the effective energy gap is determined by the recombination between electrons in ZnS layers and holes in ZnTe layers, and therefore their effective energy gap tends to be small compared with those of ZnS and ZnTe with increasing layer thickness.

These two SLSs are common-cation superlattices so, in order to produce SLS structures, the molecular beams for Se sources and for S (Te) sources were alternated, while that for the Zn source was kept on during the entire growth.

The growth of ZnSe–ZnS SLSs was demonstrated for the first time in MOMBE by using DEZn, cracked DESe and cracked DES [17]. The formation of superlattice structures was confirmed by observing some satellite peaks in X-ray diffraction patterns. Figure 5.18 shows the photoluminescence spectrum measured at 4.2 K of the SLSs consisting of 30 Å ZnS and 5 Å ZnSe (denoted (30–5) ZnS–ZnSe SLS in the figure) [17]. The wavelength of the emission peak at 402 nm was shorter than that corresponding to the energy gap of ZnSe, and this emission was attributed to the recombination of electron–hole pairs between the quantized levels in the ZnSe well layer. Figure 5.19 shows the relationship between the thickness of the ZnSe well layer and the photoluminescence peak energy [17]. The thickness of the ZnS barrier layer was fixed at 30 Å for all samples. It is shown that the photoluminescence peak shifts towards higher energy as the thickness of the ZnSe well layer decreases. The solid line represents the calculated photoluminescence peak assuming the valence band offset between ZnS and ZnSe to be 800 meV. The conduction band offsets between ZnS and ZnSe for different ZnSe layers are also shown in the figure, and it has been concluded that the conduction band offset for ZnS–ZnSe SLSs increases with an increase in strain.

Growth of lattice-matched ZnSe–ZnS SLSs onto GaAs substrates was also investigated [21, 22]. In SLSs, not only the bandgap but also the average lattice constant of the layers can be designed by controlling the thickness of each layer. In the case of ZnSe–ZnS SLSs, ZnSe has a larger lattice constant and ZnS a smaller one than does GaAs. So, the average lattice constant of the SLS which is parallel to the interface can be lattice matched to GaAs, when

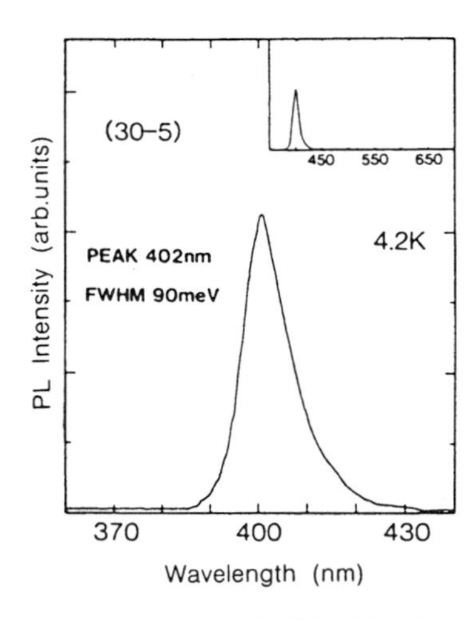

Fig. 5.18 Photoluminescence spectrum of a (30–5) ZnS–ZnSe SLS with 50 periods.

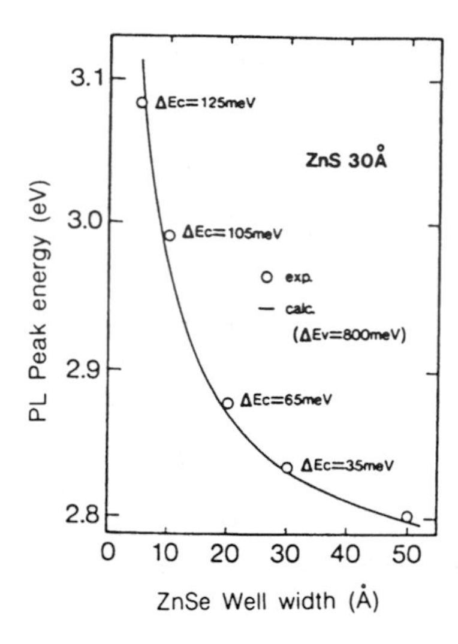

Fig. 5.19 Relationship between ZnSe well layer thickness and photoluminescence peak energy shift. The solid curve was calculated by the Krönig–Penny model.

each layer thickness is appropriate. That is, the average lattice constant of the SLSs, which is parallel to the interface, is determined by the ratio of each layer (d_{ZnS}/d_{ZnSe}), and when it is 0.05 the lattice constant of SLSs matches that of GaAs [21, 22]. Hence, utilizing the ZnSe–ZnS superlattice structure, the problem arising from lattice mismatch in widegap II–VI epilayers can be eliminated.

Figure 5.20 shows the X-ray diffraction pattern of the ZnSe–ZnS SLSs

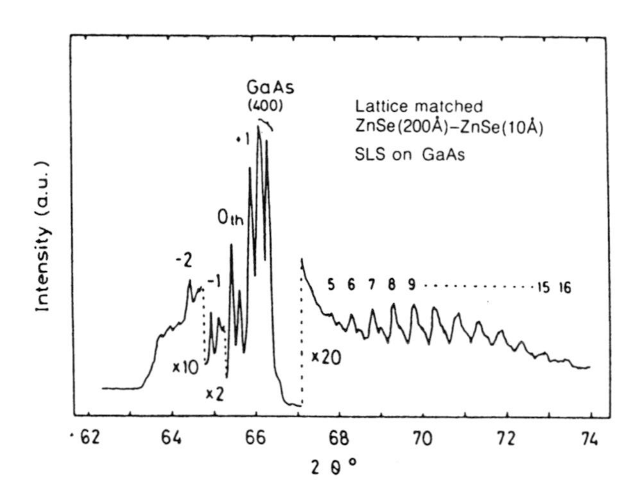

Fig. 5.20 X-ray diffraction pattern for a (200–10) ZnSe–ZnS SLS which is lattice-matched to GaAs.

consisting of a 200 Å ZnSe and a 10 Å ZnS layer in one period, which should be lattice-matched SLSs with (100)GaAs [21]. The SLSs were grown by a conventional MOMBE system using cracked DMZn, cracked H_2Se and H_2S as source materials. The number of total periods was 50. Since the critical thickness for coherent growth of ZnS onto ZnSe was smaller than 13 Å [33], the thickness of ZnS layer was chosen to be 10 Å to eliminate the introduction of misfit dislocations in the SLSs. It is shown in the figure that the satellite peaks up to 16th order are clearly observed. This indicates that high-quality SLSs with extremely flat interfaces are grown. Figure 5.21 shows the photoluminescence spectrum measured at 18 K for the lattice-matched ZnSe–ZnS SLSs having 25 periods [21]. Quite strong and narrow free-excitonic emission is observed and the intensity of the deep-level emission is extremely weak. These results also indicates that high-quality SLSs are obtained.

In order to confirm whether these SLSs are lattice-matched to GaAs or not, the excitonic peak position in photoluminescence for several film thicknesses has been investigated. The results are shown in Fig. 5.22(a) for the SLSs and in Fig. 5.22(b) for ZnSe layers on GaAs [21]. For ZnSe films, it is shown that the peak position shifts toward the longer-wavelength side as the film thickness increases owing to the effect of lattice accommodation on the energy gap [34]. That is, the peak shift with increasing film thickness indicates that the accommodation through misfit-dislocation formation occurs in ZnSe layers on GaAs. In contrast to the case of the ZnSe layer, no peak shift is observed in the case of the SLSs on GaAs. This suggests that the accommodation through misfit-dislocation formation is prevented in the SLSs as expected.

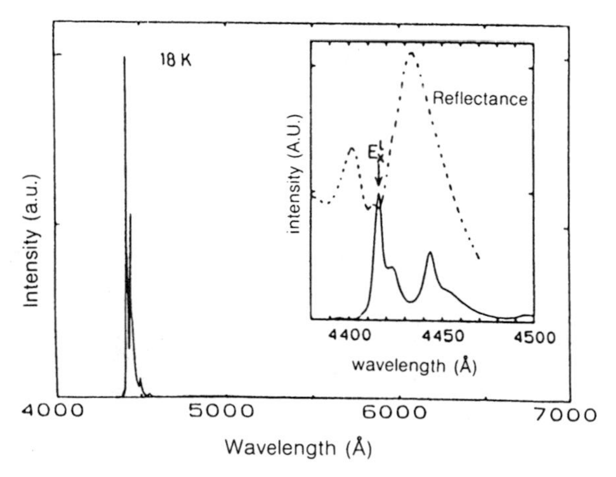

Fig. 5.21 Photoluminescence spectrum of a lattice-matched ZnSe–ZnS SLS with GaAs.

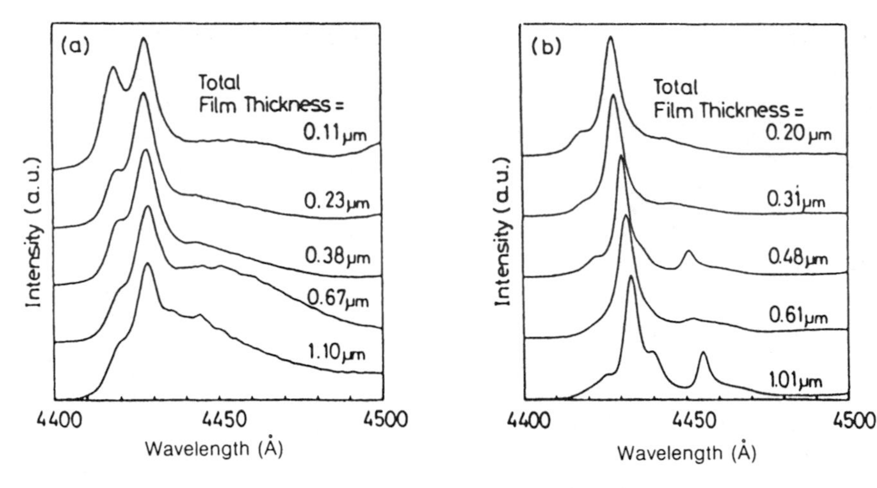

Fig. 5.22 The total layer thickness dependence of the excitonic peak position in photoluminescence; (a) for lattice-matched ZnSe–ZnS SLSs and (b) for ZnSe films on GaAs.

In order to confirm further a reduction of misfit dislocations in the lattice-matched ZnSe–ZnS SLSs, electron-beam-induced current (EBIC) image observations for Schottky diodes fabricated on the layer surface were carried out. Figures 5.23(a) and 5.23(b) show the EBIC images for a ZnSe film on GaAs and for the SLSs respectively [21]. Each sample thickness was about 2500 Å. In the ZnSe film, cross-hatched patterns which run along the ⟨110⟩ direction were observed as shown in Fig. 5.23(a). These patterns correspond to the dislocations which can be caused by misfit dislocations, suggesting that misfit dislocations are actually generated in ZnSe layers on GaAs. On the other hand, these patterns could not be observed for the SLSs, indicating that the generation of misfit dislocations was actually reduced in the lattice-matched SLSs.

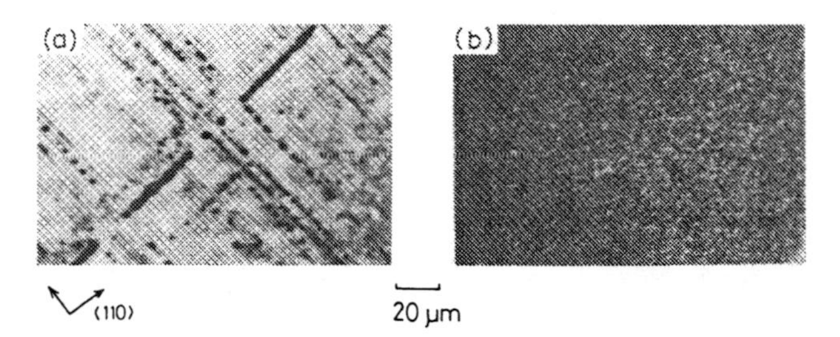

Fig. 5.23 EBIC images: (a) for a ZnSe film and (b) for a lattice-matched ZnSe–ZnS SLS on GaAs.

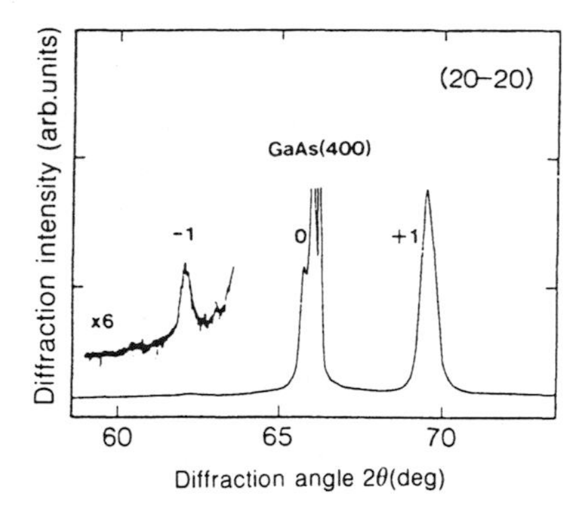

Fig. 5.24 X-ray diffraction pattern of a (20–20) ZnS–ZnTe SLS with 200 periods on a (100)GaAs substrate.

Lastly, results for the ZnS–ZnTe SLSs are quite briefly shown. Figure 5.24 shows the X-ray diffraction pattern for 200 periods (20 Å–20 Å) ZnS–ZnTe SLSs grown on (100)GaAs by using DEZn, cracked DESe and cracked DETe as source materials [17]. The formation of a superlattice structure was confirmed by observing the satellite peaks as shown in Fig. 5.24. As for the photoluminescence properties of the SLSs, a broad blue emission peaking at about 460 nm was observed. Although this did not originate from the quantized levels in the SLSs, the intensity was very strong and observed even at 400 K.

5.6 SUMMARY

The present status of MOMBE growth and properties of widegap II–VI compounds were discussed. First, features of MOMBE for the growth of widegap II–VI compounds were briefly summarized. Further, the growth mechanism for ZnSe and ZnS layers especially was investigated, and the effects of the source-gas cracking on the growth kinetics were discussed. As for the electrical and optical properties of the layers, it was shown that MOMBE was a promising technique for controlling these properties of the epilayers. Also, it was shown that p–n junction ZnSe diodes could be fabricated by MOMBE. Furthermore, ALE of ZnSe and ZnS layers in the MOMBE system and growth and properties of novel structure II–VI compound layers such as SLSs were also discussed.

Compared with the case of III–V compounds, the number of papers reporting on the growth and properties of widegap II–VI compounds is very small at present. Further, the MOMBE apparatuses used until now for the growth of II–VI compounds are not as good as those for III–V compounds. Therefore, the properties of the MOMBE-grown widegap II–VI compound layers stated in this chapter may not be good enough at present for achieving conductivity control, including conduction type. Considering that these results are preliminary, they are still hopeful data for us. I believe that MOMBE is essentially a promising low-temperature epitaxy method for widegap II–VI compounds, which can utilize the merits of both MBE and MOVPE.

REFERENCES

1. Veuhoff, E., Pletschen, W., Balk, P. and Luth H. (1981) *J. Cryst. Growth*, **55**, 30.
2. Morris, F.J. and Fukui, H. (1974) *J. Vac. Sci. Technol.*, **11**, 506.
3. Panish, M.B. and Sumski, S. (1984) *J. Appl. Phys.*, **55**, 3571.
4. Yoshida, S., Misawa, S. and Itoh, A. (1975) *Appl. Phys. Lett.*, **26**, 461.
5. Tsang, W.T. (1987) *J. Cryst. Growth*, **81**, 261.
6. Ando, H., Taike, A., Kimura, R., Konagai, M. and Takahashi, K. (1986) *Jpn. J. Appl. Phys.*, **25**, L279.
7. Ando, H., Taike, A., Konagai, M. and Takahashi, K. (1987) *J. Appl. Phys.*, **62**, 1251.
8. Oniyama, H., Yamaga, S., Yoshikawa, A. and Kasai, H. (1988) *J. Cryst. Growth*, **93**, 679.
9. Yoshikawa, A., Oniyama, H., Yasuda, H., Yamaga, S. and Kasai, H. (1989) *J. Cryst. Growth*, **94**, 69.
10. Yoshikawa, A., Oniyama, H., Yamaga, S. and Kasai, H. (1989) *J. Cryst. Growth*, **95**, 572.
11. Kobayashi, N., Shinoda, Y. and Kpobayashi, Y. (1988) *Jpn. J. Appl. Phys.*, **27**, L1728.
12. Kobayashi, N. (1988) *Jpn. J. Appl. Phys.*, **27**, L1597.
13. Migita, M., Taike, A., Shiiki, M. and Yamamoto, H. (1990) *J. Cryst. Growth*, **101**, 835.
14. Taike, A., Migita, M. and Yamamoto, H. (1990) *App. Phys. Lett.*, **56**, 1989.
15. Kaneda, S., Satou, S., Setoyama, T., Motoyama, S., Yokoyama, M. and Ota, N. (1985) *Extended Abstracts of the 17th Conference on Solid State Devices and Materials, Tokyo*, p. 225.
16. Kaneda, S., Satou, S., Setoyama, T., Motoyama, S., Yokoyama, M. and Ota, N. (1986) *J. Cryst. Growth*, **76**, 440.
17. Teraguchi, N., Takemura, Y., Kimura, R., Konagai, M. and Takahashi, K. (1988) *Cryst. Growth*, **93**, 720.
18. Yoshikawa, A., Okamoto, T., Yasuda, H., Yamaga, S. and Kasai, H. (1990) *Growth*, **101**, 86.
19. Wu, Y., Toyoda, T., Kawakami, Y., Fujita, Sz. and Fujita, Sg. (1990) *Jpn. J. Appl. Phys.*, **29**, L727.
20. Konagai, M., Kobayashi, M., Kimura, R. and Takahashi, K. (1988) *J. Cryst. Growth*, **86**, 290.
21. Oniyama, H., Yamaga, S. and Yoshikawa, A. (1989) *Jpn. J. Appl. Phys.*, **28**, L2137.
22. Oniyama, H., Yamaga, S. and Yoshikawa, A. (1990) *Mater. Res. Soc. Symp. Proc.*, **161**, 187.
23. Nishizawa, J. and Kurabayashi, T. (1983) *J. Electrochem. Soc.*, **130**, 413.
24. Mullin, J.B., Irvine, S.J.C. and Ashen, D.J. (1981) *J. Cryst. Growth*, **55**, 92.
25. Oniyama, H. (1990) *Doctoral Thesis*, Chiba University.
26. Ohki, A., Shibata, N., Ando, K. and Katsui, A. (1988) *J. Cryst. Growth*, **93**, 692.
27. Suntola, T. (1984) *Extended Abstracts of the 16th Conference on Solid State Devices and Materials, Kobe*, p. 647.
28. Yao, T. and Takeda, T. (1986) *Appl. Phys. Lett.*, **48**, 160.

29. Yao, T. (1986) *Jpn. J. Appl. Phys.*, **25**, L544.
30. Dosho, S., Takemura, Y., Konagai, M. and Takahashi, K. (1989) *J. Cryst. Growth*, **95**, 580.
31. Konagai, M., Takemura, Y., Kimura, R., Teraguchi, N. and Takahashi, K. (1990) *Mater. Res. Soc. Symp. Proc.*, **161**, 177.
32. Yasuda,. H. (1990) *Master's Thesis*, Chiba University.
33. Yokogawa, T., Sato, H. and Ogura, M. (1988) *Appl. Phys. Lett.*, **52**, 1678.
34. Mitsuhashi, H., Mitsuishi, I., Mizuta, M. and Kukimoto, H. (1985) *Jpn. J. Appl. Phys.*, **24**, L578.

6

Quantum-sized microstructures of wide bandgap II–VI semiconductors

M. Kobayashi, R.L. Gunshor and L.A. Kolodziejski

6.1 INTRODUCTION

The past several years have seen dramatic advances in epitaxial growth techniques which have led the way to an improved understanding of the properties of II–VI materials and their alloys. The renewed interest in the family of II–VI semiconductors is thus directly related to their successful epitaxial growth by molecular beam epitaxy (MBE), atomic layer epitaxy (ALE), metal–organic chemical vapour deposition (MOCVD), and metal–organic molecular beam epitaxy (MOMBE). These growth techniques have resulted in the creation of sophisticated heterojunctions, multiple quantum wells, and strained-layer superlattices composed of II–VI/II–VI and II–VI/III–V multilayered materials. Multiple quantum well and superlattice structures composed of wide bandgap II–VI compounds were first realized in the mid-1980s. Subsequent years have seen a rapid advance in the understanding of the microstructural, optical, and electrical properties of these structures. The objective of the chapter is to provide an overview of the research activity in the area of the non-equilibrium growth of II–VI-based structures, having emphasis on MBE, and with selected recent achievements highlighted.

The properties of a number of *zincblende* MBE-grown chalcogenides have been reported to date; the list includes both binary and alloy compounds involving ZnSe, ZnS, ZnTe, CdTe, CdSe, MnSe, MnTe, and FeSe. By alloying the II–VI chalcogenides with Mn or Fe, an interesting group of materials, referred to as semimagnetic or dilute magnetic semiconductors (DMS), have been grown as epitaxial layers. The incorporation of the magnetic transition element results in a variation of the lattice parameter, the direct energy bandgap, and in some instances, a change in crystal structure. Multiple quantum well and superlattice structures incorporating layers of the diluted magnetic (or semimagnetic) semiconductors (DMS) therefore provide bandgap modulation, while also exhibiting novel phenomenon arising from the presence of the magnetic ion. Subtleties arising from the exchange interaction between the magnetic ions and band electrons in an external magnetic field provide new and additional insights into the spatial distribution of carrier wavefunctions in structures of atomic dimensions. In one example, such a novel diagnostic tool allowed determination of the valence band offset in II–VI DMS strained-layer superlattice structures. Reduced dimensionality effects in magnetic superlattices are discussed, where frustrated antiferromagnetism is observed in superlattice structures composed of metastable zincblende MnSe alternated with layers of ZnSe. With further application of the non-equilibrium growth technique of MBE, metastable zincblende MnTe has been grown with a bandgap of 3.2 eV, an energy much larger than the 1.3 eV associated with the NiAs bulk crystal, the usual crystal structure for this compound. As a further example, the metastable thin film MBE growth of FeSe has also been achieved by Jonker *et al.* A large variety of new and unique II–VI-based quantum-size structures have provided for studies of interesting phenomena which include polarized stimulated emission, exciton self-trapping, nonlinear exciton absorption, biexciton formation, modulation doping. In addition, the structures allowed a wide visible wavelength tunability via a simple variation in the layer dimensions.

In recent years a wide range of III–V/III–V and II–VI/II–VI heterostructures have been studied. Less attention, however, has been given to heterostructures fabricated by utilizing compounds having different valency on either side of the interface. For example, the implementation of the II–VI/III–V heterovalent heterostructures would greatly increase the available heterojunction combinations having both technological potential and scientific importance. Three such structures which have received attention are ZnSe/GaAs, ZnTe/(Al,Ga)Sb, and CdTe/InSb. The common property shared by all three is a close lattice match. The implication of the near lattice match is the opportunity to form a dislocation-free interface by requiring one of the pair to be strained (pseudomorphically) such that the lattice spacings match in the plane of the interface. The ZnSe/GaAs structure is attractive for certain applications where the wide gap II–VI serves as a pseudoinsulator or passivating layer to the III–V semiconductor. The ZnTe/(Al,Ga)Sb structure is primarily of interest for light-emitting applications, but one can also envision field-effect devices. The motivation for considering the CdTe/InSb heterostructure is somewhat different; the primary interest here is to realize InSb quantum well structures. The issue addressed by the latter II–VI/III–V heterojunction is the absence of suitable III–V compounds to serve as the barrier layers in conjunction with InSb quantum wells; there are no available wide bandgap III–V compounds having lattice parameters compatible with InSb.

6.2 ZnSe-BASED SUPERLATTICES, MULTIPLE QUANTUM WELLS AND MICROSTRUCTURES

The primary motivation for the study of ZnSe-based layered structures revolved around potential commercial and military applications of visible blue/blue–green opto-electronic devices. The room temperature direct energy bandgaps of ZnSe, ZnTe, and associated heterostructures, covers the aforementioned spectral regions. In addition, the lattice parameters of these II–VI semiconductors are quite compatible with technologically important III–V compounds, thus facilitating the integration of the two families of compound semiconductors. Although considerable investigation of II–VI materials was previously carried out in the 1960s and 1970s [1, 2], the current excitement and renewed research interest is a result of the significant advantages which are offered by utilizing the new non-equilibrium crystal growth technologies. Many of the chapters included in this book discuss the recent results in the MBE, MOCVD, MOMBE and ALE growth of the wide bandgap II–VI compounds. The material properties of epitaxial layers have seen remarkable improvement when fabricated with the aforementioned growth techniques. The following section discusses the optical, micro-structural, and electronic properties of layered structures composed of wide bandgap ZnSe-based II–VI semiconductors, and advanced structures containing II–VI/III–V heterovalent interfaces, such as ZnSe-on-GaAs.

6.2.1 ZnSe/(Zn,Mn)Se superlattice and multiple quantum well structures

The random incorporation of the magnetic transition metals (Mn, Fe, and Co) into ZnSe, results in a group of materials known as diluted magnetic (DMS) or semimagnetic semiconductors [3–5]. In the case of (Zn, Mn)Se, the characteristic band structure of the host II–VI semiconductor ZnSe is not directly modified by the presence of Mn since the two *s*-electrons of the outer shell replace those of Zn and become part of the band electrons in extended states. The five electrons in the unfilled 3*d* shell of Mn give rise to localized magnetic moments which are partially aligned in an external magnetic field. The resultant magnetic moment interacts with the band electrons to cause Zeeman splitting which, at low lattice temperatures, is two orders of magnitude larger than for the host II–VI semiconductor. The presence of the magnetic ion provides a unique and useful feature to the superlattices and multiple quantum well structures in which Mn is incorporated. In one example, the magnetic field-induced changes in optical transition energies in superlattices incorporating the DMS material [6] provides additional insights into excitonic behaviour and valence band offset in strained-layer structures.

The MBE growth of the pseudobinary material system ZnSe–MnSe presented a unique opportunity to investigate the properties of metastable zincblende crystals over a large range of alloy fractions which were unavailable by conventional bulk equilibrium growth techniques. Thick (1–3 μm) epilayers of zincblende $Zn_{1-x}Mn_xSe$ have been grown by MBE over the $0 < x \leqslant 0.66$ composition range [7]. In contrast, bulk crystals of the ternary exist with pure zincblende crystal structure only up to $x < 0.10$ [8]. The ability to achieve zincblende (Zn,Mn)Se having an appreciable Mn content was crucial to the realization of quantum well structures in this system, as it was necessary to grow barrier layers with a high Mn fraction to achieve sufficient band offset for carrier confinement.

The study of the optical properties of ZnSe/(Zn,Mn)Se structures has focused mainly on excitonic transitions. In comparison with CdTe/(Cd,Mn)Te heterostructures, two principal differences arise. First, the increase of the bandgap of $Zn_{1-x}Mn_xSe$ with Mn concentration (x) is considerably less; as a result, for $x \sim 0.20$ in the barrier layer, the total bandgap difference between barrier and well layers is about 150 meV (as against nearly 400 meV for the $Cd_{1-x}Mn_xTe$ structure). Second, the uniaxial component of the lattice mismatch strain is opposite in sign, so that the uppermost valence band has a light-hole character (parallel to superlattice axis at $k = 0$). At the same time, the exchange effect on hole states near band extreme by the Mn-ion d-electron spins is larger than that in (Cd,Mn)Te.

Figure 6.1 shows the photoluminescence excitation spectrum near the $n = 1$ exciton ground state from a (1 0 0) oriented MQW sample of $ZnSe/Zn_{1-x}Mn_xSe$ ($x = 0.23$), with a well width of 67 Å [6]. For such typical parameters, only the $n = 1$ transition is seen, consisting of the light-hole (LH) and heavy-hole (HH) excitonic resonances. The identification of the resonances is made through magneto-optical studies [6] where both LH and HH transitions are

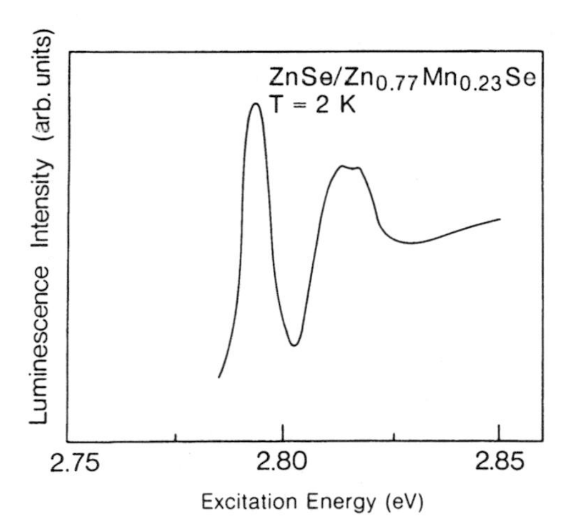

Fig. 6.1 PL excitation spectrum for a ZnSe/(Zn,Mn)Se MQW structure. The $n = 1$ light- and heavy-hole resonances are the dominant features in the spectrum.

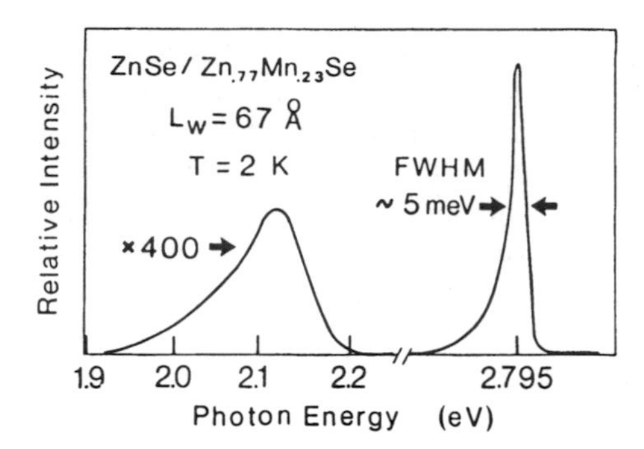

Fig. 6.2 Comparison of blue and yellow photoluminescence emission for a ZnSe/(Zn,Mn)Se MQW sample. Note the difference in amplitude between the two.

found to exhibit large Zeeman splittings due to a finite exciton wavefunction overlap with the (Zn,Mn)Se barrier layers.

Apart from increasing the direct bandgap at $k = 0$, the incorporation of Mn into ZnSe leads to additional features in the optical spectrum due to the d-electron transitions internal to the Mn-ion. The lowest transition corresponds to absorption at about 2.1 eV, and the zero-phonon line in luminescence is at about 2.0 eV. In heterostructures containing (Zn,Mn)Se, these 'yellow' resonances can compete for electronic excitation with the 'blue' resonances which are associated with band edge transitions. That there are efficient energy transfer paths from the band edge exciton states in (Zn,Mn)Se directly into the Mn-ion internal excitation, can be graphically demonstrated by comparing thin films of (Zn,Mn)Se with ZnSe/(Zn,Mn)Se quantum wells [9, 10] through time-resolved luminescence spectroscopy [11].

Figure 6.2 shows luminescence spectra from a ZnSe/(Zn,Mn)Se MQW sample at $T = 2$ K under cw excitation above the barrier bandgap [11]. The ZnSe well thickness was 67 Å and the Mn-ion concentration in the $Zn_{1-x}Mn_xSe$ barriers was $x = 0.23$. The spectrally sharp (blue) exciton recombination dominates the broad yellow Mn-ion recombination from the barriers. In contrast, the blue recombination in a thin film of (Zn,Mn)Se is dwarfed by the now dominant yellow contribution; the blue/yellow intensity ratio is about 4×10^{-3}. It is seen that the quantum well structure efficiently collects electrons and holes before any significant energy transfer of band edge excitation to the d-electron states of the Mn-ion occurs.

Subsequent to capture in a ZnSe quantum well, the photoenergetic electrons and holes relax by optical phonon emission to the $n = 1$ confined particle states. The relaxation step is fast (likely below 1 ps in such polar material). The subsequent exciton formation and further energy relaxation (localization by quantum well width fluctuations) is a slower process which can be time-resolved through the use of picosecond pulsed laser excitation and a streak camera [11]. Figure 6.3 (left panel) shows the transient

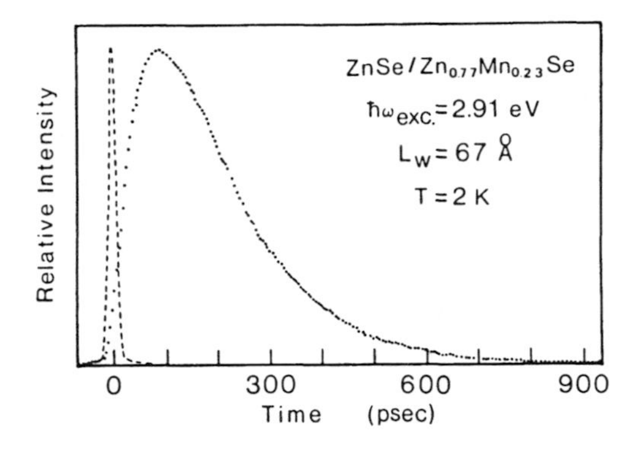

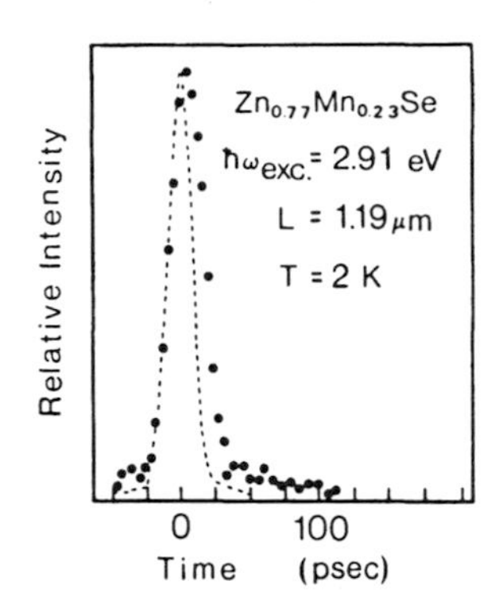

Fig. 6.3 Time-resolved exciton luminescence at the fundamental edge for (left) a $ZnSe/Zn_{0.77}Mn_{0.23}Se$ MQW and (right) a $Zn_{0.77}Mn_{0.23}Se$ epilayer. Broken lines indicate the time response of the detection (monochromator–streak camera) system.

luminescence (dotted line) from the ZnSe/(Zn,Mn)Se MQW sample when excited above the $n = 1$ exciton resonance. The rise time is approximately 90 ps; this time constant shortens to approximately 20 ps under resonant excitation, thus yielding a direct measure of the exciton formation step. Recombination lifetime is, of course, also obtained [12] from the data (~ 200 ps). In contrast, in the MQW case, the 'blue' exciton decay in a (Zn,Mn)Se thin film is very fast (~ 15 ps) as shown in Fig. 6.3 (right panel); this gives a direct measure of the rate of energy conversion from the exciton state to the *d*-electron excitation of the Mn-ion.

Nonlinear excitonic absorption was measured in both ZnSe epilayers. and ZnSe/(Zn,Mn)Se MQWs at 77 K [13–15] and room temperature [14–16]. As an example, Fig. 6.4 shows the change in excitonic absorption in a ZnSe film (1.3 μm) and a $ZnSe/Zn_{1-x}Mn_xSe$ MQW sample at $T = 77$ K [13]. The quantum well thickness was 73 Å with $x = 0.51$ in the barrier layer. For an exciton Bohr orbit diameter of about 60 Å (bulk ZnSe), the $n = 1$ light-hole exciton transition is therefore influenced by confinement effects in the well. One can obtain a rough quantitative measure for the saturation of absorption from the data in Fig. 6.4, by assuming the case of a homogeneous lineshape. When compared with the thin epitaxial film, the saturation intensity decreases from approximately 10.7 kW cm^{-2} to 1.3 kW cm^{-2} for the MQW sample [13].

Using equilibrium statistics to estimate the relative free electron-hole pair and exciton densities corresponding to these experimental conditions, one obtains some insight into the mechanisms which contribute to the observed absorption saturation. It should be noted that the relatively short lived exciton (order of 200 ps) causes estimates based on thermal equilibrium to be somewhat uncertain. Considering the Mott screening of excitons in the Debye limit, one estimates for the bulk ZnSe a critical density $n_c = 3 \times 10^{17}$ cm^{-3}

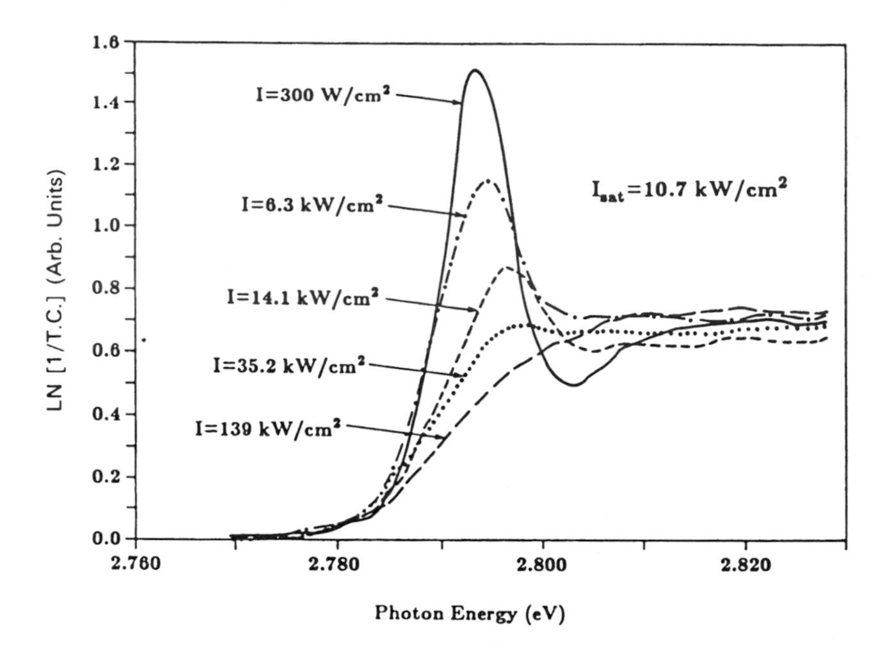

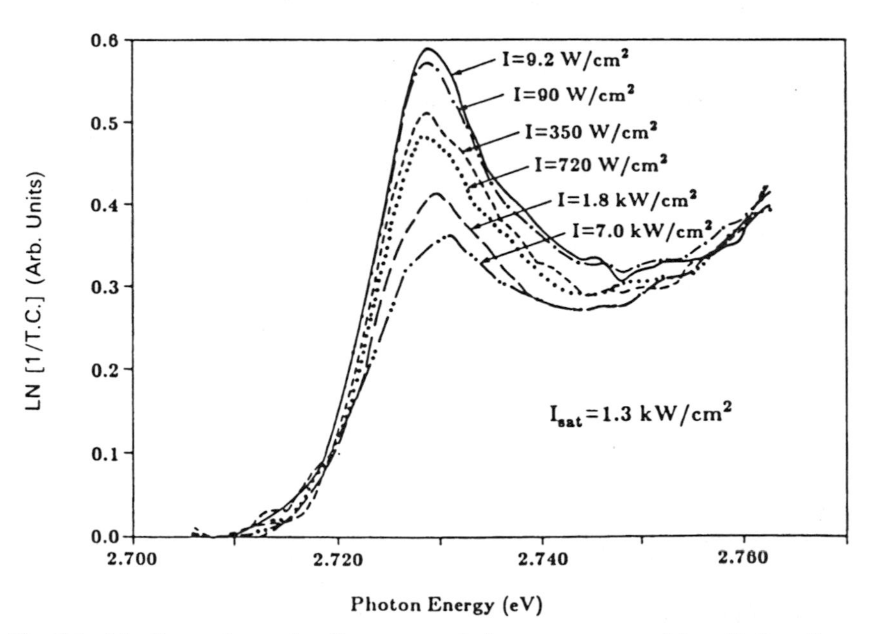

Fig. 6.4 Nonlinear absorption (from transmission measurements) for a ZnSe epilayer (upper) and a ZnSe/(Zn,Mn)Se MQW structure ($L_W = 73$ Å) (lower) at 77 K for the ground state exciton resonance. (The abscissa is ln(1/transmission coefficient).)

at 77 K, while the experimentally generated electron–hole pair density at observed saturation is estimated at $9 \times 10^{16}\,cm^{-3}$. For the MQW sample, the estimated screening of excitons occurs at $n_c = 7 \times 10^{16}\,cm^{-3}$, while experimentally measured saturation corresponds to a pair density of $2 \times 10^{16}\,cm^{-3}$. It is inferred that Coulomb screening appears to be a dominant mechanism in both cases, although there is some additional evidence in thinner ZnSe quantum well samples that the phase space filling becomes an important issue.

(a) Stimulated emission and lasing in quantum wells

Laser oscillations have been observed in ZnSe-based structures with both optical excitation and electron beam pumping. It appears that the first report of lasing in an MBE-grown ZnSe structure involved experiments with ZnSe/(Zn,Mn)Se multiquantum wells [17]. The experiments were performed on cleaved cavities after removal of the GaAs substrate. Gain spectra were measured, and thresholds of stimulated emission determined for various emission wavelengths. Optically pumped lasers were fabricated from these multiple quantum well structures and found to operate in the blue; lasing was observed at a temperature of 80 K. The stimulated emission thresholds of these first (Zn,Mn)Se MQWs exhibited an order of magnitude improvement over previously reported results [18] using single crystal ZnSe grown from a melt. Further improvements in the thresholds of QW-based lasers in this material system are expected with the addition of cladding layers to provide optical confinement.

(b) Polarization-dependent luminescence in quantum well structures

Of considerable significance is the observation that quantum well structures have anisotropic optical properties for light polarized parallel to the layers (TE) and perpendicular to the layers (TM), even though the constituent materials are isotropic in bulk form [19]. An explanation for the anisotropy, neglecting band-mixing effects, is as follows: the relative oscillator strength for the heavy-hole conduction band transition is 3 for the TE polarization and 0 for the TM polarization, while for the light-hole conduction band transition it is 1 for the TE polarization and 4 for the TM polarization. In bulk isotropic materials the heavy-hole and light-hole bands are degenerate at the band edge, so that the absorption and gain depend on the sum for the two bands, a value which is the same for both polarizations. However, in quantum well structures the valence band degeneracy is removed, providing for optical anisotropy. In the case of (Ga,Al)As quantum wells, in which the lattice constants of the well and barrier materials are closely matched, the splitting of the heavy- and light-hole bands occurs primarily from size quantization. The size quantization causes the heavy-hole band to have a smaller blue shift relative to the light-hole band. Consequently, when valence band-splitting is principally a result of size quantization, the optical properties near the band edge are dominated by heavy-hole conduction band transitions, leading to a greater absorption and gain for the TE polarization relative to the TM polarization.

Optical gain and absorption spectra for TE and TM modes propagating in the plane of the layers of (1 0 0) oriented (Zn,Mn)Se MQW structures have been measured [20]. In these structures, contrasting the case of (Ga,Al)As MQWs, there is significant strain in both well and barrier layers due to the lattice constant mismatch. The strain plays an important role in the splitting of heavy- and light-hole bands in addition to the usual size quantization effect observed. In the zincblende (Zn, Mn)Se material system structure, the lattice constant is found to increase with increasing Mn mole fraction [7]. As a result, the ZnSe well regions are subjected to a compressive uniaxial strain in the growth direction. The compressive uniaxial strain acts to lower the energy of the heavy-hole band while raising that of the light-hole band. Thus in (Zn,Mn)Se quantum wells the strain acts in an opposite sense to the effect of quantum confinement on the shifting of the valence bands. In particular MQWs, the ground state transition energy in the ZnSe well region is actually red-shifted with respect to bulk ZnSe [9]. In measurements of absorption spectra for a (Zn,Mn)Se MQW structure, the relative positions of the TE and TM absorption edges are found to be opposite to that observed in (Ga,Al)As MQW structures; the TM absorption edge is at a lower energy than the TE absorption edge [20]. The greater oscillator strength of the TM mode near the band edge is also reflected in the gain spectra (Fig. 6.5). In fact only the TM gain spectra are measured; no TE signal can be detected. These widegap II–VI quantum well structures are the first to exhibit TM polarized stimulated emission originating from a MQW structure [21]. In the usual case, including (Cd,Mn)Te MQWs, the TE mode absorption and gain dominate. (The gain spectra for the (Cd,Mn)Te MQW is seen in Fig. 6.6.). The opposite behaviour of (Cd,Mn)Te and (Zn,Mn)Se MQWs is attributed to the opposite sense of the uniaxial strain.

6.2.2 ZnSe/MnSe superlattices

The existence of zincblende MnSe is a consequence of the kinetic (non-equilibrium) nature of crystal growth by MBE; bulk crystals of MnSe grown under conditions of equilibrium have the NaCl crystal structure [22]. The extrapolated lattice constant and bandgap data, obtained for zincblende (Zn,Mn)Se epilayers, predicts a bandgap and lattice constant for zincblende MnSe of 3.4 eV (6.5 K) and 5.93 Å, respectively.

When grown as a 'thick' epilayer (400 Å), the zincblende crystal structure persisted in the presence of strain-relieving misfit dislocations. Thin epilayers of the metastable zincblende MnSe were incorporated with ZnSe in a number

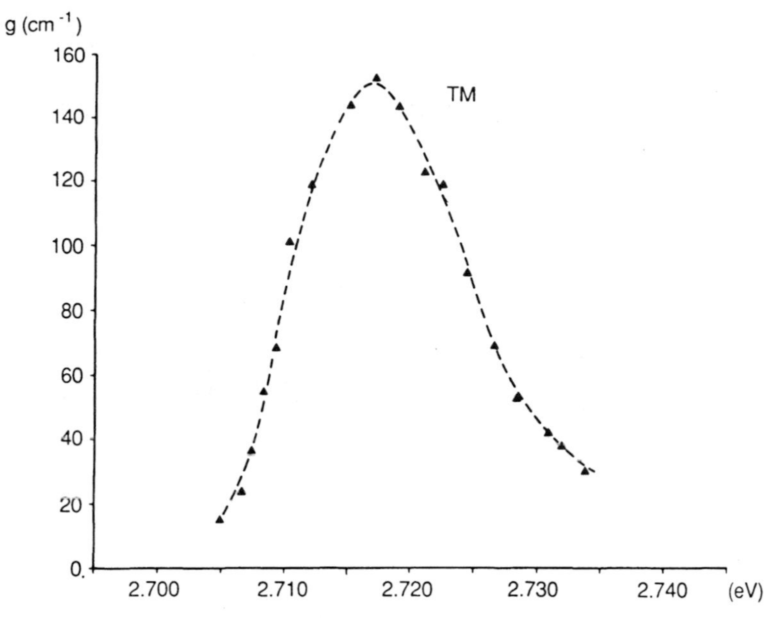

Fig. 6.5 Polarization-dependent gain spectra for a (Zn,Mn)Se MQW structure at 6 K. The excitation density was 4.9×10^5 W cm^{-2}. No TE signal could be detected.

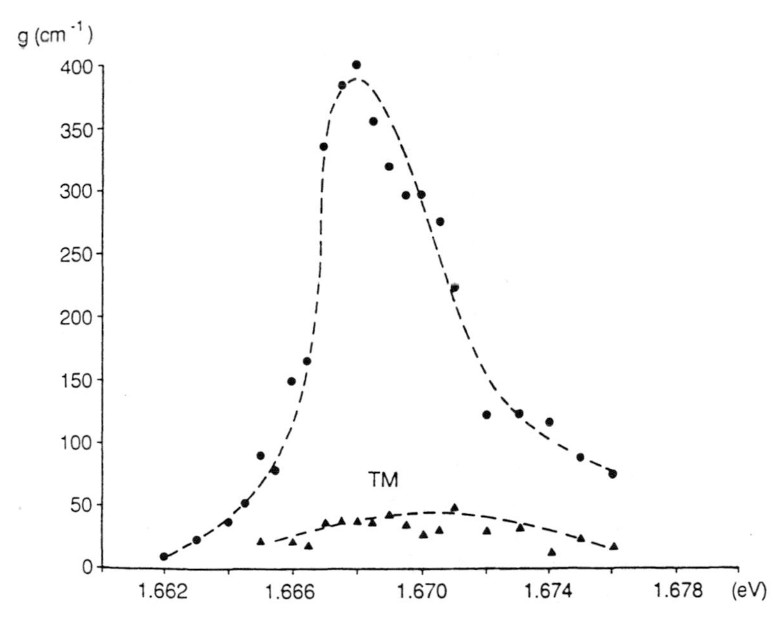

Fig. 6.6 Polarization-dependent gain spectra for the (Cd,Mn)Te MQW structure at 25 K. The excitation density was 1.6 W cm^{-2}.

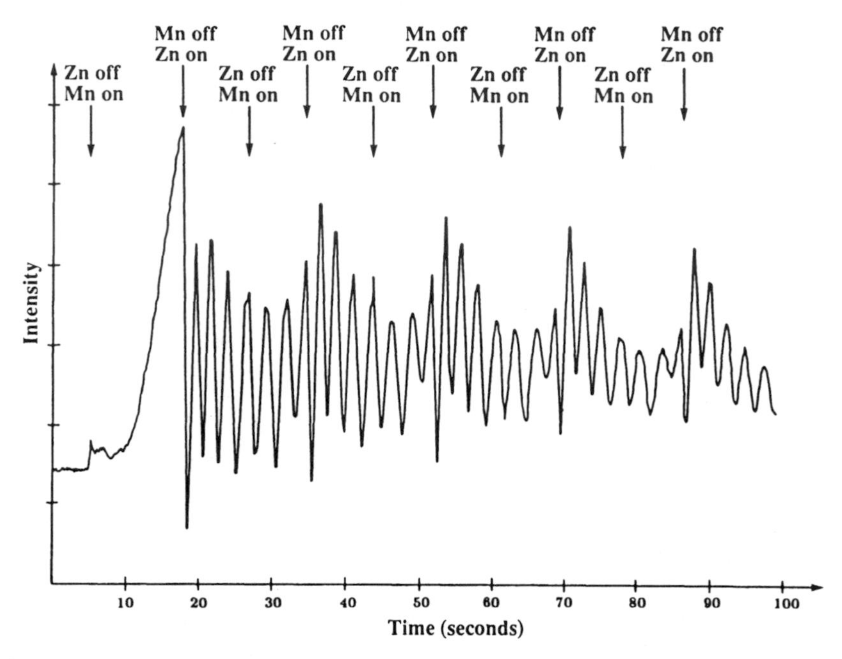

Fig. 6.7 RHEED intensity oscillations during growth of a ZnSe/MnSe binary superlattice. (The first MnSe layer was nucleated on a ZnSe buffer.)

of superlattice structures where the zincblende MnSe functioned as barrier layers. By controlling the thickness of, and spacing between layers of MnSe, an opportunity was presented for a study of magnetic ordering as dimensionality was varied from 3D to quasi-2D. A variety of superlattice structures composed of ZnSe layered with MnSe (3–32 Å) have been grown by MBE. During the growth of these superlattices the layer thickness of the MnSe (with monolayer resolution) was controlled by the use of RHEED intensity oscillations [23–25]. Figure 6.7 shows the effect of alternating the cation species during the growth of a superlattice composed of MnSe/ZnSe. The layer thicknesses were controlled by counting the number of oscillation periods, and subsequently confirmed by the HREM imaging of cross-sectional specimens. The intensity oscillations persisted throughout the entire growth of the five-period superlattice.

A series of 'comb-like' superlattices consisting of 30–100 periods were grown with MnSe layer thicknesses of one, three, and four monolayers; the MnSe layers were separated by approximately 45 Å of ZnSe. A fourth related superlattice structure had 30 periods, each containing 10 monolayer thick MnSe alternated with 24 Å of ZnSe. Photoluminescence measurements performed in the presence of an external magnetic field (up to 5 T) have been used to deduce the magnetic behaviour of these magnetic superlattices; complementary information was acquired through magnetization measurements using a SQUID magnetometer [26–28]. In photoluminescence, the observed magneto-optical shifts of the ground state excitonic transition originated from the exchange interaction between electron–hole states of the superlattice and the magnetic moments of the Mn-ions in the thin MnSe layers. In the absence of a magnetic field the observed optical transition energies were in agreement with predictions of a Kronig–Penney model.

Since a spectral red shift in the presence of an external magnetic field was not anticipated for a superlattice structure containing antiferromagnetic MnSe, the degree of paramagnetic behaviour actually observed was considered significant. The largest Zeeman shifts occurred for superlattices containing MnSe layer thicknesses approaching the monolayer limit; no Zeeman shift was observed for the superlattice containing MnSe layers of 10 monolayer thickness. Although further studies were necessary to rigorously identify the origin of the frustrated antiferromagnetism, the interpretation of the experimental observations was that the origin of the tendency for spins to align in an external magnetic field arose from 'loose' spins at the MnSe/ZnSe heterointerfaces. The thinner the MnSe layer, the greater was the influence of the heterointerface, whereas for thicker MnSe layers, the antiferromagnetic 'inner core' began to dominate the magnetic behaviour [28].

Complementary information about the electronic and magnetic properties was acquired through optical studies near the superlattice bandgap [29], including superlattices containing the monolayer limit of MnSe. The efficient photoluminescence emission originating from such a superlattice transition is illustrated in Fig. 6.8 (at T = 1.8 K in zero and 5 T external magnetic fields in a Faraday geometry). As expected, the zero field exciton (at 2.844 eV) is distinctly blue-shifted from that in bulk ZnSe. The Zeeman effect immediately suggests that a large magnetization is induced by an external field in the

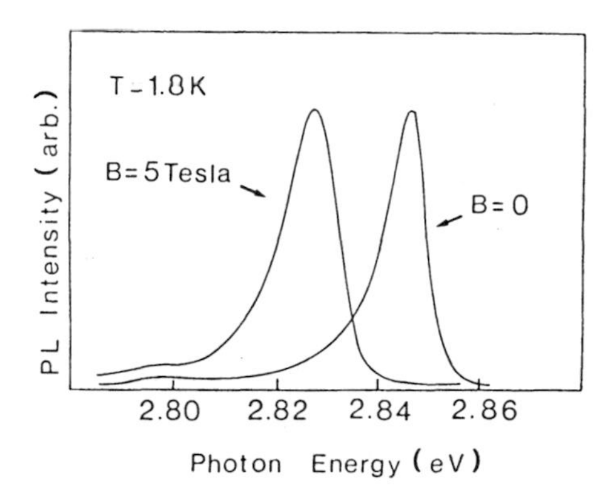

Fig. 6.8 Exciton luminescence at the superlattice bandgap for a 3 ML sample, and circularly polarized emission in a 5 T magnetic field. Measurements were taken at 1.8 K.

MnSe layers; this can by indirectly inferred by considering the strong exchange interaction of the band edge states with the Mn-ion *d*-electrons. In Fig. 6.8, the strong circularly polarized character of the emission verifies the transition as connecting the spin-split $|1/2, -1/2\rangle$ conduction and $|3/2, -3/2\rangle$ valence bands. In its simplest description, the Zeeman effect is proportional to the product of the exchange constant (*s–d* and *p*–d), the Mn spin system magnetization, and the exciton wavefunction overlap with MnSe barriers. For samples containing one and three monolayers of MnSe per superlattice period, two types of structure were fabricated: (i) those of normal growth and (ii) structures where the MBE growth was interrupted at each MnSe/ZnSe interface. It was found that very little change in the magnitude of the Zeeman shifts resulted from the finite growth interruption. Overall, the largest Zeeman shifts occurred in the limit of single monolayer MnSe barrier layers, as shown in Fig. 6.9.

The direct magnetization and magneto-optical measurements cross-check

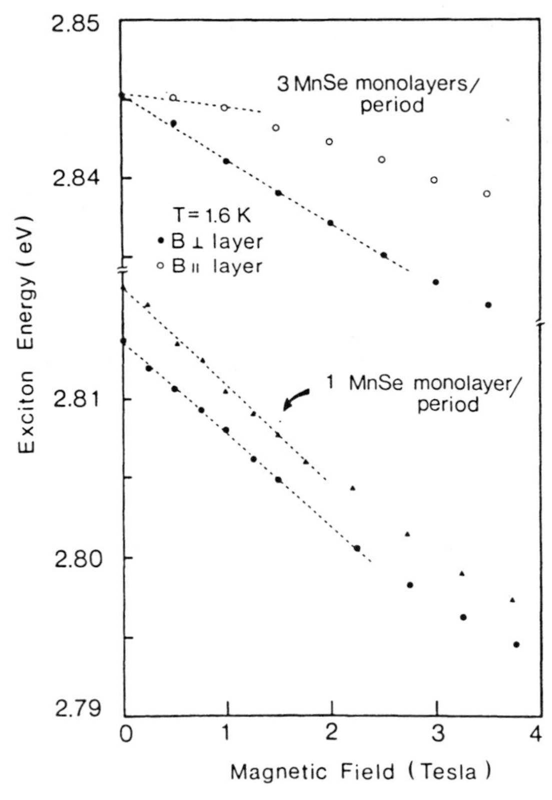

Fig. 6.9 Zeeman shifts for 1 ML and 3 ML superlattices at $T = 1.6$ K. Triangles and dark circles refer to 1 ML structures with and without growth interruption. Field anisotropy is shown for the 3 ML case. (Broken lines are to guide the eye.)

each other, and display strikingly large low temperature magnetizations which can be induced in the ultrathin layer MnSe samples. For example, optical information ensures that these originate from the superlattice portion of the samples. The key conclusion which emerged from these experiments was that the anomalous, nearly paramagnetic low temperature contributions originated from the heterointerface regions in the MnSe/ZnSe superlattices. That is, the magnetic moments at the interfaces appear to be relatively free from expected AF couplings.

The magnetic 'probes' examine microstructure on a scale on the order of chemical bond lengths. This follows from the short range nature of the superexchange paths in insulators such as MnSe (nearest neighbour Mn-ions coupled through the intervening Se anion), and give magnetic measurements an enhanced sensitivity to deviations from 'perfect' atomic arrangements (against ideal bulk) within a monolayer or so at the interface region. Some qualitative arguments can be applied concerning the roles of intrinsic and extrinsic microstructure effects at the MnSe/ZnSe heterointerfaces which underline the frustration of AF interactions seen in our experiments. Diffusion or chemical intermixing effects can provide regions of diluted Mn-ion concentration, but would also cause significant linewidth broadening of the luminescence beyond what is observed. On the other hand, while two-dimensional growth is characteristic of this SL system [24, 25], we cannot eliminate the possibility of incomplete layer growth during the heteroepitaxy. Finite size islands at the interfaces, i.e. two-dimensional clusters, can be effective in frustrating antiferromagnetic ordering. At present the answers to these questions remain largely a matter of guesswork. The issue presented is important because of the need to improve understanding of heterointerface formation in contemporary efforts to generate sophisticated artificial semiconductor layered structures.

6.2.3 ZnSe/ZnTe microstructures

The difficulty in obtaining *p*-type ZnSe to serve as an injector of holes was the primary motivating factor leading to the growth of ZnSe/ZnTe superlattice structures. The band offsets predicted by the electron affinity rule would suggest a type II superlattice where holes and electrons are confined in separate layers; the resultant decrease in the oscillator strength of optical transitions could pose a problem for light-emitting devices. An important consideration is the large lattice constant mismatch between ZnSe and ZnTe (7.4%); however, strained-layer superlattice structures are still possible with layer thicknesses restricted to a few tens of angstroms. For structures containing primarily ZnSe, a reasonable lattice match to GaAs is still possible, whereas for structures containing approximately equal amounts of ZnSe and ZnTe, the average lattice constant approaches the lattice constant of InP. ZnSe/ZnTe superlattice structures have been grown by a variety of techniques including hot wall epitaxy [30], MBE [31, 32], atomic layer epitaxy (ALE) [31] and by a combination of ALE and MBE [33]. Photoluminescence studies [34] of the MBE-grown superlattice structures show a wide wavelength tunability; photoluminescence emission is observed from the red to the green portion of the visible spectrum.

Zn(Se,Te) is of particular interest due to the observation that the photoluminescence yield of the alloy can be significantly enhanced over that of bulk ZnSe crystals [35] and epitaxial layers, due to localization of excitons in the random alloy. MBE growth of the Zn(Se,Te) mixed crystal is complicated, however, by a difficulty in controlling the composition. In the work reported by Yao *et al.* [36] over the entire range of Te fraction, a Te-to-Se flux ratio of 3 to 10 was required. To circumvent the problems associated with controlling the alloy concentrations, ZnSe-based structures consisting of ultrathin layers of ZnTe spaced by appropriate dimensions of ZnSe to approximate Zn(Se,Te) with low or moderate Te compositions were grown. As an illustration, such a 'pseudo-alloy' was incorporated in ZnSe/(Zn,Mn)Se multiple quantum well structures [33]. In these structures, either one or two ZnTe ultrathin layers were placed in the centre of each ZnSe well; the well thickness ranged was 44–130 Å. In an effort to optimize the interface abruptness of the ZnTe monolayers, the ALE of ZnTe was performed on a recovered ZnSe surface which made up, for example, the first half of a quantum well, whereas the remainder of the structure was grown by MBE. Although the configuration of these structures was substantially different from a bulk alloy, their optical transitions, as viewed in photoluminescence, were dominated by features which were quite similar to those found in the bulk Zn(Se,Te) alloy crystals at low to moderate Te composition. These luminescence features arise from the capture and strong localization of excitons at the isoelectronic Te sites. Fig. 6.10 dramatically illustrates exciton trapping at the ZnTe monolayer sheets present in the ZnSe quantum well. For comparison purposes, Fig. 6.10a shows the low temperature photoluminescence spectrum of a ZnSe/(Zn,Mn)Se MQW structure in the absence of ZnTe [6]. The luminescence was dominated by the sharp (FWHM < 5 meV), bright blue exciton recombination at the $n = 1$ (light-hole) quantum well transition. As a striking contrast to Fig. 6.10a, the photoluminescence from a ZnSe/(Zn,Mn)Se MQW, which now incorporates the ZnTe 'sheets' inserted into each quantum well, is shown in Fig. 6.10b. The broad luminescence features at lower energy originated in exciton localization at the ZnTe/ZnSe heterointerfaces and were similar to those features observed from bulk Zn(Se,Te) mixed crystals [35]. Figure 6.10c, showing a photoluminescence excitation (PLE) spectrum, indicates that the position of the lowest energy exciton transition has not been significantly shifted by the presence of the ZnTe sheets [37].

The excitation spectrum indicates that in the absorptive process the confined states of the ZnSe/(Zn,Mn)Se quantum well valence and conduction

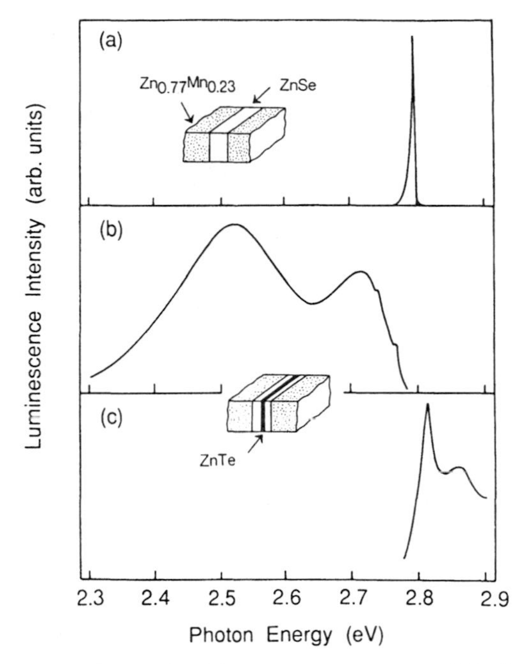

Fig. 6.10 Comparison of photoluminescence spectra at $T = 2$ K of a ZnSe/(Zn,Mn)Se MQW sample (a) with that of a similar structure but with the insertion of monolayer sheets of ZnTe in the middle of the quantum well (b). (The amplitude of emission in (a) has been reduced to bring the peak to scale.) The photoluminescence excitation spectrum of sample (b) is shown in the bottom panel (c).

band are not significantly perturbed by the addition of the ultrathin ZnTe layer. The luminescence spectrum [33, 37] however, reflects the energetics of the relaxed exciton, that is, the strong localization of the hole at the ZnTe layer in the quantum well. The self-trapping is initiated by potential energy lowering at the isoelectronic Te sites for valence band states, and followed by strong local lattice relaxation effects so that the final hole Bohr orbit is below 10 Å. The two primary features in the PL spectra of Fig. 10a and b are associated with exciton trapping at single Te and double Te sites, respectively. Calculations show that, for a one monolayer intermixing of the anion, the Te distribution is dominated by the double sites and single sites, followed by larger clusters. It was concluded that the ability to incorporate the Te isoelectronic trap centres in a planar and spatially controllable way had provided a substantial new insight into the capture process. Considerable microscopic understanding of the exciton trapping process has been obtained from the spectroscopy in magnetic fields and measurement of the exciton kinetics through time-resolved spectroscopy in such isoelectronically δ-doped ZnSe/(Zn,Mn)Se quantum wells [37–39]. The demonstration that the recombination spectra at low and moderate lattice temperatures in these specially designed ZnSe/ZnTe heterostructures is dominated by very pronounced exciton trapping effects at the heterointerfaces suggests that this phenomena should be taken into account when considering the possible use of ZnTe/ZnSe-based heterostructures for light-emitting purposes.

In an experimental extension of this work, the optical properties of such unique, isoelectronically doped quantum wells have been studied by increasing the ZnTe layer thickness in monolayer steps up to four. The aim has been to use the ZnSe/ZnTe system to approach the subject of band offset formation from the isoelectronic centre point of view. That is, while the intermixed δ-doped ZnTe monolayer case clearly shows the effects of isolated centers, in the opposite limits of a ZnTe/ZnSe quantum well, type II band offset is expected from bulk considerations. Recent spectroscopic results suggests that the transition in behaviour from one limit to the other is a continuous one where a structure containing four monolayers of ZnTe in ZnSe shows a type II-like behaviour in the absorption spectra at this limit; the spectrum is still partially influenced by the exciton trapping processes in the recombination spectra [40].

6.2.4 ZnSe/Zn(S,Se) superlattices

The MBE growth of Zn(S,Se) is relatively difficult due to the problems associated with incorporating a high vapour pressure, and highly reactive element such as S in an ultrahigh vacuum environment. Cammack *et al.* [41] circumvented some of the problems by utilizing a ZnS_xSe_{1-x} compound as the source of S in various ZnSe/Zn(S,Se) superlattice structures. In superlattice structures which are heavily strained, the RHEED patterns developed diagonal streaking occurring between spots; such RHEED patterns were believed to arise from the presence of defects, such as stacking faults and twins, in the heavily strained layers [42]. ZnSe/Zn(S,Se) superlattices [41] have shown lasing under electron beam pumping. Room temperature lasing of an electron-beam-pumped ZnSe epilayer was obtained with a threshold of 5 A cm^{-2} whereas the superlattice structures has a 12 A cm^{-2} threshold current density [41].

Low loss, short wavelength optical waveguide structures using ZnSe–ZnS strained-layer superlattices were grown by MOVPE [43]. The propagation loss of the three-dimensional optical waveguide at a 0.633 μmTE fundamental mode was measured to be 0.71 cm^{-1} for a superlattice structure composed of 80 periods of ZnSe (50 Å)–ZnS (50 Å). The loss measured for the TM mode was much larger than that for the TE mode; the difference in the loss values may be related to the anisotropy of the refractive index. This behaviour can make these waveguides very useful in polarization sensitive devices, such as polarizing optical waveguides, to be incorporated into integrated optics systems.

6.2.5 ZnSe/(Zn,Cd)Se quantum well structures

Very recently, optical pumped laser oscillations of ZnSe/$Zn_{0.86}Cd_{0.14}Se$ quantum well laser structures with appropriately designed optical waveguiding have been obtained at room temperature under low duty cycle excitation. Lasing was observed at temperatures up to 100 K under 'quasi-continuous' high repetition rate operation [44]. The room temperature threshold was measured to be 500 kW cm^{-2} while at $T = 77$ K the value was 30 kW cm^{-2}. These lasers exhibit low threshold behaviour and suggest that the goal of a room temperature continuous-wave optically pumped laser action may be attainable. In general, electron beam and optically pumped lasers are quite useful because heterostructure laser designs can be constructed and tested without dealing with the still serious difficulties of achieving both *p*- and *n*-type material in widegap II VI compounds.

6.3 CdTe-BASED MICROSTRUCTURES

CdTe is one of the most technologically important of the II–VI compound semiconductors. Apart from potential opto-electronic applications, CdTe is a key substrate for the fabrication of focal plane arrays based on (Hg,Cd)Te alloys and superlattices. As for ZnSe, the alloying of CdTe with Mn results in a dilute magnetic semiconductor. The growth of epitaxial layers of (Cd,Mn)Te by MBE presents an opportunity for developing integrated optical isolators, modulators, and switches in the visible and near-infrared portions of the spectrum. Experiments indicate that $Hg_{0.08}Cd_{0.67}Mn_{0.25}Te$ is an attractive material for isolator applications in the 820 nm range, while $Cd_{0.55}Mn_{0.45}Te$ is applicable as an isolator at the somewhat shorter wavelengths of interest to optical disc storage applications.

6.3.1 MBE growth of CdTe on GaAs

A continuing problem for the epitaxial growth of II–VI compounds is the difficulty in obtaining high quality substrates for homo- and heteroepitaxy. In the case of CdTe, for example, bulk crystals exhibiting sufficiently high crystalline quality have only recently become available, but are limited to relatively small areas. The alloying of bulk CdTe material with Zn [45] or Se [46] as a means of improving the crystalline quality and providing a variable lattice parameter, has recently been implemented. Even with suitable available II–VI substrates however, problems associated with the preparation of the surface for epitaxy remain. Substrates of many of the III–V compounds, such as GaAs, InP, GaSb, and InSb, can be chemically prepared such that a native oxide is formed and is thermally desorbed in vacuum prior to the start of growth. Because GaAs is readily available at low cost and high quality, it is an attractive candidate for use as a substrate for the heteroepitaxy of CdTe. Furthermore, a suitable surface of GaAs can be obtained without impingement of the group V element flux during the thermal desorption of the oxide. Having noted the many advantages of GaAs as a substrate for CdTe, the presence of a 14.6% lattice constant mismatch would appear to be a formidable obstacle to achieving high quality, single crystalline epitaxial growth. Nevertheless, high quality CdTe epilayers have been grown on GaAs. A fascinating consequence of the large lattice mismatch is the nucleation of two orientations of CdTe, (1 1 1) and (1 0 0), on (1 0 0) oriented GaAs substrates. During the early activity associated with the growth of CdTe on GaAs, some groups reported the observation of consistent nucleation of only one orientation or the other, while others experienced the seemingly random occurrence of (1 1 1) CdTe on (1 0 0) GaAs or (1 0 0) CdTe on (1 0 0) GaAs under what appeared to be identical growth conditions. Given the potential importance of CdTe-on-GaAs as an alternative substrate to bulk CdTe, a great deal of effort was directed at determining the factors that contributed to the selective orientation of either (1 1 1) or (1 0 0) CdTe on (1 00) GaAs.

Several groups [47–49], using growth techniques apparently similar to those described above, reported the observation of (1 0 0) CdTe on (1 0 0) GaAs; the relevant factors controlling the orientation were unknown. The first insight into the details of the CdTe/GaAs interface was provided by Otsuka *et al.* [50, 51] using TEM. In this study the (1 1 1) oriented CdTe layer was grown at Purdue whereas the (1 0 0) oriented CdTe layer was grown at North Carolina State University. The HREM revealed the presence of a thin (10 Å) residual GaAs oxide at the interface between the (1 0 0) CdTe and (1 0 0) GaAs materials. The occurrence of the residual oxide could be attributed to the much lower temperature (500–550 °C) used for the oxide desorption step. Although it is interesting that the presence of a thin oxide layer can affect the orientation of CdTe on GaAs, an oxide layer is not necessary to achieve parallel epitaxy. Two specific growth techniques can be used to nucleate the (1 0 0) epitaxial orientation on 'clean' (1 0 0) GaAs. One technique involves the nucleation of the (1 0 0) CdTe at an elevated substrate temperature, whereas the second technique involves the prior nucleation of $Cd_{0.60}Zn_{0.40}Te$. Faurie *et al.* [52] have observed a range of Zn mole fractions which nucleate (1 0 0) (Cd, Zn)Te on (1 0 0) GaAs. For all growth techniques of (1 0 0) nucleation employed, RHEED observations show the occurrence of an initially spotty pattern (indicating three-dimensional nucleation) which eventually elongates into streaks of uniform intensity; this contrasts with the two-dimensional nucleation observed with the (1 1 1) orientation.

Using the aforementioned select growth techniques to nucleate the (1 0 0) CdTe layer on (1 0 0) GaAs, (1 0 0) CdTe/(1 0 0) GaAs interfaces have been prepared without an interfacial (oxide) layer and have been studied with TEM [53–55]. Fig. 6.11 shows an HREM image (obtained with the 1 MV TEM microscope) of a CdTe/GaAs interface, resulting from nucleating the (1 0 0) CdTe epilayer at an elevated temperature to induce the occurrence of

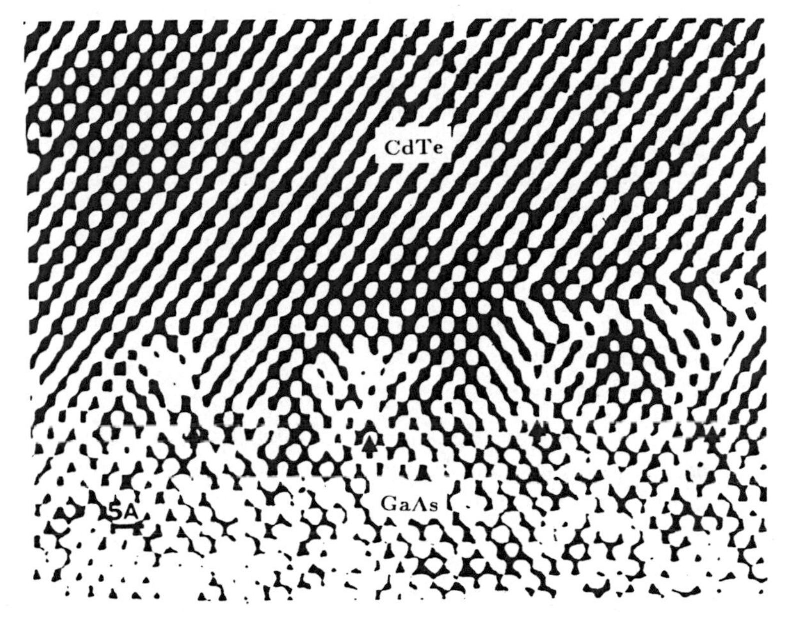

Fig. 6.11 High resolution electron micrograph of the (1 0 0) CdTe–(1 0 0)GaAs interface. The arrows indicate the presence of purely edge-type misfit dislocation with Burgers vector of 1/2 [0 1 1].

the (1 0 0) epitaxial relationship. The images show the presence of a misfit dislocation array which exists in the plane of the (1 0 0) CdTe/(1 0 0) GaAs interface. The presence of additional (1 1 1) and $(1\,1\,\bar{1})$ fringes in the image indicates that these misfit dislocations have a Burgers vector of 1/2[0 1 1] and are pure edge dislocations [54,55]. Figure 6.11 is representative of both perpendicular directions ([0 1 1] and $[0\,\bar{1}\,1]$). The occurrence of these misfit dislocations tend to generate threading dislocations which propagate perpendicular to the interface. These threading dislocations are not present in HREM imaging but are readily visible in dark field imaging. The density of these dislocations are quite high near the interface (10^{12}–10^{11} cm^{-2}), but are dramatically reduced as the CdTe film thickness is increased. A dislocation line density of 10^4 cm^{-2} has been reported at the surface of a 6.6 μm MBE-grown CdTe epilayer [48].

Many researchers now agree on the conditions required to achieve the nucleation of a (1 1 1) CdTe film on a (1 0 0) GaAs substrate. More elusive however, is the identification of the factors relevant to the nucleation of (1 0 0) CdTe on (1 0 0) GaAs. It is apparent that the (1 0 0) epitaxial occurrence is associated with a perturbation of the surface 'state' of the (1 0 0) GaAs substrate. Recently several groups have performed experiments to identify different types of perturbation and their effects on the epitaxial orientation.

Feldman and Austin [56] have studied the various surface structures which result from the adsorption of Te on a (1 0 0) GaAs surface; it is believed that both the surface structure and the interaction of the group II element with the surface affects the resultant orientation of the CdTe. The effect of the surface stoichiometry of the GaAs substrate on the epitaxial orientation of the CdTe has also been studied [57]; both As-stabilized and Ga-stabilized GaAs surfaces were found to control the orientation such that (1 0 0) CdTe nucleated on an As-stabilized surface, while (1 1 1) nucleation occurred on the Ga-stabilized surface.

6.3.2 CdTe/(Cd,Mn)Te superlattice and multiple quantum well structures

Alloying CdTe with the transition element Mn results in epitaxial layers of the dilute magnetic semiconductor. Incorporation of Mn in the host CdTe lattice increases the direct energy bandgap of CdTe while decreasing the lattice parameter. $Cd_{1-x}Mn_xTe$ has been grown with Mn mole fractions between $0 < x \leqslant 0.53$ on GaAs substrates [59,60]. Using the growth techniques described above for CdTe, both (1 1 1) and (1 0 0) oriented (Cd, Mn)Te epilayers have been grown on (1 0 0) GaAs substrates; the (1 0 0) oriented structures have been grown on (1 0 0) CdTe buffer layers on (1 0 0) GaAs. The uniformity of the Mn concentration was investigated after film growth using Auger electron spectroscopy and depth profiling. The concentrations of the Cd and the Mn were very uniform throughout the film. X-ray diffraction performed on the (Cd, Mn)Te epitaxial layers to determine the lattice constant [58] indicated the absence of any MnTe or $MnTe_2$ phases. (Cd, Mn)Te has also been grown using the related growth technique of atomic layer epitaxy [59].

In contrast to the (Zn, Mn)Se system, the lattice constant of (Cd, Mn)Te decreases as the Mn fraction increases [58]. As a result, the strain subjects the well material to expansive uniaxial strain normal to the interface and compressive strain parallel to the interface. The hydrostatic component of the strain increases the optical gap, while the uniaxial component acts to remove the valence band degeneracy at $k = 0$ so that the heavy-hole band (in the direction of the superlattice axis) now moves up in energy relative to the light-hole band.

The first superlattices in the CdTe/(Cd, Mn)Te material system were grown by MBE on GaAs substrates with superlattice interfaces parallel to (1 1 1) planes [60, 61]. Unusual optical and magneto-optical properties exhibited by these (1 1 1) strained-layer superlattices were attributed to the existence of interface-localized excitons associated with the presence of (1 1 1) interfacial planes [62]. The ability to grow either (1 1 1) or (1 0 0) CdTe on (1 0 0) GaAs provided a unique opportunity to compare directly the effect of orientation on the optical properties of the superlattices.

A number of superlattice configurations have been fabricated with both (1 0 0) and (1 1 1) orientations. For the (1 1 1) orientation, both CdTe and $Cd_{1-x}Mn_xTe$ have been used as the quantum well material, with a range of Mn mole fractions in the $Cd_{1-x}Mn_xTe$ barrier material [60, 61, 63]. For superlattices having the (1 0 0) orientation, only CdTe has so far been used as the well material. These superlattice structures were at first grown on 1–2 μm (Cd, Mn)Te or CdTe buffer layers on (1 0 0) GaAs substrates [53] and subsequently grown on CdTe [64] and InSb [65] substrates.

TEM observations have been made for both (1 0 0) and (1 1 1) oriented superlattice structures. High resolution images of the superlattice interfaces were observed, and no discontinuity in lattice fringes at the interfaces was found. The implication of the interface coherence is that the lattice mismatch between superlattice layers is accommodated by elastic strain rather than by misfit dislocation networks, resulting in strain-layer superlattices. [66].

The low temperature photoluminescence (Fig. 6.12) from a number of (1 1 1) oriented MQW structures of various well thicknesses resulted in the discovery that the intense low temperature recombination observed originated from localized excitons, particularly so in narrower well (< 100 A) samples. Such

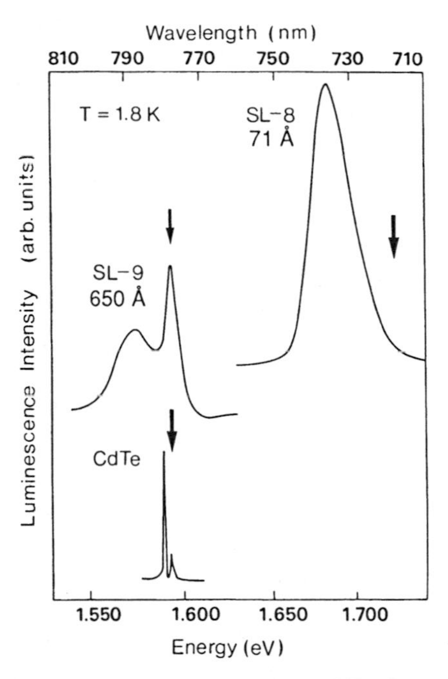

Fig. 6.12 Photoluminescence spectra at $T = 1.8$ K for two (1 1 1) oriented $CdTe/Cd_{0.74}Mn_{0.26}Te$ MQW samples (well widths 71 Å and 650 Å) in comparison with high quality bulk CdTe (bottom trace). The arrows denote the assigned free exciton energies for the (1 1 1) oriented MQW and the bulk sample as determined from reflectance spectra.

localized exciton emission luminescence spectra were found readily to red-shift in external magnetic fields [67]. While spectral shifts were expected from considering the effects by spin exchange for the finite penetration of the exciton wavefunction into the (Cd, Mn)Te barriers, they were unexpectedly large and also showed pronounced anisotropy regarding the direction of the applied field with respect to the superlattice axis. The measured low field shifts for the narrow well MQWs showed that as much as half of the hole wavefunction lies within the (Cd, Mn)Te layers. This is consistent with a small valence band offset. The anisotropy has been used to argue that the heavy hole wavefunction is substantially two-dimensional [62, 67, 68]. Furthermore, magneto-optical data at very high magnetic fields (>20 T), following the evident paramagnetic saturation of the contributing Mn-ion spins, indicated little change in the exciton ground state energy, thereby supporting the notion that the exciton is rather two-dimensional [67]. The combination of large penetration of the hole wavefunction into the Mn-containing region of the MQWs, together with the substantially two-dimensional character of the hole and the exciton has provided an empirical insight which strongly suggests that the recombining excitons are physically localized at the (1 1 1) CdTe/(Cd, Mn)Te heterointerface [62, 69].

The ability to grow the (Cd, Mn)Te superlattices in both (1 1 1) and (1 0 0) crystalline orientations [53] provided an opportunity to study possible differences in associated electronic characteristics in these strained-layer structures for the two polar interfaces. The contrast between the two growth orientations was pronounced in luminescence excitation spectra [70]. While a strong ground state exciton absorption peak was evident for the (1 0 0) case (which also showed strain split valence band and excited state structure), this resonance was almost completely broadened out in the (1 1 1) structures. It should be noted that excitation spectra taken on (1 1 1) structures at considerably lower Mn concentrations ($x = 0.06$) has shown how well-defined transition features do indeed emerge even for this orientation [71].

Raman scattering has provided clear evidence for acoustic phonon zonefolding in a narrow well/barrier, (1 1 1) oriented MQW structure, in good agreement with an elastic continuum model, and consistent with a high degree of crystalline quality [72]. Similar Raman measurements show an unexpected contrast between (1 1 1) and (1 0 0) oriented CdTe/(Cd, Mn)Te MQWs in studies of the longitudinal optical (LO) phonon dispersion in samples comparable to those discussed above. These Raman data show LO phonon confinement occurring in the separate layers of the (1 0 0) structures, but it is not evident in the (1 1 1) case. It is possible that the differences seen in the Raman experiments [73] have a common underlying physical origin to that which produces the large difference in the excitation spectra between the two orientations.

Further evidence of the inherent optical quality of the material system is supported by the demonstration of stimulated emission in both

CdTe/(Cd, Mn)Te [20, 74] and $Cd_{1-x}Mn_xTe/Cd_{1-y}Mn_yTe$ quantum well structures [75]. Laser samples were prepared using a selective chemical etch [76] to remove the GaAs substrate; the resultant free-standing epilayers were cleaved and mounted on a copper heat sink [77] for optical pumping experiments. For lasers structured with (1 1 1) CdTe as the active material, lasing was obtained at wavelengths of 763–766 nm at 25 K with a threshold power density of $1.35 \times 10^4\,W\,cm^{-2}$ [74]. When (1 1 1) $Cd_{1-x}Mn_xTe$ ($x = 0.19$) was the active quantum well material, lasing occurred in the red spectral region at 665–670 nm at 15 K; the threshold for laser action was $2.0 \times 10^4\,W\,cm^{-2}$ [75]. The exchange interaction occurring in the DMS material provided a shift in the energy of the quantum well states such that the application of a magnetic field allowed for the tuning of the output energy of the DMS laser (Fig. 6.13). For the DMS lasers studied, a magnetic field tuning rate of $3.4\,meV\,T^{-1}$ was obtained at 1.9 K [78]. This magnetic-field-induced shift was approximately one fifth that obtained from bulk $Cd_{1-x}Mn_xTe$ having a comparable x value. (1 0 0) oriented quantum well

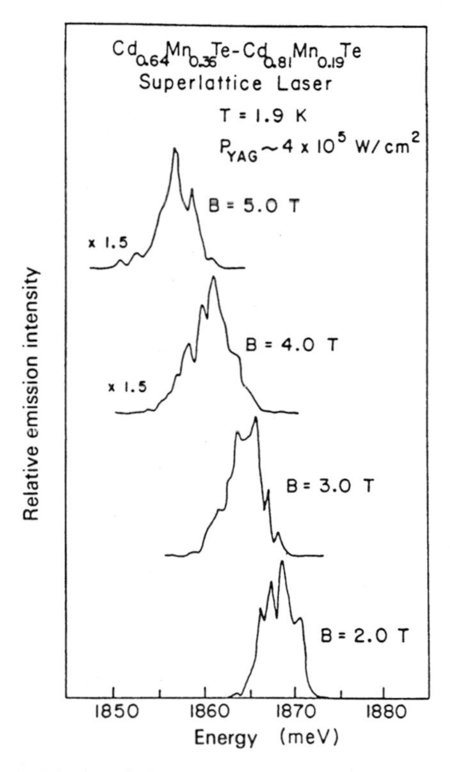

Fig. 6.13 Stimulated emission from a CdMnTe/CdMnTe superlattice structure as a function of applied magnetic field illustrating magnetic field tunability due to the presence of the Mn-ion in the active laser region [78].

structures with CdTe as the active well material, have exhibited stimulated emission up to 119 K [20]. The effects of strain in these (1 0 0) strained-layer superlattices results in TE-polarized stimulated emission from the sample edge, and has been compared with the edge emission from oppositely strained (Zn, Mn)Se multiple quantum well structures [20].

(*a*) *Modulation doping*

A long-standing problem associated with II–VI compounds is their propensity for defect generation, and the attendant self-compensation, as dopant impurity species are incorporated into the material. When indium atoms are incorporated during conventional MBE growth of CdTe (using a CdTe compound source), the photoluminescence is degraded, while little evidence of activation is obtained. Bicknell *et al.* [79] have employed a technique of photoassisted MBE to overcome this tendency for self-compensation. By illuminating the growing film with a low intensity beam from an argon laser (150 mW cm^{-2}), a high degree of dopant activation was obtained. A similar success has been achieved for MBE-grown CdTe doped with Sb [80]. Comparisons of photoluminescence and transport properties have shown dramatic improvements when laser illumination is employed for both *n*- and *p*-doping. The laser assisted doping technique has also been used during dopant incorporation in CdTe/(Cd, Mn)Te superlattices where the DMS barrier layer is doped with indium [64]. In the case of relatively wide CdTe wells, the measured mobility values exceeded that obtained in single layer, indium-doped CdTe films. As the wells became narrower, however, the mobilities tended to decrease, suggesting that the interfaces may play a role in transport parallel to the superlattice layers. Having achieved the controlled substitutional doping of MBE-grown CdTe, a variety of devices, such as *pn* diodes and metal–semiconductor field-effect transistors, have been fabricated and studied [81]. Although details of the mechanism by which the photons interact with the growing surface layer during photoassisted MBE is not completely understood, recent experiments suggest that the photon illumination selectively desorbs Te [82] to compensate for the greater desorption rate of Cd at the growth surface [83]. For the readers convenience, references 82 and 84–89 discuss the molecular beam epitaxy of II–VI semiconductors in the presence of photons, but will not be detailed here.

6.3.3 CdTe/ZnTe superlattice and quantum well structures

The binary superlattice CdTe/ZnTe [90] has been investigated as a possibly useful pseudomorphic structure. The configuration, however, involves a very large lattice mismatch (approximately 6.4%). Initial optical studies had suggested a reasonable agreement between experimentally derived superlattice bandgap values and calculations in the 'free standing' superlattice limit (assuming also the absence of a finite valence band offset) [91]. A small valence band offset would follow the situation already established with CdTe/(Cd, Mn)Te and ZnSe/(Zn, Mn)Se, in that the finite lattice mismatch strain (through hydrostatic and uniaxial components) may be the main factor determining the actual valence band offsets, thus making them dependent on individual sample parameters. Apart from the band offset issue, subsequent resonant Raman scattering experiments have, however, cast doubts on the arguments on attaining the free-standing superlattice limit in the CdTe/ZnTe system [92].

One important consequence of a small band offset in a real, highly-strained system subject to small but finite structural irregularities has been established by using the lowest interband exciton resonances as indicators in photoluminescence experiments. In particular, time-resolved and resonantly excited spectroscopies have shown that excitons exhibit unusual localization in the CdTe/ZnTe system [93]. Qualitatively, variations in the layer thickness on a monolayer scale in a highly-strained structure are sufficient to produce significant fluctuations in the local strain about some mean value. In the absence of strong confinement (i.e. small offset), the associated random potential fluctuations may be quite efficient in capturing electronic quasiparticles at low or moderate lattice temperatures.

Low threshold, optically pumped lasers, emitting in the yellow-orange portion of the visible spectrum and fabricated with $Cd_{0.25}Zn_{0.75}Te$/ZnTe superlattice structures, have been reported by Glass *et al.* [94]. The binary CdTe/ZnTe superlattices are heavily strained, thus the laser structures incorporated alloys of (Cd, Zn)Te as the well layer in an effort to reduce the strain. In the superlattice structures studied, the lasing wavelength increased from 575 nm at 8 K to 602 nm at 310 K. At low temperatures, the threshold pump intensity was found to be quite low at 7 kW cm^{-2}. At room temperature the threshold pump intensity increased to ~55 kW cm^{-2} and represented the first report of room temperature, optically pumped lasing in a II–VI superlattice.

Transmission measurements demonstrating room temperature excitonic saturation were recently reported from (Cd, Zn)Te/ZnTe quantum well structures [95]. In this work, strong absorption peaks from $Cd_{0.25}Zn_{0.75}$-Te/ZnTe quantum wells (50 Å/100 Å) at room temperature were reported. The absorption coefficient of the sample was 6×10^4 cm^{-1}, a value three times higher than observed from bulk $Cd_{0.14}Zn_{0.86}Te$. In addition, the absorption coefficient per well was larger than that typically found in GaAs/AlGaAs MQWs. Saturation intensities were calculated with parameters giving the best fit to experimental data. The data taken with the sample having a well dimension of 50 Å were fitted to a saturation intensity of 31 kW cm^{-2}, while the data from the sample with 100 Å wells were fitted to a saturation intensity of 15 kW cm^{-2}. These values were significantly larger than those of previously reported III–V compound semiconductor MQWs.

6.3.4 CdTe/InSb quantum well structures and superlattices

One of the few zinc blende semiconductors available with appropriate lattice parameter and bandgap to serve as a barrier with InSb is CdTe. Theoretical band offset predictions [96] as well as photoelectron spectroscopy data [97] indicate that the resultant quantum well structure will be type I. The degree of lattice match (0.05%) approaches that of GaAs and (Al, Ga)As, hence the effects of strain should be minimized. The relatively small effective mass for electrons and holes tends to amplify quantum shifts; a 75 Å quantum well dimension doubles the ground state transition energy of bulk InSb [98]. Easily achievable well dimensions allow access to a 2–5.5 μm wavelength range. The high carrier mobilities and band non-parabolicity suggest interesting optical and electronic device characteristics.

The realization of proposed device structures has been hampered by the significant material problems associated with these II–VI/III–V heterostructures. The optimum temperatures for the growth of high-quality InSb and CdTe are reported to be quite different (approximately 400 °C [98] and 200 °C [99], respectively). Secondly, Raman spectroscopy provided evidence of the formation of interfacial compounds (such as In_2Te_3) under various growth conditions [100]. It was recently demonstrated that the chemical reaction at the interface could be suppressed by growing CdTe with a Cd-enhanced flux ($J_{Cd}/J_{In} = 3$) [101, 102]. To achieve high-quality InSb material, the use of Sb_2 was employed [103] at Purdue with the anticipation of similar results as those obtained for the low temperature growth of GaAs using As_2 [104, 105] in place of As_4. In addition, very low growth rates (0.18 Å s^{-1}) were also used for the growth of InSb layers in various heterostructure configurations. In these experiments, two relatively lattice-matched substrates were used, InSb and CdTe, for the various double heterostructures and multiple quantum wells. Figure 6.14 shows a four-crystal

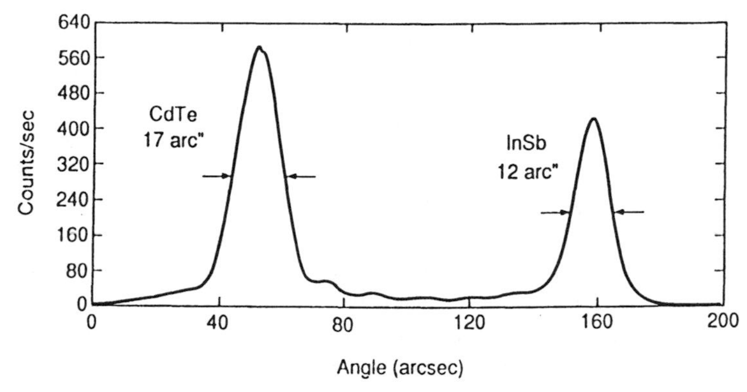

Fig. 6.14 X-ray rocking curve of a heterostructure consisting of a 1.1 μm CdTe epilayer grown on an InSb substrate, with an intermediate (0.36 μm) InSb buffer layer.

X-ray rocking curve of a structure consisting of a 1.1 μm CdTe layer grown on an InSb homoepitaxial buffer layer at a temperature of 310 °C. The narrowness of the peaks, as well as the diffraction interference effects observed in the region between the peaks, are indicative of a high degree of crystalline quality. Single quantum well, multiple quantum well (MQW), and InSb/CdTe superlattice structures have been grown at substrate temperatures between 280 and 330 °C.

For the growth of InSb/CdTe double heterostructures, the Purdue group utilized two interconnected MBE chambers. InSb (1 0 0) homoepitaxial layers were grown at 300–330 °C using InSb (as a source of antimony) and elemental indium sources. A characteristic $(2\sqrt{2} \times 2\sqrt{2})\ 45°$ reconstructed surface was observed by RHEED during growth indicating an Sb-stabilized surface [106]. Unlike the RHEED patterns observed during the nucleation of (1 0 0) CdTe on (1 0 0) InSb, the nucleation of (1 0 0) InSb on (1 0 0) CdTe was seen to occur by a three-dimensional process, as evidenced by an initially spotty RHEED pattern. The formation of streaks was seen to occur after ~75 s (80 Å) of growth. Following InSb layer growth, samples were transferred back to the CdTe chamber where CdTe capping layers 2000 Å thick were nucleated and grown at 240 °C; during growth the CdTe again showed a (2×1) reconstructed surface in the RHEED patterns.

Infrared photoluminescence data has been taken at 10 K with a YAG laser and cooled InSb infrared detector. The luminescence from a 5600 Å InSb active layer, confined between CdTe layers, is plotted with luminescence from an InSb substrate in Fig. 6.15. In this case, the double heterostructure is

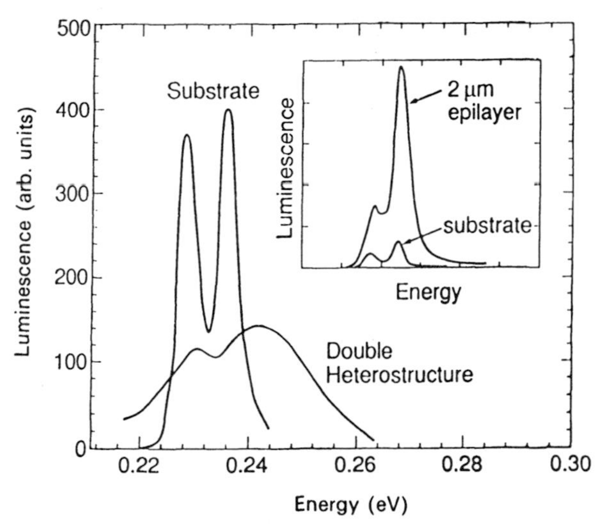

Fig. 6.15 Infrared luminescence of a CdTe/InSb/CdTe double heterostructure at 10 K compared to that from an InSb substrate. Inset: Low temperature luminescence of a 2 μm homoepitaxial MBE-grown InSb layer compared to an InSb bulk substrate.

based on an InSb substrate with a 0.42 μm InSb buffer layer, a 1.63 μm CdTe buffer layer, the InSb active layer, and a 0.22 μm CdTe capping layer. Two peaks are seen in the luminescence spectra for the heterostructure, with the higher energy feature being assigned to band-to-band recombination [107]. Spectra for the double heterostructure are seen to be broader (to the high energy side) than that observed for the InSb substrate. For additional comparison where a CdTe/InSb interface is not involved, luminescence features from a Cd-doped InSb substrate and a 2 μm InSb epilayer are seen to be more intense than for the InSb substrate by a factor of ~ 7.

For the growth of multilayer structures of more than a few periods, the substrate transference method was deemed to be unrealistic. Given the need to grow multilayer structures, an antimony cracker and an indium source were installed in the chamber designated for CdTe growth. As a base for multiple quantum well growth on InSb substrates, InSb buffer layers were grown at 330 °C using the antimony cracker. Alternately, CdTe substrates were prepared for epitaxy by *in situ* heating to 325 °C followed immediately by dropping the substrate temperature to 200 °C. At 200 °C, CdTe buffer layers were nucleated and stepped up by 1 °C min^{-1} to a growth temperature of 240 °C.

Before the end of the CdTe buffer layer growth the substrate temperature was raised by 2 °C min^{-1} to a value of 280 °C. After a short stabilization period, growth of InSb/CdTe multilayer structures was commenced without interruption of epitaxy. InSb layers were grown at very low rates (0.2 Å s^{-1}) with different antimony cracking temperatures (850–1040 °C). In addition to the double heterostructures, a number of multiple quantum well and superlattice structures have been grown at 310 °C. The X-ray rocking curve for a 15 period superlattice is shown in Fig. 6.16. Of the two main peaks

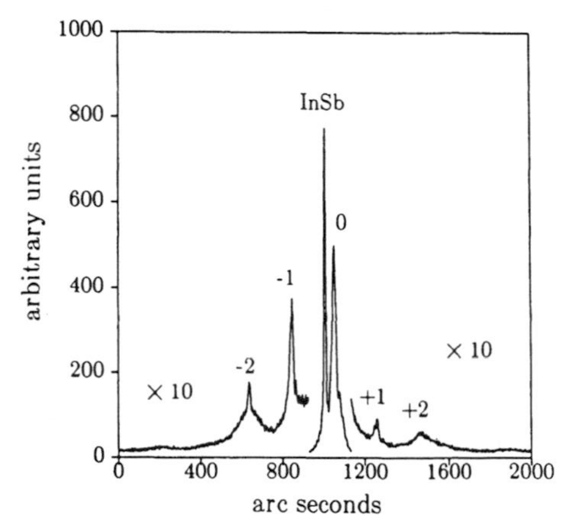

Fig. 6.16 X-ray rocking curve of a 15 period superlattice (period = 870 Å) grown at 310 °C.

in the diffraction spectrum, the peak at the higher angle side (FWHM = 22 arc second) was attributed to the zero order diffraction peak of the multilayer structure. Satellite peaks, having a separation of approximately 208 arc second, were observed in a symmetrical angular distribution about the zero order peak. The spacing between satellite peaks corresponded to a periodicity of about 870 Å, in good agreement with the dimensions (883 ± 10 Å) determined from TEM images. The additional high intensity peak seen in the figure, having a FWHM value of 11 arc second, was attributed to the (0 0 4) reflection from the InSb buffer/substrate.

Since the lattice spacing in the growth direction for the InSb substrate/buffer should be smaller than the average lattice plane spacing associated with the periodic structure, the zero order diffraction peak from the superlattice was expected to lie at a lower angle than the peak corresponding to InSb. The observation that the positions of the peaks were reversed could be explained by assuming that an ultrathin, and highly strained layer of some interfacial compound, such as zincblende In_2Te_3, was incorporated at the interfaces [100, 101]. (Zincblende In_2Te_3 has a lattice constant of 6.14 Å, a dimension significantly smaller than that of InSb/CdTe.) A simple calculation involving the minimization of strain energy in the strained layer structure, with interfacial layers included as a component of the superlattice, was used to predict the effect of such interfacial layers on the average lattice dimension of the superlattice in the growth direction. The calculation was found to predict the measured angular positions of the diffraction peaks provided several monolayers (~ 5) of the assumed interfacial In_2Te_3 were distributed in some fashion between the two interfaces associated with each period of the structure. In the calculation, the elastic constants for the assumed interfacial layer, taken as consisting of the zincblende phase In_2Te_3 (where every third site of the metal sublattice is vacant) [108], were approximated as an average of the elastic constants of CdTe and InSb. The multiple quantum well structures studied also revealed X-ray diffraction features which could be explained by the presence of interfacial layers of In_2Te_3.

The microstructure of the MQW and superlattice structures were examined by the TEM observation of cross-sectional specimens. In the preparation of TEM specimens, iodine ions were used in the final stage of thinning during which specimens were cooled with liquid nitrogen. As reported by Cullis *et al.* [109], the radiation damage of II–VI semiconductors, and the formation of indium particles on InSb surfaces, are significantly reduced by the use of iodine ions instead of argon ions. The TEM observations have shown that the interfaces in the MQW structures have degrees of roughness that varied from sample to sample; variation in the amount of roughness was also observed within the same sample. In all cases, InSb-on-CdTe interfaces exhibited greater degrees of roughness than CdTe-on-InSb interfaces. Figure 6.17 is a [1 1 0] HREM image of a 20 period MQW structure having

Fig. 6.17 [1 1 0] HREM image of a CdTe/InSb MQW structure.

135 Å CdTe and 180 Å InSb layers. Darker bands are CdTe barrier layers and brighter ones InSb well layers. The InSb-on-CdTe interfaces had roughness amplitudes of 20 Å (six monolayers), while the width of CdTe-on-InSb interfaces appeared to be less than 10 Å. (These MQW structures appear to have considerably smoother interfaces when compared to those of an earlier report [1 1 0].) Both interfaces exhibited dark line contrast which appear to be characteristic of the interfaces between these two semiconductors [1 1 1]. Despite the roughness and dark line contrast, InSb and CdTe crystals maintain a coherent relation as seen in the continuation of {1 1 1} lattice fringes at the interfaces. From the present observation, neither misfit dislocations nor incoherent particles were found at the interfaces of these MQW structures.

As discussed above, the results of X-ray diffraction measurements suggest that both superlattice and MQW structures incorporate very thin In_2Te_3 layers at their interfaces. The possible existence of the In_2Te_3 layers is examined by analyzing changes of spacing and orientation of lattice fringes in HREM images. If thin pseudomorphic layers of stoichiometric In_2Te_3 exist at the interfaces, the spacing of the (0 0 2) lattice fringes in the In_2Te_3 layer are expected to be smaller than those of InSb and CdTe by 0.27 Å; also the {1 1 1} lattice fringes of the layer should change their orientations from those of InSb and CdTe by 2.8° in [1 1 0] HREM images.

HREM images of two MQW structures, taken at 10 different areas for each sample, were examined. Due to the extremely small difference in the

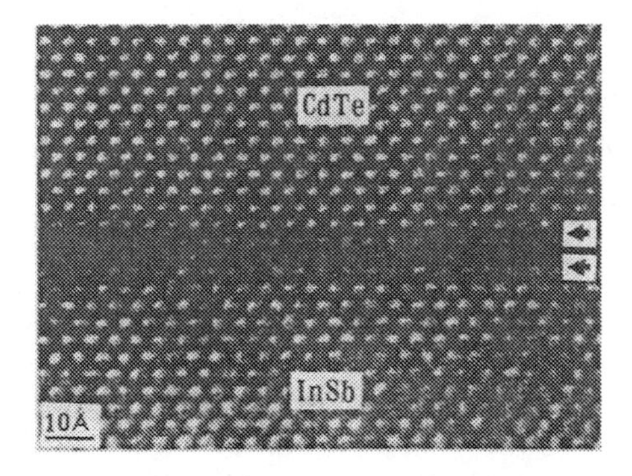

Fig. 6.18 Magnified [1 1 0] HREM image of a CdTe on InSb interface.

spacing and orientation of lattice planes, the changes in lattice fringes were detected only from areas which appeared to have more than two In_2Te_3 monolayers. Figure 6.18 is a [1 1 0] HREM image of one of such interface area. In the image, (0 0 2) lattice fringes in the range between the two arrows had slightly smaller spacings than those in the CdTe and InSb layers. Also both (1 1 1) and 1 $\bar{1}$ 1) lattice fringes showed small shifts of their positions at the interface, as a result of the change of their orientations in the narrow range. Nearly one half of the interface areas examined had similar features; no significant change in the spacing or shift of lattice fringes were observed in other areas. In all cases where changes of lattice fringes at an interface were detected, they corresponded to the small spacing of (0 0 2) lattice planes, which suggested that the changes were not due to artifacts caused by the imaging conditions. In rough interface areas, gradual changes in the spacing of (0 0 2) lattice fringes over distances of several monolayers were detected by overlapping pieces of HREM images taken from portions of InSb or CdTe layers.

Raman spectroscopy was employed to explore the issue of a mixed interface by focusing on the optical phonon region of the spectrum. Typically, strong, bulk-like InSb LO phonons were seen in the CdTe/InSb heterostructures examined. Apart from these bulk-like LO phonons in the Raman spectra, occasional weak features were observed (in certain isolated samples) which are apparently related to lattice vibrations reported [100] for chemically intermixed InSb/CdTe interfaces (In_2Te_3). For the majority of samples grown, such features, if present, were below our current level of detection.

6.3.5 CdTe/MnTe single quantum well structures

This section describes the growth and evaluation of quantum structures incorporating the metastable zincblende phase of MnTe. The MBE growth technique enables single crystal growth of the zincblende phase [112, 113], whereas bulk-grown crystals of MnTe exhibit the hexagonal NiAs crystal structure [114]. The difference in bandgap energy between the two crystal structures is dramatic; the NiAs phase has an effective optical bandgap of 1.3 eV while the bandgap of the zincblende phase is 3.18 eV at 10 K [112, 115], which is in the near-ultraviolet portion of the spectrum. The motivation for the research reported here is the development of a suitable widegap semiconductor to serve as the barrier layer to quantum well structures in the lattice constant range of InSb and CdTe. Epilayers of zincblende MnTe were grown up to a thickness of 0.5 μm, while a series of strained single quantum wells were fabricated with MnTe forming widegap barrier layers for quantum wells of CdTe, ZnTe, and InSb.

The structures having CdTe quantum wells were grown on InSb substrates with a buffer layer of InSb. The InSb buffer layer was grown in a III–V growth chamber of the modular MBE system, and transferred under ultrahigh vacuum to a second chamber for the growth of the MnTe and CdTe epilayers. The CdTe layers were grown using a compound source, while the MnTe was grown from elemental sources at an approximately unity cation/anion flux ratio. During the sequential growth of barrier and well layers of the MnTe/CdTe structures, the RHEED patterns appeared to be virtually unchanged as the layers were alternated, with each exhibiting a (2×1) reconstruction suggesting an anion-stabilized surface [116]. TEM measurements were used to determine the MnTe barrier layer thickness.

TEM observations of the single quantum well structures were made using [0 1 0] and [0 1 1] cross-sectional samples. For the preparation of cross-sectional samples, iodine ions were used at the final stage of ion thinning in order to reduce damage in the CdTe crystals. Although the samples were kept at low temperature with liquid nitrogen, the MnTe layers in these structures were found to intermix rapidly with the CdTe layers during ion thinning. Single quantum well structures were observed only from those samples for which a close contact was carefully maintained with the cold stage of the ion milling machine.

Lattice fringe patterns in HREM images directly show that the MnTe layers in the single quantum well structures have a cubic zincblende structure (also confirmed in electron diffraction of cross-sectional samples), and have been grown epitaxially on the CdTe layers. In the HREM image, more distinct (2 0 0) and (0 0 2) lattice fringes are seen in areas of MnTe than those of CdTe due to the larger value of the crystal structure factor of (2 0 0)-type reflections of MnTe. No misfit dislocations are observed in bright field or dark field images of these single quantum well structures, indicating the pseudomorphic nature of MnTe/CdTe interfaces.

In the several MnTe/CdTe/MnTe single quantum well structures, spanning CdTe well thicknesses from approximately 10–56 Å, strong photoluminescence originating from the well is detected in all samples, even for excitation at energies *below* the MnTe barrier layer absorption edge. (The MnTe barrier

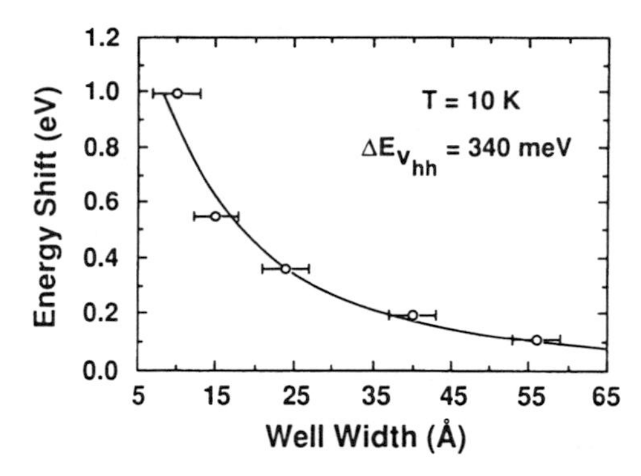

Fig. 6.19 Photoluminescence emission for series of MnTe/CdTe/MnTe single quantum well structures. (The arrow indicates the location of the bulk CdTe excitonic bandgap.)

layer thicknesses are kept constant at approximately 35 Å.) Separate optical reflectance measurements on relatively thick MnTe films yield an approximate value of 3.2 eV for the *s–p* bandgap at $T = 10\,\text{K}$. Thus the (unstrained) bandgap difference in the heterostructure is about 1.6 eV which suggests the possibility of strong confinement-induced effects. Direct evidence of carrier confinement is indeed apparent in the systematic shift to higher photon energies of the PL emission with decreasing CdTe quantum well thickness. Figure 6.19 compares the experimentally determined energy shift of excitonic features with the predictions based on the calculated transmission coefficients [117] through the double barrier structure forming the quantum well. The calculation employs a heavy-hole valence band offset of 340 meV which is determined from photoluminescence and photoluminescence excitation spectroscopy (PLE) measurements [118]. The PLE provides a measure of the valence band splitting due to the combination of strain (in the barriers) and quantum confinement in the well; the configuration of the structure ensures that the well is essentially strain free. (The implied unstrained valence band offset of 270 meV [118] agrees closely with the theoretical value of 250 meV predicted by Wei and Zunger [119]. For the narrowest well sample (10 Å), an energy shift (due to confinement) of approximately 1 eV is realized, corresponding to photon emission in the blue. The PL emission is usually composed of two or more emission lines which are 40–60 meV apart. These lines, each approximately 20 meV in width, are associated with transitions between $n = 1$ confined valence band states and the $n = 1$ conduction band state. Since the spectral peaks show the same sign of circular polarization in an external magnetic field, it is likely that they are associated with the same hole state. We presently attribute the peaks to monolayer scale fluctuations in the quantum well width. Important ingredients acting to shape the photoluminescence features are the operative band offsets, including the role of the large lattice constant mismatch (2.3%), the degree and nature of exciton binding, the exchange of electron–hole states with the Mn-ion *d*-electron moments, and possible deviations from an ideal square well due to interfacial steps. We mention in passing that the temperature dependence of the PL linewidth has given us a quantitative measure of the exciton–LO phonon interaction in this quantum well system [118].

6.3.6 InSb/MnTe resonant tunnelling structures

In addition to incorporation in the CdTe quantum well structures, zincblende MnTe has also been employed to form InSb-based quantum well structures which, due to the lack of appropriate barrier materials, have eluded realization. The study of MnTe/InSb heterovalent tunnelling structures was motivated by the potential application of resonant tunnelling at high frequencies (InSb has the highest electron mobility of the conventional semiconductors), with improved device performance at room temperature. Computer simulations of MnTe/InSb double barrier resonant tunnelling has been performed using a two-band model of electron dispersion in the band gap with the inclusion of the non-parabolicity of the InSb band structure [120–122]. Calculations predicted a peak-to-valley ratio of the order of several thousand, and peak current densities of 5×10^3 A cm^{-2} from a particular InSb/MnTe resonant tunnelling structure with barrier and well dimensions of 20 Å and 50 Å, respectively. The current–voltage (I–V) characteristics were expected to remain approximately the same between 77 K and room temperature; the conduction band barriers formed by zinc-blende MnTe (with an estimated band offset of 1.86 eV) [96, 118] block most of the thermionic current.

Resonant tunnelling structures consisting of MnTe/InSb/MnTe layers were grown without growth interruption. Figure 6.20 shows a (2 0 0) dark field image of a MnTe/InSb/MnTe resonant tunnelling structure. The barrier and well thicknesses were measured to be 35 Å and 70 Å, respectively, which agreed

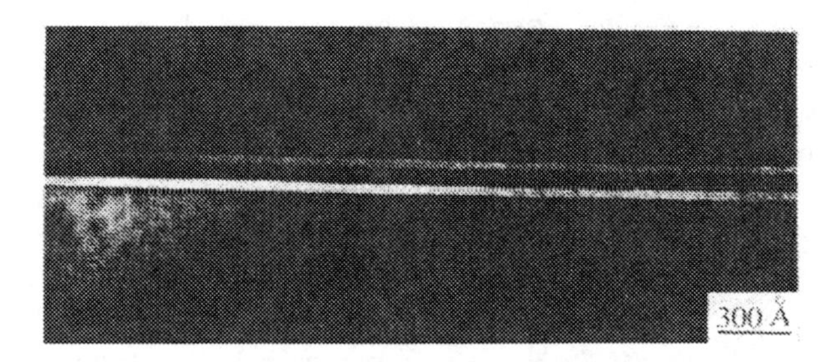

Fig. 6.20 (2 0 0) dark field electron microscope image of the MnTe/InSb resonant tunnelling structure.

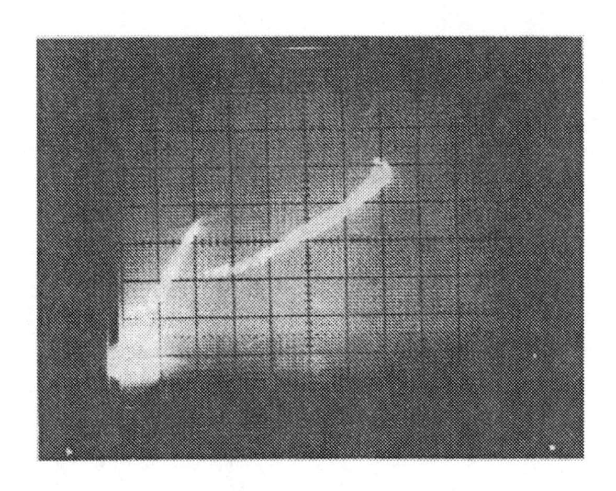

Fig. 6.21 *I–V* characteristic of the MnTe/InSb resonant tunnelling structure. The measurement is carried out at 77 K.

with the dimensions estimated from the growth rate. The image reveals the high microstructural quality of the resonant tunnelling structure with abrupt and smooth interfaces. Dislocations were not observed in any imaged area of the sample.

Diodes are fabricated by evaporating Ti/Au on the InSb cap layer, followed by the etching of mesa structures. The *I–V* characteristics are measured at various temperatures. Preliminary *I–V* results employing pulse measurement (to avoid sample heating) are shown in Fig. 6.21. A peak-to-valley ratio of 1.7:1 is observed at 77 K with a peak current density of $980\,A\,cm^{-2}$. Asymmetry about the origin in the *I–V* characteristics is observed with the resonant peak appearing at a higher voltage for the reverse bias condition compared to the forward bias condition. For these initial samples tested, the negative differential resistance becomes less pronounced with increasing temperature, and disappears at about 140 K. The origin of the deviation of the *I–V* characteristics from theoretical predictions is not yet known, and is still under investigation. One could speculate about such factors as interface scattering. (MnTe/InSb interfaces have not been previously reported.) Another possibility is leakage current paths along the side wall of the etched mesas. In any case, to our knowledge, the resonant tunnelling represents the first report of a quantized state in InSb.

6.4 CONCLUSIONS

In the contemporary development of the epitaxy of II–VI microstructures, the first and necessary stages are now completed. The generally optimistic attitude of researchers in the field suggests the beginning of a new era addressing II–VI device development. There is increasing evidence that the non-equilibrium growth property of MBE (and MOCVD) will provide the means to surmount the many obstacles of the past which have held back the development of the II–VI compound family. Significant understanding of the epitaxy process, and the optical behaviour of quantum well and superlattice structures incorporating wide bandgap II–VI compounds has been gained, with new results concerning the doping and transport properties of the II–VIs under intense investigation. In the case of MBE growth of ZnSe, the ability to control the resistivity was shown to depend primarily on the intrinsic purity of the source materials. Two directions which could be heavily impacted by similar developments are the MBE growth of ZnS-based configurations, as well as the growth of wide bandgap II–VI semiconductors by chemical beam epitaxy. There is expected to be an increased use of techniques which exploit the nature of the new non-equilibrium epitaxial growth techniques to 'engineer' the surface conditions during dopant incorporation. Some initial attempts to use the surface accessibility to improve doping involved the planar doping of ZnSe [123, 124] and include recent speculations on means for the reduction of compensating defects by adjustment of the Fermi level at the growth surface [125]. As is apparent throughout this chapter, the II–VI semiconductor compound family and associated superlattice systems are emerging as viable materials of extreme interest to the opto-electronic community.

6.5 ACKNOWLEDGEMENTS

The authors have attempted to review the latest progress made in the research of II–VI semiconductor microstructures. Undoubtedly relevant work has been overlooked. However, the many references cited in the chapter should provide sufficient additional access to the literature of the field to more than compensate for our inadvertent omissions. We especially thank the many graduate students and colleagues who contributed to the work at Purdue and Brown. We thank the authors who supplied figures, reprints, and preprints describing their work. The research at Purdue and Brown Universities was supported by the Office of Naval Research, the Air Force Office of Scientific Research, the Defense Advanced Research Projects Agency, the National Science Foundation, and the Naval Research Laboratory.

REFERENCES

1. Jonker, B.T., Krebs, J.J., Qadri, S.B. *et al.*, (1988) *J. Appl. Phys.*, **63**, 3303.
2. Aven, M. and Prener, J.S. (1967) *Physics and Chemistry of II-VI Compounds*, North Holland Publishing Company, Amsterdam.

3. Hartman, H., Mach, R. and Selle, B. (1982) *Current Topics in Materials*, **9**, (Ed. E. Kaldis), North Holland Publishing Company, Amsterdam, p. 1.
4. Galazka, R.R. (1979) *Inst. Phys. Conf. Ser.*, **43**, 133.
5. Gaj, J.A. (1980) *J. Phys. Soc. Japan*, **49**, suppl. A, 797.
6. Furdyna, J.K. (1982) *J. Appl. Phys.*, **53**, 7637.
7. Hefetz, Y., Nakahara, J., Nurmikko, A.V. *et al.* (1985) *Appl. Phys. Lett.*, **47**, 989.
8. Kolodziejski, L.A., Gunshor, R.L., Otsuka, N. *et al.* (1986) *IEEE J. Quantum Electronics*, **QE-22**, 1666.
9. Twardowski, A., Dietl, T. and Demianuk, M. (1983) *Solid State Commun.*, **48**, 845.
10. Kolodziejski, L.A., Gunshor, R.L., Bonsett, T.C. *et al.* (1985) *Appl. Phys. Lett.*, **47**, 169.
11. Gunshor, R.L., Kolodziejski, L.A., Otsuka, N. and Datta, S. (1986) *Surface Sci.*, **174**, 522.
12. Hefetz, Y., Goltsos, W.C., Nurmikko, A.V. *et al.* (1986) *Appl. Phys. Lett.*, **48**, 372.
13. Hefetz, Y., Goltsos, W.C., Lee, D. *et al.* (1986) *Superlattices and Microstructures*, **2**, 455.
14. Andersen, D.R., Kolodziejski, L.A., Gunshor, R.L. *et al.* (1986) *Appl. Phys. Lett.*, **48**, 1559.
15. Andersen, D.R. (1986) 'Nonlinear optical properties of II-VI semiconductor compounds grown by molecular beam epitaxy', Ph.D. Thesis, Purdue University.
16. Peyghambarian, N., Park, S.H., Koch, S.W. *et al.* (1988) *Appl. Phys. Lett.*, **52**, 182.
17. Gunshor, R.L. and Kolodziejski, L.A. (1988) *IEEE J. Quantum Electronics*, **24**, 1744.
18. Bylsma, R.B., Becker, W.M., Bonsett, T.C. *et al.* (1985) *Appl. Phys. Lett.*, **47**, 1039.
19. Catalano, I.M., Cingolani, A., Ferrara, M. and Lugara, M. (1982) *Solid State Commun.*, **43**, 371.
20. Weiner, J.S., Chemla, D.S., Miller, D.A.B. *et al.* (1985) *Appl. Phys. Lett.*, **47**, 664.
21. Bonsett, T.C., Yamanishi, M., Gunshor, R.L. *et al.* (1987) *Appl. Phys. Lett.*, **51**, 499.
22. Bylsma, R.B. (1986) 'Photoluminescence and stimulated emission of the diluted magnetic semiconductor $Zn_{1-x}Mn_xSe$, Ph.D Thesis, Purdue University.
23. Pajaczkowska, A. (1978) *Prog. Crystal Growth Charact.*, **1**, 289–326.
24. Kolodziejski, L.A., Gunshor, R.L., Nurmikko, A.V. and Otsuka, N. (1987) *Thin Film Growth Techniques for Low Dimensional Structures*, Series B, Physics, Vol. **163**, (Eds. R.F.C. Farrow, S.S.P. Parkin, P.J. Dobson, J.H. Neave, and A.S. Arrott), Plenum Press, New York, pp. 247–260.
25. Kolodziejski, L.A., Gunshor, R.L., Otsuka, N. *et al.* (1987) *J. Crystal Growth*, **81**, 491.
26. Gunshor, R.L., Kolodziejski, L.A., Otsuka, N. *et al.* (1987) *Superlattices and Microstructures*, **3**, 5.
27. Lee, D., Mysyrowicz, A., Nurmikko, A.V. and Fitzpatrick, B.J. (1987) *Phys. Rev. Lett.*, **58**, 1475.
28. Lee, D., Chang, S.-K., Nakata, H. *et al.* (1987) *Mat. Res. Soc. Symp. Proc.*, **77**, 253.
29. Chang, S.-K., Lee, D., Nakata, H. *et al.* (1987) *J. Appl. Phys.*, **62**, 4835.
30. Nurmikko, A.V., Lee, D., Hefetz, Y. *et al.* (1986) *Proceedings of the 18th International Conference on the Physics of Semiconductors*, Stockholm, p. 775, Proc. Ed. O. Engstrom, World Scientific Publishing Co. Singapore.
31. Fujiyasu, H., Mochizuki, K., Yamazaki, Y. *et al.* (1986) *Surface Science*, **174**, 543.
32. Takeda, T., Kurosu, T., Lida, M. and Yao, T. (1986) *Surface Science*, **174**, 548.
33. Kobayashi, M., Mino, N., Konagai, M. and Takahashi, K. (1986) *Surface Science*, **174**, 550.
34. Kolodziejski, L.A., Gunshor, R.L., Fu, Q. *et al.* (1988) *Appl. Phys. Lett.*, **52**, 1080.

35. Kobayashi, M., Mino, N., Katagiri, H. *et al.* (1986) *Appl. Phys. Lett.*, **60**, 773.
36. Lee, D., Mysyrowics, A., Nurmikko, A.V. and Fitzpatrick, B.J. (1987) *Phys. Rev. Lett.*, **58**, 1475.
37. Yao, T., Makita, Y. and Maekawa, S. (1978) *J. Crystal Growth*, **45**, 309.
38. Fu, Q., Lee, D., Nurmikko, A.V. *et al.* (1989) *Phys. Rev.*, **B39**, 3173.
39. Lee, D., Fu, Q., Nurmikko, A.V. *et al.* (1989) *Superlattices and Microstructures*, **5**, 345.
40. Trzeciakowski, W., Hawrylak, P., Aers, G. and Nurmikko, A.V. (1989) *Solid State Comm.*, **71**, 653.
41. Ding, J., Fu, Q., Pelekanos, N. *et al.* (1990) *Proc. of 20th Int. Conf. on Physics of Semiconductors*, Thessaloniki (Greece); World Scientific Publishing Co. Pte. Ltd. Singapore (Eds E.M. Anastassakis and J.D. Joannopoulos), p. 1198.
42. Cammack, D.A., Dalby, R.J., Cornelissen, H.J. and Khurgin, J. (1987) *J. Appl. Phys.*, **62**, 3071.
43. Cornelissen, H.J., Cammack, D.A. and Dalby, R.J. (1988) *J. Vac. Sci. Technol.*, **B6**, 769.
44. Yokogawa, T., Ogura, M. and Kajiwara, T. (1988) *Appl. Phys. Lett.*, **52**, 120.
45. Jeon, H., Ding, J., Nurmikko, A.V. *et al.* (1990) *Appl. Phys. Lett.*, **57**, 2413.
46. Qadri, S.B., Skelton, E.F., Webb, A.W. and Kennedy, J. (1985) *Appl. Phys. Lett.*, **46**, 257.
47. Dean, B.E. and Johnson, C.J. (1987) *Proc. of the Third Intern. Conf. on II–VI Compounds*, Monterey, July 12–17.
48. Nishitani, K., Ohkata, R. and Murotani, T. (1983) *J. Elect. Mater.*, **12**, 619.
49. Bicknell, R.N., Yanka, R.W., Giles, N.C., *et al.* (1984) *Appl. Phys. Lett.*, **44**, 313.
50. Faurie, J.P., Sivananthan, S., Boukerche, M. and Reno, J. (1984) *Appl. Phys. Lett.*, **45**, 1307.
51. Otsuka, N., Kolodziejski, L.A., Gunshor, R.L. *et al.* (1985) *Appl. Phys. Lett.*, **46**, 860.
52. Otsuka, N., Kolodziejski, L.A., Gunshor, R.L. *et al.* (1985) *Mater. Res. Soc. Symp. Proc.*, **37**, 449.
53. Faurie, J.P., Shu, C., Sivanathan, S. and Chu, X. (1986) *Surface Science*, **168**, 473.
54. Kolodziejski, L.A., Gunshor, R.L., Otsuka, N. *et al.* (1985) *Appl. Phys. Lett.*, **47**, 882.
55. Kolodziejski, L.A., Gunshor, R.L., Otsuka, N. and Choi, C. (1986) *J. Vac. Sci. Technol.*, **A4**, 2150.
56. Ponce, F.A., Anderson, G.B. and Ballingall, J.M. (1986) *Surface Science*, **168**, 564–570.
57. Feldman, R.D. and Austin, R.F. (1986) *Appl. Phys. Lett.*, **49**, 954.
58. Srinivasa, R., Panish, M.B. and Temkin, H. (1987) *Appl. Phys. Lett.*, **50**, 1441.
59. Kolodziejski, L.A., Bonsett, T.C., Gunshor, R.L. *et al.* (1984) *Appl. Phys. Lett.*, **45**, 440.
60. Kolodziejski, L.A., Gunshor, R.L., Datta, S. *et al.* (1985). *J. Vac. Sci. Technol.*, **B3**, 714.
61. Kolodziejski, L.A., Sakamoto, T., Gunshor, R.L. and Datta, S. (1984) *Appl. Phys. Lett.*, **44**, 799.
62. Herman, M.A., Jylha, O. and Pessa, M. (1984) *J. Cryst. Growth*, **66**, 480.
63. Bicknell, R.N., Yanka, R.W., Giles-Taylor, N.C. *et al.* (1984) *Appl. Phys. Lett.*, **45**, 92.
64. Zhang, X.-C., Chang, S.-K., Nurmikko, A.V. *et al.* (1985) *Phys. Rev.*, **B31**, 4056.
65. Harwit, A., Ritter, M.B., Hong, J.M. *et al.* (1989) *Appl. Phys. Lett.*, **55**, 1783.
66. Bicknell, R.N., Giles, N.C. and Schetzina, J.F. (1987) *Appl. Phys. Lett.*, **50**, 691.
67. Ashenford, D.E., Lunn, B., Davis, J.J. *et al.* (1989) *J. Crystal Growth*, **95**, 557.

68. Zhang, X.-C., Chang, S.-K., Nurmikko, A.V. *et al.* (1985) *Solid State Commun.*, **56**, 255.
69. Nurmikko, A.V., Zhang, X.-C., Chang, S.-K. *et al.* (1985) *J. Luminescence*, **34**, 89.
70. Petrou, A., Warnock, J., Bicknell, R.N. *et al.* (1985) *Appl. Phys. Lett.*, **46**, 692.
71. Chang, S.-K., Nurmikko, A.V., Kolodziejski, L.A. and Gunshor, R.L. (1986) *Phys. Rev.*, **B33**, 2589.
72. Warnock, J., Petrou, A., Bicknell, R.N. *et al.* (1985) *Phys. Rev.*, **B32**, 8116.
73. Venugopalan, S., Kolodziejski, L.A., Gunshor, R.L. and Ramdas, A.K. (1984) *Appl. Phys. Lett.*, **45**, 974.
74. Suh, E.-K., Bartholomew, D.U., Ramdas, A.K. *et al.*, (1987) *Phys. Rev.*, **B36**, 4316.
75. Bicknell, R.N., Giles-Taylor, N.C., Schetzina, J.F. *et al.* (1985) *Appl. Phys. Lett.*, **46**, 238.
76. Bicknell, R.N., Giles-Taylor, N.C., Blanks, D.K. *et al.* (1985) *Appl. Phys. Lett.*, **46**, 1122.
77. Williams, G.M., Cullis, A.G., Whitehouse, C.R. *et al.* (1989) *Appl. Phys. Lett.*, **55**, 1303.
78. Holonyak, N. and Scifres, D.R. (1971) *Rev. Sci. Instrum.*, **12**, 1885.
79. Isaacs, E.D., Heiman, D., Zayhowski, J.J. *et al.* (1986) *Appl. Phys. Lett.*, **48**, 275.
80. Bicknell, R.N., Giles, N.C. and Schetzina, J.F. (1986) *Appl. Phys. Lett.*, **49**, 1095.
81. Bicknell, R.N., Giles, N.C. and Schetzina, J.F. (1986) *Appl. Phys. Lett.*, **49**, 1735.
82. Dreifus, D.L., Kolbas, R.M., Harris, K.A. *et al.* (1987) *Appl. Phys. Lett.*, **51**, 931.
83. Benson, J.D. and Summers, C.G. (1988) *J. Crystal Growth*, **86**, 354.
84. Bicknell-Tassius, R.N., Waag, A., Wu, Y.S. *et al.* (1990) *J. Crystal Growth*, **101**, 33.
85. Farrell, H.H., Nahory, R.E. and Harbison, J.P. (1988) *J. Vac. Sci. Technol.*, **B6**, 779.
86. Benson, J.D., Rajavel, D., Wanger, B.K. *et al.* (1989) *J. Crystal Growth*, **95**, 543.
87. Harper, Jr., R.L., Hwang, S., Giles, N.C. *et al.* (1988) *J. Vac. Sci. Technol.*, **A6**, 2627.
88. Harper, Jr., R.L., Hwang, S., Giles, N.C. *et al.* (1989) *Appl. Phys. Lett.*, **54**, 170.
89. Ohnishi, M., Saito, H., Okano, H., Ohmori, K. (1989) *J. Crystal Growth*, **95**, 538.
90. Gunshor, R.L., Kolodziejski, L.A., Nurmikko, A.V. and Otsuka, N. (1990) Molecular-Beam Epitaxy of II—VI Semiconductor Microstructures. in: Semiconductors and Semimetals (T. Pearsall, ed.), Vol. 33, pp. 337–409, Academic Press, San Diego.
91. Monfroy, G., Sivananthan, S., Chu, X. *et al.* (1986) *Appl. Phys. Lett.*, **49**, 152.
92. Miles, R.H., Wu, G.Y., Johnson, M.B. *et al.* (1986) *Appl. Phys. Lett.*, **48**, 1383.
93. Menendez, J., Pinczuk, A., Valladares, J.P. *et al.* (1987) *Appl. Phys. Appl. Phys. Lett.*, **50**, 1101.
94. Hefetz, Y., Lee, D., Nurmikko, A.V. *et al.* (1986) *Phys. Rev.* **B34**, 4423.
95. Glass, A.M., Tai, K., Bylsma, R.B. *et al.* (1988) *Appl. Phys. Lett.*, **53**, 834.
96. Lee, D., Zucker, J.E., Johnson, A.M. *et al.* (1990) *Appl. Phys. Lett.*, **57**, 1132.
97. Tersoff, J. (1986) *Phys. Rev. Lett.*, **56**, 2755.
98. Mackey, K.J.G., Allen, P.M., Herrenden-Harker, W.G. *et al.* (1986) *Appl. Phys. Lett.*, **49**, 354.
99. Noreika, A.J., Greggi, Jr., J. Takei, W.J. and Francombe, M.H. (1983) *J. Vac. Sci. Technol.*, **A1**, 558.
100. Feng, Z.C., Mascarenhas, A., Choyke, W.J. *et al.* (1985) *Appl. Phys. Lett.*, **47**, 24.
101. Mackey, K.J., Zahn, D.R.T., Allen, P.M.G. *et al.* (1987) *J. Vac. Sci. Technol.*, **B5**, 1233.
102. Golding, T.D., Martinka, M. and Dinan, J.H. (1988) *J. Appl. Phys.*, **64**, 1873.
103. Golding, T.D., Greene, S.K., Pepper, M. *et al.* (1990) *Semicond. Sci. Technol.*, **5**, S311.

104. Glenn, Jr., J.L., O., Sungki, Kolodziejski, L.A. *et al.* (1989) *J. Vac. Sci. Technol.*, **B7**, 249.
105. Neave, J.H., Blood, P. and Joyce, B.A. (1980)*Appl. Phys. Lett.*, **36**, 311.
106. Missous M. and Singer, K.E. (1987) *Appl. Phys. Lett.*, **50**, 694.
107. Noreika, A.J., Francombe, M.H. and Wood, C.E.C. (1981) *J. Appl. Phys.*, **52**, 7416.
108. Pehek, J. and Levinstein, H. (1965) *Phys. Rev.*, **140**, A576.
109. Woolley, J.C. and Smith, B.A. (1958) *Phys. Soc. of London*, **72**, 862.
110. Cullis, A.G., Chew, N.G. and Hutchison, J.L. (1985) *Ultramicroscopy*, **17**, 203.
111. Williams, G.M., Whitehouse, C.R., Cullis, A.G. *et al.* (1988) *Appl. Phys. Lett.*, **53**, 1847. (In this publication the roughness of reported interfaces was ~ 100 Å.)
112. Williams, G.M., Whitehouse, C.R., Chew, N.G. *et al.* (1985) *J. Vac. Sci. Technol.*, **B3**, 704.
113. Durbin, S.M., Han, J., O., Sungki *et al.* (1989) *Appl. Phys. Lett.*, **55**, 2087.
114. Durbin, S.M., Kobayashi, M., Fu, Q. *et al.* (1990) *Surface Science*, **228**, 33.
115. Allen, J.W., Lucovsky, G. and Mikkelsen, J.C. Jr. (1977) *Solid State Commun.*, **24**, 367.
116. Lee, Y. and Ramdas, A.K. (1988) *Phys. Rev.*, **B38**, 10600.
117. Benson, J.D., Wagner, B.K., Torabi, A. *et al.* (1986) *Appl. Phys. Lett.*, **49**, 1034.
118. Gunshor, R.L., Kobayashi, M., Kolodziejski, L.A. *et al.* (1990) *J. Crystal Growth*, **99**, 390.
119. Pelekanos, N., Fu, Q., Ding, J. *et al.* (1990) *Physical Review*, **B41**, 9966.
120. Wei, S.-H. and Zunger, A. (1988) *Phys. Rev.*, **B37**, 8958.
121. McLennan, M.J. and Datta, S. *SEQUAL 2.1 User's manual, Purdue University*, (TR-EE 89-17, 1989).
122. Yu, E.T. and McGill, T.C. (1988) *Appl. Phys. Lett.*, **53**, 60.
123. van Welzenis, R.G. and Ridley, B.K. (1984) *Solid State Electronics*, **27**, 113.
124. de Miguels, J.L., Shilbi, S.M., Tamargo, M.C. *et al.* (1988) *Appl. Phys. Lett.*, **53**, 2065.
125. Venkatesan, S., Pierret, R.F., Qiu, J. *et al.* (1989) *J. Appl. Phys.*, **66**, 3656.
126. Woodall, J.M., Hodgson, R.T. and Gunshor, R.L. (1990) *Appl. Phys. Lett.*, **58**, 379.

Part Two: Materials Characterization

7

Optical properties of the wide bandgap II–VI semiconductors

H. Mar

7.1 INTRODUCTION

The various effects and phenomena observed during the interaction of light with matter in an optical experiment are unique to the particular material system under investigation. The manner in which a material system responds to the impingement of light reveals details of the optical, electronic and microstructural characteristics of the material. Light in the form of photons incident on a semiconductor material interacts with the individual electrons that constitute the bound states of the host atoms. In the process, an electron may be excited to a high energy state in the conduction band from which it rapidly thermalizes to the lowest unfilled states within the band. In the process, a net positive charge (hole) is left in the valence band. In addition to exciting electrons across the bandgap, photons interact with a variety of quasi-particles such as electron–hole pairs (excitons), polaritions and plasmons. It is also well known that light interacts with the host atoms through the phonon–photon interaction process in a wide variety of scattering phenomena such as Raman and Brillouin scattering events in the host material.

In this chapter our discussions will be directed to the optical properties of the II–VI semiconductors with particular emphasis on the interaction of visible and near-IR radiation with these materials. The response of the semiconductor to excitation by light of various wavelengths, or photons of various energies, is unique to the material. The interaction process reveals subtle variations in the physical characteristics of these materials. This is the basis for the widespread use of optical techniques for precise characterization of the quality of the materials so important today in the evaluations of films and structures grown by a variety of modern single-crystal deposition techniques. Optical techniques have proven to be very sensitive tools for studying growth processes and structural defects in these materials. In addition to the use of these techniques for material growth characterization, the basic understanding of the underlying physical and optical properties of materials and material structures sheds light on new phenomena with tremendous potential for device applications. Thus, the interaction of light with superlattice structures requires extensive investigation and many new effects are just now being observed in these structures.

Other areas of application of the wide bandgap II–VI materials are in the incorporation of these materials with II–V heterostructures making use of the large bandgaps and refractive indices of these materials. Photoluminescence and other optical investigative techniques such as Raman scattering are proving to be powerful tools in the task of characterizing these structures and understanding the processes when the materials are used in opto-electronic devices.

The purpose of this chapter is to introduce the reader to the extensive investigative literature on the structural characterization and closely related optical and electrical properties of the wide bandgap materials grown by the leading-edge single-crystal deposition techniques presented by the other contributors. The discussions will be centred principally on ZnSe, ZnTe, CdS and CdSe. These materials have been investigated extensively because of their relative ease of growth. However, ZnSe is of singular importance because of the technological importance of this material. Other wide bandgap materials which have been as intensely studied in recent years are CdTe and ZnS. Of these CdTe has attracted the most interest because it is readily alloyed with HgTe to form the ternary $Cd_{0.2}Hg_{0.8}Te$ which is a very important material for 8–12 μm remote sensing applications. ZnS is also of significant interest because it readily alloys with ZnSe to form a ternary suitable for the blue–green device applications.

There have been a number of extensive early reviews of the chalcogenides of Cd and to a lesser extent of Zn. An extensive review by Zanio [1] of the early work on CdTe was published in 1978 and includes extensive details of the major advances that have occurred in that material. Most of the work presented there still represents the current understanding of CdTe, but the results, concepts and experimental techniques are applicable to a large extent to the wide bandgap II–VI semiconductors. That work is an excellent reference for new researchers in the field of the II–VI semiconductors. Other significant references are the optical studies on CdS and ZnS by Halsted [2] and by Curie and Prener [3] in the *Physics and Chemistry of II–VI Compounds*, and they are still of major relevance today.

It is worth noting at this point that, whereas ZnSe, CdS, CdSe and ZnS are readily made with n-type conductivity and are extremely difficult to make with p-type conductivity, the reverse is the case with ZnTe. The reasons for this may prove to be very revealing of the nature of the doping problems in the wide bandgap II–VI semiconductors. We shall present the discussions that appear in the literature on this topic. The research carried out on these binaries will be discussed to present a consistent picture of current research in this field.

In the past ten years there has been tremendous insight derived from research in the Zn chalcogenides, in particular from work done on ZnTe and ZnSe from photoluminescence and electrical studies [4–11] and research conducted using magneto-optical techniques [10]. In addition, new growth techniques such as vapour phase epitaxy (VPE), organometallic chemical vapour deposition (OMCVD) and molecular beam epitaxy (MBE) developed within the past 15 years have provided the additional impetus to a greater understanding and significant developments in these materials. Witness particularly the major advances in CdTe [12–14] over the past 15 years using these techniques.

7.2 THE WIDE BANDGAP II–VI SEMICONDUCTORS: CdSe, CdS, ZnTe AND ZnSe

The II–VI materials have always been an attractive family from the very early days of semiconductor materials research. Indeed, they are at present of substantial commercial importance albeit in the polycrystalline form as phosphors, as well as in electroluminescence and thin film electronic devices. Today, however, interest in the wide bandgap semiconductors has been extended to the single-crystalline form for solid-state opto-electronic device applications. Early contributions to the understanding of the wide bandgap II–VI materials was derived from the work of Henry and colleagues [15–17]. These materials are direct bandgap semiconductors with T_d lattice symmetry. During optical excitation bound valence electrons are excited into the conduction band with only a small change in the momentum of the electron; that is without the need for phonon involvement in an inelastic scattering process. Of more importance for opto-electronic device application is the recombination rate which is high since phonon involvement is not required in the recombination process.

The wide bandgap II–VI materials CdSe, CdS as well as ZnS were investigated extensively [15–19] especially prior to 1968. The results of optical studies on these materials will be examined. However, the main thrust of the chapter will be on the growth and doping studies of ZnSe. The technological importance of ZnSe as well as the limited space available here dictate that our discussions will be centred on this material. One section will also be devoted to a brief discussion of the p-type conductivity in ZnTe. ZnTe seems to exhibit anomalous doping characteristics in that, whereas p-type conductivity is easy to obtain in this material, the reverse is true for the other wide bandgap semiconductors ZnSe, ZnS, CdS and CdSe. This anomalous behaviour will be discussed in some detail in section 7.3.3.

7.2.1 Photoluminescence, adsorption and optical scattering

By about 1968 good quality single crystals of the wide bandgap II–VI semiconductors CdSe, CdS and ZnSe could be grown in the laboratory.

Indeed, the materials were of sufficiently high quality that optical techniques could be employed to study details of the electronic radiative transitions in these materials. The early work of Henry *et al.* [17] was centred on the defect characterization and p-type conductivity of CdS and CdSe, and resulted in a fundamental understanding of the physical nature of the defects and the thermodynamic principles that controlled the concentrations in these materials. Also, the various complexes formed among the native defects as well as the extrinsic defects (impurity atoms) were well understood.

The principal techniques employed in these early studies were photoluminescence and optical adsorption. A discussion of the results of the earlier work and the influence on current studies on ZnSe, ZnTe and CdTe will be presented later. Suffice it to say at this point that photoluminescence and the related techniques became powerful tools for characterization of the optical and electronic nature of the defects in these materials and, by extension, determination of layer quality. Optical absorption experiments were used to provide corroboratory data to the results obtained by photoluminescence. Nonetheless, the technique also provided useful information on the properties of free carriers including scattering cross-sections and carrier concentrations, parameters that are vital for a complete understanding of the material properties.

In photoluminescence, samples, cooled to 4.2 K by immersion in a liquid helium bath or suspended in the He vapour above the bath, are excited with light of energy greater than the bandgap of the semiconductor with the subsequent creation of excited electron–hole (e–h) pairs. The e–h pairs generated by the optical excitation thermalize and in the process may become trapped as a single entity at donor or acceptor sites [2, 5, 15–17], or dissociate to form separate electrons and holes. The e–h pair may be thought of conceptually as being a single entity with unique characteristics and is referred to as the exciton [20]. Unless dissociated during a scattering event the e–h pair moves about freely within the host lattice until recombination occurs with the annihilation of the exciton and the subsequent emission of a photon with an energy equal to the e–h binding energy. It should be mentioned that excitons may be coupled with photons within the crystal to form mixed modes called polaritons. These exciton–polaritons have particle- as well as light-like characteristics and have been reportedly observed in the II–VI materials. For the interested reader refs 21–23 are key works on the topic. Alternatively, the exciton may be localized at neutral or ionized crystal defects which may be native point or extended defects, or substitutional or interstitial impurity atoms. The mechanism of exciton localization at defects is discussed in detail by Dean and Herbert [24], Hopfield [25] and Halsted [2]. For the particular case of ZnSe, the details of the recombination process are reflected in the energy and shape of the radiation emission peaks around 440 nm. At liquid helium temperatures and under low excitation levels (typically less than 50 $mW\,cm^{-2}$) thermal broadening effects on the emission peaks are very small, and the emission spectrum reflects in great detail the structural imperfections present within the crystalline semiconductor.

Emission bands with substantially lower intensities also appear around 500 nm in the low-temperature luminescence spectrum of ZnSe, and these result from the recombination of electron and hole pairs captured at acceptor and donor sites respectively. Recombination of this type reflects the energy difference between the donor and acceptor levels involved in the recombination process involving distant donor–acceptor (D–A) pairs. However, for D–A pairs separated by interatomic distances fine structure is observed in the spectrum. For nearest neighbours, nearest neighbours once removed, etc., the separation between donor and acceptor atoms occurs only at well-defined discrete distances on 'shells'. The energy of the emission peak is given approximately by

$$h\nu(\text{D–A}) = E_g - (E_A + E_D) + e^2/cr$$

where r takes on discrete values and c is the dielectric constant. From this equation it is evident that for small r, that is small separation between the donor and acceptor atoms, the energy of the peak is substantially different from that of the distant pairs [7, 26–28]. This gives rise to the fine structure observed on the high-energy skirt of the relatively broad no-phonon D–A pair emission peak of the distant pairs. For the zinc blende lattice it has been shown [12–14] that certain discrete peaks may be absent in the fine structure, and this depends on whether both the donor and the acceptor impurity atoms substitute on the same sublattice or on different sublattices. D–A pair fine structure emission when combined with selective excitation [7, 27, 28] has proven to be a powerful technique for doping studies in the wide bandgap II–VI semiconductor materials. The broad emission bands centred around 610 nm are a consequence of the recombination of e–h pairs at deep acceptor levels sites.

Early work by Thomas and Hopfield [18] on CdS developed and applied the concepts of the acceptor and donor bound excitonic states first introduced by Lampert [21] for semiconductors. In a comprehensive review of the recombination processes in the near-band-edge region Halsted [2] noted that the bound states play a significant role in near-band-edge transitions. This was established through experiments using the Zeeman effect [10]. Subsequent photoluminescence studies by Henry *et al.* [16] of shallow acceptors in CdS and CdSe, in particular Li and Na, helped to establish photoluminescence as a powerful tool for studying impurity- and defect-related acceptors not only in these materials but also in the family of II–VI semiconductors as well as that of the III–V semiconductors. In an earlier study Thomas and Hopfield [18] showed that shallow acceptors in CdS give rise to the I_1 lines which are excitons bound to neutral acceptors, and the I_2 and I_3 which are associated with excitons bound to neutral and ionized donors respectively.

Another powerful tool which has been used in the characterization of the materials is optical scattering; in particular Raman scattering by phonons [30–32]. Recently, a study was undertaken using Brillouin scattering [33] to determine the elastic constants of ZnSe. Other less common optical investigative tools used in the studies include scattering by surface plasmons, and magneto-optical measurements [10, 34].

In what follows, we shall elaborate on and discuss the results of photoluminescence and related techniques and absorption experiments made on ZnSe, ZnTe and, to a lesser extent, CdTe. As argued earlier these are currently of greater technological importance, a fact reflected in the number of publications devoted to investigations of these materials. In section 7.4, we shall discuss more recent research on CdSe, CdS and ZnTe. However, the emphasis is on the characterization of ZnSe by means of luminescence techniques, particularly photoluminescence, as a powerful tool to investigate the crystal quality and concentration of unwanted impurities in a material grown by various growth techniques such as discussed elsewhere in this monograph.

7.2.2 ZnSe

The interest shown in recent years in ZnSe derives from the potential of this material for use in the fabrication of blue-light-emitting devices. This semiconductor has a direct bandgap of about 2.7 eV at room temperature. The lattice constant of this material (5.6676 Å) is relatively closely matched with GaAs (5.6531 Å) and so it has been identified as having potential for use in the monolithic integration of blue opto-electronic devices with III–V electronic and opto-electronic technology.

Early research [15, 35–37] in the growth of ZnSe by bulk techniques suggested a fundamental difficulty in obtaining low defect concentration in ZnSe. Henry *et al.* [15, 16] have suggested that ZnSe grown under near-thermal-equilibrium conditions by bulk techniques would always contain high concentrations of native point defects. These point defects give rise to donor levels in very high concentrations and make it virtually impossible to obtain p-type conductivity in this material. However, by 1970 research on CdS as well as on CdS and ZnSe [15–17] leads to the suggestion that the problem was in fact, not one of compensation by intrinsic donors but rather a problem of extrinsic donor impurities being incorporated into the material during growth. Indeed, it was suggested that donor impurities such as Ga, In, Al and Cl are readily incorporated into these wide bandgap semiconductors and are found in relatively large concentrations in the source materials and the growth equipment.

Early work by researchers in the field of the II–VI materials revealed an apparent relationship between the ease of growing semiconducting material with both n- and p-type conductivity and the degree of ionicity in the covalent bonds of the host atoms. There appears to be a trend of increasing difficulty in producing p-type conductivity material (or ease of growing n-type conductivity material) with an increase in the ionicity of the bonds. Thus, one finds that for CdTe it is relatively easy to grow material of either conductivity type [1]. On the other hand, early researchers investigating the growth, and optical and electrical characteristics of ZnSe [15–19] were not able to obtain ZnSe with p-type conductivity although it was readily available with low-resistivity n-type conductivity. Similar results were obtained for the II–VI compounds CdS, CdSe. Because of this but, perhaps more importantly, because of the potential application of CdTe and HgTe in IR remote sensing technology, the early efforts in the II–VI materials were directed to the growth of the CdTe and HgTe and alloys of these two materials.

The ease of obtaining both conductivity types in the narrower bandgap II–VI binaries and their ternaries combined with the technological importance of (Cd, Hg)Te for IR applications in the 8–12 μm range of wavelengths resulted in tremendous advances in II–VI research in this field. Optical techniques such as photoluminescence and absorption spectroscopy were combined with electrical and chemical techniques to provide tremendous insight into the nature and characteristics of the defects that determined the opto-electronic properties of the narrow bandgap II–VI materials. The optical studies revealed that, in addition to native point defects such as vacancies and interstitials, there is almost a plethora of complex structures involving two or more point defects as well as extrinsic impurities present in the crystal [2, 35–37]. A good understanding of the thermodynamics of native defects and complexes involving these defects was developed. By proper heat treatment of the crystals in excess pressures of the more volatile constituent element of the binary, it was possible to predict and control the concentrations of the native defects and concomitant donor–acceptor levels in the bandgap. A good example is CdTe which could be made n-type by annealing in Cd vapour at 1 atm pressure and a temperature of about 600 °C. However, this was not possible for most of the other II–VI materials, particularly the wide bandgap semiconductors. As will be discussed later, ZnTe seemed to be a notable exception.

Unintentionally doped ZnSe could not be made low-resistivity p-type by heat treatment although the degree of n-type conductivity could be controlled by this process [16]. Furthermore, as appeared to be the case for CdS and CdSe [5, 15–17], ZnSe intentionally doped with Li, Na or N was found to be of high-resistivity n-type or fully compensated. Mandel *et al.* [6] and Kroger *et al.* [6] argued that compensation was due to the native donors controlled by the thermodynamics of the material. Furthermore, there was speculation [6] that the solubility limits of acceptors in these materials were less than the background donor concentrations, that acceptors located on different sites may act as shallow donors and that acceptors may form electrically inactive complexes with native defects.

In 1971 Henry *et al.* [15] carried out interesting optical investigations of shallow acceptors in CdS and CdSe grown by vapour phase sublimation in a rapid argon stream. High purity CdSe and CdS were the source materials and alkali-metal dopants were introduced in the form of sulphates or carbonates. Photoluminescence measurements taken at liquid helium temperatures of the Li- and Na-doped CdS and CdSe revealed very narrow peaks associated with excitons bound to shallow Li and Na acceptor levels. They suggest that the shallow acceptor that forms during doping is a complex and not a simple substitutional acceptor. They conclude that the evidence from photoluminescence measurements supports the hypothesis that a native double donor exists in CdS but that Li and Na are compensated by ordinary donor impurities and not by native double donors. They cited the work done by Smith [2] on temperature electrical measurements which also suggests that compensation by isolated native donors is unimportant in CdSe and ZnSe.

Lithium doping studies [15–17, 38, 39] suggest that at high Li concentrations the principal means of compensation is by Li interstitials. Thus, it appeared highly unlikely that Zn and Cd sulphide or selenide could be made low-resistivity p-type by doping with these two elements.

7.3 HIGH QUALITY ZnSe AND ZnTe

One of the areas of considerable interest to researchers in the field of wide bandgap II–VI semiconductors is the problem of obtaining both n-type and p-type conductivity in these materials because of the potential for fabricating p–n junction devices capable of emitting in the blue–green region of the visible spectrum. ZnSe and ZnTe, both direct-gap semiconductors, are prime candidates. Furthermore, these two compounds are miscible over a wide range of compositions, giving rise to the possibility of obtaining a variety of wavelengths for various applications. For this reason the emphasis will be on the results of optical studies of these two materials. Furthermore, the discussions on optical studies of the more recent results will be confined to studies on heteroepitaxial layers grown by VPE, MOCVD and MBE. Nonetheless, studies of other II–VI semiconductors will be discussed as the need arises.

Unlike the III–V compounds and, even more so, the elemental semiconductors, the consequences of the presence of defects in the II–VI semiconductors are more pronounced because of the ionic nature of the bonding between the group II and group VI host atoms. This property and the materials characteristics that result from this have had far-reaching consequences on the advances in fundamental understanding of the properties of the II–VI semiconductors. In the following section the discussion will be centred on the results of photoluminescence studies of the wide bandgap II–VI semiconductors with particular emphasis on ZnSe.

7.3.1 Photoluminescence of ZnSe: high purity, high crystalline quality

Heteroepitaxial ZnSe layers deposited on a variety of substrates using VPE, OMCVD and MBE have been characterized by a number of techniques including optical [8–10, 40, 41], X-ray diffraction [42, 43] and electron diffraction [43]. In heteroepitaxy the principal characteristics one is typically concerned with are high purity and high crystalline quality. In the context of crystalline quality the concerns relate to low intrinsic point defect concentrations and low density of dislocations which arise from strain due to lattice mismatch at the heterointerface. Of equal importance for studies in the wide bandgap II–VI semiconductors is the identification of the defects or impurities responsible for donor and acceptor levels in these materials. Ultimately, one is interested in using these materials in devices and so the desire is to develop an understanding of the processes involved in the incorporation of impurity atoms, and the characterization of the energy levels associated with these impurities. Optical techniques have proven to be very powerful tools in the characterization of these layers. In particular, high-resolution photoluminescence at liquid helium temperatures has been developed into a powerful tool for elucidating some of the long-standing questions of the shallow donors and acceptors associated with these materials. Magneto-optic [10] and electron resonance measurements have also been used to corroborate the findings of the photoluminescence studies.

Figure 7.1 shows the photoluminescence spectrum taken at 4.2 K of a 1 μm thick layer of ZnSe on GaAs(1 0 0) [44] and is typical of high-quality layers grown by MBE at 330 °C. For this layer, growth was from 6N purity Zn and Se independent sources. Although the spectrum is specific to ZnSe the characteristics are, in general, applicable to the direct wide bandgap II–VI semiconductors and, indeed, to the direct bandgap III–V semiconductors as well. The measured emission peaks are associated with the radiative recombination processes that occur through trapping levels within the bandgap of the semiconductor. These processes become prominent under appropriate levels of excitation and at the very low temperatures. At liquid helium temperatures typical of photoluminescence studies discussed here, electron–acoustic phonon scattering events are sufficiently rare that the recombination process results in very narrow emission lines.

For ZnSe of high crystalline quality and high purity, the visible spectrum from about 440 nm to 900 nm is normally dominated by the so-called near-edge emission. The extremely narrow near-edge peaks are associated with the annihilation of free excitons [41], or excitons that are localized at shallow donor or acceptor traps [23–25]. In Fig. 7.1 the emission line at 2.798 eV is associated with the annihilation of an exciton trapped at a neutral donor site; in this case the donor is believed to be Ga impurities [41] substituting at a Zn site in ZnSe. The full width at half-maximum is seen to be about 0.56 meV and represents material of high crystalline quality and

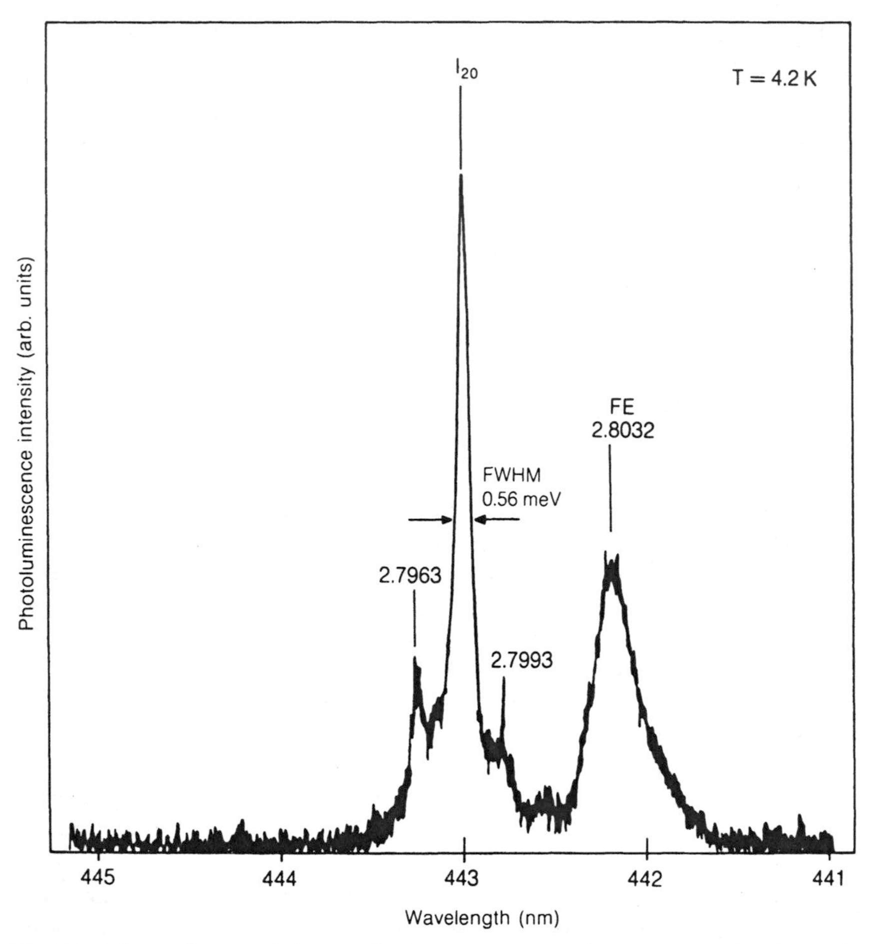

Fig. 7.1 The exciton emission peaks from 441 nm to about 445 nm dominate the visible spectrum, indicative of high crystalline quality ZnSe grown by MBE. The peak at 443 nm, possibly due to an exciton bound to a neutral Ga atom, has a narrow linewidth of 0.56 meV. The free-exciton (FE) peak at 2.8052 eV is clearly visible.

low impurity concentration. Werkhoven *et al.* [45] also reported similar results for homoepitaxial ZnSe grown by liquid phase epitaxy.

The low-temperature photoluminescence spectrum of the ideal very high purity material would show a dominant free-exciton emission peak. Because the recombination of localized excitons bound at donor or acceptor sites is a much more efficient process than free-exciton annihilation, emission peaks associated with bound excitons are readily observed even for impurity concentrations at or below the typical detection level of secondary ion mass spectrometry (SIMS) (about $10^{14}\,cm^{-3}$ depending on the impurity and the matrix).

In 1984 Yoneda *et al.* [46] reported the growth of undoped, high purity, low resistivity ZnSe layers grown on GaAs(1 0 0) by MBE. In low temperature photoluminescence only the free-exciton emission line was observed to be dominant, and all the bound exciton emission lines were undetectable, characteristic of very-high-purity ZnSe. These results were achieved using extremely high-purity Se source material refined by a sublimation process. The thickness of the typical layer was about 3 μm.

Park *et al.* [41] reported the growth of very-high-purity ZnSe on GaAs(100) by MBE. The source materials reportedly used in their experiments were of standard 6N purity but grown at very slow rates of about 0.1 μm h^{-1} and high Zn to Se effective beam pressure ratios. The measured heteroepitaxial layer thickness was typically 0.5 μm. The 4.2 K photoluminescence spectrum is shown in Fig. 7.2 and is seen to be dominated by the exciton–polariton emission peaks at 2.8008 eV and 2.803 eV. The deep-level emission peaks at about 550 nm and 610 nm were observed to be quite small. No peaks clearly identifiable with shallow native donors were detected. These results indicate that it is possible to grow by MBE layers of very high purity. Indeed, they seem to support the earlier contention that compensation by extrinsic donor impurities is a major barrier to obtaining p-conductivity ZnSe.

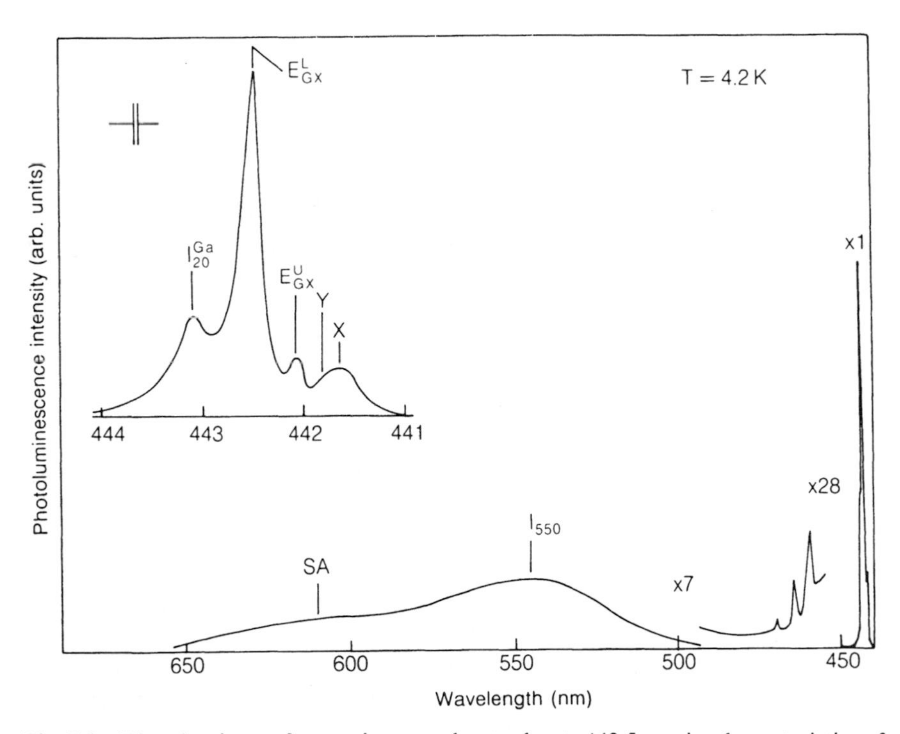

Fig. 7.2 The dominant free-exciton peak at about 442.5 nm is characteristic of high-quality, high-purity ZnSe grown on GaAs(100) by MBE.

Kolodziejski *et al.* [47] demonstrated the possibility of growing heteroepitaxial layers having both very high purity and high crystalline quality. They reported that the heteroepitaxial growth of ZnSe deposited on GaAs(1 0 0) buffer layers by MBE occurred via a two-dimensional growth mechanism. The low-temperature (8 K) photoluminescence spectrum of a 0.1 μm thick pseudomorphic ZnSe layer grown on the GaAs buffer was found to be dominated by emission peaks at 2.8064 eV and 2.8178 eV. They note that the peak observed at 2.7997 eV is normally associated with a neutral donor bound exciton, but suggest that the higher energy peaks are due to the annihilation of the free-exciton (2.8064 eV) and the first excited free-excitonic state (2.8178 eV), both shifted to higher energies because of the bi-axial compressive strain resulting from the pseudomorphic growth of mismatched lattices. Park *et al.* [41] and Yoneda *et al.* [46] observed the free-excitonic peaks at lower energies presumably because the thicker ZnSe layers were partially or fully relaxed with the formation of mismatch dislocations. Interestingly, Tamargo *et al.* [48] demonstrated that on an As-stabilized GaAs surface, as occurs on high quality GaAs buffers, growth of the ZnSe heteroepitaxial layer exhibited two-dimensional characteristics. On the other hand, on a Ga-stabilized surface, growth of the ZnSe was three dimensional.

7.3.2 Doping studies in ZnSe

The technological importance of ZnSe for potential application in blue solid state light-emitting devices (LEDs) and laser diodes (LDs) is reflected in the research effort into resolving the problem of p-type conductivity in the material. The problems appear to be substantially similar to those of the other wide bandgap II–VI semiconductors, namely ZnS, CdS and CdSe, and for this reason the research done on these materials could play an important role in achieving the goal of p-type conductivity in ZnSe. It is also interesting to note at this point that ZnTe exhibits anomalous doping behaviour vis-à-vis the other wide bandgap II–VI semiconductors. A discussion of the anomaly is presented later in section 7.3.3.

The early work of doping CdS, CdSe as well as ZnSe by Henry *et al.* [15–17] left little doubt that the most viable candidates for shallow acceptors in these materials were Li and Na substituting on the Zn sublattice and N substituting on the Se sublattice. The other group V elements, P, As and Sb, were all found [49] to result in deep acceptor levels when incorporated as substitutional impurities. Photoluminescence and Hall mobility studies of Li- and Na-doping revealed that these impurities were soluble in the materials to concentrations in excess of $10^{18}\,\text{cm}^{-3}$. Henry *et al.* [15–17] argued that nitrogen would be a shallow acceptor but, based on the large difference in the covalent radii of N and Se, it would be insoluble in ZnSe.

The barriers to obtaining p-type conductivity in these materials, and in

ZnSe in particular, are (a) the large ionization energies for most of the possible group I (on Zn sublattice) and group V (on Se sublattice) elements, (b) the apparent low solubility of these impurities in the materials, and (c) the problem of compensation by donors, both native and extrinsic impurities, in these semiconductors. With this background, researchers in the field entered into detailed studies to understand fully the mechanism of impurity incorporation and compensation in ZnSe. In this section we shall discuss the efforts to overcome these difficulties.

(*a*) *Li and Na doping*

In refs 15 and 16 on doped CdS and CdSe, the photoluminescence spectra taken at 4.2 K are characterized by the presence of two narrow I_1 peaks which are closely spaced. These narrow peaks correlate with Li and Na doping and confirmed the results of earlier work [17]. For lightly doped material the full width at half-maximum linewidths were typically 0.4 meV and this broadens with increasing acceptor concentration. Of singular importance for ZnSe was the conclusion from optical experiments [15, 16] that, although a native double donor exists in CdS, compensation is by ordinary donors and not the double-donor native complexes. The analysis of the results of Smith's high temperature Hall measurements [4, 52] confirms that compensation by isolated native donors is unimportant in CdTe, CdSe and ZnSe. These experiments also reveal that the principal means of compensation in Li- and Na-doped II–VI semiconductors is by Li and Na that occupy interstitial sites, giving rise to shallow donor levels in the semiconductor. For concentrations in excess of $10^{17}\,cm^{-3}$ interstitial concentrations are significant in causing self-compensation of the substitutional Li and/or Na impurities. They concluded that low-conductivity p-type sulphides and selenides of Zn and Cd are unlikely to be produced by doping with Li and Na, and contended that, since these are the only soluble substitutional impurities which act as shallow acceptors in these materials, the possibility of ever obtaining p-type conductivity material is very remote.

ZnSe grown by various new deposition techniques such as OMCVD and MBE as well as the more established techniques, VPE and liquid phase epitaxy (LPE), were doped with Li and Na acceptor impurities at low concentrations, and the layers characterized extensively by photoluminescence and, to a lesser extent, by Hall mobility measurements.

Merz *et al.* [28] investigated the excitonic emission peaks in ZnSe by making measurements of the low-temperature donor bound excitonic peaks for Ga, In and Al on the Zn sublattice, and Cl on the Se sublattice. Low levels of the impurities were deliberately introduced into the samples by means of back doping. The two-electron transition peaks characteristic of the specific donors were also investigated. D–A pair analysis of fine structure by Dean *et al.* [8–10], Neumark *et al.* [38, 39], and Merz *et al.* [5, 28] showed that in Li- and Na-doped ZnSe the donors involved in the D–A pair spectrum were on Zn sites. In these studies the group III donors Ga, In and Al were deliberately introduced in these samples. In their study, Dean *et al.* made use of selective D–A pair fine structure to identify the sites occupied by the impurities. In selective D–A pair excitation monochromatic light from a tunable dye laser is used to excite the sample at wavelengths corresponding to impurity-specific donor bound excitonic emission lines in the luminescence spectrum. The results were interpreted to mean that the donors Ga, In and Al deliberately introduced into the sample occupied substitutional sites on the same sublattice as the Li and Na acceptor impurities. No other donor peaks were observed in these materials. However, from Hall mobility measurements the material was found to be highly resistive and highly compensated. The absence of other donor peaks according to photoluminescence was taken to mean that compensation was by donor impurities incorporated unintentionally during growth. This was particularly true for samples doped with Li and Na to concentrations less than $10^{17}\,cm^{-3}$. The source of the impurities could be the walls of the growth chamber, source materials and dopant sources. However, compensation by non-optically active donor levels still remained a possibility.

Werkhoven *et al.* [45] investigated the possible source of these impurities. ZnSe was grown by LPE on ZnSe substrates using Sn as the solvent because of its very high purity. The procedure was to grow a buffer layer and to use higher growth rates to shorten the duration of the growth cycle. Photoluminescence measurements taken at 5.2 K revealed narrow well-resolved peaks at 2.7983, 2.7977 and 2.7973 eV. Acceptor bound excitons of Li and Na are also observed. The results seem to support the contention that compensation is by extrinsic donor impurities. The energy level of the Na and Li acceptors in ZnSe has been found to be around 114 meV.

Li and Na ion-implantation studies were also performed by Rosa and Streetman [51] and Yoda and Yamashita [52]. Although the authors claimed that p-conductivity ZnSe was obtained by this technique, the results remain inconclusive.

(*b*) *Nitrogen doping*

Studies on As, P, Au and Ag [50, 53] revealed that only deep levels are formed which makes them unsuitable for p-type conductivity at room temperature. P seems to form a shallow acceptor level but this level is now believed [49] to be associated with a complex involving a P atom. The ionization energy of the substitutional P acceptor atom is about 400 meV. Henry *et al.* [15–57] and Neumark [54] pointed out that, from all other possible acceptors such as As, P, Sb, Au, Ag and N, nitrogen appears to be the most likely candidate as a shallow acceptor in the wide bandgap II–VI semiconductors.

Henry *et al.* [15, 16] showed that the activation energy for nitrogen is about 90 meV as well. However, they suggest it has a low solubility limit and is perhaps even insoluble. Nonetheless, a detailed photoluminescence study of ZnSe grown by LPE [55, 56] and OMCVD using NH_3 as the source of nitrogen [55, 57] revealed that in fact nitrogen can be incorporated in sufficient concentrations to be detected by photoluminescence. A rough estimate of the nitrogen concentration associated with the emission line is 10^{11}–$10^{12}\,cm^{-3}$.

Nitrogen ion implantation studies were undertaken by Wu *et al.* [58] in an effort to increase the number of active nitrogen atoms incorporated in the ZnSe. The I_1^N acceptor BE peak was observed at 2.7916 eV together with the longitudinal optical (LO) phonon replicas. However, a post-implantation anneal was required to remove the damage caused by the implantation and, as Dean [59] has pointed out, there were insufficient control experiments performed to separate the effects of lattice damage from those of the incorporated nitrogen atoms. Fitzpatrick *et al.* [56] investigated nitrogen doping of ZnSe grown by LPE and Stutius [57] used NH_3 as a nitrogen source for doping of OMCVD grown ZnSe. In each of these studies an acceptor bound exciton peak as well as strong D–A pair emission peaks associated with the presence of incorporated nitrogen were observed. From the energy of the zero-phonon D–A pair peak, E_A was determined to be around 110 meV. In all cases carrier concentrations were too low to be measured by electrical means ($\rho \gg 10^5\,\Omega\,cm$) although an acceptor bound excitor peak at 2.7916 eV, shown to be due to nitrogen incorporation, was quite pronounced in the spectrum.

Park *et al.* [60] investigated the incorporation of nitrogen into ZnSe during MBE growth on GaAs. The low-temperature photoluminescence spectrum of unintentionally doped ZnSe was shown to be dominated by free exciton emission and indicative of high-purity material. An acceptor bound exciton peak at 2.7916 eV is shown in Fig. 7.3 for ZnSe grown under the same conditions but exposed to a flux of N_2 or NH_3. The I_1^N peak is observed in the spectrum indicating the presence of nitrogen. Further, the D–A pair emission peaks were observed as a dominant feature of the full spectrum. The presence of such strong D–A pair emission under relatively low excitation levels suggests a high concentration of donors. This is evidenced by the presence of the I_2 peak at 2.7969 eV which is probably Ga out-diffusing from the GaAs substrate [Mar]. Hall mobility measurements suggest that the resistivity of the material was $\gg 10^5\,\Omega\,cm$. The conductivity type could not be determined from the measurements but one suspects that with the high Ga concentration it is probably n-type.

A potential difficulty for N-incorporation is the low sticking coefficient of N on the II–VI semiconductors. In the N-doping studies undertaken by Park *et al.* [60] the nitrogen overpressure required in the MBE growth chamber before incorporation could be detected by photoluminescence exceeded

10^{-5} Torr. This result suggests that the sticking coefficient of molecular nitrogen on ZnSe is very low but, as noted above, Henry *et al.* [15, 16] pointed out the possibility of a very low solubility limit for nitrogen in this material.

In an attempt to increase the incorporation of nitrogen in ZnSe, Mitsuyu *et al.* [61] irradiated the growing ZnSe surface with a flux of ionized NH_3 molecules from a low-energy saddle field source. The ion energy selected was 10 eV at a current density that varied from 10 to 1000 nA cm^{-2}. The reason given for choosing an ion energy of 10 eV was in order to minimize radiation damage. However, it has been pointed out [62] that, for molecular nitrogen ions with an energy of 10 eV, dissociation of the molecule would occur on impact with a surface to form the more reactive atomic species. It was suggested there that the sticking coefficient of the atomic species on ZnSe could be high. This has not yet been confirmed experimentally. Mitsuyu *et al.* reported that the photoluminescence results suggest a significant increase in the concentration of acceptors. From the D–A pair spectrum the activation energy of the acceptor was calculated to be 110 eV which they

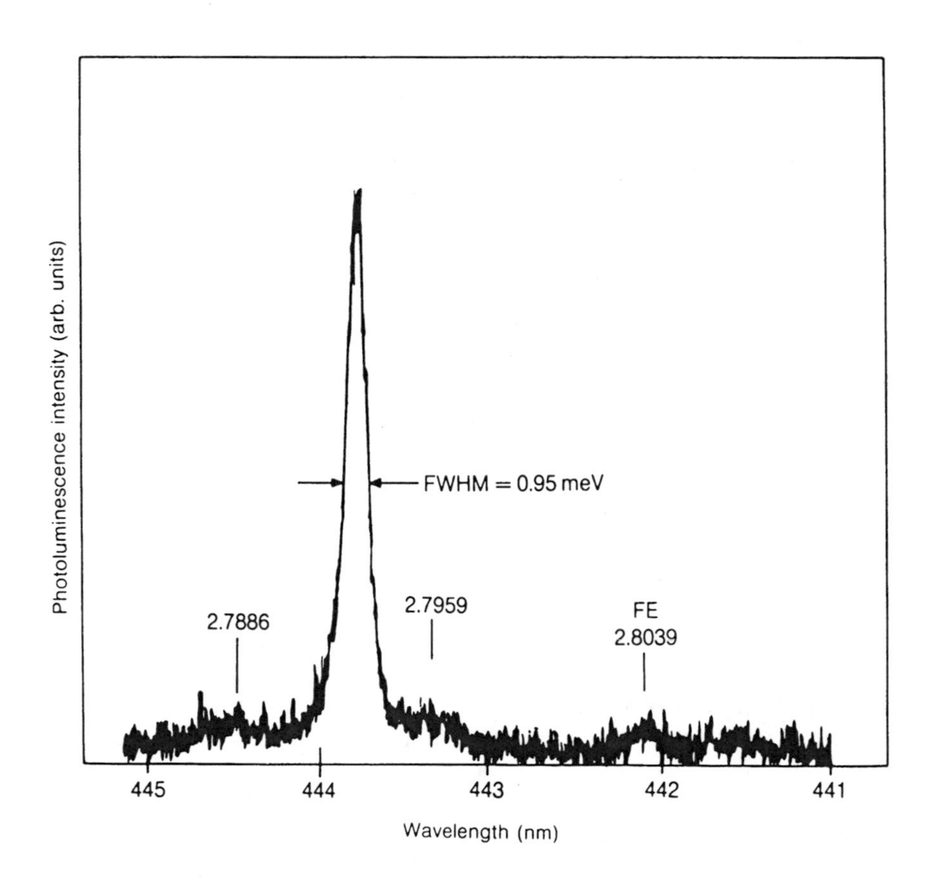

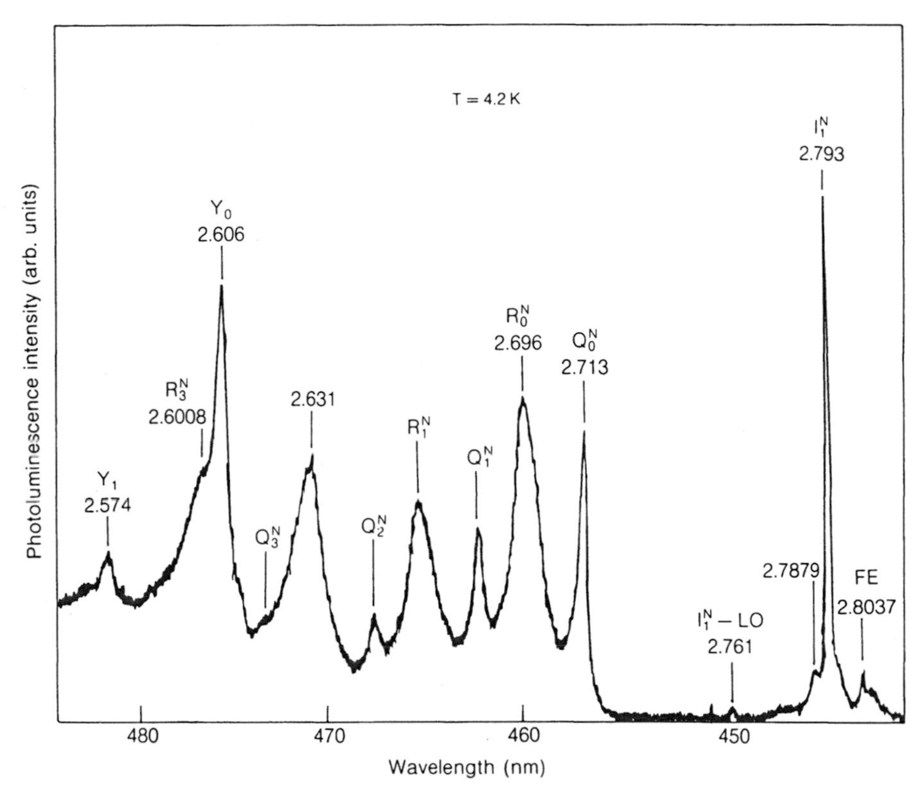

Fig. 7.3 Photoluminescence spectra taken at liquid helium temperatures of nitrogen-doped MBE-grown ZnSe showing (a) a dominant I_1^N peak in the excitonic emission band (note the relatively weak free exciton peak at 2.8037 eV) and (b) strong donor–acceptor pair emission lines associated with the nitrogen acceptor observed between 455 nm and 475 nm.

attribute to nitrogen incorporation. However, the electrical resistivity was reported to be greater than $10^4\,\Omega$ cm, and it was not possible to observe p-conductivity by Hall measurements.

Kosai *et al.* [63] investigated N-doped ZnSe grown by LPE. From the D–A pair spectra they estimated E_A to be about 90 meV. Electrical measurements showed that incorporation of N had not yielded a high-conductivity p-type semiconductor. Rather, the resistivity was greater than $10^5\,\Omega$ cm and mobility was less than $5\,\mathrm{cm^2\,V^{-1}\,s^{-1}}$. The ability to measure the mobility was in itself an achievement. The electrical measurements were from photocapacitance experiments on Schottky diodes.

The results discussed so far suggest that nitrogen can be incorporated in ZnSe in concentrations in excess of $10^{12}\,\mathrm{cm^{-3}}$, but so far the studies have failed to provide an indication of the mechanisms responsible for the apparent low incorporation. One of the principal difficulties is the paucity of electrical data on these materials. All evidence for the presence of the acceptor levels in the II–VI semiconductors is from the low-temperature photoluminescence studies. Towards correcting this imbalance, Leigh and Wessels [64] investigated nitrogen-doped ZnSe grown by VPE using optically excited deep-level transient spectroscopy. The results of the measurements revealed the presence of three electron traps at energies $E_v + 0.085\,\text{eV}$, $E_v + 0.10\,\text{eV}$ and $E_v + 0.16\,\text{eV}$. The former two traps are consistent with results of photoluminescence measurements reported by Fitzpatrick *et al.* [56] on LPE-grown ZnSe ($E_A = 85\,\text{meV}$), Stutius [57] on OMCVD ZnSe ($E_A = 110\,\text{meV}$), Dean *et al.* [55] on LPE and OMCVD ZnSe ($E_A = 114\,\text{meV}$) and Park *et al.* [41] on MBE ZnSe ($E_A = 110\,\text{meV}$). However, Leigh and Wessels note that the trap at $E_A + 0.16\,\text{eV}$ has not been observed by photoluminescence. By extension the question one may ask is, why is it that only one of the other two traps is observed for any given sample in the photoluminescence measurement? The range of traps indicates that the acceptor levels associated with nitrogen as an impurity atom in ZnSe may not simply involve a single substitutional defect but may also involve complexes comprising either native defects or impurities. The optical deep-level transient spectroscopy study also revealed the presence of a deep donor at $E_c - 0.35\,\text{eV}$ that seems to be derived from the nitrogen. This suggests that some self-compensation of nitrogen may occur in N-doped ZnSe. The structures of these defects remain to be resolved.

7.3.3 Photoluminescence in ZnTe

Unlike the wide bandgap II–VI Se and S chalcogenides of Zn and Cd, As-grown ZnTe is always of p-type conductivity. The bandgap of ZnTe is about 2.39 eV at room temperature so that p–n junctions fabricated from this material could be important for green-light-emitting devices. Unfortunately, practical n-type conductivity in ZnTe is very difficult to achieve [59, 65–67] and, indeed, successful devices have not yet been fabricated. Nonetheless, Venghaus and Dean [67] have pointed out that ZnTe can be made n-type only at very high concentrations of Al ($2 \times 10^{20}\,\text{cm}^{-3}$) but it has not been possible to obtain moderate doping levels. The questions remain: why is As-grown ZnTe of p-type conductivity, and why is it difficult to grow n-type conductivity material?

Native defects and complexes involving native defects in ZnTe have a profound effect on the electrical and optical characteristics of the material, and in this sense ZnTe resembles the other wide bandgap chalcogenides of Zn and Cd. Park and Shin [68] showed that complexes involving native defects and impurities can give rise to shallow acceptor levels in these materials. However, unlike the case of P in ZnSe, detailed studies by Venghaus and Dean [67], Dean [59] and Dean *et al.* [65] using photoluminescence excitation spectroscopy have shown not only that Li and Na form shallow acceptors at cation sites, as occurs in ZnSe as well as CdS and CdSe, but also that there is a large number of shallow anion site acceptors in ZnTe. These include P, As and Cu, all of which are known, from low-temperature photoluminescence spectroscopy, to form deep substitutional acceptor levels in ZnSe. The formation of these deep levels has been attributed to the large, trigonal, static Jahn–Teller distortion of the tetrahedral cluster of P (As) and nearest-neighbour Zn atoms [69] which reduces the ZnSe lattice symmetry in the vicinity of the P atom from T_d to C_{3v}. Although similar studies have not been conducted for CdS and CdSe, substitutional P and As are known to give rise to deep acceptor levels. The reason why the trigonal, static Jahn–Teller distortion occurs in ZnSe and not ZnTe is not yet known. Neither are the reasons clear why ZnTe cannot be moderately doped n-type.

7.4 DISCUSSIONS

Great strides have been made in determining that extrinsic donors play a major role in compensation in the II–VI materials. The identification of impurities which are likely to be shallow acceptors in these materials also represents a significant advance in this field. However, the goal that still remains elusive is the development of a full understanding of the compensation processes by extrinsic donors, and the various impurity–native defect complexes not revealed by photoluminescence that evidently still appear to have a dominating effect on p-type conductivity in these materials. Unlike LPE and melt-growth techniques, VPE, OMCVD and particularly MBE are highly non-equilibrium single-crystal growth techniques. Indeed, MBE is characterized by kinetic processes. Perhaps because of it, new microstructures revealed in photoluminescence as the I_1^{Deep} and I_x peaks are more commonly observed in MBE-grown ZnSe.

Annealing studies by Yao and Taguchi [70] on ZnSe in Se vapour suggest that for the MBE-grown ZnSe the I_1 line associated with a neutral acceptor bound exciton becomes more intense and narrow with a concomitant decrease in the I_x line as the Se overpressure during the anneal is increased. They suggest that the characteristics of the principal BE lines observed in MBE-grown ZnSe may be related to changes in native defect concentrations. There is no evidence to suggest that these native defects in MBE-grown ZnSe act as compensating donors even though the material is found to be of n-type conductivity. Nonetheless, these results appear to negate earlier arguments that extrinsic impurities rather than native unassociated donors compensated acceptor impurities such as shallow Li, Na and N acceptors. Although these defects appear to be specific to the MBE non-equilibrium growth process and may yet be controlled under the appropriate growth conditions, compensation by Li and Na interstitials still remains a problem.

Having said that, Yasuda *et al.* [71] have reported the growth of low-resistivity p-type ZnSe by MOVPE. The acceptor impurity was lithium and low-temperature photoluminescence measurements of heavily doped ZnSe:Li reveal a broad peak centred around 2.79 eV that becomes dominant in the spectrum with increasing Li concentration. The authors report Hall measurements made in a van der Pauw configuration showing p-type conductivity with carrier concentrations ranging from 3×10^{16} to $9 \times 10^{17}\,cm^{-3}$. An activation energy of 80 meV was estimated from the slope of the n versus $1/T$ straight line. This is smaller than the 114 meV reported [55, 57, 60] but similar to the 85 meV determined by Fitzpatrick *et al.* [56] from photoluminescence studies and by Leigh and Wessels [64] from optically excited deep-level transient spectroscopy studies. Since these initial results, Cheng *et al.* [72] have reported observing a dominant, broad emission peak in photoluminescence corresponding to an Li acceptor bound exciton in intentionally Li-doped MBE-grown ZnSe. They have also reported p-type conductivity in the material as determined by Hall measurements.

These results represent significant advances in the quest for p-conductivity ZnSe. However, one still has to contend with stability of the Li_{Zn} acceptor, as well as the interstitial Li impurity atoms which tend to drift in the presence of high electric fields such as exist in the practical p–n junctions. This is a fundamental property of the system and would appear to negate the advance made in obtaining p-type conductivity in the material.

The efforts with N in LPE appeared equally unpromising. The problem with nitrogen appears to be the low level of incorporation or even something as fundamental as a low solubility limit of N in the wide bandgap II–VI semiconductors as suggested by Henry *et al.* The lack of progress in the research perhaps rests with an over-reliance on photoluminescence data and general lack of electrical data. Photoluminescence techniques have provided significant insight into the problems of doping in II–VI semiconductors but any non-radiative levels associated with defects would remain undetected. These could only be probed by electrical means. One suspects that much more research on doping, and characterization by optical as well as other means including electrical measurements, is required before the full picture is revealed.

REFERENCES

1. Zanio, K. (1978) *Semiconductors and Semimetals*, Vol. 13, Academic Press, New York.
2. Halsted, R.E. (1967) In *Physics and Chemistry of II–VI Compounds* (eds M. Aven and J.S. Prener), Chapter 8.
3. Curie, D. and Prener, J.S. (1967) In *Physics and Chemistry of II–VI Compounds* (eds M. Aven and J.S. Prener), Chapter 9.
4. Smith, F.T.J. (1969) *Solid State Commun.*, **7**, 1757.
   Smith, F.T.J. (1970) *Solid State Commun.*, **8**, 263.
   Smith, F.T.J. (1970) *Metall. Trans.*, **1**, 617.
5. Merz, J.L., Kukimoto, H., Nassau, K. and Shiever, J.W. (1972) *Phys. Rev. B*, **6**, 545.
6. Hite, G.E., Marple, D.T.F., Aven, M. and Segall, B. (1967) *Phys. Rev.*, **156**(3), 850.

7. Tews, H., Venghaus, H. and Dean, P.J. (1979) *Phys. Rev. B*, **19**(10), 5178.
8. Dean, P.J., Herbert, D.C. and Lahee, A.M. (1980) *J. Phys. Soc. Jpn.*, **49** (Suppl. A), 185.
9. Dean, P.J., Herbert, D.C., Pfister, J.C., Schaub, B. and Marine, J. (1978) *J. Lumin.*, **16**, 363.
10. Dean, P.J., Herbert, D.C., Werkhoven, C.J., Fitzpatrick, B.J. and Bhargava, R.N. (1981) *Phys. Rev. B*, **23**(10), 4888.
11. Dean, P.J. and White, A.M. (1978) *Solid State Electron.*, **21**, 1351.
12. Farrow, R.F.C. (1981) *J. Vac. Sci. Technol.*, **19**(2), 150.
13. Faurie, J.-P., Million, A. and Jacquier, G. (1982) *Thin Solid Films*, **90**, 107.
14. Myers, T.H., Lo, Y., Schetzina, J.F. and Jost, S.R. (1982) *J. Appl. Phys.*, **53**, 9232.
    Myers, T.H., Lo, Y., Bicknell, R.N. and Schetzina, J.F. (1983) *Appl. Phys. Lett.*, **42**, 247.
15. Henry, C.H., Nassau, K. and Shiever, J.W. (1971) *Phys. Rev. B*, **4**(8), 2453.
16. Henry, C.H., Nassau, K. and Shiever, J.W. (1970) *Phys. Rev. Lett.*, **24**, 820.
17. Henry, C.H., Faulkner, R.A. and Nassau, K. (1969) *Phys. Rev.*, **183**, 798.
18. Thomas, D.G. and Hopfield, J.J. (1962) *Phys. Rev.*, **128**, 2135.
19. Thomas, D.G. (1961) *J. Appl. Phys.*, **32S**, 2298.
20. Cho, K. (1979) *Topics in Current Physics: Excitons*, Springer, Berlin.
21. Hopfield, J.J. (1958) *Phys. Rev.*, **112**, 1555.
22. Lang, M. (1970) *Phys. Rev. B*, **2**(10), 4022.
23. Hopfield, J.J. (1966) *Proceedings of the International Conference on Physics of Semiconductors, Kyoto, 1966*, p. 77.
24. Dean, P. J. and Herbert, D.C. (1979) in *Excitons*, (ed. K. Cho), Springer, Berlin, Chapter 2.
25. Hopfield, J.J. (1964) *Proceedings of the International Conference on Physics of Semiconductors, Paris*, p. 725.
26. Prener, J.S. and Williams, F.E. (1956) *J. Electrochem. Soc.*, **103**, 342.
27. Dean, P.J. and Merz, J.L. (1969) *Phys. Rev. B*, **178**, 1310.
28. Merz, J.L., Nassau, K. and Shiever, J.W. (1972) *Phys. Rev. B*, **8**, 1444.
29. Lampert, M.A. (1958) *Appl. Phys. Lett.*, **1**, 450.
30. Pinzcuk, A. and Burstein, E. (1983) in *Light Scattering in Solids I* (ed. M. Cardona), Springer, Berlin, p. 23.
31. Nakashima, S., Kojima, H. and Hattori, T. (1975) *Solid State Commun.*, **17**, 689.
32. Hattori, T., Nakashima, S. and Mitsuishi, A. (1976) *Proceedings of the 13th International conference on the Physics of Semiconductors, Rome, 1976*, p. 635.
33. Lee, S., Hillebrands, B., Stegeman, G.I., Cheng, H., Potts, J.E. and Nizzoli, F. (1988) *J. Appl. Phys.*, **63**(6), 1914.
34. Cheng, C.H. and Nassau, K. (1970) *Phys. Rev. B*, **2**(4), 997.
    Reynolds, D.C., Litton, C.W. and Collins, T.C. (1967) *Phys. Rev.*, **156**, 881.
35. Mandel, G., Moorehouse, F.F. and Wagner, P.R. (1964) *Phys. Rev.*, **136**, A826.
    Mandel, G. (1964) *Phys. Rev.*, **134**, A1073.
36. Kroger, F.A. (1965) *J. Chem. Phys. Solids*, **26**, 1717.
    Watt, R.K. (1973) *J. Mater. Sci.*, **8**, 1201.
37. Aven, M. (1967) in *II–VI semiconducting Compounds* (ed. D.G. Thomas), Benjamin, New York, p. 1232.
38. Neumark, G.F. and Herko, S.P. (1982) *J. Cryst. Growth*, **59**, 189.
39. Neumark, G.F., Herko, S.P., McGee III, T.F. and Fitzpatrick, B.J. (1984) *Phys. Rev. Lett.*, **53**(6), 604.
40. Dean, P.J. (1984) *Phys. Status Solidi A*, **81**, 625.
41. Park, R.M., Mar, H.A. and Salansky, N.M. (1985) *Appl. Phys. Lett.*, **46**(4), 386.
42. Park, R.M., Kleiman, J. and Mar, H.A. (1987) *Proc. Soc. Photo-Opt. Instrum. Eng.*, **796**, 86–90.

43. Kleiman, J., Park, R.M. and Mar, H.A. (1988) *J. Appl. Phys.*, **64**(3), 1201.
    Park, R.M., Mar, H.A. and Kleiman, J. (1988) *J. Cryst. Growth*, **86**, 335.
44. Mar, H.A. and Park, R.M. Unpublished.
45. Werkhoven, C., Fitzpatrick, B.J., Herko, S.P. and Bhargava, R.N. (1981) *Appl. Phys. Lett.*, **34**(7), 540.
46. Yoneda, K., Hishida, Y., Toda, T., Ishii, H. and Niina, T. (1984) *Appl. Phys. Lett.*, **45**(12), 1300.
47. Kolodziejski, L.A., Gunshor, R.L., Melloch, M.R., Vaziri, M., Choi, C. and Otsuka, N. (1987) *Proc. Soc. Photo-Opt. Instrum, Eng.*, **796**.
48. Tamargo, M.C., De Miguel, J.L., Hwang, D.M. and Farrell, H.H. (1988) *J. Vac. Sci. Technol. B*, **6**(2), 784.
49. Reinberg, A.R., Holton, W.C., de Wit, M. and Watts, R.K. (1971) *Phys. Rev. B*, **3**(2), 410.
    Adachi, S. and Machi, Y. (1976) *Jpn. J. Appl. Phys.*, **15**(8), 1513.
    Watts, R.K., Holton, W.C. and de Wit, M. (1977) *Phys. Rev. B*, **3**(2), 404.
50. Smith, F.T.J. (1970) *Metall Trans.*, **1**, 617.
    Smith, F.T.J. (1970) *Solid State Commun.*, **8**, 263.
51. Rosa, A.J. and Streetman, B.G. (1975) *J. Lumin.*, **10**, 211.
52. Yoda, T. and Yamashita, K. (1988) *Appl. Phys. Lett.*, **53**(24), 2403.
53. Dean, P.J., Fitzpatrick, B.J. and Bhargava, R.N. (1982) *Phys. Rev. B*, **26**(4), 2016.
54. Neumark, G. (1980) *J. Appl. Phys.*, **51**, 3383.
55. Dean, P.J., Stutius, W., Neumark, G.F., Fitzpatrick, B.J. and Bhargava, R.N. (1983) *Phys. Rev. B*, **27**(4), 2419.
56. Fitzpatrick, B.J., Werkhoven, C.J., McGee, T.F., III, Harnack, P.M., Herko, S.P., Bhargava, R.N. and Dean, P.J. *IEEE Trans. Electron Devices* **28**, 440.
57. Stutius, W. (1982) *Appl. Phys. Lett.*, **40**(3), 246.
58. Wu, Z.L., Merz, J.L., Werkhoven, C.J., Fitzpatrick, B.J. and Bhargava, R.N. (1982) *Appl. Phys. Lett.*, **40**(4), 345.
59. Dean, P.J. (1979) *Inst. Phys. Conf. Ser.*, **46**, 100.
60. Park, R.M., Mar, H.A. and Salansky, N.M. (1985) *J. Appl. Phys.*, **58**(2), 1047.
61. Mitsuyu, T., Ohkawa, K. and Yamazaki, O. (1986) *Appl. Phys. Lett.*, **49**(20), 1348.
    Ohkawa, K., Mitsuyu, T. and Yamazaki, O. (1987) *3rd International Conference on II–VI Compounds, Monterey, 1987.*
62. Mar, H. Unpublished 3M Internal Report.
63. Kosai, K., Fitzpatrick, B.J., Grimmeiss, H., Bhargava, R.N. and Neumark, G.F. (1979) *Appl. Phys. Lett.*, **35**(2), 194.
64. Leigh, W.B. and Wessels, B.W. (1984) *J. Appl. Phys.*, **55**(6), 1614.
65. Dean, P.J., Venghaus, H., Pfister, J.C., Schaub, B. and Marine, J. (1978) *J. Lumin.*, **16**, 363.
66. Bensahel, D., Magnea, N., Pautrat, J.L., Pfister, J.C. and Revoil, L. (1979) *Inst. Phys. Conf. Ser.*, **46**, 421.
67. Venghaus, H. and Dean, P.J. (1980) *Phys. Rev. B*, **21**(4), 1596.
68. Park, Y.S. and Shin, B.K. (1974) *J. Appl. Phys.*, **45**, 1444.
69. Watts, R.K., Holton, W.C. and de Wit, M. (1971) *Phys. Rev. B*, **3**(2), 404.
    Reinberg, A.R., Holton, W.C., de Wit, M. and Watts, R.K. (1971) *Phys. Rev. B*, **3**(2), 410.
70. Yao, T. and Taguchi, T. (1985) Characteristics of p type impurities and the effect of heat-treatment in high-purity ZnSe grown by MBE, p. 1221.
71. Yasuda, T., Mitsuishi, I. and Kukimoto, H. (1988) *Appl. Phys. Lett.*, **52**(1), 57.
72. Cheng, H., Depuydt, J.M., Potts, J.E., and Smith, T.L. (1988) *Appl. Phys. Lett.*, **52**, 147.
73. Mar, H.A. and Park, R.M. (1986) *J. Appl. Phys.*, **60**(3), 1229.
74. Park, R.M., Mar, H.A. and Salansky, N.M. (1985) *J. Vac. Sci. Technol. B*, **3**(2), 676.

8

Transport properties of widegap II–VI compounds

H.E. Ruda

8.1 INTRODUCTION

Widegap II–VI materials have now become a focus of attention for researchers around the globe. This may be attributed to two facts: firstly, the inherent properties of these materials for advanced opto-electronic device applications and, secondly, the rapid advances that have been made in materials preparation.

One of the key characteristics of these materials is their large direct bandgap at room temperature. This underscores the potential of these materials for optoelectronic devices in the visible to ultraviolet spectral range.

Transport properties of semiconductors have come to encompass a broad variety of phenomena originating from the transport of carriers in these materials, in applied fields. These may include electric fields, magnetic fields, optical and thermal fields, independently or combined. Furthermore, the influences of direct and alternating fields are important, as well as low- and high- field conditions, which should also be considered. An important recent development has been studies of transport in low-dimensional structures, where quantization changes the number of degrees of freedom in the system.

Transport property studies provide two important categories of information. Firstly, they help to establish an understanding of the materials. For example, information may be extracted on the number and types of carriers in these materials, as well as band structure and carrier-scattering mechanisms. In this way it is possible to separate the effects of crystal perfection and impurities from the pure lattice. Secondly, such studies provide a quantitative benchmark for assessing materials quality and hence the suitability of the material for device applications. The exhaustive materials characterization efforts of many research groups have certainly contributed to the rapid recent progress in achieving high quality n- and p-type widegap II–VI materials of controlled properties.

This chapter will concentrate on the electrical transport characteristics of the widegap II–VI materials with particular emphasis on ohmic transport in the bulk materials. However, a.c. transport with particular reference to free-carrier absorption, high-field transport, and 2D transport phenomena will also be briefly reviewed.

8.2 SCATTERING MECHANISMS AND TRANSPORT PARAMETERS

8.2.1 Band structure

All of the widegap II–VI compounds crystallize into the zinc blende crystal structure. Since the energy band structure is determined by the crystal structure, the Brillouin zone is defined for these materials. It should be noted that, for problems of carrier transport, we are principally concerned with the band extrema. An obvious exception would be the case of high-field transport.

The Brillouin zone for these materials has the shape of a truncated octahedron, or a so-called tetrakaidecahedron, having fourteen plane faces; six square faces are along the $\langle 1\,0\,0 \rangle$ directions and eight hexagonal faces are along $\langle 1\,1\,1 \rangle$ directions. The zone centre is denoted the Γ point, and the $\langle 1\,1\,1 \rangle$ and $\langle 1\,0\,0 \rangle$ directions are called Λ and Δ directions respectively, with their intersections with the zone boundaries called the L and the X points respectively. The energy band relations, commonly denoted $E(\boldsymbol{k})$, may be expressed about band extrema. In particular, for the widegap II–VI compounds referenced to the Γ point, we may write

$$\mathrm{E(k)} = \frac{\hbar^2 k^2}{2m^*}\left[A \pm \left(B^2 + \frac{C^2}{6}\right)^{1/2}\right] \tag{8.1}$$

The band parameters for these materials are listed in Table 8.1. A schematic of the band structure of ZnSe about the Γ point is given in Fig. 8.1 for illustration. Notice how the effect of band edge non-parabolicity is essentially negligible in these materials. That is, the term in parentheses is approximately unity. Notice also that for these widegap materials the effective masses are relatively high, and therefore these materials tend to remain non-degenerate for moderate carrier concentrations. Another important feature of the band structure of these materials relates to the valence band structure. The large values of Δ_0 compared with $k_b T$ at room temperature (and below) justify neglecting the split-off hole band in transport analysis of these materials. Thus for hole conduction in p-type widegap II–VI compounds, only the two degenerate bands (at the zone centre) hh and lh need to be considered.

In order to establish the scattering probabilities for various carrier scattering mechanisms, it is necessary to describe carrier eigenfunctions in initial and final states. Following Kane [3] who pioneered the $\boldsymbol{k}\cdot\boldsymbol{p}$ interaction

Table 8.1 Energy band parameters[a] of widegap II–VI materials

Parameter	*ZnTe*	*ZnSe*	*ZnS*
Lattice constant, a (Å)	6.1037	5.6687	5.4093
Energy gap at O K, E_g (eV)	2.38	2.94	3.84
dE_g/dT (10^{-4} eV K^{-1})	−5	−8	−4.6
Spin–orbit splitting, Δ_0 (eV)	0.91	0.43	0.27
Electron effective mass, m_e^*/m_0	0.16	0.17	0.31
Heavy hole effective mass, m_{hh}^*/m_0	NA	0.60	NA
Light hole effective mass, m_{lh}^*/m_0	0.154	0.149	0.230
Density of state hole mass, m_{dh}^*/m_0	NA	0.649	NA
Valence band parameters:			
A	−3.29	−2.12	−3.06
B	−0.08	−0.48	−0.53
C^2	0.60	4.90	2.18

[a]Data are derived from refs 1 and 2 and references cited therein.

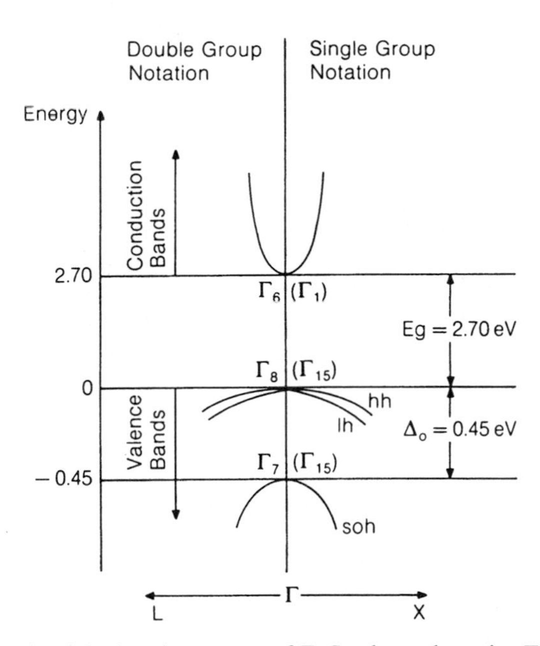

Fig. 8.1 Schematic of the band structure of ZnSe about the point Γ in the Brillouin zone. Single-group (no spin–orbit interaction) and double-group notations are used to label the symmetry points at the zone centre.

approach to this problem, eigenfunctions may be written as

$$\begin{aligned}
\psi^{i}_{k\{\alpha,\beta\}} &= u^{i}_{k,\{\alpha,\beta\}} \exp(i\boldsymbol{k}\cdot\boldsymbol{r}):\ i = \text{cb, hh, lh, soh bands} \\
u^{\text{cb}}_{k,\alpha} &= [\mathrm{i}S\downarrow];\ u^{\text{cb}}_{k,\beta} = [\mathrm{i}S\uparrow] \\
u^{\text{hh}}_{k,\alpha} &= (1/\sqrt{2})[(X + \mathrm{i}Y)\uparrow] \\
u^{\text{hh}}_{k,\beta} &= -(1/\sqrt{2})[(X + \mathrm{i}Y)\downarrow] \\
u^{\text{lh}}_{k,\alpha} &= -(1/\sqrt{6})[(X - \mathrm{i}Y)\uparrow] - \sqrt{(2/3)}[Z\downarrow] \\
u^{\text{lh}}_{k,\beta} &= -(1/\sqrt{6})[-(X + \mathrm{i}Y)\downarrow] - \sqrt{(2/3)}[Z\uparrow] \\
u^{\text{soh}}_{k,\alpha} &= (1/\sqrt{3})[(X - \mathrm{i}Y)\uparrow] - \sqrt{(1/3)}[Z\downarrow] \\
u^{\text{soh}}_{k,\beta} &= (1/\sqrt{3})[-(X + \mathrm{i}Y)\downarrow] - \sqrt{(1/3)}[Z\uparrow]
\end{aligned} \tag{8.2}$$

where the symbols $\uparrow$ and $\downarrow$ denote spin-up and spin-down states respectively, and X, Y, and Z refer to the periodic parts of the Bloch functions which transform under the tetrahedral group like p_x, p_y, and p_z atomic functions.

The probability of transition from an initial state $\boldsymbol{k}$ to final state $\boldsymbol{k}'$ is proportional to a parameter $G_{r,s}(\boldsymbol{k}, \boldsymbol{k}')$ [1,4–7]. This parameter is a measure of the overlap between the cell periodic block functions of equation (8.2), and may be written as

$$G_{r,s}(\boldsymbol{k}, \boldsymbol{k}') = \frac{1}{2} \sum_{\mu_1,\mu_2} \left| \int u^{*}_{s,\mu_2,k'}(\boldsymbol{r}) u_{r,\mu_1,k}(\boldsymbol{r}) \mathrm{d}\boldsymbol{r} \right|^2 \tag{8.3}$$

where the sum extends over spin states, and the functions $u_{k,i}(\boldsymbol{r})$ are as defined in equation (8.2). G therefore measures the overlap between initial and final states. For spherically symmetric s-like electron eigenfunctions, the overlap will be unity for all $\boldsymbol{k}$ and $\boldsymbol{k}'$. However, for anisotropic eigenfunctions such as the hole eigenfunctions, there is a reduced overlap between eigenfunctions owing to the strong p-component. Wiley [5] has derived expressions for the overlap functions for inter- and intraband hole scattering and gives the following approximate forms, which he argues provide an excellent approximation in the case of wide bandgap semiconductors:

$$\begin{aligned}
G_{11}(\gamma) &= G_{22}(\gamma) = (1/4)(1 + 3\cos^2\gamma) \\
G_{12}(\gamma) &= G_{21}(\gamma) = (3/4)\sin^2\gamma
\end{aligned} \tag{8.4}$$

where γ refers to the angle subtended between the initial and the final state wavevectors $\boldsymbol{k}$ and $\boldsymbol{k}'$ respectively. The subscripts i and j refer to the initial- and final-states combination, where 1 and 2 denote the hh and the lh bands respectively.

8.2.2 Doping and transport

(*a*) *General*

In analysing transport phenomena in widegap materials one of our main requirements is a knowledge of the relationship between carrier concentration and Fermi energy. Under thermal equilibrium this is determined by the materials band structure, lattice temperature and the concentration of impurity atoms. In general, an expression may be derived between Fermi energy and the concentration and ionization energy of impurities. However, in these materials a number of factors often complicate this approach. There may be a number of different impurity atom types, each of which can produce a number of energy levels corresponding to different states of excitation, and each level may be degenerate to a different degree. The energy positions of these levels are also affected by external fields, for example an applied magnetic field or pressure. More importantly, however, is the general case for much of the experimental data reported on these materials to date: there tends to be a wide assortment of impurities, apart from those intentionally introduced for doping, which are often present in large concentrations. These unintentional impurities may be of acceptor and donor types and invariably result in material which is compensated to some degree. An *a priori* evaluation of the Fermi energy is thus difficult, even though analysis of transport phenomena in these materials relies on a knowledge of ionized and neutral impurity concentrations, respectively. In moderately doped material, the band structure of the host lattice is essentially unperturbed, owing to the wide separation between impurity atom potential wells. Thus free-electron motion is generally only very weakly disturbed by the presence of these impurities. The result is that discrete hydrogenic (Bohr) states are produced separated from the band extrema by an ionization energy, E_{I}. Since the eigenfunctions for the weakly bound outermost electron–hole at the impurity centre cover a large number of host crystal atoms, the permittivity value used to describe the centre is that of the host crystal, ε. Similarly, bandedge effective masses m^* should be used to describe these carriers. Thus we may write

$$E_{\text{I}} = m^* e^4 (128\pi^4 \varepsilon^2 h^2)^{-1} \tag{8.5}$$

As the impurity concentration increases, the impurity potential wells overlap, transforming the discrete levels discussed above into energy bands. However, in such highly doped material there will be a large concentration of ionized centres. For example in n-type material the attractive potential of these ionized donors will lower the conduction bandedge. The same argument applies for acceptors. As a result the conduction and valence bandedges merge into the impurity state bands resulting in so-called tail states.

(b) Statistics and impurity descriptions

Electrons and holes in the bands of these widegap semiconductors behave as nearly free particles. However, when such particles are in thermal equilibrium with the lattice they become distributed among available energy levels in accord with Fermi–Dirac statistics. The resulting concentrations of electrons and holes in the material may be written as

$$n = N_c \mathscr{F}_{1/2}(\eta)$$
$$p = N_v \mathscr{F}_{1/2}(-\varepsilon_i - \eta) \quad (8.6)$$

where

$$N_{c,v} = 4.831 \times 10^{15}[T(m^*_{eh}/m_0)]^{3/2}\ \text{cm}^{-3}$$
$$\eta = (E_f - E_C)/k_b T$$
$$\varepsilon_i = E_g/kT$$
$$\mathscr{F}_{1/2}(\eta) = \frac{2}{\sqrt{\pi}} \int_0^\infty \frac{\varepsilon^{1/2}\,d\varepsilon}{1 + \exp(\varepsilon - \eta)}$$

where all terms are as defined in ref. 8. The intrinsic carrier concentration may therefore be written as

$$n_i = \sqrt{N_c N_v \mathscr{F}_{1/2}(\eta) \mathscr{F}_{1/2}(-\varepsilon_i - \eta)} \quad (8.7)$$

The room temperature values for n_i for the widegap materials listed in Table 8.1 are all below about $10^2\ \text{cm}^{-3}$. Thus all of these materials are dominated by extrinsic effects for all temperatures of interest.

Critical electron and hole concentrations n_{cr} and p_{cr} respectively can be established in these materials to describe the degree of degeneracy. These are given as

$$n_{cr} = N_c \mathscr{F}_{1/2}(0)$$
$$p_{cr} = N_v \mathscr{F}_{1/2}(-\varepsilon_i) \quad (8.8)$$

Thus at room temperature $n_{cr} \gg p_{cr}$, with n_{cr} varying from about $1.3 \times 10^{18}\ \text{cm}^{-3}$ for ZnTe to about $2.7 \times 10^{18}\ \text{cm}^{-3}$ for ZnS. At 77 K these values drop to around $1.8 \times 10^{17}\ \text{cm}^{-3}$ and $3.5 \times 10^{17}\ \text{cm}^{-3}$ for ZnTe and ZnS respectively. Therefore, in general, when analysing transport properties of these materials, generalized Fermi–Dirac statistics should be used.

From the above analysis it is clear that impurity effects dominate transport characteristics of the widegap II–IV materials. In our previous discussions of impurity level influences, we have highlighted many of the complications in trying to understand impurity effects in real materials. However, for simplicity, transport properties are usually described in terms of the concentration and energy of the dominating impurity. From this assumption, together with the net residual ionized shallow level concentration, the Fermi energy may simply be deduced. We illustrate this by reference to n-type material, where the predominant donor level is assumed present in a concentration, N_d, at an energy level E_d below the conduction band edge, with a degeneracy, β. Thus,

$$n \approx N_d\{1 + \beta \exp[(E_f - E_d)/k_b T]\}^{-1} - N_a \tag{8.9}$$

where N_a is the total concentration of acceptors. In practice all of these will be ionized as the Fermi energy must lie above midgap in n-type material. The factor β represents donor level degeneracy, and multiplies the number of ways that the donor level can be occupied. For a simple donor state β has a value of 2. N_d is the sum of the number of ionized and neutral donors, N_d^+ and $N_d^\times$, respectively. Fermi statistics gives the relation between these two quantities as

$$N_d^+/N_d^\times = \{\beta \exp[(E_f - E_d)/k_b T]\}^{-1} \tag{8.10}$$

Recalling the dependence of n on Fermi energy, equation (8.9) may be solved explicitly for Fermi energy and hence carrier concentration. Such an approach is commonly used to analyse temperature dependent Hall data for these materials. The importance of neutral and ionized impurity effects in transport phenomena cannot be understated: hence the attention to these parameters, as discussed above. However, the total ionized impurity concentration must include both N_a and N_d^+. Bearing this in mind, a frequently used description of these widegap materials, based on compensation, may be discussed. The usual description of compensation is given in terms of the parameter, θ, the compensation ratio. θ is defined as follows:

$$\theta = N_a^-/N_d^+ \tag{8.11}$$

where for n-type material, $N_a^- = N_a$, the ionized acceptor concentration. A useful form for expressing the carrier concentration in terms of the total ionized impurity concentratin, N_I, can thus be written

$$n = N_I \frac{1-\theta}{1+\theta} \tag{8.12}$$

Immediately it is apparent that the presence of compensating impurities has a dramatic influence on the net carrier concentration. The higher the degree of compensation, the more ionized impurities are required to generate a similar number of free carriers. Ideally, widegap II–VI materials with $\theta = 0$ are desirable, giving $n = N_I$.

The arguments presented above apply equally well to p-type material in their analogous forms.

Table 8.2 presents materials parameters of widegap II–VI materials and Table 8.3 collects together values of the depths of various impurity levels below their band edges in the widegap II–VI materials. Also given are the hydrogenic levels in these materials, derived using equation (8.5) and the data

Table 8.2 Materials parameters[a] of widegap II–VI materials

Parameter	*ZnTe*	*ZnSe*	*ZnS*
High frequency dielectric constant, ε_∞	7.28	6.20	5.13
Low frequency dielectric constant, ε_0	9.67	9.20	8.32
Optical phonon energy, hw_l (meV)	26.0	31.4	43.6
Polar phonon Debye temperature, θ_d (K)	297	364	503
Optical mode coupling constant, α	0.30	0.47	0.77
Longitudinal elastic constant, C_l (10^{10} N m^{-2})	8.41	10.34	12.82
Transverse elastic constant, C_t (10^{10} N m^{-2})	2.48	3.29	12.82
Intervalley deformation potential, E_1 (eV)	3.50	4.20	4.90
Piezoelectric interaction parameter, h_{14} (10^9 V m^{-1})	0.31	0.61	2.26

[a]Data are derived from refs 1 and 8–10 and references cited therein.

of Tables 8.1 and 8.2. The apparent inconsistencies between data from various authors and the wide variations in reported ionization energies for different impurities in p-ZnSe have been discussed in the literature, and similar arguments are applicable here. For example, photoluminesce spectra have been interpreted in the following ways. Wu *et al.* [23] have used acceptor bound exciton to free-exciton separations and invoked Haynes' rule [24] to determine the acceptor level depth. That is,

$$E_A \approx 10[E_x - E(I_1^A)] \tag{8.13}$$

That is, however, only a very approximate empirical relationship. A more common approach has been to use the donor–acceptor pair bands, [1], and writing the energy of the no-phonon peak as Q_0^A gives

$$Q_0^A = E_g - (E_A + E_D) + e^2/\varepsilon_0\langle r\rangle \tag{8.14}$$

The coloumbic term $E_c = e^2/\varepsilon_0\langle r\rangle$ is defined in terms of $\langle r\rangle$, the average separation between donor and acceptor impurities in the pair from which the recombination band originates. Clearly E_c will depend on N_I and may approximately be written

$$\langle r\rangle \approx (3/4\pi N_I)^{1/3} \tag{8.15}$$

At higher carrier concentrations the screening of the pair potential by free carriers will tend to lower E_c. These considerations highlight some of the problems associated with assigning energies to impurity levels in these materials.

As a point of caution, some of the levels given in Table 8.3 appear to be lower than those estimated for hydrogenic levels. There are a number of possible reasons that may be used to reconcile the discrepancies. Firstly, as we have discussed, there are most certainly experimental problems with level

Table 8.3 Impurity ionization energies[a] in widegap II–VI

	ZnTe	*ZnSe*[b]	*ZnS*
Donors			
Al	18.5[c]	26.3[d]	180[e]
Ga		27.9[d]	
In	[c]	28.9[d]	
Cl	20.1[c]	190[f]	240[g]
Br		210[h]	
I			140[i]
F		28.2[j]	
Hydrogenic level[k]	24.7	27.3	53.0
Acceptors			
Li	60.5[l]	113–114	150[m]
Na	160[i]	92–97	190[m]
Cu	150[f]	650, 750	1250[m]
Ag	110[f]	431	720[m]
Au	220[f]	550	
Tl		35	
N		80–90, 109–111	
P	78[o]	84–90, 600	
As	79[l]	100–110	
Hydrogenic level[k]	87.3	96.4	117.9

[a] All energies are given in units of meV.
[b] Acceptor energies are from ref. 1.
[c] Reference 11.
[d] Reference 12.
[e] Reference 13.
[f] Reference 14.
[g] Reference 15.
[h] Reference 16.
[i] Reference 17.
[j] Reference 18.
[k] Calculated from the data of Tables 8.1 and 8.2
[l] Reference 19.
[m] Reference 20.
[n] Reference 21.
[o] Reference 22.

assignment. In addition, only quite recently has it been feasible to prepare ultrahigh quality material in which the number and concentration of extraneous (unintentional impurity centres) impurities are controlled to levels far below those of intentionally introduced species. In addition, there are obvious instances where unusual and perhaps complex phenomena are occurring. One case in point would be that of Tl [25] which was reported to form a 35 meV shallow acceptor level in ZnSe whereas the Tl_{Zn} species would rather be expected to act as a shallow donor. Another obvious reason for the discrepancies is the uncertainty in the calculated hydrogenic level energies themselves. Clearly these values are only as good as the band structure and dielectric constant data available. Again, the rigorous applicability of a simple hydrogenic level approach, especially in the case of acceptors, should certainly not be implicitly assumed. It is therefore suggested that the data in Table 8.3 should serve as a useful guide for the relative positions of the different impurity energy levels in these widegap II–VI materials, taking note of the aforementioned cautions.

8.3 DESCRIPTION OF TRANSPORT PHENOMENA

8.3.1 General

In this chapter we shall concentrate on three basic areas of transport phenomena in the widegap II–VI materials: firstly, bulk electron and hole transport under low field (ohmic) d.c. conditions; secondly, the possibility of using free carrier absorption (low-field a.c perturbation) to probe these materials; thirdly we shall briefly discuss quantized (2D) interfacial transport in these materials. In the first two instances, the electron and hole scattering processes that are considered are similar. In the third case, additional mechanisms, particular to this case, are discussed.

8.3.2 Scattering mechanisms

In these materials, there are three distinct factors that define scattering transitions; initial states, final states and the mechanism which induces the transition. Possible transitions to consider would be inter- and intraband, and inter- and intravalley transitions. Only the former applies to electrons, whereas both types apply for hole transport. Within this framework, phonon and impurity carrier scattering mechanisms must be included. For holes, phonon scattering mechanisms include polar and non-polar phonon scattering, acoustic deformation potential and piezoelectric–acoustic phonon scattering. Non-polar phonon scattering is unique to hole scattering, since for electrons the s-like eigenfunctions do not couple to the non-polar optical phonon modes (vanishing matrix elements). Neutral and ionized impurity scattering both come under the umbrella of impurity carrier scattering. Carrier–carrier scattering is generally of very limited importance in the case of the widegap II–VI materials, as it need only be considered at exceedingly high carrier densities. Furthermore, the difficulties [26] in formally treating this mechanism in transport analysis are generally responsible for its omission. A major reason for this is that this mechanism results in a redistribution of energy and momentum among carriers, and hence a change in electron distribution function. Typically the distribution function is far from equilibrium under these circumstances, often assuming a maxwellian form [27–30], and necessitating a complicated self-consistent scheme for analysing transport behaviour.

A number of features are specific to the case of 2D quantized transport [31–33]. For example since the band structure is quite distinct from its 3D counterparts, intra- and intersubband scattering must be included for each of the mechanisms. Furthermore, the nature of the interfacial potential and the quantum condition itself impose specific analogues of the phonon scattering mechanisms in this case. In addition, both remotely located and 'background' (3D-like) impurity scattering events occur. Finally, apart from the obvious differences in carrier statistics, screening of charges in 2D and 3D are dissimilar.

8.4 RESULTS ON TRANSPORT PHENOMENA

Devlin [34] offers an excellent review of the state of the art in this field up to 1967. The present review is aimed at providing supplementary information defining the present state of the art. This section discusses results on each of the widegap II–VI materials considered in turn.

8.4.1 ZnTe

Much of the early work in this material concentrated on bulk material of relatively poor quality with large background concentrations of a variety of impurities. Normally the material was p-type, and early doping studies with impurities such as Cu, Ag and Au helped to identify impurity levels and to achieve relatively high carrier concentrations. Aven and Segal [14] reported achieving room temperature hole concentrations in the range of mid 10^{16} to mid $10^{17}\,cm^{-3}$, by varying the partial pressure of tellurium in annealing experiments. However, such material appeared to be heavily compensated with θ values in excess of 0.9. Hole mobility values were typically less than $100\,cm^2\,V^{-1}s^{-1}$ at room temperature.

Since that time, the research thrust concentrated more on epitaxial technologies to achieve growth at lower temperatures, and in purer and more controlled conditions. Lower growth temperatures appeared to be the solution to minimizing native defect-related problems, as well as impurity incorporation and diffusion. Kitagawa and Takahashi [35] were among the first researchers to investigate molecular beam epitaxy (MBE) growth of ZnTe on a variety of substrates. Their electrical characterization focused on resistivity measurements. They investigated the dependence of film resistivity on Zn:Te flux ratio, designated F_1/F_2 in Fig. 8.2. Notice that both homoepitaxial and heteroepitaxial (GaAs substrates) were investigated, with the same basic dependence. Clearly, it was possible within their experiments

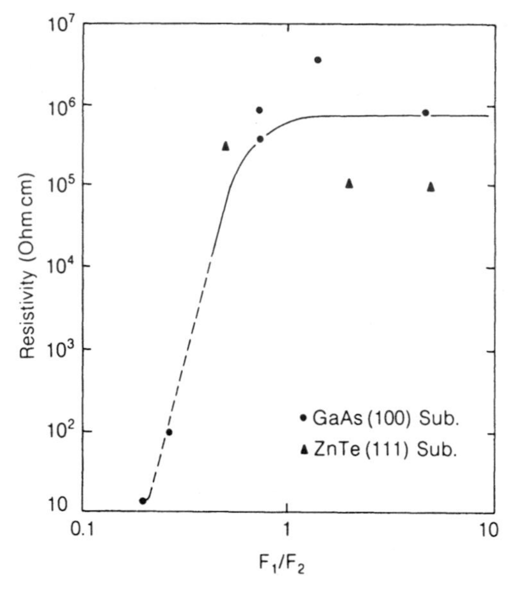

Fig. 8.2 Resistivity of homo- and heteroepitaxial MBE ZnTe films versus flux ratio, (F_1/F_2). (Reproduced from ref. 35 and published with kind permission of F. Kitagawa and K. Takahashi.)

to use stoichiometry to change samples from conducting ($\sim 10\,\Omega\,\text{cm}$) to highly resistive films ($\sim 10^6\,\Omega\,\text{cm}$). The carrier concentration (and possibly mobility) of these films was therefore strongly dependent on native defect equilibria.

Nishio *et al.* [36] studied the effects of HCl on the open-tube vapour phase growth of ZnTe. These authors studied source depletion rate dependence on growth temperature and particularly the input molar fraction of HCl. They performed a series of experiments studying the hole concentration and mobility dependence on the HCl input mole fraction and found a strong correlation between transport parameters and source depletion. Hole mobilities of $102\,\text{cm}^2\,\text{V}^{-1}\,\text{s}^{-1}$ for hole concentrations of $2.0 \times 10^{14}\,\text{cm}^{-3}$ were reported. Hishida *et al.* [22] studied MBE-grown phosphorus-doped ZnTe films on GaAs substrates, with the aim of using this intentional acceptor impurity to control hole concentrations precisely. As can be seen in Fig. 8.3, mobilities and carrier concentrations of these films could be varied by varying the phosphorus dopant flux. Hole densities in the range of approximately 10^{16}–$10^{18}\,\text{cm}^{-3}$ were achieved, with mobilities in the range of about 20–$30\,\text{cm}^2\,\text{V}^{-1}\,\text{s}^{-1}$.

Although ZnTe is normally p-type as grown, reports of n-ZnTe go back to the original reports of over two decades ago [37, 38]. For example Fisher *et al.* [38] prepared n-ZnTe crystals by Al-doping melt-grown material, grown from three zinc-rich melts. These authors reported electron mobilities of

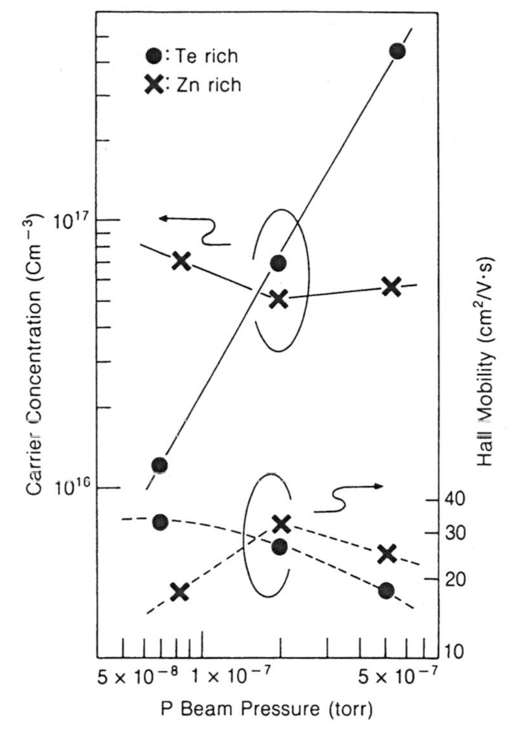

Fig. 8.3 The dependence of carrier concentration and Hall mobility of P-doped ZnTe films grown under Te-rich conditions and Zn-rich conditions on P beam pressure. (Reproduced from ref. 22 and published with the kind permission of Y. Hishida, H. Ishii, T. Toda and T. Niina.)

70–350 $cm^2 V^{-1} s^{-1}$ and corresponding resistivities of 10^5–10^7 Ω cm. Hou *et al.* [37] used fluorine ion implantation of bulk p-ZnTe to effect n-type conversion, and hence p–n junction information.

Recently Ruda [9] established a transport model for n-ZnTe. This model solved the Boltzmann transport equation exactly using a variational method, and included all major scattering mechanisms, and screening. The model predicted the dependences of carrier transport on carrier density, compensation ratio and temperature. Figure 8.4 shows the dependence of electron mobility on carrier concentration in uncompensated material at room temperature. The dominant lattice scattering mechanism is clearly polar optical phonon scattering. The other competing scattering mechanism is ionized impurity scattering. The curve for $\theta = 0$ originates from the requirement that a minimum number, $N_I = n$, of impurities are required to provide the indicated carrier concentrations. Clearly this figure sets a maximum or inherent limit to electron mobilities in ZnTe at room temperature. This maximum value is listed in Table 8.4. Compensation effects

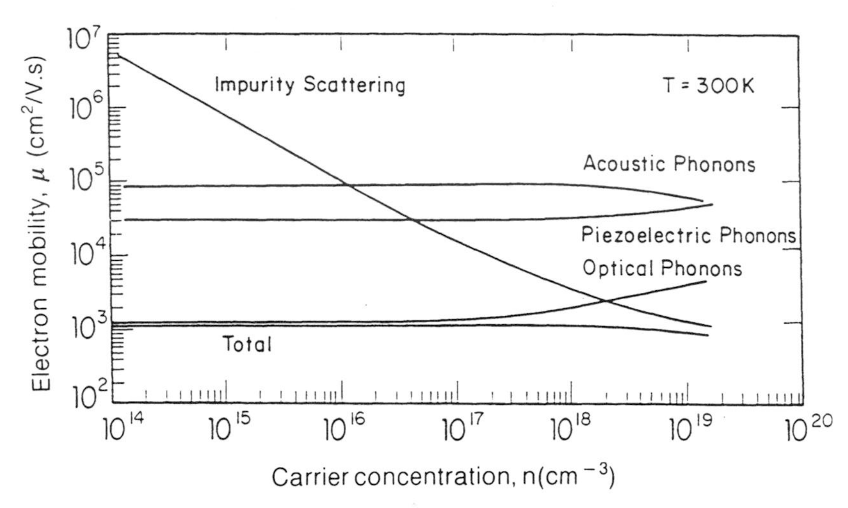

Fig. 8.4 Calculated component contributions to the total electron mobility in ZnTe at $\theta = 0$ versus carrier concentration at 300 K.

Table 8.4 Maximum inherent mobilities in widegap II–VI materials

Parameter	*ZnTe*	*ZnSe*	*Zns*
% ionic character[a]	6	15	19
μ^{3D}_{max} (300 K)(cm^2 V^{-1} s^{-1})	1 500	800	230
μ^{3D}_{max} (77 K)(cm^2 V^{-1} s^{-1})	20 000	10 000	3000
μ^{2D}_{max} (77 K)(cm^2 V^{-1} s^{-1})	—	100 000	—

[a] Based on Pauling electronegativities.

are apparent in Fig. 8.5. With increasing θ, more impurities are required to maintain a given n value and the family of mobility curves shown in this figure results. Figure 8.6 is an important figure which shows how polar optical phonon scattering is the dominant lattice scattering mechanism for temperatures in excess of about 30 K, where piezoelectric–acoustic phonon scattering plays an ever more important role. These curves also serve to illustrate how the inherent mobility limits for this material depend on temperature.

8.4.2 ZnSe

A number of attempts [14, 34, 39, 40] have been made to understand electron transport characteristics of ZnSe. Prior to 1986 there were no significant

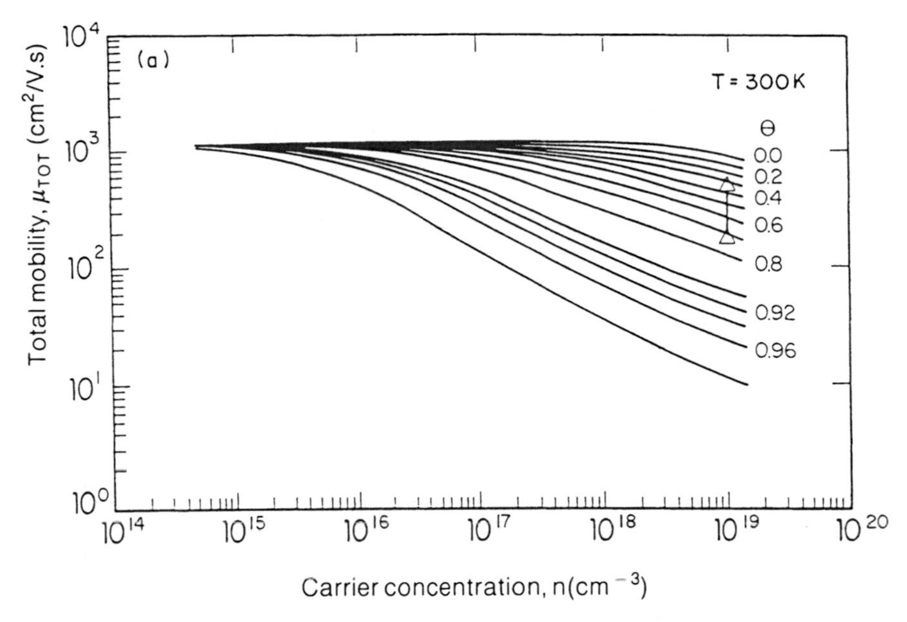

Fig. 8.5 Total electron mobility in ZnTe at different θ versus carrier concentration at 300 K. Data of ref. are indicated by triangles.

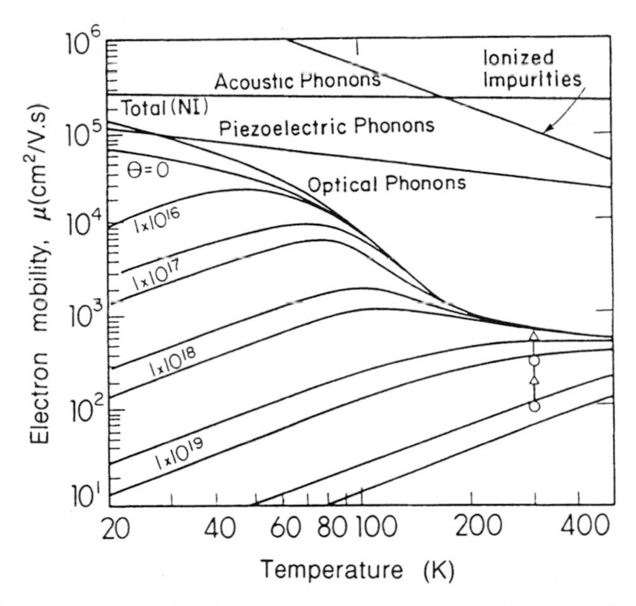

Fig. 8.6 Component contributions to the total electron mobility in ZnTe at different N_I, versus temperature. The $\theta = 0$ limit curve is also plotted. The experimental data of ref. 38 are indicated by circles and of ref. 37 by triangles.

changes in these authors' treatment of the problem, and the pioneering scheme of Aven and Segall [14] was followed or modified. These works focused on the relaxation time approximation, even for polar optical phonon–electron scatterings, and restricted analysis to non-degenerate electron statistics. Mattheissen's rule was invoked to combine relaxation rates for individual scattering mechanisms, thus providing an approximate solution to the Boltzmann transport equation. Naturally this approach removed all dependence of electron transport characteristics on carrier concentration by energy averaging. In 1986, Ruda [8] proposed an exact treatment for transport characteristics in n-ZnSe. All major electron scattering processes were included in the treatment, together with a proper treatment of screening for all of these processes. No artificial limitations on electron statistics were imposed as generalized Fermi–Dirac statistics were employed. The Boltzmann transport equation was solved exactly using a variational procedure to combine the individual electron scattering processes. As was stated in that paper, the necessity to resort to that type of an approach was dictated by the fact that longitudinal optical phonons contribute significantly to overall electron scattering rates. However, polar mode scattering is highly inelastic and thus cannot be defined within the universal relaxation time approximation. Figure 8.7 gives the results of these calculations for electron mobility versus carrier concentrations at room temperature in uncompensated material. Contrary to previously reported average energy calculations, there is a distinct concentration dependence apparent from the figure. Average energy calculations would predict straight lines parallel to the concentration axis. The essential feature of transport behaviour seen here is a dominance of electron scattering by the polar optical phonon scattering mechanism up to high concentrations of free carriers, whereupon ionized impurity scattering dominates. The essential critical concentration which partitions the behaviour is the critical degeneracy electron concentration of about $1.3 \times 10^{18}\,\mathrm{cm}^{-3}$. A further important feature of transport behaviour in this material is the apparent significance of screening effects for electron concentrations in excess of about $1 \times 10^{17}\,\mathrm{cm}^{-3}$. Note how deformation potential scattering, both acoustic and piezoelectric–acoustic, provides negligible perturbations to overall scattering rates at room temperature. Optical phonons cannot arbitrarily exchange small quantities of energy. Since the characteristic optical phonon temperature is about 364 K from Table 8.2, at 300 K and below, absorption of optical phonons is the dominating behaviour. The inherent mobility limit for ZnSe predicted for uncompensated material by the model was $800\,\mathrm{cm}^2\,\mathrm{V}^{-1}\,\mathrm{s}^{-1}$ at 300 K.

In practice, material is not uncompensated ($\theta \neq 0$). Indeed, to generate a given free carrier concentration many times that number of impurities are required, further increasing the dominance of ionized impurity scattering over other competing scattering mechanisms. Figures 8.8 and 8.9 show the results of model calculations at 300 K and 77 K respectively for mobility

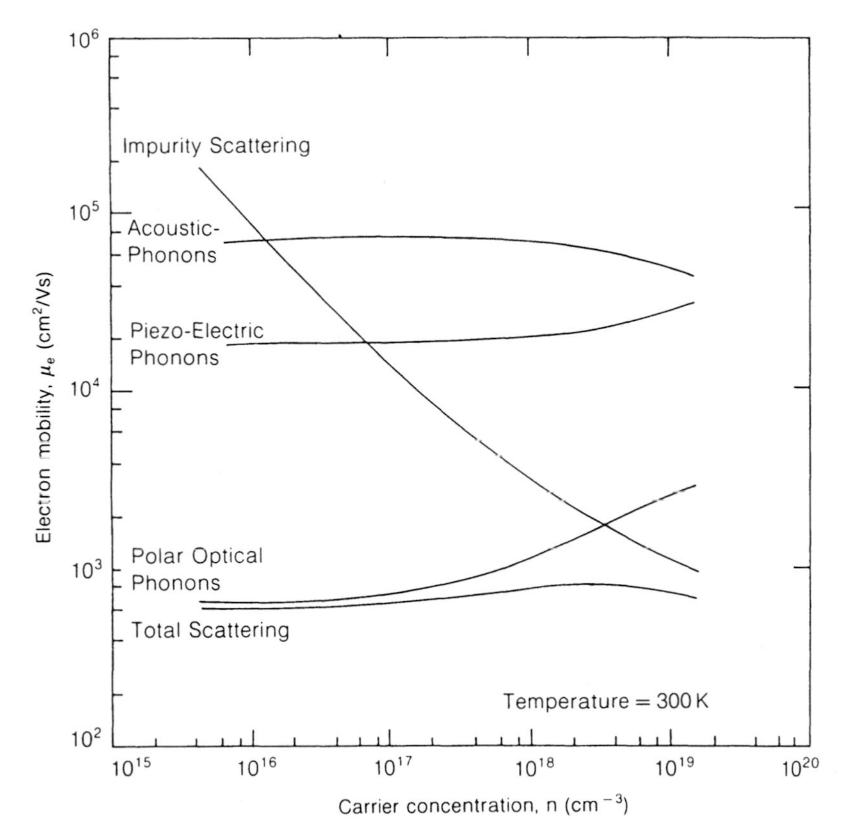

Fig. 8.7 Calculated component contributions to the total electron mobility in ZnSe at $\theta = 0$ versus carrier concentration at 300 K.

dependence on θ, compared with experimental data collected from the literature. The uppermost curves in both cases correspond to $\theta = 0$ curves. The insert to Fig. 8.8 illustrates the range of characteristics for high-quality n-ZnSe collected at that time. Region A corresponds to typical high-quality bulk and MBE-grown material with ionized impurity densities typically below about $5 \times 10^{18}\,\text{cm}^{-3}$, while region B corresponds to that of typical high-quality metal–organic chemical vapour deposition (MOCVD) grown material with ionized impurity densities in the range 5×10^{18}–$5 \times 10^{19}\,\text{cm}^{-3}$. Open and closed symbols on these figures correspond to experimental data for intentionally and unintentionally doped material respectively. Growth techniques for data points are designated as follows; squares (bulk), circles (MOCVD), diamonds (vapour phase epitaxy (VPE)), inverted triangles(MBE), triangles (liquid phase epitaxy (LPE)). The numerals adjacent to data points refer to reference numbers given in ref. 8.

Inspection of Fig. 8.8 for 300 K shows how the contribution of impurity

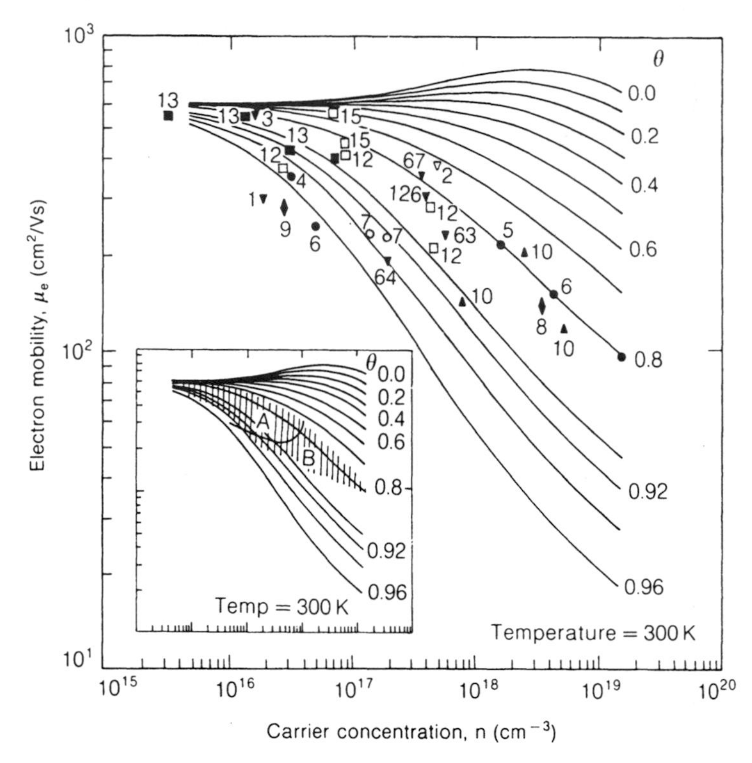

Fig. 8.8 Total electron mobility in ZnSe at different θ versus carrier concentration at 300 K. The numerals adjacent to each data point correspond to references cited in ref. 8.

scattering becomes increasingly important as θ increases, being of approximately equal importance for θ values of about 0.3. Reported mobility values are all well below the $\theta = 0$ limit. Some of the best quality material reported in the literature [14] is bulk chlorine-doped material having $\theta \sim 0.6$ at 300 K. Most material reported typically has $\theta > 0.7$ at 300 K. Thus material quality is some way from that of GaAs where $\theta = 0.22$ [41] is attainable. There are definite trends in material quality grown by various techniques, as illustrated in Fig. 8.8. Region A tends to be dominated by high-quality bulk and MBE-grown material with θ ranging from 0.70 to 0.94, while region B is dominated by high-quality MOCVD material with θ around 0.80. The very wide spread in material quality cannot be attributed to a single phenomenon. However, there are a number of factors that should be considered. One important factor is the growth technique. In the case of melt or solution growth techniques, impurities may be incorporated via a liquid phase, in which diffusion is fast. Also in melt growth and solution growth, growth temperatures tend to be much higher than in vapour growth techniques, resulting in higher impurity solubilities, coupled with higher

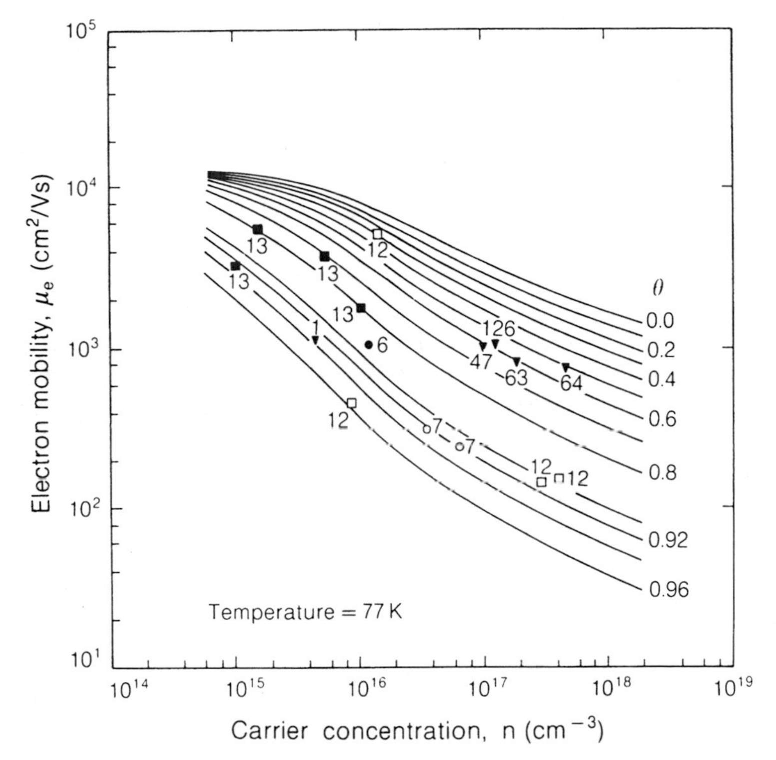

Fig. 8.9 The same as Fig. 8.8 at 77 K.

concentrations of native point defects owing to a wider existence region at higher temperatures. However, in the case of melt growth, the growth velocities are orders of magnitude faster than for MBE or MOCVD; for example, the rapidly advancing growth interface in this case can act kinetically to limit impurity incorporation. Thus the widespread use [14] of a post-growth firing in molten Zn to reduce drastically excess native point defects, coupled with the above considerations, can explain the high electronic quality attainable from bulk ZnSe. In the case of MBE, it is the combination of the low growth temperatures (typically around 300–400 °C) and the ultrahigh vacuum conditions (around 10^{-11} Torr) during growth that ensures high-quality material. In this growth technique, it is the ability to control stoichiometry by independent control over component fluxes that determines both native defect generation and also impurity incorporation. One of the most important contributions to the wide variation in electronic quality of current MBE and MOCVD material is stoichiometry. For example, the GaAs substrates are believed to be a major source of impurities [42], providing gallium which out-diffuses into the epilayers during growth, and acts as a shallow donor (substitutionally on the zinc side). Since the number of native zinc vacancies into which gallium can diffuse are stoichiometry controlled, the impurity concentration in the epilayer will itself be stoichiometry controlled. In the case of MOCVD, although growth temperatures are comparable with those in MBE, it is the combination of much higher background pressures together with the poor purity of currently available MO sources (with concomitant side reaction products that can be produced) that results in much higher levels of impurity incorporation than in MBE.

Figure 8.10 shows the calculations [8] of total mobility as a function of temperature for uncompensated material. Lowering of the temperature results in a decrease in the phonon occupation number and coincident rise in the scattering transition rate for optical phonon absorption. This phenomenon generally results in higher lattice mobilities with decreasing temperature, re-emphasizing the importance of impurity scattering. This interplay is most apparent in Fig. 8.10 and results in a peak in the electron mobility with decreasing temperature. The actual temperature for the peak mobility decreases with decreasing impurity concentration, whilst the peak mobility value itself increases. A 77 K inherent mobility limit of about $1 \times 10^4\ cm^2\ V^{-1}\ s^{-1}$ was determined for n-ZnSe from this theory.

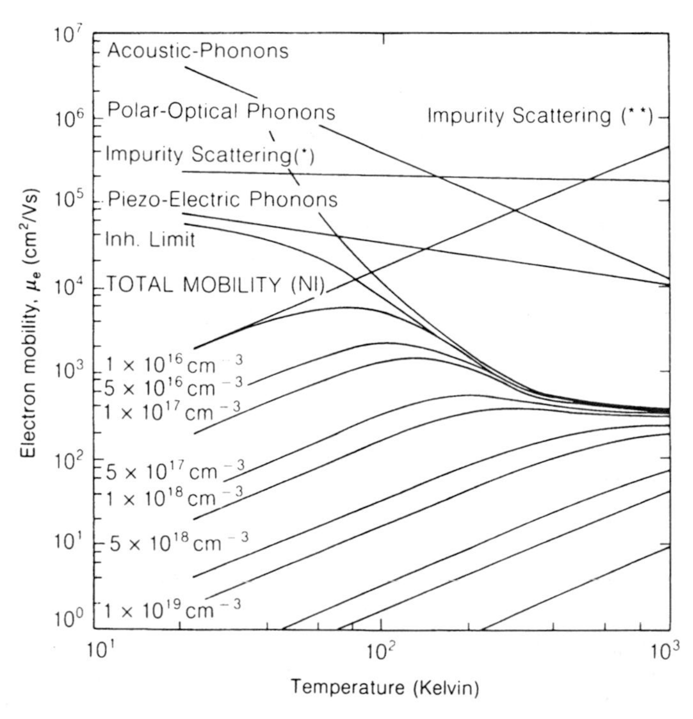

Fig. 8.10 Component contributions to total electron mobility in ZnSe at different N_I versus carrier concentration. The $\theta = 0$ limit curve is also plotted.

A recent paper by Morimoto [43] investigates the influence of stoichiometry on electron transport characteristics of ZnSe. The films were grown on (100)GaAs substrates by low pressure MOCVD using DMZ and H_2Se sources. The films of Fig. 8.11 were grown at a substrate temperature of 350 °C. The figure shows the dependence of carrier concentration and mobility on selenium:zinc flux in the vapour. By varying the stoichiometry, the carrier concentration varied by about two orders of magnitude from about 10^{15} to 10^{17} cm^{-3} with room-temperature electron mobilities varying correspondingly. By analysis of these results using the theoretical model of Ruda [8] the impurity density in these films varied by over an order of magnitude with stoichiometry. These results point to the subtlety of engineering n-ZnSe materials characteristics. Even for a given growth technique, there are many factors that must be strictly controlled to control transport properties.

Kamata *et al.* [44] report on some of the highest quality material prepared to date. These authors used MOCVD to prepare films, and introduced Cl

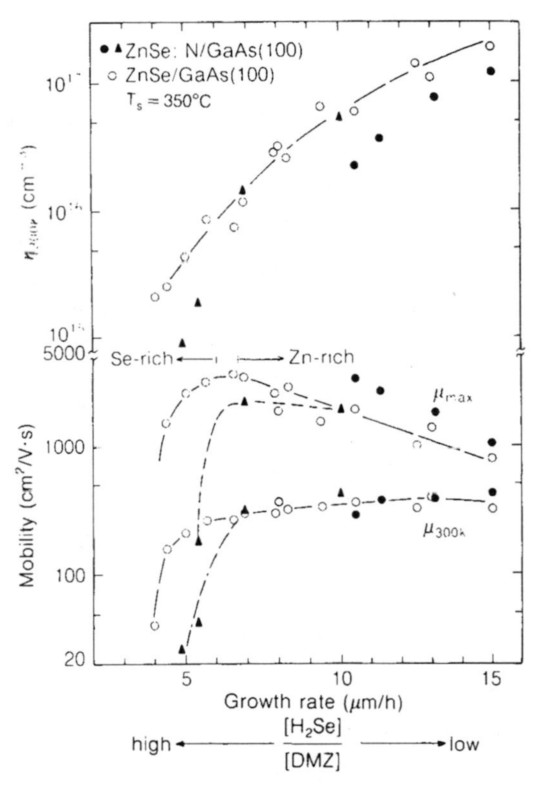

Fig. 8.11 The dependence of carrier concentration and electron mobility in ZnSe at 300 K on gas composition. (Reproduced from ref. 43 and published with the kind permission of K. Morimoto.)

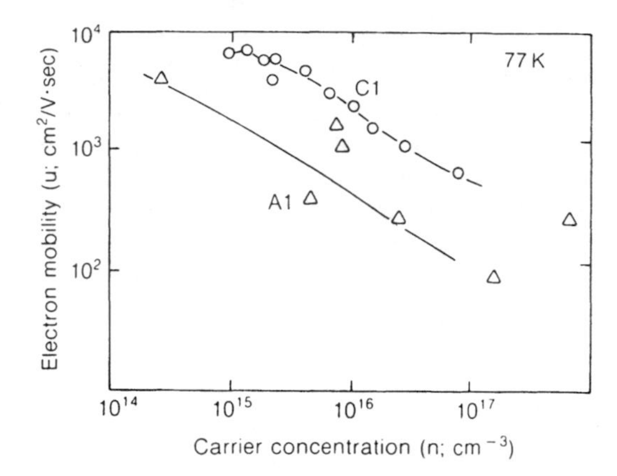

Fig. 8.12 The dependence of electron mobility on carrier concentration at 77 K for Al- and Cl-doped ZnSe samples. Samples were grown at a VI:II ratio of 0.9. (Reproduced from ref. 44 and published with the kind permission of A. Kamata, T. Uemoto, M. Okajima, K. Hirahara, M. Kawachi and T. Beppu.)

and Al (using a TEAl source) to permit n-type doping. Figure 8.12 is taken from their paper. Analysis of these data suggests that chlorine is a much more effective donor dopant than aluminium in ZnSe. For ZnSe:Cl, the best quality material is grown with lower carrier concentrations, having θ values of about 0.6 and impurity concentrations of about mid $10^{15}\,cm^{-3}$. At higher carrier concentrations, θ approaches about 0.9 and impurity concentrations may exceed $10^{19}\,cm^{-3}$. Calculating the donor doping efficiency (n/N_d^+) for this material reveals that at low concentrations the efficiency of doping is over 25%, whereas at higher doping concentrations this values sinks below 10%.

Preliminary studies of hole transport in ZnSe were reported by Kranzer [45]. This work was followed by a rigorous treatment by Ruda [1] which included all major hole scattering mechanisms and importantly accounted for the p-symmetry of valence band states. The Boltzmann transport equation was then solved for these partially coupled bands (for light and heavy holes) applying a variational method. The relaxation time approximation was not invoked and generalized Fermi–Dirac statistics was applied throughout. Both intra- and interband hole scattering processes were included. Figure 8.13 shows the calculated component hole mobilities as a function of carrier concentration at 300 K. Clearly, heavy hole scattering is the dominant contribution to the total scattering rate. Heavy hole polar optical phonon scattering and ionized impurity scattering are the two most important scattering mechanisms in this material, determining a room-temperature inherent mobility limit of about $110\,cm^2\,V^{-1}\,s^{-1}$.

Figures 8.14 and 8.15 provide data on calculated dependences of total hole mobility on carrier concentration at different levels of compensation. Figure 8.16 is taken from ref. 1 and plots all experimental data available at

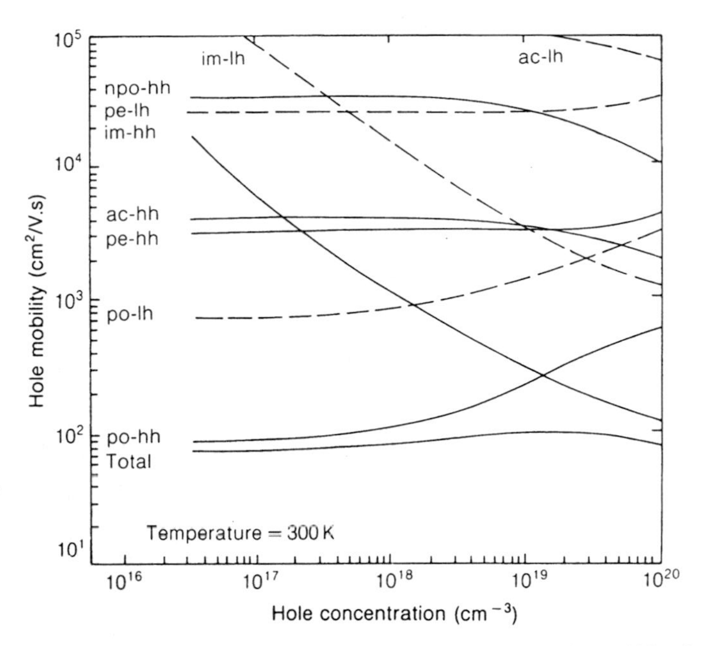

Fig. 8.13 Calculated component contributions to the total hole mobility in ZnSe at $\theta = 0$ versus carrier concentration at 300 K.

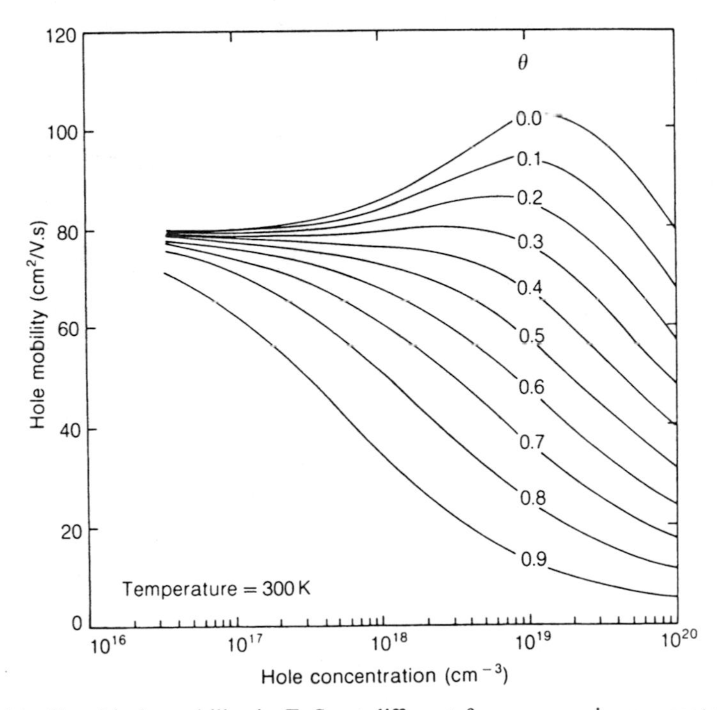

Fig. 8.14 Total hole mobility in ZnSe at different θ versus carrier concentration at 300 K.

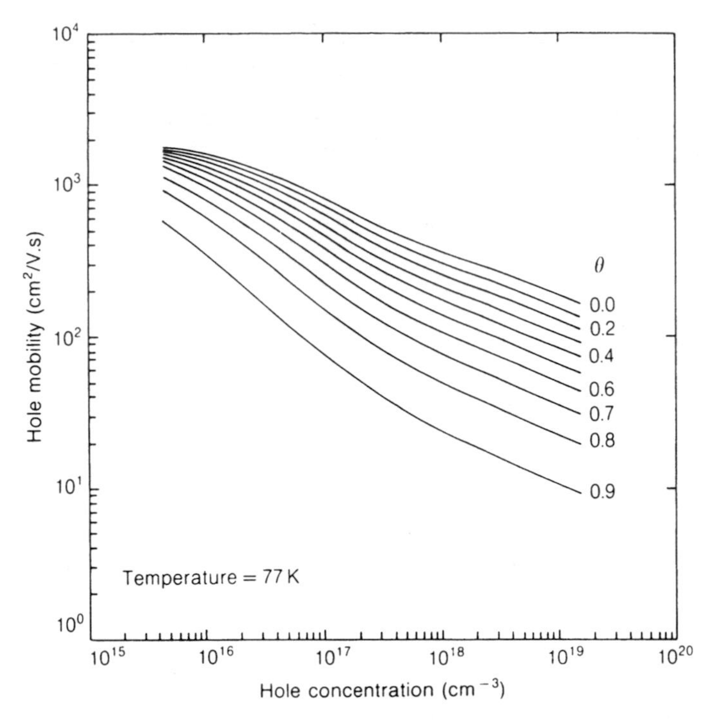

Fig. 8.15 The same as Fig. 8.14 at 77 K.

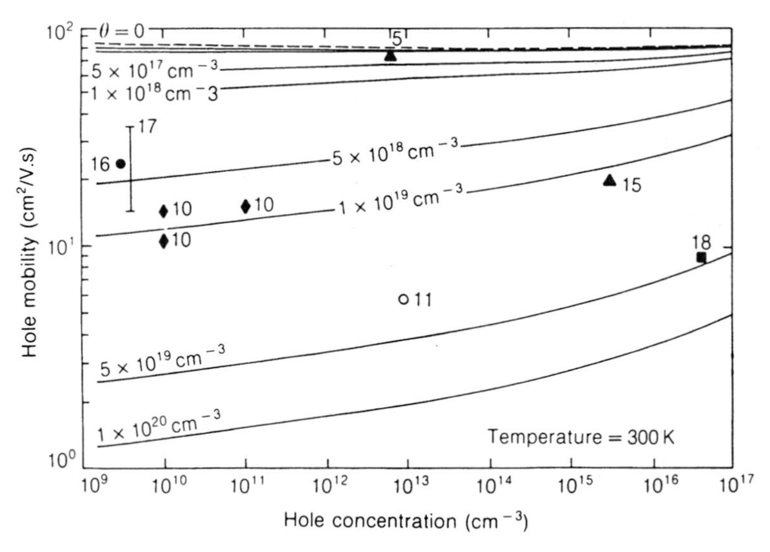

Fig. 8.16 Total hole mobility in ZnSe at different N_I versus carrier concentration at 300 K. Data points are those given in ref. 1.

that time on p-type ZnSe. Note how almost all of the samples plotted correspond to impurity densities of about $5 \times 10^{18}\,cm^{-3}$ or worse. The work of Nishizawa [46] (reference point 15 on the figure) is an exception to this pattern, with exceedingly high hole mobility and very low ionized impurity density (just above about $1 \times 10^{17}\,cm^{-3}$). Figure 8.17 shows calculated dependences of hole mobility on temperature [1] compared with the experimental data of Park *et al.* [47]. The excellent agreement between theory and experiment point to the usefulness of this type of modelling for understanding hole transport in ZnSe.

There have recently been two noteworthy contributions to studies of p-type ZnSe electronic properties. The first is the paper by Yasuda *et al.* [48] who report on low-resistivity lithium-doped p-ZnSe layers grown by MOCVD. These authors report on the lowest resistivity to date of $0.2\,\Omega\,cm$ and material with $9 \times 10^{17}\,cm^{-3}$ carrier concentration. The second exciting development in this field is the paper by Potts *et al.* [49] on group I doping of ZnSe by MBE. This is a first report of successful p-type doping of ZnSe by MBE and opens the way to novel light-emitting device developments in this important materials system.

Free-carrier absorption (FCA) has recently been suggested [50] as a useful

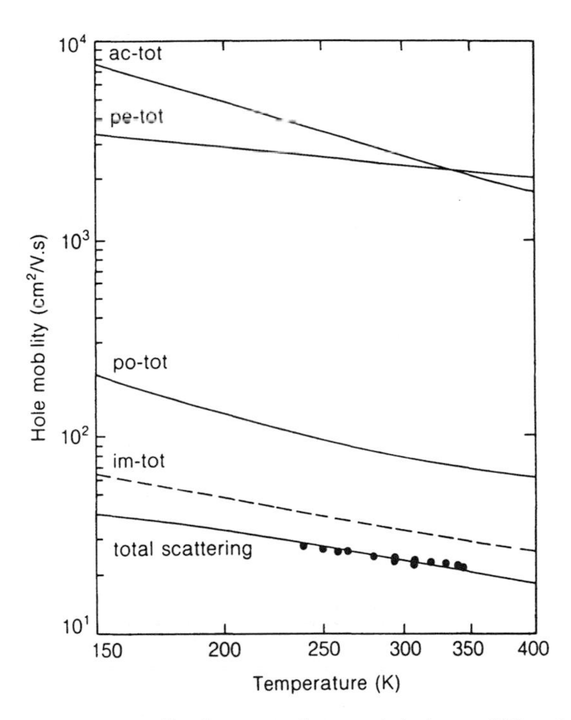

Fig. 8.17 Component contributions to the total hole mobility in ZnSe versus temperature. The experimental data points are those cited for Fig. 12 of ref. 1.

means of characterizing and understanding electronic transport properties of ZnSe. Ruda [50] has formulated a theory for FCA, and studied the influence of impurity concentration and compensation on FCA characteristics. Inherent and absolute FCA coefficients were derived for probe wavelengths of 3.3, 5.0 and 10 μm. A new parameter, *p*, the logarithmic absorption coefficient, was shown to be a powerful tool in analysing FCA behaviour. Figure 8.18 shows the contributions of various scattering mechanisms to FCA as a function of carrier concentration for uncompensated material. Notice how polar optical phonon scattering again dominates the picture. Figure 8.19 shows calculated FCA curves for different θ values versus carrier concentration at 3.3 μm and 5.0 μm. Superimposed on this figure are the experimental data of Dutt *et al.* [51] for aluminium-doped ZnSe. The ease of deducing θ values from these data is quite apparent.

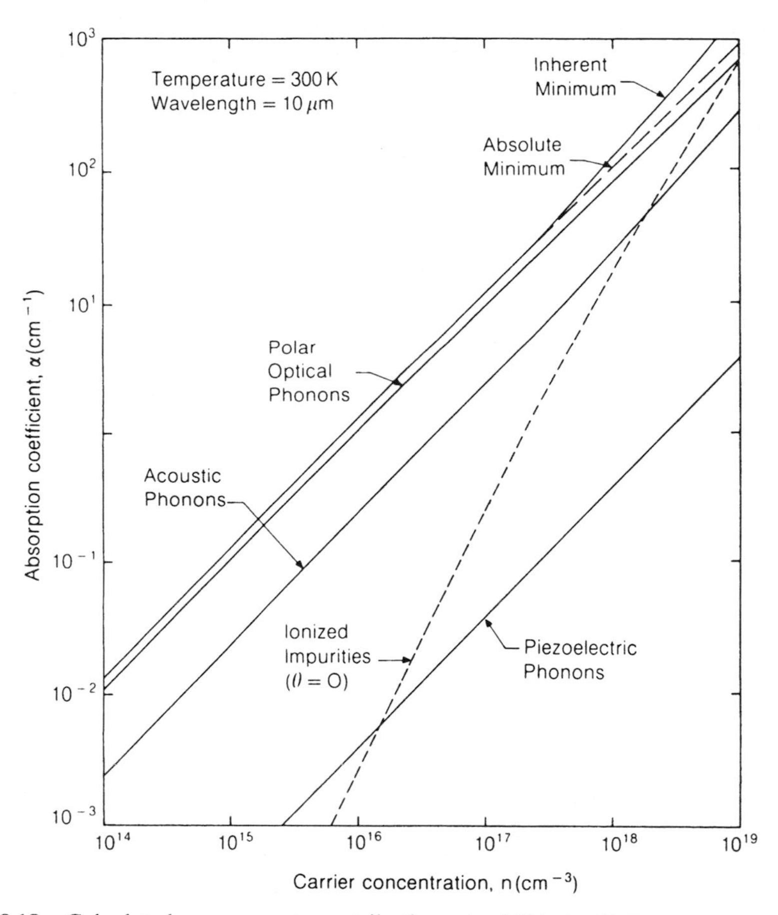

Fig. 8.18 Calculated component contributions to FCA in ZnSe versus carrier concentration at 10 μm and 300 K, for $\theta = 0$.

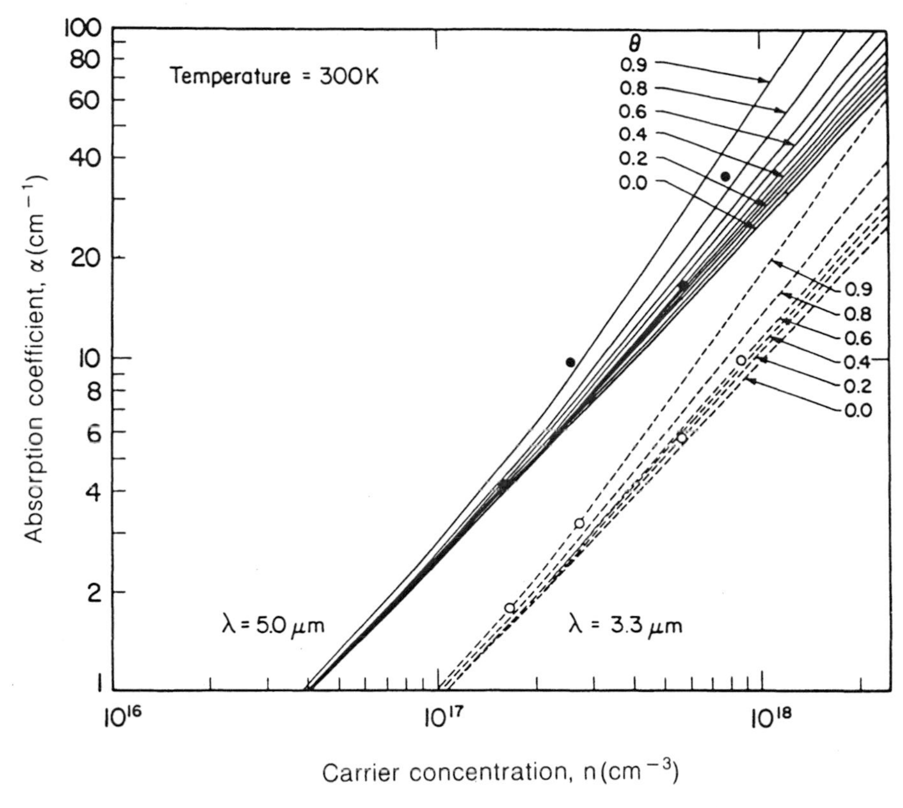

Fig. 8.19 Calculated total FCA in ZnSe versus carrier concentration as a function of θ for 3.3 μm and 5.0 μm.

Brennan [52] has recently published a paper on a theory of high-field electronic transport in ZnSe. The calculations were based on an ensemble Monte Carlo method under conditions of high applied field strengths. Full details of conduction band structure and electron–phonon interactions were included in this author's analysis. Figure 8.20 shows the results of these calculations, giving electron drift velocity versus applied field at 300 K. Such results are clearly applicable in helping to understand the behaviour of electroluminescent devices.

Other developments in ZnSe transport include the work by Ruda [33] on quantum transport in a ZnSe 2D electron gas confined at a lattice-matched Zn(S, Te) heterointerface. The proposed heterostructure is shown schematically in Fig. 8.21. Carriers are transferred from the selectively doped ternary alloy across an undoped spacer. The purpose of this proposed scheme was to minimize coulombic scattering from the tails of these remote ionized impurity centres. Indeed, the confinement of a high density of free carriers spatially separated from their parent ionized donors is a unique characteristic

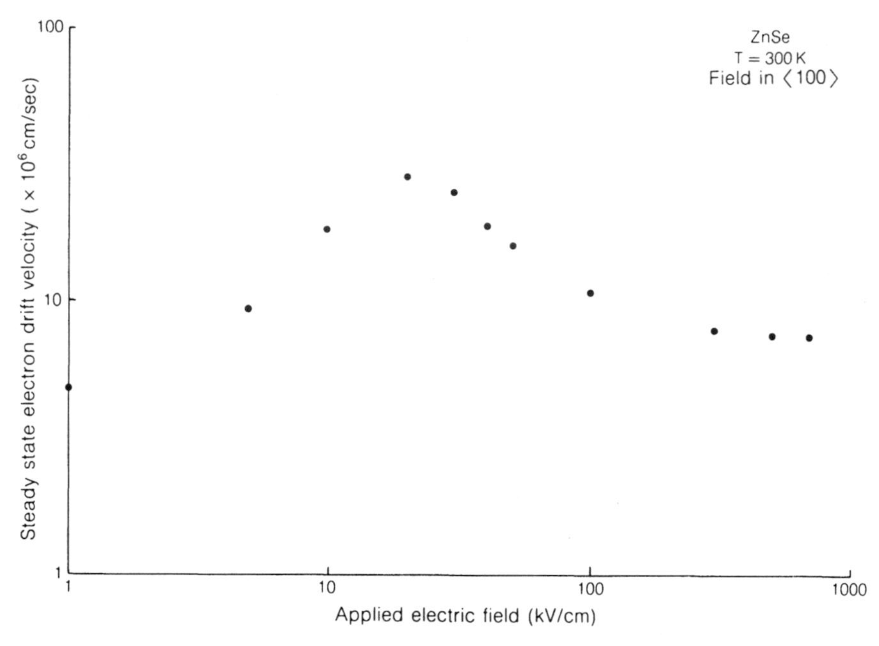

Fig. 8.20 Steady state electron drift velocity as a function of applied electric field in ZnSe at 300 K. The field is applied in the ⟨100⟩ direction. The threshold field for intervalley transfer is found to be roughly 25 kV cm^{-1}. (Reproduced from ref. 52 and published with the kind permission of K. Brennan.)

of these structures. This is responsible for providing a significant enhancement in transport characteristics compared with conventional bulk transport. The calculated component mobilities for electrons in the ZnSe channel are given in Fig. 8.22. This figure is based on a 100 Å spacer width, and investigates a variety of alloy scattering potentials for electrons whose wavefunctions seep

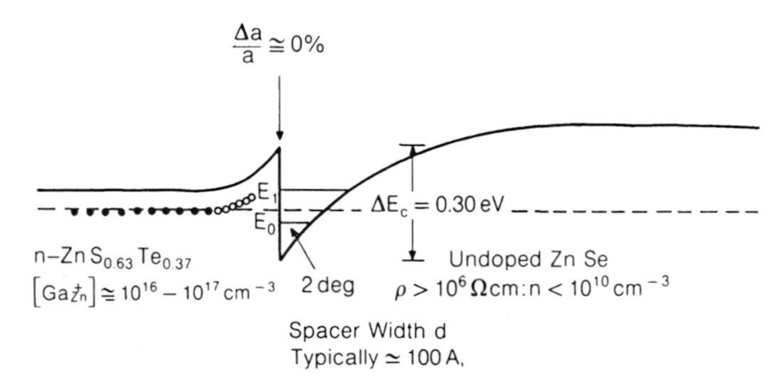

Fig. 8.21 Schematic of 2DEG (2d electron gas) heterostructure with electron gas confined in ZnSe at a lattice-matched interface to a selectively doped Zn(S, Te) layer.

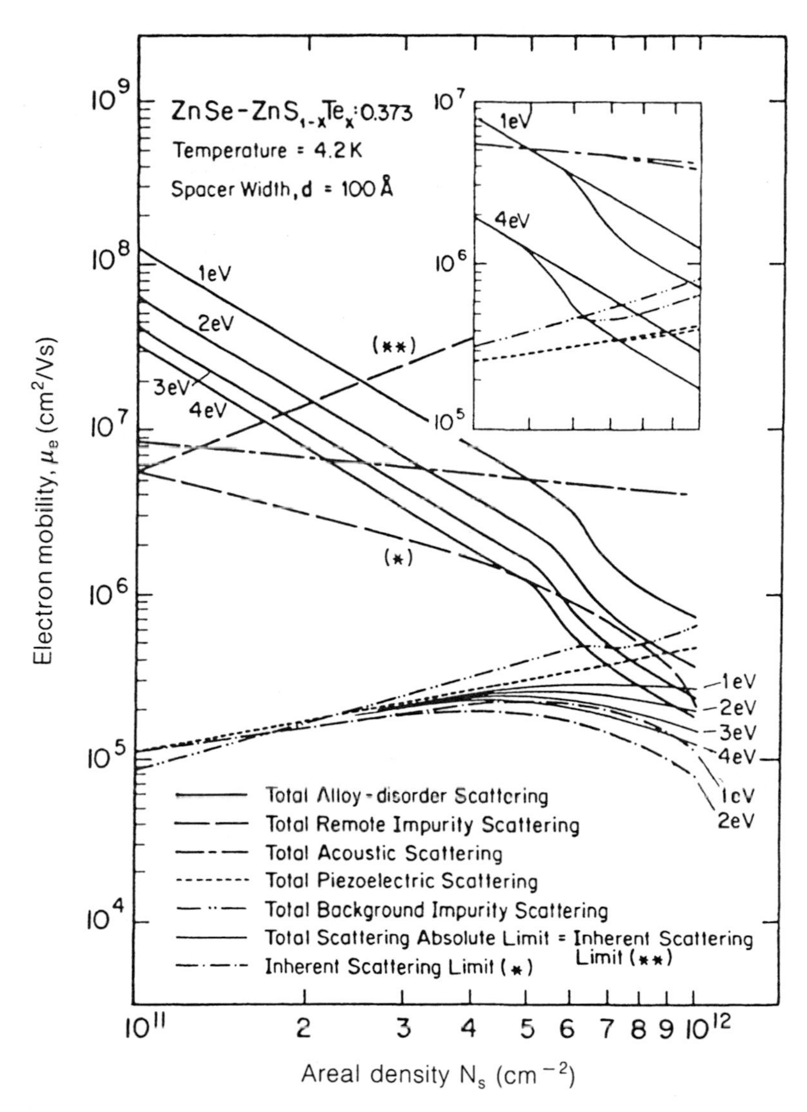

Fig. 8.22 Calculated component contributions to electron mobility in a ZnSe 2DEG, as shown schematically in Fig. 8.21. (Taken from ref. 33).

through the well walls and are scattered by the alloy disorder scattering mechanism in the ternary. Figure 8.23 shows the calculated inherent mobility limits for such structures at 77 K and 4.2 K, of about $1 \times 10^4\,cm^2\,V^{-1}\,s^{-1}$ and $3 \times 10^5\,cm^2\,V^{-1}\,s^{-1}$ respectively. This 77 K mobility is about an order of magnitude higher than that for bulk ZnSe at the same temperature.

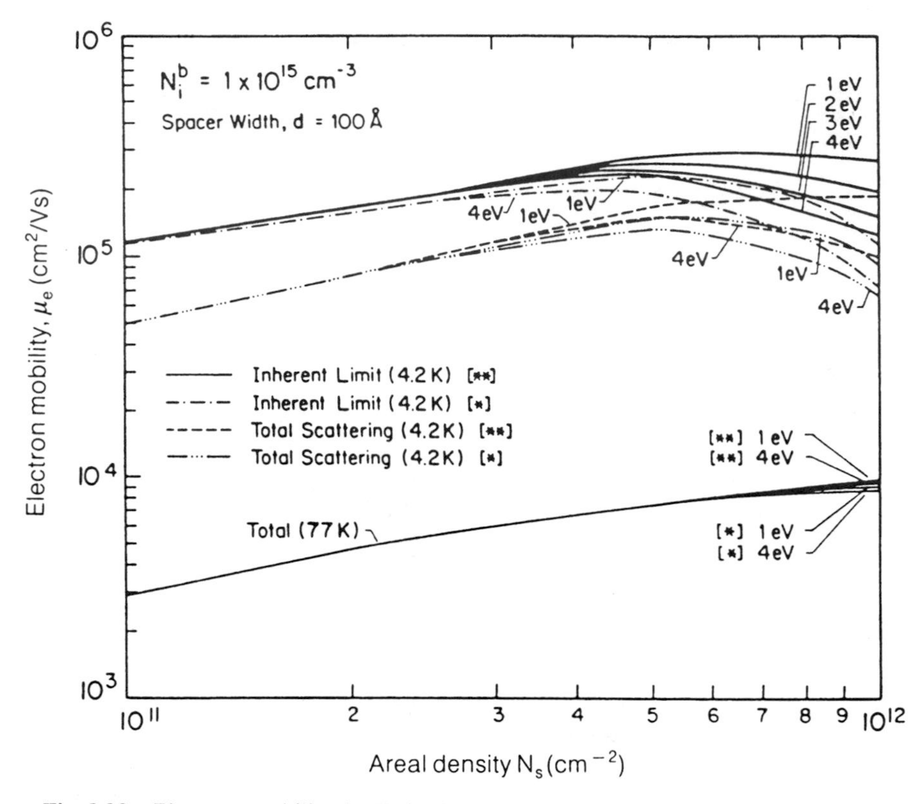

Fig. 8.23 Electron mobility in ZnSe 2DEG versus sheet carrier concentration at 4.2 K and 77 K. (Taken from ref. 33.)

8.4.3 ZnS

Recent attempts have been made to incorporate shallow donor and acceptor impurities in ZnS, leading to junction formation. Until recently only a single attempt at modelling transport characteristics of ZnS was made. This was an exploratory work by Aven and Mead [13] of similar scope and approach to the one they reported for ZnSe. Since that early work some 25 years ago, there has only been a single other work by Ruda and Lai [10]. Their paper outlines an exact approach to the problem following the same guidelines as earlier reported works by these authors. Figure 8.24 shows the results of their calculations for component mobility contributions to the total scattering rate at 300 K versus carrier concentration. The transport characteristics are very similar to those reported for ZnSe [8], for example. Polar optical phonon scattering and polar interactions in general are much more pronounced in this material, reflecting an increased ionicity of the zinc blende lattice. Furthermore, typically mobilities are some four to five times lower in this

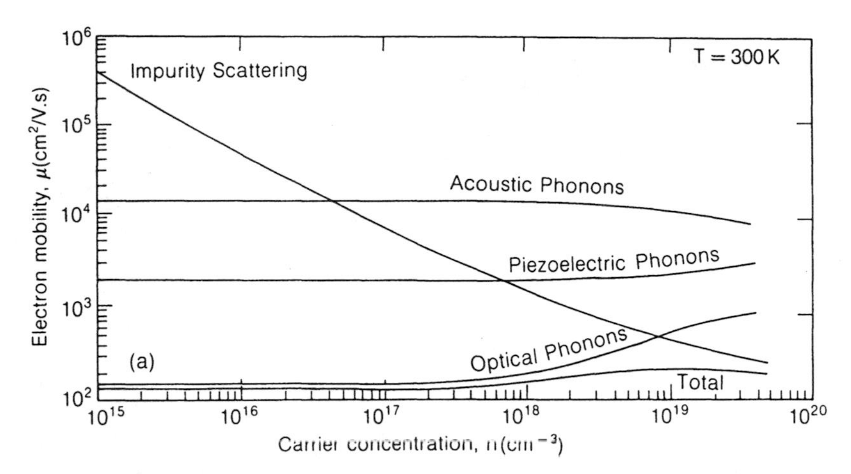

Fig. 8.24 Calculated component contributions to the total electron mobility in ZnS at $\theta = 0$ versus carrier concentration at 300 K.

material system. Figures 8.25 and 8.26 show the calculated total mobilities versus carrier concentration as a function of compensation ratio at both 300 K and 77 K respectively. Available experimental data from the literature have been plotted on these figures. The error of the old maxim of equating high quality material with high mobilities can clearly be seen by reference

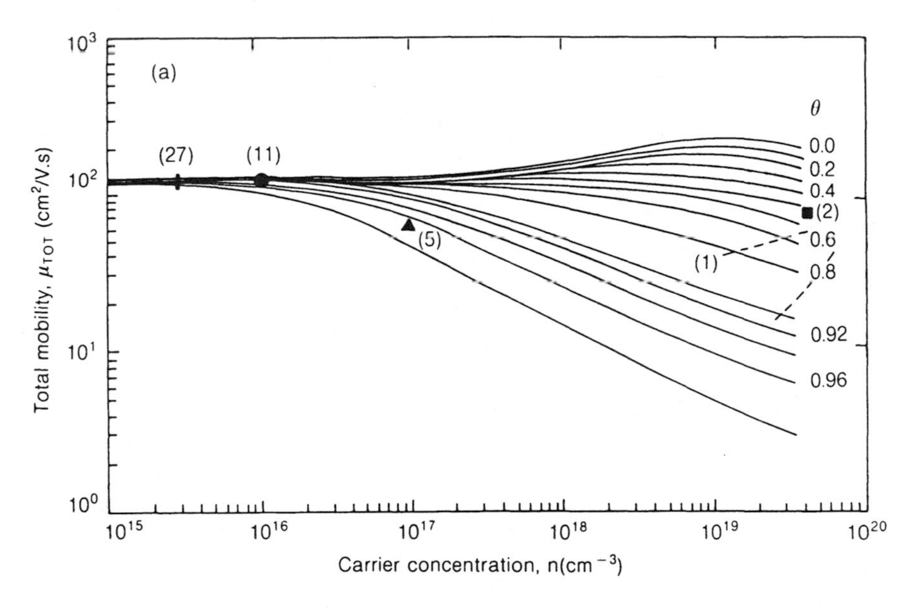

Fig. 8.25 Total electron mobility in ZnS at different θ versus carrier concentration at 300 K. The numerals adjacent to each data point correspond to the references cited in ref. 10.

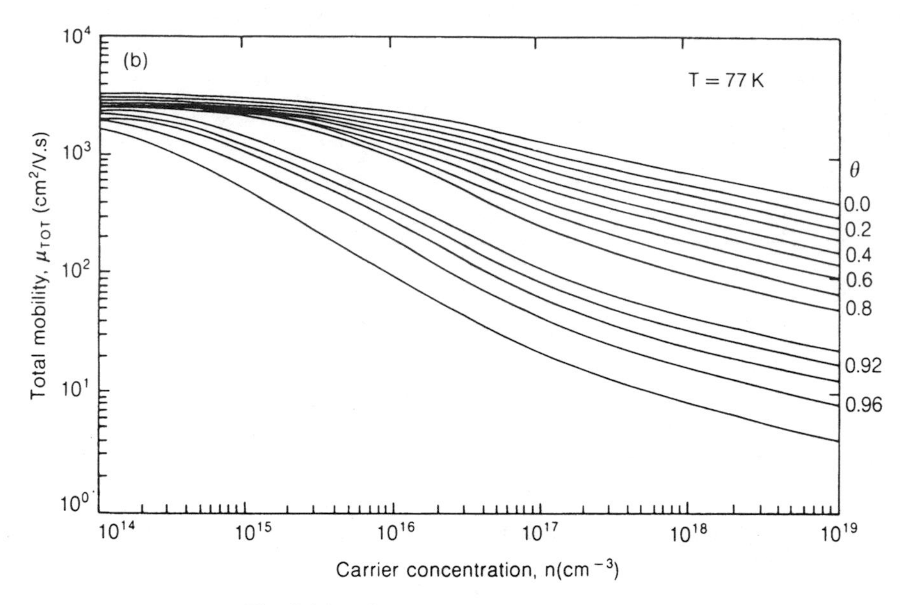

Fig. 8.26 The same as Fig. 8.25 at 77 K.

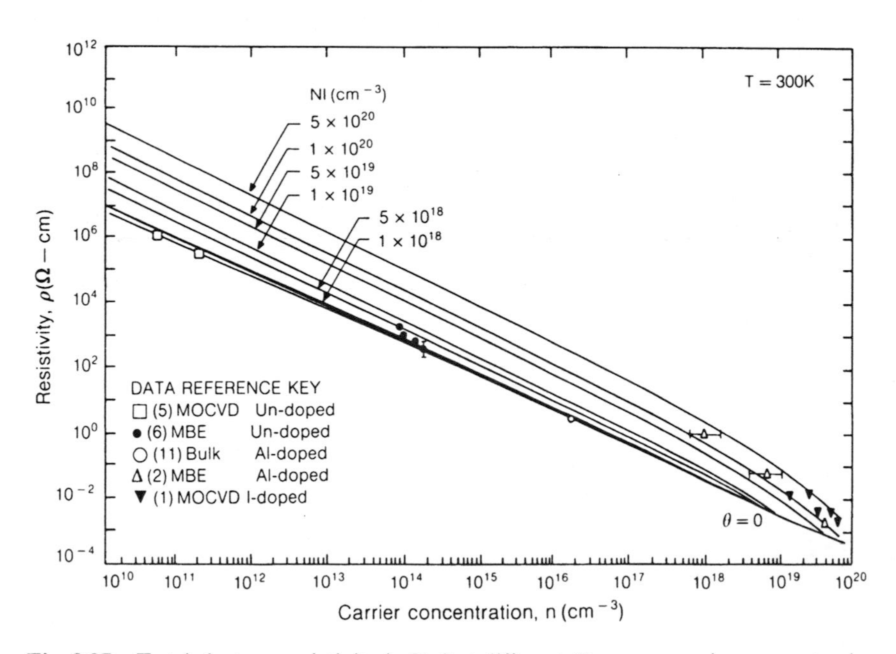

Fig. 8.27 Total electron resistivity in ZnS at different N_I versus carrier concentration at 300 K. The data points are those cited in ref. 10.

to Fig. 8.25. Data points 2 and 5 exhibit very similar mobilities, but quite different levels of compensation; that is, $\theta \sim 0.55$ and 0.97 respectively. These figures also, interestingly, indicate that present state-of-art ZnS is of high quality (low θ and N_I), approaching theoretical limit predictions in some cases.

Figure 8.27 taken from ref. 10 shows total resistivity versus carrier concentration for different N_I values, at 300 K. Data points from a variety of experiments are superimposed on this figure. The theory is able to account for the observed sample resistivities over a ten decade range of carrier concentrations, for both doped and undoped samples grown by a variety of different growth techniques. Undoped ZnS is semi-insulating as might be expected given its wide bandgap at room temperatures; however, resistivity can be made to vary over at least four orders of magnitude. Doping with either iodine or aliminium is effective in lowering resistivities to the mid $10^{-3}\,\Omega$cm range by introducing shallow donors.

The data of Fig. 8.28 were derived from the experimental results of iodine-doped ZnS grown by MOCVD [17]. Measured resistivities and

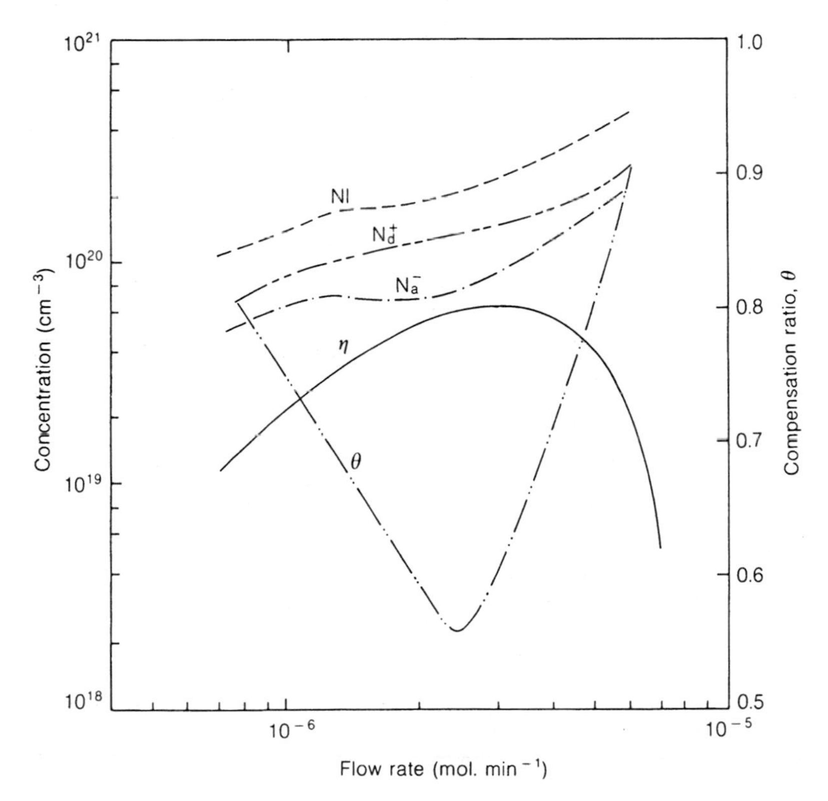

Fig. 8.28 Calculated dependence of electron concentration and compensation ratio on iodine dopant flux taken from the data of ref. 17, and analysed in ref. 10.

mobilities were reported for samples grown at constant stoichiometry; zinc and sulphur gas flows were fixed, as was the GaAs substrate temperature. Thus mobility and carrier concentration versus iodine dopant flux could be derived. Using the model of Ruda and Lai [10], θ, N_I, N_d^+, and N_a^- values were calculated. As can be seen from Fig. 8.28, the ionized donor concentration increases approximately in proportion to the dopant flux, indicative of substitutional incorporation of iodine on sulphur sites.

Unexpectedly, at intermediate iodine flow rates there is a plateau in the ionized acceptor concentration, resulting in a sharp rise in the net carrier density. Thus θ has a minimum of about 0.56 with iodine flow rate. Off-optimum θ values of about 0.9 are possible, indicating the usefulness of such models for guiding growth experiments for optimum materials preparation.

Figure 8.29 shows calculated component mobility contributions to total mobility as a function of temperature, together with the experimental data of Aven and Mead [13]. Interestingly, the role of different scattering mechanisms is somewhat different from that in the other widegap II–VI materials. At successively lower temperatures piezoelectric–acoustic phonon scattering takes over as the dominant scattering mechanism from polar optical phonon scattering. The agreement between theory and experiment is excellent. Inherent mobility limits were derived for ZnS as about 230 $cm^2 V^{-1} s^{-1}$ and over 3000 $cm^2 V^{-1} s^{-1}$ at 300 K and 77 K respectively.

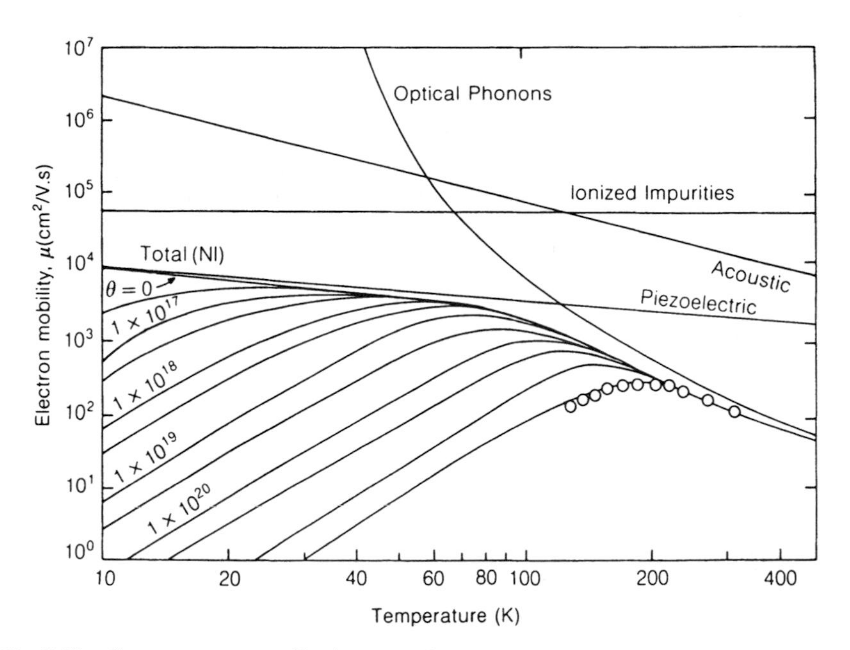

Fig. 8.29 Component contributions to the total electron mobility in ZnS versus temperature.

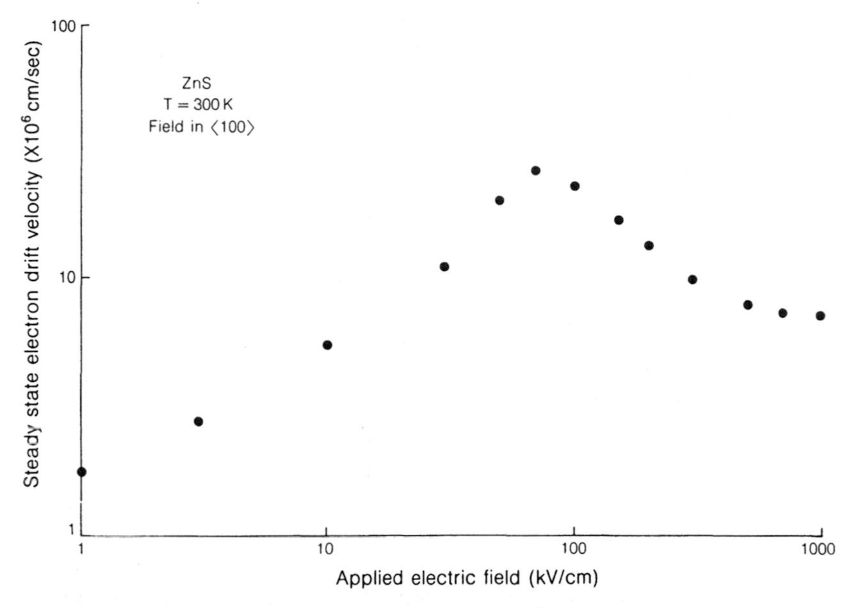

Fig. 8.30 Steady state electron drift velocity in ZnS as a function of applied field. The threshold field for intervalley transfer is about 70 kV cm^{-1}; this is much greater than for ZnSe owing to greater confinement of electrons in the valley. (Reproduced from ref. 52 and published with the kind permission of K. Brennan.)

Figure 8.30 shows the results of calculations of high-field transport behaviour of ZnS, from the work of Brennan [52]. These calculations were based on an ensemble Monte Carlo method under conditions of high applied field strengths. Electron drift velocity versus field is shown in this figure. These results are very pertinent to the development of high field a.c. electroluminescent cells.

8.4.4 Ternary alloys

Although there has been considerable work on the optical properties of the ternary alloys, there has been very little work on transport properties. Aven and Garwacki [53] measured carrier concentration and mobilities of several n- and p-type samples of Zn(Se, Te). They found deep donors ($E_D \sim 0.01$ eV) and deep acceptors ($E_A \sim 0.37$ eV).

8.5 CONCLUSIONS

A key thread through transport characteristics of II–VI materials considered in this review is the influence of bond ionicity. In all cases the dominant intrinsic scattering mechanisms reflect the polar nature of these lattices. Thus polar optical phonon scattering and to a lesser extent piezoelectric–acoustic deformation potential scattering are especially important. These considerations naturally follow through to all the transport properties discussed. These include low-field and high-field transport under d.c. conditions, as well as a.c. transport (FCA), and low-dimensional transport.

A second major point that emerges is the need for an awareness of the importance of controlling and understanding impurity effects. Compensation is a major problem that antagonizes our efforts to dope these materials effectively. Control of stoichiometry and background unintentional residual impurities will continue to be problems.

Although significant progress has been made in achieving amphoteric doping of these materials, suitable species and incorporation conditions remain to be established.

As a final note, the possibility of engineering structures by advanced growth techniques such as MOCVD and MBE is only just beginning to occur. Reference to Table 8.4 shows a summary of the influence of ionicity on mobility limits in these materials. However, it also shows the inherent advantage of 2D transport systems over 3D systems. This area, coupled with examining multicomponent systems which are lattice matched to selected substrates, should prove a fertile new area for achieving tunable optical materials with enhanced transport characteristics.

ACKNOWLEDGEMENTS

Research on the transport properties of widegap II–VI materials has been supported by the Defence Advanced Research Projects Agency under ONR Grant number N00014-85-C-0052, by 3M Corporation, and by the Natural Sciences and Engineering Research Council of Canada.

REFERENCES

1. Ruda, H.E. (1986) *J. Appl. Phys.*, **59**, 3516.
2. Nag, B.R. (1980) *Electron Transport in Compound Semiconductors*, Springer Series in Solid-State Sciences 11, Springer, Berlin.
3. Kane, E.O. (1966) In *Semiconductors and Semimetals* (eds Willardson, R.K. and Beer, A.C. Academic Press, New York, Vol. 1.
4. Ehrenreich, H. (1957) *J. Phys. Chem. Solids*, **2**, 131.
   Ehrenreich, H. (1959) *J. Phys. Chem. Solids*, **8**, 130.
   Ehrenreich, H. (1959) *J. Phys. Chem. Solids*, **9**, 129.
5. Wiley, J.D. (1971) *Phys. Rev. B*, **4**, 2485.
6. Matz, D. (1967) *J. Phys. Chem. Solids*, **28**, 373.
   Matz, D. (1968) *Phys. Rev*, **168**, 843.
7. Vassell, M.O., Ganguly, A.K. and Conwell, E.M. (1970) *Phys. Rev. B*, **2**, 948.
8. Ruda, H.E. (1986) *J. Appl. Phys.*, **59**, 1220.
9. Ruda, H.E. (1991) *J. Phys. D*, **24**, 1158.
10. Ruda, H.E. and Lai, B. (1990) *J. Appl. Phys.*, **68**, 1714.

11. Magnea, N., Saminadayar, K., Galland, D., Pautrat, J.L. and Triboulet, R. (1982) In *Extended Abstracts of Electrochemical Society Meeting, Minneapolis, MN, 1982*, p. 404.
12. Merz, J.L., Kukimoto, H., Nassau, K. and Shiever, J.W. (1972) *Phys. Rev. B*, **6**, 545.
13. Aven, M. and Mead, C.A. (1965) *Appl. Phys. Lett.*, **7**, 8.
14. Aven, M. and Segall, B. (1963) *Phys. Rev.*, **130**, 81.
15. Kroger, F.A. (1956) *Physica*, **22**, 637.
16. Bube, R.H. and Lind, E.L. (1958) *Phys. Rev.*, **110**, 1040.
17. Kawazu, Z., Kawakami, Y., Taguchi, T. and Hiraki, A. (1989) *Mater. Sci. Forum*, **38–41**, 555.
18. Dean, P.J., Herbert, D.C., Werkhoven, C.J., Fitzpatrick, B.J. and Bhargava, R.N. (1981) *Phys. Rev., B*, **23**, 4888.
19. Magnea, N., Bansahel, D., Pautrat, J.L. and Pfister J.C. (1979) *Phys. Status Solidi B*, **94**, 627.
20. Gezci, S. and Woods, J. (1980) *J. Appl. Phys.*, **51**, 1866.
21. DePuydt, J.M., Smith, T.L., Potts, J.E., Cheng, H. and Mohapatra, J.M. (1988) *J. Cryst. Growth*, **86**, 318.
22. Hishida, Y., Ishii, H., Toda, T. and Niina, T. (1989) *J. Cryst. Growth*, **95**, 517.
23. Wu, Z.L., Merz, J.L., Werkhoven, C.J. and Fitzpatrick, B.J. (1982) *Appl. Phys. Lett.*, **40**, 345.
24. Halstead, R.E. and Aven, M. (1965) *Phys. Rev. Lett.*, **14**, 64.
25. Nakau, N., Fujiwara, J., Yoshitake, S., Takenoshita, H., Itoh, N. and Okuda, M. (1982) *J. Cryst. Growth*, **59**, 196.
26. Bacchelli, L. and Jacobini, C. (1972) *Solid State Commun.*, **10**, 71.
27. Spitzer, L. and Harm, R. (1953) *Phys. Rev.*, **89**, 977.
28. Luong, M. and Shaw, A.W. (1971) *Phys. Rev. B*, **4**, 2436.
29. Bate, R.T., Baxter, R.D., Reid, F.J. and Beer, A.C. (1965) *J. Phys. Chem. Solids*, **26**, 1205.
30. Appel, J. (1961) *Phys. Rev.*, **122**, 1760.
31. Ruda, H.E., Walukiewicz, W., Lagowski, J. and Gatos, H.C. (1984) *Phys. Rev. B*, **29**, 4818.
32. Ruda, H.E., Walukiewicz, W., Lagowski, J. and Gatos, H.C. (1984) *Phys. Rev. B*, **30**, 4571.
33. Ruda, H.E. (1986) *Appl. Phys. Lett.*, **49**, 35.
34. Devlin, S. (1967) In *Physics and Chemistry of II–VI Compounds* (eds Aven M. and Prener, J.S.) North-Holland, Amsterdam, Chapter 11.
35. Kitagawa, F. and Takahashi, K. (1977) *Electr. Eng. Jpn.*, **97**, 1.
36. Nishio, M., Nakamura, Y. and Ogawa, H. (1983) *Jpn. J. Appl. Phys.*, **22**, 1101.
37. Hou, S.L., Beck, K. and Marley, J.A. (1969) *Appl. Phys. Lett.*, **14**, 151.
38. Fisher, A.G., Carides, J.N. and Dresner, J. (1964) *Solid State Commun.*, **2**, 157.
39. Aven, M. (1971) *J. Appl. Phys.*, **42**, 1204.
40. Ray, A.K. and Kröger, F.A. (1979) *J. Appl. Phys.*, **50**, 4208.
41. Walukiewicz, W., Lagowski, J., Jastrebski, L., Lichtensteiger, M. and Gatos, H.C. (1975) *J. Appl. Phys.*, **50**, 899.
42. Park, R., Mar, H.A. and Salansky, N.M. (1985) *J. Vac. Sci. Technol. B*, **3**, 676.
43. Morimoto, K. (1989) *J. Appl. Phys.*, **66**, 4206.
44. Kamata, A., Uemoto, T., Okajima, M., Hirahara, K., Kawachi, M. and Beppu, T. (1988) *J. Cryst. Growth*, **86**, 285.
45. Kranzer, D. (1973) *J. Phys. C*, **6**, 2977.
46. Nishizawa, J., Itoh, K., Okuno, Y. and Sakurai, F. (1985) *J. Appl. Phys.*, **57**, 2210.
47. Park, Y.S., Hemenger, P.M. and Chung, C.H. (1971) *Appl. Phys. Lett.*, **18**, 45.
48. Yasuda, T., Mitsuishi, I. and Kukimoto, H. (1987) *Appl. Phys. Lett.*, **52**, 57.

49. Potts, J.E., Cheng, H., DePuydt, J.M. and Haase, M.A. (1989) *Proceedings of the 4th International Conference on II–VI Compounds, Berlin, September 17–22, 1989.*
50. Ruda, H.E. (1987) *J. Appl. Phys.*, **61**, 3035.
51. Dutt, B.V., Kim, O.K. and Spitzer, W.G. (1977) *J. Appl. Phys.*, **48**, 2110.
52. Brennan, K. (1988) *J. Appl. Phys.*, **64**, 4024.
53. Aven, M. and Garwacki, W. (1964) *Appl. Phys. Lett.*, **5**, 160.

9

Transmission electron microscopy of layered structures of widegap II–VI semiconductors

N. Otsuka

9.1. INTRODUCTION

Widegap II–VI semiconductors have long been considered as the leading candidate of materials for development of light-emitting devices operating in the blue and green spectral ranges. Difficulty in obtaining both p-type and n-type conductive forms of these materials, however, has prevented progress towards the realization of such opto-electronic devices. In the past decade, new low-temperature epitaxial growth techniques represented by molecular beam epitaxy (MBE) and metal–organic chemical vapour deposition (MOCVD) have led to a resurgence in the research on widegap II–VI semiconductors with the expectation that the advantage of employing these epitaxial growth techniques may enable us to overcome the aforementioned material problem. In the last several years, a variety of layered structures of widegap II–VI semiconductors have been grown by MBE and MOCVD, stimulating new research activities in the US, Japan and Europe. During the course of this recent development, transmission electron microscopy (TEM) has been utilized as one of the main tools for characterization of layered structures. With this technique, we can directly observe microstructures in epitaxial layers and, hence, can provide valuable feedback to growth experiments. In addition, TEM, especially high-resolution transmission electron microscopy (HRTEM), is capable of revealing interfacial atomic structures which are known to have critical effects on electrical and optical properties of layered structures. This chapter will provide a review of recent TEM studies on layered structures of widegap II–VI semiconductors, citing primarily the studies made at Purdue University where the TEM program has in recent years had close collaborations with a number of MBE programs pursuing the development of opto-electronic devices of widegap II–VI semiconductors.

9.2 THIN SAMPLE PREPARATION

The preparation of thin samples is an important part of a TEM characterization of layered structures, as microstructural information which can be derived from the characterization depends, to a large extent, on the type and quality of thin samples. Two types of thin samples are commonly utilized for the TEM characterization of semiconductor layered structures. A cross-sectional sample, which is prepared by cutting an epitaxial layer and a substrate into a thin slice vertically in the growth direction, enables us to observe directly interfaces in the layered structure and to examine defect distributions along the growth direction. A plan-view sample, on the other hand, prepared by thinning a sample from the back side of a substrate, is suitable for examination of lateral distribution of lattice defects in a layered structure, particularly for the analysis of misfit dislocations at a heteroepitaxial interface.

Thinning with high energy argon ions is normally employed at the final stage of the preparation of cross-sectional and plan-view samples. By utilizing this technique, we can obtain thin samples of most III–V and elemental semiconductors which preserve As-grown microstructures. High-energy argon ions, however, cause severe damage in certain materials including widegap II–VI semiconductors. Damage created by high-energy ions results in the formation of small dislocation loops and short segments of stacking faults in thin samples. Such damage also enhances atomic diffusion, causing a significant degree of intermixing of layered structures. In order to avoid such intermixing, a sample containing a widegap II–VI semiconductor needs to be kept at low temperatures with liquid nitrogen during the ion thinning. An important contribution was made by Cullis *et al.* for overcoming the problem of ion thinning damage [1,2]. Their studies demonstrated that the use of iodine ions, instead of argon ions, markedly reduces the degree of damage in thin samples of widegap II–VI semiconductors. Figures 9.1(a) and 9.1(b) are HRTEM images of a ZnSe epitaxial layer containing a periodic arrangement of ZnTe layers with thicknesses of one monolayer. This layered structure was grown by MBE on the (100) surface of a GaAs substrate [3]. The image (a) was taken from a cross-sectional sample prepared by argon ion thinning, and the image (b) was obtained from a cross-sectional sample prepared by iodine ion thinning. Two images were taken under identical conditions with the incident beam in the [010] direction. Clear images of ultrathin ZnTe layers can be seen in image (b), while the existence of these ZnTe layers is obscured by inhomogeneous dark and bright contrasts resulting from damage in image (a). From these two images, we can see that the preparation of damage-free thin samples by utilizing iodine ions greatly enhances the ability of TEM to reveal atomic structures of samples containing widegap II–VI semiconductors.

Another important aspect of thin samples used for the TEM

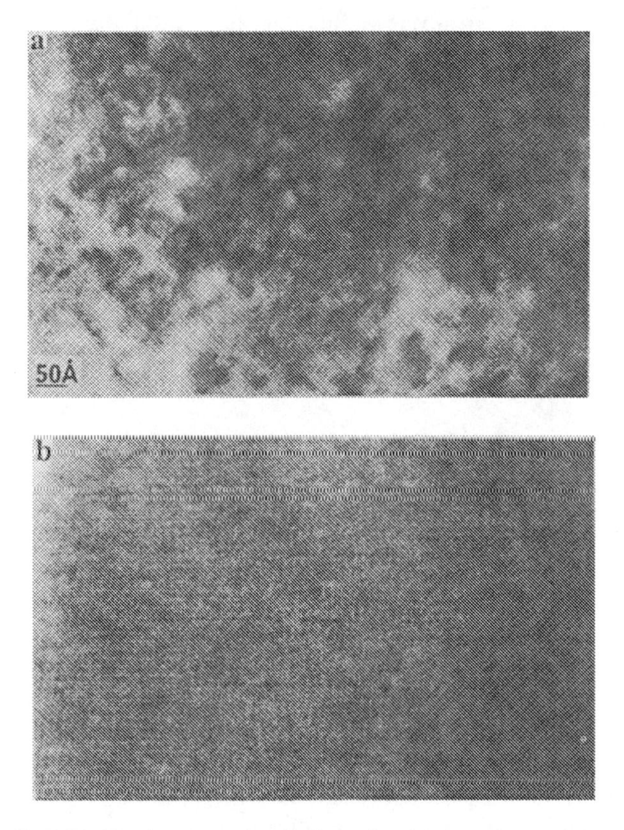

Fig. 9.1 [0 1 0] HRTEM images of a ZnSe epitaxial layer containing a periodic arrangement of ultrathin ZnTe layers. The images (a) and (b) were taken from cross-sectional samples prepared by the argon ion thinning and iodine ion thinning respectively.

characterization of layered structures, which is not explicitly stated in most of the published reports, is the extremely small sizes of areas that can be examined by this technique. In one cross-sectional sample, we can normally observe an area of only a few tens of square microns in an epilayer with a thickness of 1 or 2 μm. A plan-view sample provides a larger area than does a cross-sectional sample, but the area ranges typically a few thousands of square microns. Because of this limitation, it is important to examine a number of samples grown under similar conditions, if one wishes to draw a significant conclusion from the TEM characterization of layered structures.

9.3 LATTICE DEFECTS AND PSEUDOMORPHIC LAYERS

Up to now, the majority of epitaxial layers of widegap II–VI semiconductors have been grown on III–V semiconductor substrates. The use of III–V semiconductors as substrates is mainly due to the difficulty in obtaining large high-quality single crystals of widegap II–VI semiconductors. Lattice defects in epitaxial layers grown on III–V semiconductor substrates, especially misfit dislocations existing at II–VI/III–V semiconductor heteroepitaxial interfaces, have been the main subject of a number of recent TEM studies [4–9]. One of the findings from these studies is the existence of two types of misfit dislocations at (100) heteroepitaxial interfaces. These two types of misfit dislocations are called Lomer edge dislocations and 60° dislocations based on the orientations of Burgers vectors with respect to those of dislocation lines. The former type is found mainly in the systems having large lattice mismatches such as CdTe/GaAs [4–6], while the latter type is commonly observed in the systems with small lattice mismatches including ZnSe/GaA [7, 8].

A detailed analysis of misfit dislocations at (100)ZnSe–GaAs interfaces was carried out by *Choi et al.* [8]. For this study, ZnSe epitaxial layers with a thickness of 1 μm were grown on (100)GaAs surfaces at 400 °C by MBE. The lattice mismatch between ZnSe and GaAs crystals is 0.28% at the growth temperature, from which a critical thickness of the pseudomorphic state is expected to be 1800 Å according to the model proposed by Matthews and Blakeslee [10]. The thickness of ZnSe layers examined in this study, therefore, is considerably greater than the critical thickness. Figure 9.2(a) is a bright field image of a plan-view sample of a ZnSe/GaAs heterostructure, which shows the existence of a highly developed network of misfit dislocations. Most of the misfit dislocations in the network are aligned in the [011] or $[0\bar{1}1]$ orientations with a few curved dislocations which are also considered as a form of misfit dislocation. The Burgers vectors of misfit dislocations were determined by using the weak beam dark field imaging technique [11] along with calculations of image contrasts of different types of dislocations. Through the analysis, most of the misfit dislocations were identified as either 60° dislocations or pairs of Shockley partial dislocations, which are considered to be in the same family of misfit dislocations because a pair of Shockley dislocations can be formed through dissociation of a 60° dislocation. There are a few misfit dislocations which were identified as pairs of Shockley and Frank partial dislocations. Figure 9.2(b) is a weak beam dark field image of misfit dislocations at the ZnSe–GaAs interface which was taken by using a $02\bar{2}$ reflection. In the image, 60° dislocations and pairs of Shockley partial dislocations, all of which appear as bright lines, are indicated by B and C respectively. The existence of a large number of partial misfit dislocations, which are not commonly observed in heteroepitaxial interfaces of III–V and II–VI semiconductors, is attributed to the low stacking fault energy of ZnSe [12]. The dissociation of a 60° dislocation results in the formation of a stacking fault between two Shockley partial dislocations [13].

Petruzzello *et al.* studied the evolution of a network of misfit dislocations at the (100)ZnSe–GaAs interface by observing a series of heterostructures

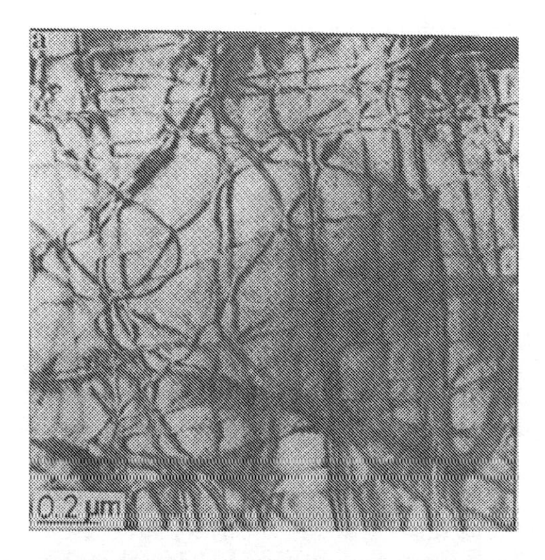

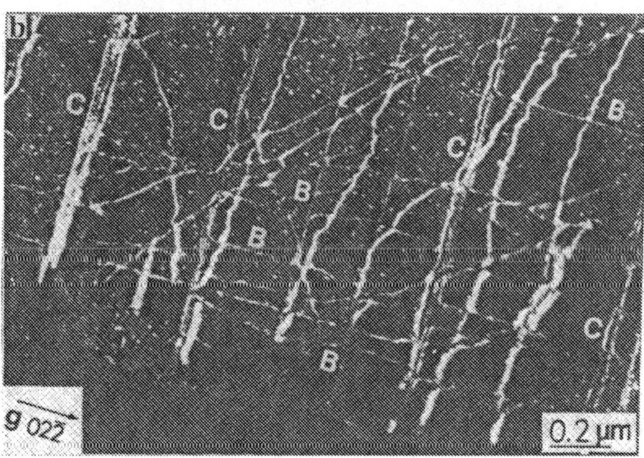

Fig. 9.2 (a) Plan-view bright field image of a network of misfit dislocations at a (100)ZnSe–GaAs interface (b) Plan-view weak beam dark field image of a network of misfit dislocations at a (100)ZnSe–GaAs interface. The image was taken by using a $02\bar{2}$ reflection.

with different thicknesses of ZnSe layers which ranged from 500 Å to 4.9 μm [7]. The ZnSe/GaAs heterostructures examined in this study were grown by MBE on GaAs substrates. From the study, an increase of the number of Lomber edge misfit dislocations with the thickness of the ZnSe layer was found, while the majority of misfit dislocations were identified as 60° dislocations. This trend is explained as a result of the formation of Lomer edge misfit dislocations through reactions of 60° misfit dislocations which had been initially introduced into the interface [7].

In addition to dislocations, stacking faults and microtwins are commonly observed in epitaxial layers of widegap II–VI semiconductors as expected from low stacking fault energies in these materials [12, 14]. The formation of these planar defects is strongly dependent on the growth direction of an epitaxial layer. More stacking faults and microtwins are normally observed in (111) epitaxial layers than in epitaxial layers of other growth orientations. Stutius and Ponce studied planar defects in ZnSe epitaxial layers grown on GaAs substrates by MOCVD [15]. In their study, planar defects in (100), (110) and (111)B ZnSe layers were examined by TEM, and results of the examination were compared with electrical properties of the ZnSe layers.

One important observation for further improvement of the structural quality of epitaxial layers of widegap II–VI semiconductors was made by TEM characterizations of ZnSe epitaxial layers grown directly on GaAs substrates and those grown on GaAs epitaxial layers. TEM images of the ZnSe layers grown directly on GaAs substrates show many lattice defects which appear to have formed at the substrate surfaces. These defects, most of which are stacking faults, are seen even in layers with thicknesses smaller than the critical thickness [7]. They become sources of misfit dislocations when the layer thickness exceeds the critical value. ZnSe epitaxial layers grown on GaAs epitaxial layers, on the other hand, were found to be free from such lattice defects [16]. Figure 9.3 is a bright field image of a cross-sectional sample of a (100) pseudomorphic ZnSe layer grown on a (100)GaAs epitaxial layer. For the preparation of the heterostructure, a GaAs layer was first grown on a GaAs substrate in one MBE system and transferred to another MBE system for the growth of a ZnSe layer with a thickness of 1000 Å [16]. An arsenic layer was deposited for the protection of the surface

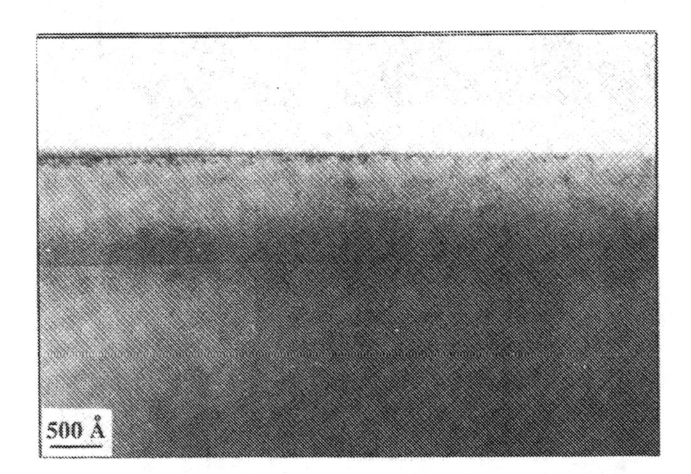

Fig. 9.3 Bright field image of a cross-sectional sample of a pseudomorphic ZnSe layer grown on a (100)GaAs layer.

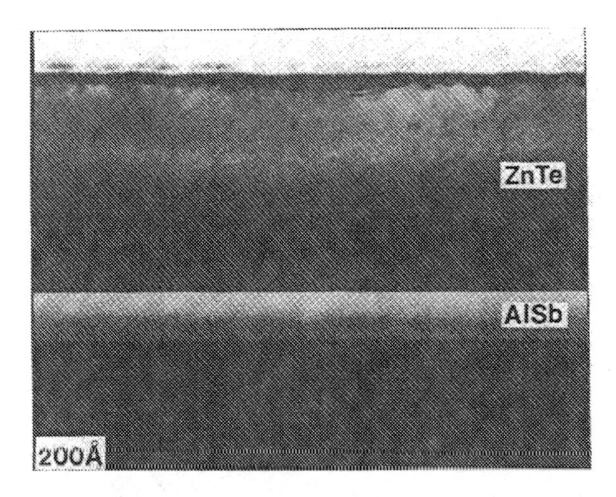

Fig. 9.4 Bright field image of a cross-sectional sample of the layered structure with 1380 Å thick ZnTe and 300 Å thick AlSb layers grown on a GaSb layer.

of the GaAs layer and desorbed prior to the growth of the ZnSe layer. In this way, the ZnSe layer was grown on an atomically flat and clean surface of the GaAs layer. In the bright field image, the ZnSe–GaAs interface appears as a clean defect-free plane. No dislocations or stacking faults were found in the observation of cross-sectional or plan-view samples of this heterostructure, which indicates that the defect density in the ZnSe layer is very low, probably comparable with that of the GaAs substrate crystal. Similar results were obtained from the growth of a pseudomorphic ZnTe epitaxial layer on an AlSb epitaxial layer [17]. Figure 9.4 is a bright field image of a cross-sectional sample of a heterostructure consisting of a pseudomorphic ZnTe layer, a pseudomorphic AlSb layer, and a GaSb layer grown on a (100)GaSb substrate. From the aforementioned observations and those of pseudomorphic III–V semiconductor layers grown on III–V semiconductor substrates, it appears that the substrate surface tends to become a source of lattice defects more easily in the growth of a II–VI semiconductor layer on a III–V semiconductor substrate than in the growth of a III–V semiconductor layer on a III–V substrate. This tendency may be attributed to atomic structures of II–VI/III–V semiconductor interfaces which will be discussed further in section 9.5.

9.4 SUPERLATTICES AND QUANTUM WELL STRUCTURES

In attempts to develop opto-electronic devices of widegap II–VI semiconductors, a variety of superlattices and quantum well structures have been grown by MBE and MOCVD in the last several years. One group of superlattices which have to date attracted much attention is that consisting of layers of widegap II–VI semiconductors and layers of their ternary alloys called diluted magnetic semiconductors (DMSs) [18–21]. Representatives of this group are ZnSe–$Zn_{1-x}Mn_xSe$ superlattices and CdTe–$Cd_{1-x}Mn_xTe$ superlattices. The addition of manganese increases bandgaps of host II–VI semiconductors, so that DMS layers serve as barriers in these superlattices. The incorporation of manganese also changes lattice parameters significantly, resulting in considerably large lattice mismatches between superlattice layers. Figure 9.5, which is a bright field image of a ZnSe–$Zn_{0.67}Ma_{0.33}Se$ superlattice, shows an example of the microstructures of this group of superlattices. The superlattice was grown on a (100) surface of a GaAs substrate with a ZnSe buffer layer [19]. The lattice parameter of $Zn_{0.67}Mn_{0.33}Se$ is by 0.95% greater than that of ZnSe. In the image, ZnSe and $Zn_{0.67}Mn_{0.33}Se$ layers appear as bright and dark bands respectively. In the image, it is seen that the lattice mismatch between ZnSe and $Zn_{0.67}Mn_{0.3}{}^3Se$ layers is accommodated in the form of a strained layer superlattice, leaving interfaces free from misfit dislocations. Dislocations seen in the image have threaded from the substrate surface. In electron diffraction patterns, satellite spots resulting from the superlattice structure are seen up to the sixth and seventh orders, indicating the existence of atomically abrupt interfaces. One common feature found from TEM observations of superlattices containing DMS layers is the existence of a large number of stacking faults with a high concentration of manganese in DMS layers. This trend is explained by the tendency of DMS materials to form wurtzite structures as stable phases in a range of high manganese concentrations [22]. The wurtzite structure can be considered as a periodic arrangement of stacking faults in a zinc blende structure. With relatively low manganese concentrations, on the other hand, superlattices appear as highly regular structures as seen in Fig. 9.5.

Fig. 9.5 Bright field image of a cross-sectional sample of a ZnSe–$Zn_{0.67}Mn_{0.33}Se$ superlattice.

Most of the superlattices consisting of widegap II–VI semiconductors, especially those which possess great significance for possible device applications, suffer from large lattice mismatches between superlattice layers. One such superlattice is that consisting of ZnSe and ZnTe layers. Because of the tendency of ZnSe and ZnTe to become only n-type and p-type respectively, strained layer superlattices consisting of these two compounds are considered as an attractive candidate with which we may be able to develop light-emitting devices of widegap II–VI semiconductors [23]. The very large lattice mismatch between ZnSe and ZnTe, being nearly 7%, however, presents a difficult problem in the growth of such superlattices. Extremely short-period superlattices need to be grown in order to accommodate the lattice mismatch in the form of strained layer superlattices, which requires highly controlled growth processes. In addition, a very large elastic strain resulting from the lattice mismatch may easily cause instability of the growth plane of the superlattice. Recent TEM observations, along with X-ray diffraction, confirmed successful growth of strained layer superlattices of ZnSe and ZnTe [24], but observed images indicates the need of further significant improvement of the microstructure in order to achieve the structure quality acceptable for the device application.

Figure 9.6(a) is a dark field image of a $(ZnSe)_3(ZnTe)_3$ superlattice. The image was taken from a cross-sectional sample by using a 200 reflection. The superlattice was grown at 250 °C on a (100) surface of an InP substrate, the lattice parameter of which is nearly equal to the average of those of ZnSe and ZnTe. In order to control layer thicknesses as well as to obtain atomically abrupt interfaces, the atomic layer epitaxy (ALE) technique was employed for the growth. In the dark field image, individual ZnSe and ZnTe layers are clearly seen as dark and bright bands respectively. Figure 9.6(b) is an HRTEM image of the superlattice which was taken with an incident beam of [011] orientation. In the image, ZnSe and ZnTe layers appear as bright and dark bands respectively. Tetragonal distortions of ZnSe and ZnTe crystals which result from the lattice mismatch are directly seen from spacings of (200) lattice fringes. No misfit dislocations were found in HRTEM images as expected from thicknesses of superlattice layers. As seen in both dark field image and HRTEM image, however, superlattice layers exhibit a significant degree of waviness. This waviness is seen from the bottom to the upper part of the superlattice, the total thickness of which is 2500 Å. In addition to the wavy superlattice layers, a large number of dislocations were observed in the superlattice. These dislocations were found to have threaded from the surface of the InP substrate. It appears from the TEM observations that the wavy superlattice layers resulted from the instability of the growth plane which was caused by the threading of a large number of dislocations during the growth of the superlattice.

A number of recent studies have demonstrated the growth of epitaxial layers of metastable phases including those of CdS [25, 26], CdSe [25–27],

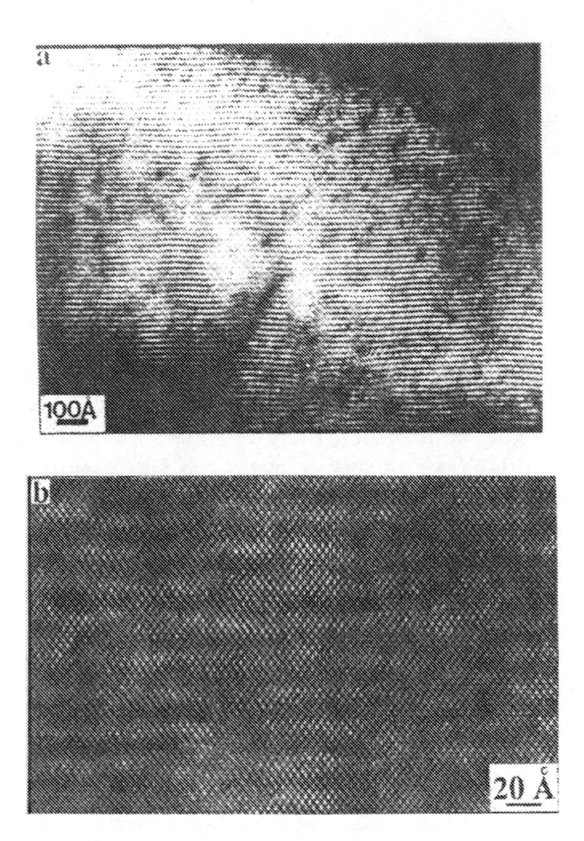

Fig. 9.6 (a) Dark field image of a cross-sectional sample of a $(ZnSe)_3(ZnTe)_3$ superlattice. The image was taken by using a 200 reflection. (b) [0 1 1] HRTEM image of a cross-section of a $(ZnSe)_3(ZnTe)_3$ superlattice.

MnSe[28] and MnTe[29] by taking the advantage of low-temperature epitaxial growth techniques. In these growth experiments, (100) planes of III–V semiconductor substrates were used, leading to the formation of epitaxial layers having metastable zinc blende structures. As revealed by TEM observations, however, these epitaxial layers were found to have a large number of stacking faults which may be attributed to their metastable nature. In attempts to explore the possibility of device applications of these metastable phases, multilayer structures including superlattices and quantum well structures were grown and examined by TEM [28, 30, 31]. By the TEM observations, some of these multilayer structures were found to have surprisingly high structural qualities despite their metastable nature. Figure 9.7 is an HRTEM image of an MnTe–CdTe–MnTe double barrier structure where metastable MnTe serves as barriers [29]. The image was taken with the incident beam in the [010] direction, showing {200}-type

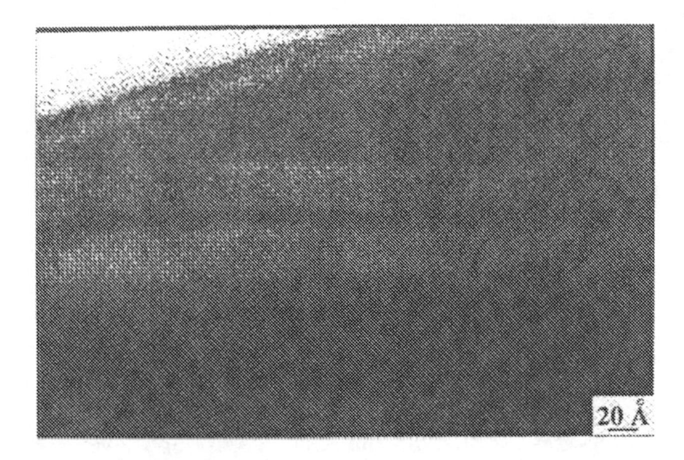

Fig. 9.7 [0 1 0] HRTEM image of a MnTe/CdTe/MnTe double-barrier structure.

lattice fringes in the MnTe layers and {220}-type lattice fringes in CdTe layers. The double-barrier structure was grown on a (100) surface of an InSb substrate by MBE. The thickness of the CdTe well layer is only 9 Å, but the existence of this layer is clearly seen in the image. MnTe layers appear as perfect zinc blende structures without having stacking faults or dislocations in these HRTEM images. Because of the high structural quality of the quantum well structure and a significantly large bandgap of metastable MnTe, a very large quantum confinement effect was observed from this sample as a blue light emission [31].

9.5 ATOMIC STRUCTURES OF II–VI/III–V SEMICONDUCTOR INTERFACES

The closing section of this chapter will describe recent TEM studies of one intriguing aspect of II–VI/III–V semiconductor interfaces, which also presents an important problem for development of opto-electronic devices of widegap II–VI semiconductors. In the last few years, many studies have focused on the structures and electrical properties of heteroepitaxial interfaces of closely lattice-matched II–VI/III–V semiconductor systems including the systems ZnSe/GaAs, ZnTe/(Al, Ga)Sb and CdTe/InSb [32–40]. The main reason for such increasing interest is the possibility of developing new devices employing closely lattice-matched II–VI/III–V semiconductor heterostructures which has been raised by the recent significant improvement of the structural quality of these heterostructures with the use of MBE and MOCVD. II–VI/III–V semiconductor heterojunctions cover a wider range of combinations of bandgaps under the restriction of close lattice matching than do III–V/III–V semiconductor heterojunctions and, hence, provide an opportunity to develop new innovative devices. In addition, the feasibility of anphoteric doping of III–V semiconductors makes the use of II–VI/III–V semiconductor heterojunctions a possible alternate approach to the development of light-emitting devices operating in the green and blue spectral ranges [17, 35]. A number of recent studies, however, indicate that the atomic structure of II–VI/III–V semiconductor interfaces may be considerably different from those of III–V/III–V or II–VI/II–VI semiconductor interfaces, presenting a possible difficulty in the control of their structures and electrical properties [16, 32, 33]. Results of spectroscopic measurements [39, 40] and electrical properties [32, 34] suggest that interfacial compound layers or interfacial alloy layers may form at II–VI/III–V semiconductor epitaxial interfaces under the growth conditions normally employed in MBE and MOCVD.

A direct observation of such an interfacial compound layer was made by Li *et al.* in their recent study of ZnSe/GaAs heterostructures [41, 42]. In addition, the capacitance–voltage measurements of ZnSe/GaAs metal–insulator–semiconductor (MIS) structures [38], which were performed with this TEM study, showed that the interface with a highly developed compound layer exhibits excellent electrical properties comparable with those of the GaAs/(Al, Ga)As system. For these studies, a set of (100)ZnSe/GaAs heterostructures were grown by using a modular MBE system having separate growth chambers which are connected by an ultrahigh vacuum (UHV) transfer tube. A GaAs epitaxial layer was first grown on a GaAs substrate and transferred through the UHV tube to another chamber for the growth of a ZnSe layer. Prior to the growth of the ZnSe layer, the surface chemistry of the GaAs epilayer was varied to a selected condition by monitoring reflection high-energy electron diffraction (RHEED) patterns. The following three heterostructures were examined by TEM: the sample grown on an As-rich surface which exhibited a c(4×4) reconstruction, the sample grown on a highly As-deficient surface which exhibited a (4×3) reconstruction, and the sample grown on a surface whose As coverage was intermediate between the above two surfaces and exhibited a (4×6) reconstruction.

Figure 9.8(a), 9.8(b) and 9.8(c) are cross-sectional dark field images of three heterostructures which were taken by using the 200 reflection under identical imaging conditions. A clear trend is seen in the images where a more distinct bright line appears at the ZnSe–GaAs interface which has formed on the more As-deficient GaAs surface, suggesting the existence of a highly developed interfacial layer at the As-deficient interface. Figure 9.9(a) is a [010] HRTEM image of the ZnSe–GaAs interface which has formed on the GaAs surface with the (4×6) reconstruction. In the image, a dark band with a width of about two monolayers is seen along the interface. Despite the appearance of the dark band, lattice fringes in the HRTEM image show perfect coherence between ZnSe and GaAs crystals at the interface. Dark field images of the interfacial layer, on the other hand, exhibited a systematic dependence of the

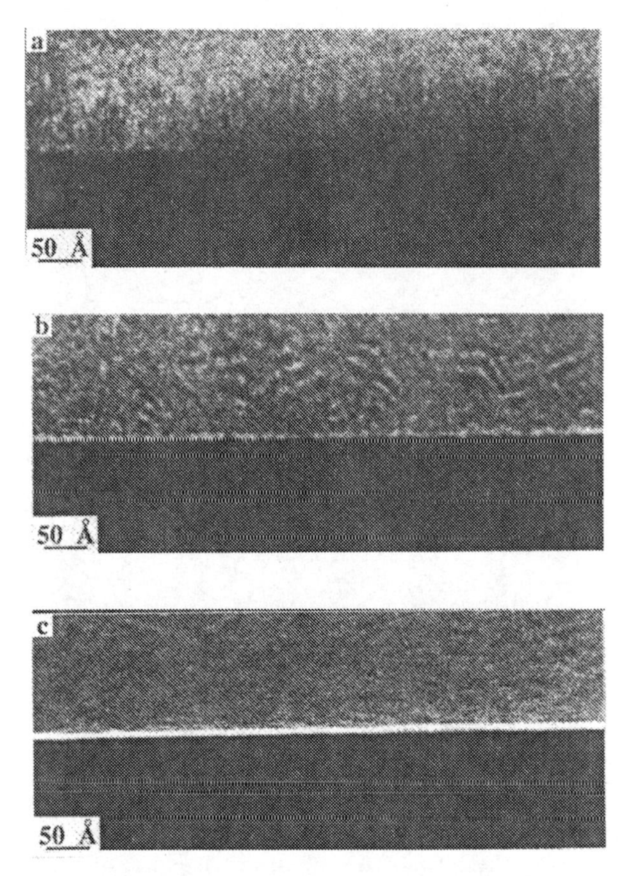

Fig. 9.8 200 dark field images of ZnSe–GaAs interfaces in the heterostructures grown on (a) the c(4 × 4) surface, (b) the (4 × 6) surface, and (c) the (4 × 3) surface. In each image, the upper side is ZnSe, and the lower side is GaAs.

contrast on the reflection used for the imaging. All 200-type dark field images as well as 420 dark field images give rise to a bright interfacial line, while all 400- and 220-type dark field images show a distinctly dark interfacial line. Figure 9.9(b) is a 400 dark field image of the ZnSe–GaAs interface which has formed on the surface with the (4 × 6) reconstruction.

The following structure model of the interfacial layer was derived from the aforementioned observations. The model is described as a thin layer having a zinc-blende-type structure which maintains a coherent interface between ZnSe and GaAs crystals. One of the f.c.c. sublattices is occupied by cations, i.e. Zn or Ga, and the other by anions, i.e. Se or As. Unlike the GaAs or ZnSe crystals, one of the f.c.c. sublattices of the thin layer has a high concentration of vacancies. Crystal structure factors of the 200 and 420 reflections of a zinc-blende-type crystal are given by a difference of scattering

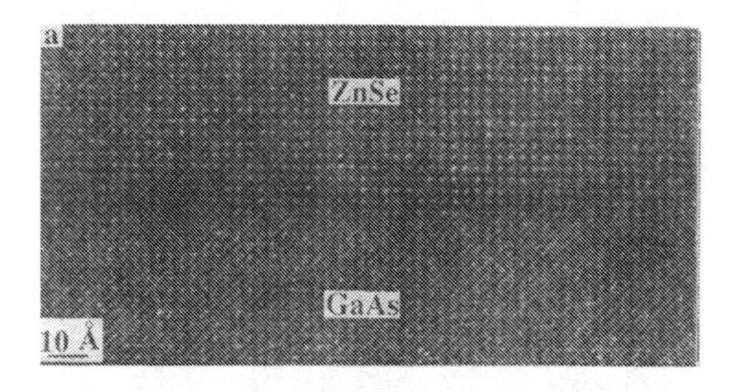

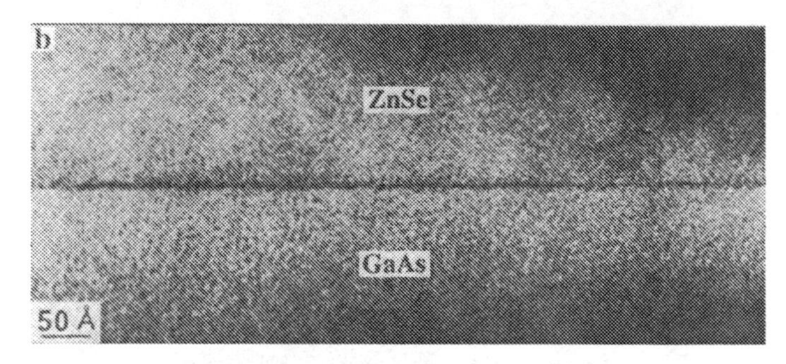

Fig. 9.9 (a) [0 1 0] high resolution transmission electron microscope image and (b) 400 dark field image of the ZnSe–GaAs interface.

factors of atoms occupying two different f.c.c. sublattices, while crystal structure factors of the 400 and 220 reflections are given by additions of scattering factors of these two types of atoms. Because of the existence of a high concentration of vacancies in one of the f.c.c. sublattices, crystal structure factors of the 200 and 420 reflections of the interfacial layer become significantly greater than those of GaAs and ZnSe, which are very small owing to nearly equal values of scattering factors of Zn, Ga, As and Se atoms. The interfacial layer, therefore, will appear as a bright line in dark field images of these reflections. In dark field images of 400 and 220 reflections, on the other hand, the interfacial layer will appear as a dark line as a result of the smaller crystal structure factors than those of GaAs and ZnSe.

Earlier studies of ZnSe–GaAs interfaces by Auger electron spectroscopy and electron energy loss spectroscopy have suggested the formation of a (Ga, Se) compound at the interface [39]. One of the stable (Ga, Se) compounds, Ga_2Se_3, is known to have a structure exactly identical to that suggested by the TEM observations [43]. It has a zinc blende structure, and one-third of the Ga sites are left as vacancies. The calculation of dark field images based on the model of an interfacial Ga_2Se_3 layer is in excellent agreement with the observed images [42, 44]. With this agreement, the existence of a thin Ga_2Se_3 layer at the ZnSe–GaAs interface which has formed on the As-deficient GaAs surface was inferred. This conclusion was supported by more recent studies which employed X-ray photoemission spectroscopy (XPS) [45] and Raman spectroscopy [46].

Similar results were obtained from the TEM observation of the CdTe–InSb epitaxial interface which is another combination of closely lattice-matched II-VI and III-V semiconductors [42]. Earlier studies by Raman spectroscopy suggested the existence of In_2Te_3 at the CdTe–InSb interface [40]. Similarly to Ga_2Se_3, In_2Te_3 forms a zinc-blende-type structure having vacancies in the cation sublattice [47]. Figure 9.10(a) and 9.10(b) are 400 and 200 dark field images respectively of a heteroepitaxial interface of a CdTe/InSb multilayer structure. The multilayer structure was grown by using a modular MBE system [37]. As expected from the crystal structure of In_2Te_3, the 200 dark field image has a bright interfacial line, and the 400 dark field image shows a dark interfacial line, supporting the results of earlier studies.

Results of TEM studies of the above two systems suggests that the formation of an interfacial compound layer may commonly occur in II–VI/III–V semiconductor systems. It is interesting to investigate such a possibility in other systems by utilizing a technique suitable for detection of the existence of an interfacial compound layer. It is also expected that the

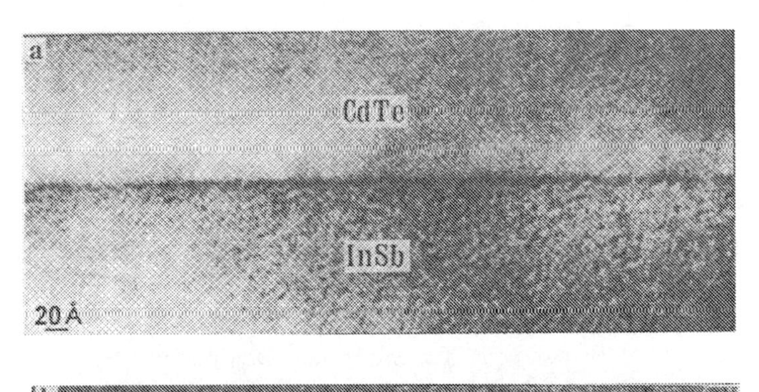

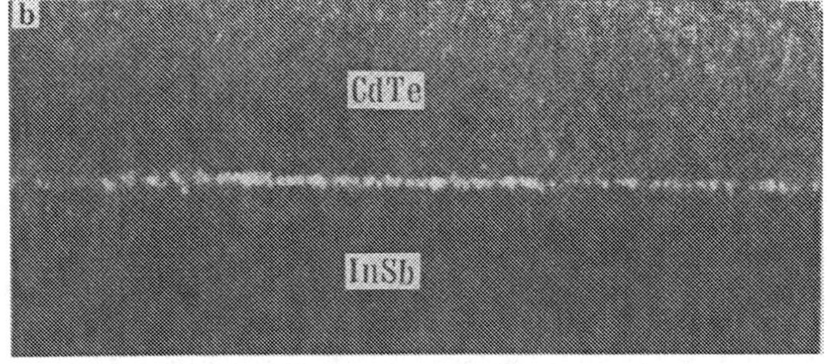

Fig. 9.10 Dark field images of the CdTe–InSb interface. The images (a) and (b) were taken by using 400 and 200 reflections.

atomic structures of II–VI/III–V semiconductor interfaces remain as an important subject for research pursuing the development devices of widegap II–VI semiconductors, because these atomic structures not only affect electrical properties of the heteroepitaxial interfaces but also play a critical role in the nucleation of II–VI semiconductor layers on III–V semiconductor substrates, influencing directly the structural quality of the epitaxial layers.

ACKNOWLEDGEMENTS

The author wishes to thank many graduate students and colleagues who have contributed to the work at Purdue University. The research has been supported by the Air Force Office of Scientific Research, National Science Foundation, Defense Advanced Project Agency.

REFERENCES

1. Cullis, A.G., Chew, N.G. and Hutchison, J.L. (1985) *J. Cryst. Growth*, **17**, 203.
2. Chew, N.G. and Cullis, A.G. (1987) *Ultramicroscopy*, **23**, 175.
3. Kolodziejski, L.A., Gunshor, R.L., Fu, Q., Lee, D., Nurmikko, A.V., Gonsalves, J.M. and Otsuka, N. (1988) *Appl. Phys. Lett.*, **52**, 1080.
4. Otsuka, N., Choi, C., Kolodziejski, L.A., Gunshor, R.L., Fischer, R., Peng, C.K., Morkoc, H., Nakamura, Y. and Nagakura, S. (1986) *J. Vac. Sci. Technol. B*, **4**, 896.
5. Ponce, F.A., Anderson, G.B. and Ballingall, J.M. (1986) *Surf. Sci.*, **168**, 564.
6. Feuillet, G., Di Cioccio, L., Million, A., Cibert, J. and Tararenko, S. (1981) *Inst. Phys. Conf. Ser.*, **87**, 135.
7. Petruzzello, J., Greenberg, B.L., Cammack, D.A. and Dalby, R. (1988) *J. Appl. Phys.*, **63**, 2299.
8. Choi, C., Otsuka, N., Kolodziejski, L.A., Melloch, M.R. and Gunshor, R.L. (1988) In *Dislocations and Interfaces in Semiconductors* (eds K. Rajan, J. Narayan and D. Ast), Metallurgical Society of AIME, Warrendale, PA, p. 141.
9. Feuillet, G. (1989) In *Evaluation of Advanced Semiconductor Materials by Electron Microscopy* (ed. D. Cherns), NATO ASI Series B, Vol. 203, p. 33.
10. Matthews, J.W. and Blakeslee, A.E. (1974) *J. Cryst. Growth*, 118.
11. Cockayne, D.J.H. (1981) *Annu. Rev. Mater. Sci.*, **11**, 75.
12. Rivaudo, G., Denanot, M.F., Garem, H. and Desoyer, J.C. (1982) *Phy. Status Solidi A*, **73**, 401.
13. Hirth, J.P. and Lothe, J. (1982) *Theory of Dislocations*, 2nd edn, Wiley, New York.
14. Takeuchi, S., Suzuki, K., Maeda, K. and Iwanaga, H. (1984) *Philos. Mag. A*, **50**, 171.
15. Stutius, W. and Ponce, F.A. (1985) *J. Appl. Phys.*, **58**, 1548.
16. Gunshor, R.L., Kolodziejski, L.A., Melloch, M.R., Vaziri, M., Choi, C. and Otsuka, N. (1987) *Appl. Phys. Lett.*, **50**, 200.
17. Mathine, D.L., Durbin, S.M., Gunshor, R.L., Kobayashi, M., Menke, D.R., Pei, Z., Gonsalves, J.M., Otsuka, N., Fu, Q., Haggerott, M. and Nurmikko, A.V. (1989) *Appl. Phys. Lett.*, **55**, 268.
18. Kolodziejski, L.A., Bonsett, T.C., Gunshor, R.L., Datta, S., Bylsma, R.B., Becker, W.M. and Otsuka, N. (1984) *Appl. Phys. Lett.*, **45**, 440.
19. Kolodziejski, L.A., Gunshor, R.L., Bonsett, T.C., Venkatasubramanian, R., Datta, S., Bylsma, R.B., Becker, W.M. and Otsuka, N. (1985) *Appl. Phys. Lett.*, **47**, 160.
20. Kolodziejski, L.A., Gunshor, R.L., Otsuka, N., Becker, W.M. and Datta, S. (1986) *IEEE Trans. Quantum Electron.*, **22**, 1666.

21. Furdyna, J.K. (1982) *J. Appl. Phys.*, **53**, 7637.
22. Twardowski, A., von Ortenberg, M. and Demianiuk, M. (1983) *Solid State Commun.*, **48**, 845.
23. Kobayashi, M., Dosho, S., Imai, A., Kimura, R., Konagai, M. and Takahashi, K. (1987) *Appl. Phys. Lett.*, **51**, 1602.
24. Takemura, Y., Nakanishi, H., Konagai, M., Takahashi, K., Nakamura, Y. and Otsuka, N. (1990) *Proceedings of 6th International Conference on Molecular Beam Epitaxy, San Diego, CA, August*, 1990.
25. Endoh, Y., Kawakami, Y., Taguchi, T. and Hiraki, A. (1988) *Jpn. J. Appl. Phys.*, **27**, L2199.
26. Cullis, A.G., Smith, P.W., Parbrook, P.J., Cockayne, B. and Wright, P.J. (1989) *Appl. Phys. Lett.*, **55**, 2081.
27. Samarth, N., Luo, H., Furdyna, J.K., Qadri, S.B., Lee, Y.R., Ramdas, A.K. and Otsuka, N. (1989) *Appl. Phys. Lett.*, **54**, 2680.
28. Kolodziejski, L.A., Gunshor, R.L., Otsuka, N., Gu, B.P., Hefetz, Y. and Nurmikko, A.V. (1986) *Appl. Phys. Lett.*, **48**, 1482.
29. Durbin, S.M., Han, J., O, S., Kobayashi, M., Menke, D.R., Gunshor, R.L., Li, D., Gonsalves, J.M. and Otsuka, N. (1989) *Appl. Phys. Lett.*, **55**, 2087.
30. Durbin, S.M., Kobayashi, M., Fu, Q., Pelekanos, N., Gunshor, R.L. and Nurmikko, A.V. (1990) *Surf. Sci.*, **228**, 33.
31. Han, J., Durbin, S.M., Kobayashi, M., Menke, D.R., Pelekanos, N., Haggerott, M., Nurmikko, A.V., Nakamura, Y. and Otsuka, N. (1990) *Proceedings of 6th International Conference on Molecular Beam Epitaxy, San Diego, CA, August 1990.*
32. Olego, D.J. (1989) *Phys. Rev. B*, **40**, 2743.
33. Tamargo, T.C., de Miguel, J.L., Hwang, D.M. and Farrall, H.H. (1988) *J. Vac. Sci. Technol B*, **6**, 784.
34. Suemune, I., Ohmi, K., Kanda, T., Yatake, K., Kan, Y. and Yamanishi, M. (1986) *Jpn. J. Appl. Phys.*, **25**, 1827.
35. McCaldin, J.O. and McGill, T.C. (1988) *J. Vac. Sci. Technol. B*, **6**, 1360.
36. Golding, T.D., Martinka, M. and Dman, J.H. (1988) *J. Appl. Phys.*, **64**, 1873.
37. Glenn, J.L., Jr., O, S., Kolodziejski, L.A., Li, D., Otsuka, N., Haggerott, M., Pelekanos, N. and Nurmikko, A.V. (1989) *J. Vac. Sci. Technol. B*, **7**, 249.
38. Qiu, J., Qian, Q.D., Gunshor, R.L., Kobayashi, M., Menke, D.R., Li, D. and Otsuka, N. (1990) *Appl. Phys. Lett.*, **56**, 1272.
39. Tu, D.W. and Kahn, A. (1985) *J. Vac. Sci. Technol., A*, **3**, 922.
40. Zahn, D.R.T., Mackey, K.J., Williams, R.H., Munadar, H., Geurts, J. and Richter, W. (1987) *Appl. Phys. Lett.*, **50**, 742.
41. Li, D., Gonsalves, J.M., Otsuka, N., Qiu, U., Kobayashi, M. and Gunshor, R.L. (1990) *Appl. Phys. Lett.*, **57**, 449.
42. Li, D., Otsuka, N., Qiu, J., Glenn, J., Jr., Kobayashi, M. and Gunshor, R.L. (1990) *Mater. Res. Soc. Symp. Proc.*, **161**, 127.
43. Hahn, H. and Klingler, W. (1949) *Z. Anorg. Chem.*, **259**, 135.
44. Otsuka, N., Li, D., Qiu, J., Kobayashi, M. and Gunshor, R.L. (1990) *Mater. Trans., JIM*, **31**, 622.
45. Qiu, J., Menke, D.R., Kobayashi, M., Gunshor, R.L., Li, D. and Otsuka, N. (1990) *Proceedings of 6th International Conference of Molecular Beam Epitaxy, San Diego, CA, August 1990.*
46. Krost, A., Richter, W., Zahn, D.R.T., Hingerd, K. and Sitter, H. (1990) *Appl. Phys. Lett.*, **57**, 1981.
47. Grzeta-Plenkovic, B., Popovic, S., Celustka, B., Ruzie-Toros, Z., Santic B. and Soldo, D. (1983) *J. Appl. Crystallogr.*, **16**, 415.

10

Self- and impurity diffusion processes in widegap II–VI materials

D. Shaw

10.1 INTRODUCTION

A knowledge of the manner and ease with which atoms can migrate in a II–VI crystal lattice is of both fundamental and technological importance. At the fundamental level the diffusion process may involve non-defect or defect mechanisms: experimental diffusivities can help to identify the type of mechanism and underpin theoretical understanding of atom movements and defects. Diffusion is important in both material and device technology through control of non-stoichiometry, impurity doping and compositional interdiffusion (via gradients in alloy composition). The introduction during the past decade of epitaxial growth techniques and their use in the fabrication of low-dimensional structures has revealed a need for diffusion studies at much lower temperatures and in much smaller spatial regions compared with bulk materials (i.e. material $\geqslant 1\ \mu m$ in extent). It is easily seen for example that in a period of 10^3 s (e.g. growth of a multiple quantum well (MQW)) diffusivities must be $\leqslant 10^{-18}\ cm^2\ s^{-1}$ if diffusional spread should not exceed ~ 10 Å. Over a period of a year diffusivities must be $\leqslant 10^{-22}\ cm^2\ s^{-1}$ to contain any spread to < 10 Å.

Recent reviews of diffusion in II–VI materials have been given by Sharma [1] and Shaw [2]. This chapter is concerned with lattice diffusion in single crystals of the Zn and Cd chalcogenide compounds and their alloys in isothermal situations. It is an update of ref. 2 with greater attention to impurity diffusion and the omission of the Hg-based chalcogenides. Some data on self- and interdiffusion in the alloy (CdMn)Te are also included. Sections 10.5–10.8 deal with diffusion in essentially bulk material. Diffusion in quantum wells and superlattices is considered separately in section 10.9.

10.2 DIFFUSION REGIMES AND THEIR DIFFUSION COEFFICIENTS

Atomic diffusion can occur under a wide variety of conditions. For meaningful (i.e. reproducible) measurements to be made it is essential that the thermodynamic parameters be controlled and specified (e.g. temperature, level of non-stoichiometry, doping or impurity content) throughout the diffusion anneal. Attention to crystal quality is also important as high concentrations of extended defects (dislocations, subgrain boundaries, grain boundaries) can distort or mask attempts to measure lattice diffusivities [3]. Diffusion regimes fall into one of two categories: conditions of chemical equilibrium or of chemical disequilibrium. In the former case there are no gradients in chemical potential so that there are no changes in chemical composition during an anneal. In these circumstances atom movement can be followed by 'tagging' atoms (e.g. by isotopic enrichment or by radiotracers) diffusing in an isotopic gradient. If solvent migration is the particular interest the regime is known as self-diffusion whereas if the focus is on solute migration (typically involving a spatially and temporally uniform doping level) the regime is an isoconcentration regime. Both regimes are basic to any investigation into the identification of diffusion mechanisms [3]. On the other hand, diffusion down gradients of the chemical potential (chemical disequilibrium) are basic to materials and device technology. Three regimes of chemical diffusion can be recognized: impurity (solute) diffusion, changes in non-stoichiometry of the solvent compound or alloy and changes in the composition of the solvent alloy. In the ternary alloy $M_{1-\Delta}(X_{1-x}Y_x)_\Delta$ where M is the cation species, X and Y represent the chalcogenide anions and Δ describes the level of non-stoichiometry ($\Delta = 1/2$ for exactly stoichiometric composition), changes in non-stoichiometry are reflected by variations in Δ and changes in alloy composition by changes in x (compositional interdiffusion).

Diffusion at chemical equilibrium gives rise to self-diffusion and isoconcentration diffusion coefficients. Chemical disequilibrium gives rise to net chemical fluxes, whatever the regime, and as this is an intermixing process it can be characterized by an interdiffussion coefficient. In order to identify the type of interdiffusion coefficient with its regime it is convenient to characterize impurity diffusion, changes in Δ and changes in x by impurity diffusion, chemical self-diffusion and interdiffusion coefficients respectively.

In the impurity diffusion regime the impurity can be introduced to the solvent lattice from an external phase (e.g. vapour, surface layer, doped epilayer) or by ion implantation. The diffusion process can be modified by prior irradiation (e.g. as in ion implantation) or by irradiation during the diffusion anneal [4]. It is also feasible that the impurity reacts chemically with the solvent anion and/or cation components, which gives rise to reactive chemical diffusion. Such surface reactions are important in making electrical contacts [5] but are beyond the scope of this chapter. There is also obvious concern to prevent the in-diffusion of a metal contact into the substrate and it has been reported that monolayers of rare earth elements will block the diffusion of contact metals into II–VI substrates [6]. Diffusion from ion-implanted sources and the effects of irradiation are complex topics where little systematic work in relation to II–VI diffusion has been attempted. These topics will not be considered further beyond noting that Sharma [1] has listed references to II–VI ion implantation.

10.3 DIFFUSION MECHANISMS

Solvent and solute atoms can migrate through a crystal lattice by means of non-defect mechanisms (e.g. ring or direct exchange processes) or of defect mechanisms (e.g. vacancy or interstitial or complexes of these point defects). Recent theoretical work [7, 8] indicates that the widely accepted view that non-defect mechanisms were energetically unfavourable relative to defect mechanisms is no longer tenable. Anion vacancies and cation interstitials are divalent donors; cation vacancies and anion interstitials are divalent acceptors [9]. In addition, the possibility of antisite defects (anion or cation occupying a cation or anion site) must also be considered. A cation (anion) occupying an anion (cation) site creates a divalent acceptor (donor) [9]. The diffusivity of a point defect will generally depend on its charge state (i.e. neutral, singly or doubly ionized) as will the diffusivities of any complexes formed from point defects (e.g. cation–anion divacancy, vacancy–interstitial pair).

In the case of self-diffusion by a defect mechanism the self-diffusivity is given by [3]

$$D^* = \sum_j f_j D_j C_j \tag{10.1}$$

where f_j, D_j and C_j are the correlation factor, diffusivity and concentration of the jth native defect. Generally C_j will be determined by the temperature, the partial pressure of the anion or cation component imposed during the anneal and, in the case of ionized defects, by the Fermi level (as determined by the electroneutrality condition (ENC). In the compound, MX, the concentrations of metal vacancies, $[V_M]$, and of metal interstitials, $[M_i]$, are proportional to $P_M^{-\gamma}$ and P_M^{γ} respectively where $\gamma = 1$ for the neutral defect and $0 \leqslant \gamma < 1$ for the ionized states, depending on the ENC. Similar proportionalities exist for the chalcogen defects. These pressure dependencies are reflected in D^* (equation (10.1)) and are the signatures of the defects involved. It should be noted that in a given set of conditions only a small number of defects are dominant in equation (10.1) [10] which helps to reduce the complexity of the situation. If M self-diffusion occurs via V_X (i.e. diffusion of M_X on the anion sublattice) or via V_M and V_X (i.e. diffusion by jumps to nearest-neighbour sites) then the associated diffusivity varies as P_M^γ with $\gamma \geqslant 1$, with similar proportionalities for X self-diffusion [11]. Obviously self-diffusion by a non-defect mechanism is independent of any Fermi level or partial pressure dependence.

In general an impurity diffusivity can be expected to be more complex because (i) the impurity, if electrically active, will influence, if not control, the local ENC which may or may not vary through the diffusion region, (ii) the impurity may be present in several configurations (e.g. Ag in CdS [12], Cu in CdTe [13], P in CdTe [14], In in CdTe [15]. As an illustration, suppose an impurity is present in two forms A and B with a total concentration $C = C_A + C_B$. If D_A and D_B are the diffusivities for A and B then the total diffusivity for the impurity is

$$D_{tot} = D_A \frac{\partial C_A}{\partial C} + D_B \frac{\partial C_B}{\partial C} \tag{10.2}$$

Introducing $K = C_A/C_B$ into equation (10.2) yields

$$D_{tot} = \frac{(C_A D_A + C_B D_B)}{C} + \frac{(D_A - D_B)}{1 + K} \cdot \frac{\partial \ln(1 + K)}{\partial C} \tag{10.3}$$

A and B could define different charge states of the same impurity defect or alternatively different point defects (e.g. substitutional and interstitial). Depending on the particular diffusion mechanism D_A and D_B may also involve the concentration of the appropriate native defect, which also may depend on C through the local ENC. It is evident from equation (10.3) that generally D_{tot} can be expected to be dependent on C and the component partial pressure. A systematic investigation of these dependencies is the basis for identifying the impurity diffusion mechanism(s).

Changes in non-stoichiometry can be described in terms of the chemical self-diffusivity $D(\Delta)$ [3]. In this case because changes in native defect concentrations are taking place it is the diffusivities of these native defects (e.g. vacancies and interstitials) that contribute to $D(\Delta)$. Clearly non-defect mechanisms can play no part in $D(\Delta)$.

Compositional interdiffusion in the ternary alloys $M_{1-x}N_xX$ and $MX_{1-x}Y_x$, where M, N and X, Y are the cation and chalcogen components respectively, is characterized by the interdiffusion coefficients, $\tilde{D}(x)$, given in ref. 16:

$$\tilde{D}(x) = xD(\mathrm{M}) + (1 - x)\, D(\mathrm{N})$$

and

$$\tilde{D}(x) = xD(\mathrm{X}) + (1 - x)\, D(\mathrm{Y}) \tag{10.4}$$

$D(\mathrm{M})$ and $D(\mathrm{N})$ are the chemical diffusion coefficients of M and N in $M_{1-x}N_xY$ and similarly for $D(\mathrm{X})$ and $D(\mathrm{Y})$. These chemical diffusivities are related to the associated self-diffusivities $D^*(\mathrm{M})$, $D^*(\mathrm{N})$, $D^*(\mathrm{X})$, $D^*(\mathrm{Y})$ at the particular alloy composition. Tang and Stevenson [16a] have shown that the general relationships are quite complex and derived expressions for $\tilde{D}(x)$ appropriate to electrically extrinsic and intrinsic conditions. Their expressions simplify in the case of a common anion ternary to

$$\tilde{D}(x) = D^*(\mathrm{M})D^*(\mathrm{N})\{(1-x)D^*(\mathrm{M}) + xD^*(\mathrm{N})\}^{-1}\theta(x) \qquad (10.5\mathrm{a})$$

for the condition $D^*(\mathrm{X}) \ll D^*(\mathrm{M})$ and $D^*(\mathrm{N})$, whereas if the equality is reversed, so that $D^*(\mathrm{X}) \gg D^*(\mathrm{M})$ and $D^*(\mathrm{N})$ then

$$\tilde{D}(x) = \{xD^*(\mathrm{M}) + (1-x)D^*(\mathrm{N})\}\,\theta(x) \qquad (10.5\mathrm{b})$$

$\theta(x)$ is the thermodynamic factor and either equation is valid for both extrinsic and intrinsic regimes. Note that equation (10.5b) is identical to that for a binary alloy system. Unfortunately we have very little quantitative information on $\theta(x)$ for the II-VI alloys. Equations (10.5a) and (10.5b) are readily adapted to the common cation by permutation of the component symbols. Equations (10.5a) and (10.5b) show that $\tilde{D}(x)$ is a function of x and, if defect diffusion mechanisms are operative, also of the local concentrations of native defects (because of the self-diffusivities).

Chemical diffusion fluxes also create effects which in turn can influence these fluxes. Two such effects are due to elastic strain and electric fields. A diffusant, because of its atomic size, can cause local dilation or contraction in the solvent lattice, so generating strain gradients which can add drift terms to the right-hand side of equation (10.2) and to the chemical diffusivities in equation (10.4) [17]. Lattice strain (uniform or non-uniform) will also change the balance between the concentrations of vacancies and self-interstitials. Results in Si indicate that diffusant concentrations $\geqslant 1\%$ are needed for significant effects to be found. In addition, if the strain gradient is large enough the strain energy can generate dislocations through plastic flow. Shockley (quoted in ref. 18) has given a criterion for the onset of plastic flow in terms of the total number of diffusant atoms per unit area. These dislocations can then provide short-circuit diffusion paths for the diffusant [19]. Dislocation climb can also be a source or sink of vacancies or self-interstitials, so disturbing the local concentrations of these defects. These changes in turn can affect diffusivities. An internal electric field is set up whenever there is a spatial gradient in $E_\mathrm{C} - E_\mathrm{F}$ and/or $E_\mathrm{F} - E_\mathrm{V}$ where E_F is the Fermi level and E_C and E_V are the energies of the conduction and valence band edges. Such gradients arise (i) with an ionized impurity concentration gradient in an otherwise homogenous semiconductor [3, 20], (ii) where there is a compositional gradient so that there is a gradient in $E_\mathrm{C} - E_\mathrm{V}$, as in compositional interdiffusion [17], and (iii) in the depletion layer of a p–n junction [21]. The electric field will add drift terms to equation (10.2) and to the chemical diffusivities in equation (10.4). The diffusivity can be enhanced or reduced depending on the polarity of the migrating species and the direction of the internal electric field. If more than one charged defect is involved the situation becomes complex and requires detailed modelling [22, 23].

For completeness we need to note that lattice strain can also arise because of lattice mismatch at interfaces such as occurs in epilayer structures [17]. Similar interfaces can also create electric fields which may extend to a significant depth [17].

10.4 MEASUREMENT OF DIFFUSION COEFFICIENTS

Most of the methods used to measure diffusion coefficients relate to a concentration profile. The most common and most reliable approach is to determine directly the concentration distribution with depth. An account of general profiling techniques can be found in a recent review by Rothman and in the references which he cites [24]. There are, however, several further techniques unique to semiconductors which are based on electrical [25, 26] or photoluminescent properties. In the former (e.g. resistivity profiling, capacitance ($C-V$) profiling, p–n junction depth) it is necessary for the diffusant to be fully ionized at both the diffusion and measuring temperatures and for it to be the only ionized species in the ENC. Whether these conditions are met in a particular case can only be decided by experiment [26]. Compensation effects can never be ruled out of consideration in compound semiconductors so that caution must be exercised in the interpretation of diffusion coefficients obtained by electrical methods. Photoluminescence (PL), based on bandgap exciton emission, can be used to measure compositional interdiffusivities in II–VI alloys, provided that the variation of bandgap with composition is known [27]. Impurity PL in transparent II–VI crystals has been widely used to obtain impurity diffusivities from visual observations of the depth of the sharp PL boundary. Similar sharp boundaries which separate a high-resistance (compensated) diffused layer from a low-resistance deeper region have been used to obtain impurity diffusivities.

These 'boundary' methods are faster than serial sectioning techniques which explains their relatively wide use and it is of some importance to examine more closely what is measured in the boundary method. If the diffusant concentration is C and its concentration dependent diffusivity is $D(C)$ then the diffusion can be described by Fick's second law:

$$\frac{\partial C}{\partial t} = \frac{\partial}{\partial x}\left[D(C)\frac{\partial C}{\partial x}\right] \tag{10.6}$$

The general solution of this equation has the form

$$C(x, t) = C_0 F(x/\lambda\sqrt{t}) \tag{10.7}$$

where C_0 is the surface concentration and λ is a constant. F is some function of the dimensionless parameter $(x/\lambda\sqrt{t})$. If the PL or resistance boundary occurs at a depth x_b corresponding to a concentration C_b then

$$C_b = C_0 F(x_b/\lambda\sqrt{t}) \tag{10.8}$$

Equation (10.8) shows that, if $C_b/C_0 = \beta$, a constant, then $x_b^2 = \lambda^2 t$. The difficulty in practice is that C_b, C_0 and λ are usually unknown and it is therefore assumed that β is constant and that $\lambda^2 = 4D$ which makes the implicit assumption that F is either an erfc or gaussian distribution (i.e. the diffusivity, D, is independent of concentration). At a given temperature 'boundary' methods generally give $x_b^2 \propto t$. At different temperatures it is possible that C_b and C_0 may vary so that β varies with temperature in which case λ will also change, in addition to any intrinsic temperature dependence λ may possess. Aven and Halstead [28] compared the PL boundary and radiotracer methods for Cu diffusion in ZnSe between 200 and 570 °C. Both methods gave agreement above ~ 300 °C but below this temperature the PL diffusivity became increasingly smaller than the radiotracer diffusivity as the temperature fell. At 200 °C the difference was a factor of 10. As the radiotracer profiles were erfcs the authors suggested that below 300 °C β became temperature dependent. Boundary methods have a useful empirical role which can give relatively quickly an order-of-magnitude impurity diffusivity. Techniques yielding complete concentration profiles, however, must be seen as providing more reliable diffusivities.

$D(\Delta)$ can be obtained directly from changes in the electrical conductivity or free-carrier concentration measured at the anneal temperature [29].

Diffusivity measurements in MQWs and superlattice structures (SLSs) are becoming of increasing interest and importance. The problems are severe because of the small diffusion lengths ($2\sqrt{Dt}$) involved which really favour non-destructive techniques. Interdiffusivities at interfaces have been measured using the intensities of the satellites in double X-ray rocking curve spectra. The method was pioneered by Fleming *et al.* [30] in III–V MQWs and has been used to measure interdiffusion in (Hg, Cd)Te MQWs [31] and Zn(Se, Te) SLSs [32]. Such X-ray spectra involve contributions from all the layers in the MQW stack and so correspond to an average for the whole of the stack. This means that it is not possible to recognise from satellite spectra whether or not there are any systematic variations in interface interdiffusion through the stack. Interdiffusivities derived from satellite data represent some kind of average across all of the interfaces in the stack. Rutherford backscattering has been used to characterize interdiffusion in a single quantum well [33] in the Si–Ge system. Notwithstanding the obvious advantage of non-destructive methods a destructive technique with a spatial resolution close to atomic layer separation has recently been described [34]. It is based on chemical lattice imaging using transmission electron microscopy and was used to measure interdiffusion in (Hg, Cd)Te MQWs. A new and novel technique has been used for measuring the diffusivity of an impurity in a single quantum well (Si in a (GaAl) As well) [35]. The impurity is incorporated during growth as a δ layer and subsequent spread of the layer with annealing is measured by C–V profiling. The diffusivity is obtained by fitting numerical simulations to the C–V data.

CdS has the würtzite crystal structure so that diffusion, in principle, could be anisotropic. Cd [36] and S [37] self-diffusion measurements carried out in directions perpendicular and parallel to the c axis, using the radiotracer technique, failed to detect any anisotropy in the self-diffusivities. Jones and Mykura found, however, that the diffusivities of the optical boundaries in CdS due to In [38] or Ga [39] impurity diffusion were clearly anisotropic with the diffusivity perpendicular to the x axis being up to a factor of 3 greater than diffusion parallel to the c axis. Anisotropy in the radiotracer diffusivity of In in CdS was, if present, much less than that of the optical boundary [38]. This seems to indicate that the anisotropy associated with the optical boundary is not directly due to the In diffusion. These radiotracer results clearly show that to within experimental error self- and impurity diffusion in CdS is isotropic and it appears reasonable to assume that the same will be the case in CdSe. Isotropic behaviour will be assumed in the following sections.

10.5 SELF-DIFFUSION

Only references not given in ref. 2 will be cited. Where ref. 2 is cited this means that the appropriate reference(s) can be found in ref. 2.

10.5.1 Chalcogen self-diffusion

On currently available data $D^*(\mathrm{X})$ exhibits the simplest overall behaviour found in the II–VI materials in that, at a given temperature, $D^*(\mathrm{X}) \propto P_{\mathrm{X}_2}^{1/2}$ [2]. Only as $P_{\mathrm{M}}(\mathrm{sat})$ is approached does $D^*(\mathrm{X})$ change over to a weak increase with rising P_{M}, varying approximately as $P_{\mathrm{M}}^{1/3}$. Strong donor doping (in CdS and in CdSe) has no effect on $D^*(\mathrm{X})$ under chalcogen-rich conditions. These features have been attributed to diffusion by the neutral self-interstitial, $\mathrm{X}_i^{\times}$, under chalcogen-rich conditions, and to a $\mathrm{V}_{\mathrm{X}}^{\cdot\cdot}$ mechanism at high P_{M}[2]. It should be noted also that the $P_{\mathrm{X}_2}^{1/2}$ dependence of $D^*(\mathrm{X})$ is not unique to diffusion by the $\mathrm{X}_i^{\times}$ defect. The same relation would be found for the neutral anti-site defect $\mathrm{X}_{\mathrm{M}}^{\times}$ diffusing by a $\mathrm{V}_{\mathrm{M}}/\mathrm{V}_{\mathrm{X}}$ mechanism [11]. Evidence of anti-site defects in the II–VIs is lacking [2] but some recent theoretical work in the tellurides suggests the occurrence of X_{M} may be preferred to that of X_i [40]. Although the absence of any effect of donor doping on $D^*(\mathrm{X})$ rules out any significant contribution of an ionized native acceptor to $D^*(\mathrm{X})$ it is quite probable that acceptor doping would increase $D^*(\mathrm{X})$ because of the enhanced $\mathrm{V}_{\mathrm{X}}^{\cdot\cdot}$ concentration. Experiments are needed to check this likelihood.

Experimentally $D^*(\mathrm{X}) = K(T)P_{\mathrm{X}_2}^{1/2}$ and $D^*(\mathrm{X}, \mathrm{X}_{\mathrm{sat}}) = D_0(\mathrm{X}, \mathrm{X}_{\mathrm{sat}}) \exp[-Q(\mathrm{X}, \mathrm{X}_{\mathrm{sat}})/kT]$ and we can therefore write $K(T) = D^*(\mathrm{X}, \mathrm{X}_{\mathrm{sat}})P_{\mathrm{X}_2}^{-1/2}(\mathrm{sat})$. Using

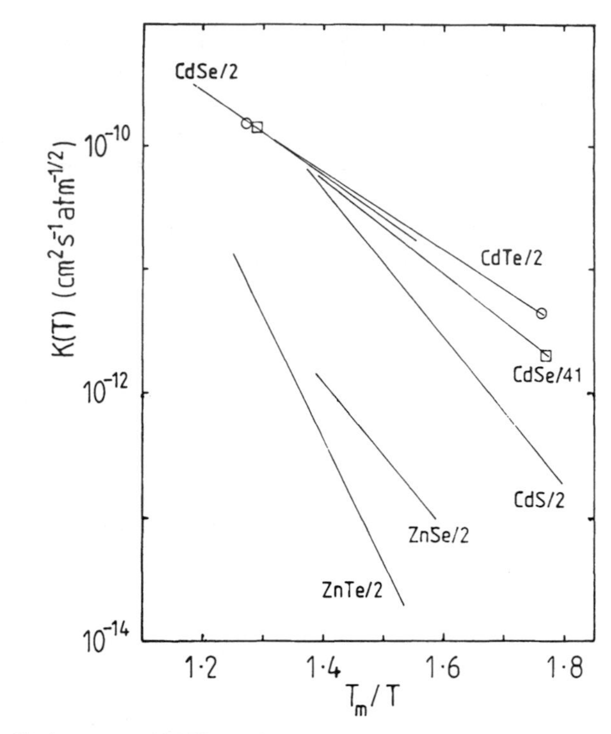

Fig. 10.1 Chalcogen self-diffusivities at $P_{X_2} = 1$ atm versus T_m/T. The source reference is given against each plot.

the tabulated results for $D_0(\mathrm{X}, \mathrm{X_{sat}})$ and $Q(\mathrm{X}, \mathrm{X_{sat}})$ [2] $K(T)$ versus T_m/T has been calculated and the results are given in Fig. 10.1 (T_m is the maximum melting point of the compound). $K(T)$ is in fact the value of $D^*(\mathrm{X})$ for $P_{X_2}^{1/2} = 1$ atm.

In ZnS, Williams [42] (see Appendix 2), using a radiotracer technique, obtained $D^*(\mathrm{S}) = 2.14 \times 10^{-3} \exp(-1.44\,\mathrm{eV}/kT)\,\mathrm{cm^2\,s^{-1}}$ between 600 and 800 °C in an argon ambient. The partial pressures should therefore approximate to the minimum total pressure condition. The corresponding $K(T)$ versus T_m/T plot lies several orders of magnitude above the CdTe/2 line in Fig. 10.1 whereas from the pattern of the other five compounds, evident in the figure, it might have been expected to lie below the Cd compounds. Measurements of $D^*(\mathrm{S})$ in ZnS versus P_{S_2} and temperature are needed.

10.5.2 Metal self-diffusion

(*a*) *Binary compounds*

Metal self-diffusion is more complex than chalcogen self-diffusion because of the effect of non-stoichiometry and of doping. The binaries divide naturally into two groups according to the variation of $D^*(M)$ with P_M in undoped material.

ZnSe, ZnTe and CdTe In this group, $D^*(M)$ is characterized by being largely independent of P_M in undoped crystals at a given temperature [2]. This feature is attributed to diffusion by a neutral associate $(M_iV_M)^\times$ or $(V_MV_X)^\times$, but a non-defect mechanism cannot be excluded. Donor and acceptor doping enhances $D^*(M)$ in ZnSe and CdTe whereas, in ZnTe, $D^*(Zn)$ is only enhanced by donor doping. These doping effects clearly demonstrate the contributions of native ionised acceptors (V'_M and/or V''_M) and ionized donors ($M_i^{\cdot}$ and/or $M_i^{\cdot\cdot}$) to $D^*(M)$ [2]. Figure 10.2 shows $D^*(M)$ versus T_m/T in undoped and donor doped material using data tabulated in ref. 2. (Note that for $D^*(Zn)$ in Al-doped ZnTe, $D^*(Zn) = 0.0116 \exp(-1.91\,\mathrm{eV}/kT)\,\mathrm{cm^2\,s^{-1}}$. The values quoted in ref. 2 are incorrect.) One can see immediately from the figure that $D^*(M)$ in CdTe is much more sensitive to donor doping than in ZnTe or ZnSe. It is also of interest to note that the Arrhenius plots for the three undoped binaries all lie within a narrow parallel band whose width is less than a factor of ~ 3.

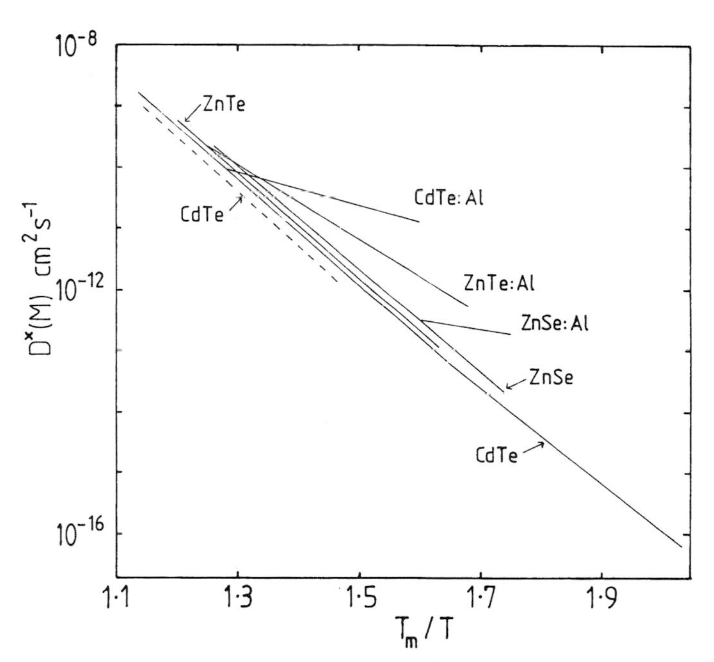

Fig. 10.2 metal self-diffusivities, $D^*(M)$, in undoped and donor doped ZnSe, ZnTe and CdTe at P_M(sat) (————) and P_{X_2}(sat) (— — —) versus T_m/T. Note that for ZnSe and ZnTe the solid and broken lines overlie each other. All the data are from ref. 2. The Al concentrations are 2×10^{19}, 5×10^{18}, $5 \times 10^{17}\,\mathrm{cm^{-3}}$ respectively for ZnTe, ZnSe and CdTe.

Astles and Blackmore [43] reported $D^*(\mathrm{Cd}) = 3 \times 10^{-13}\,\mathrm{cm^2\,s^{-1}}$ at 500 °C in a CdTe layer grown by liquid phase epitaxy (LPE) from a solution using In as solvent. The overpressure during the 500 °C anneal would approximate to the minimum total pressure. The high value of $D^*(\mathrm{Cd})$ is almost certainly attributable to In-doping of the LPE layer due to the method of growth.

ZnS, CdS and CdSe The common feature of this second group is that $D^*(\mathrm{M})$ varies with P_{M}, generally increasing, but there are substantial quantitative differences between the variations for the three members of the group.

Measurements of $D^*(\mathrm{Zn})$ in ZnS are very few. A re-assessment of Secco's [44] results (927–1073 °C at $P_{\mathrm{Zn}} = 1$ atm) yields

$$D^*(\mathrm{Zn}) = 3.0 \times 10^6 \exp(-3.90\,\mathrm{eV}/kT)\,\mathrm{cm^2\,s^{-1}} \tag{10.9}$$

An Arrhenius plot of equation (10.9) is shown in Fig. 10.3. At 1025 °C Secco also found in the pressure range 0.25–2 atm that $D^*(\mathrm{Zn})$ was approximately proportional to $P_{\mathrm{Zn}}^{1.5}$.

More information is available for CdSe where the results of Borsenberger

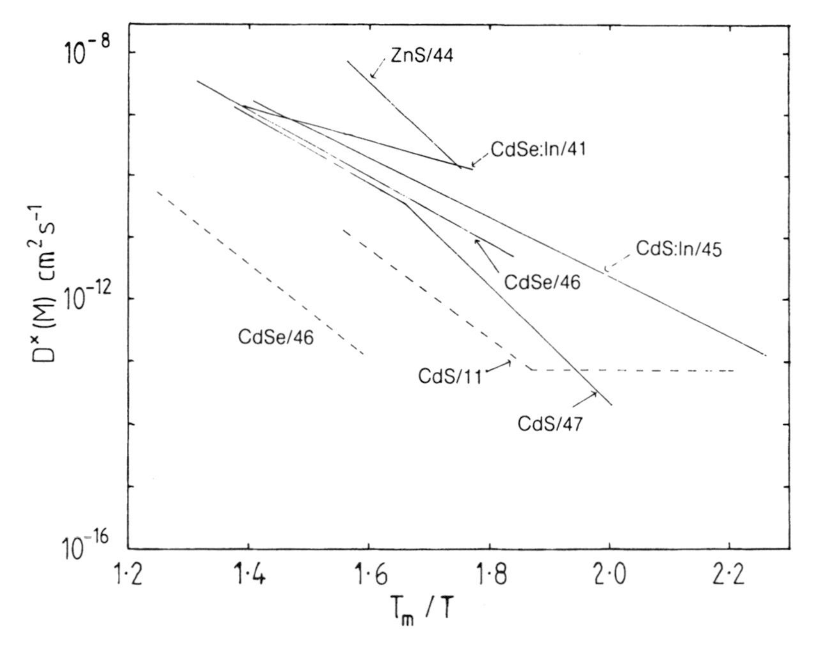

Fig. 10.3 Metal self-diffusivities, $D^*(\mathrm{M})$, versus $T_{\mathrm{m}}T$ in undoped and doped ZnS, CdS and CdSe at $P_{\mathrm{M}}(\mathrm{sat})$ (————), at minimum total pressure for CdS (— — —) and at $P_{\mathrm{Se_2}}(\mathrm{sat})$ for CdSe (— — —). The source reference is given against each plot.

et al. [46] and Zmija [41] for $D^*(\mathrm{Cd})$ at $P_{\mathrm{Cd}}(\mathrm{sat})$ and $P_{\mathrm{Se}_2}(\mathrm{sat})$ are in good agreement. Donor doping enhanced $D^*(\mathrm{Cd})$ at $P_{\mathrm{Cd}}(\mathrm{sat})$ [41]. Arrhenius plots of $D^*(\mathrm{Cd})$ at both limits of the existence range are given in Fig. 10.3. $D^*(\mathrm{Cd})$ increases continuously between $P_{\mathrm{Se}_2}(\mathrm{sat})$ and $P_{\mathrm{Cd}}(\mathrm{sat})$ [46] and the variation can be accounted for in terms of parallel diffusion by $\mathrm{Cd}_i^{\cdot}$ and $\mathrm{Cd}_i^{\cdot\cdot}$ mechanisms [2]. The observation that donor doping enhances $D^*(\mathrm{Cd})$ requires the presence of, and domination by, native ionized acceptors (e.g. $\mathrm{V}'_{\mathrm{Cd}}$ and/or $\mathrm{V}''_{\mathrm{Cd}}$) with the suppression of the ionized Cd interstitials.

Despite the greater attention devoted to measuring $D^*(\mathrm{Cd})$ in CdS the only confident conclusion that emerges is that the diffusion processes are exceedingly complex. Work to 1984 has been discussed in detail in ref. 11. The notable features are that above 675 °C in undoped CdS $D^*(\mathrm{Cd})$ increases with P_{Cd} but shows three sections: $D^*(\mathrm{Cd})$ increases with P_{Cd} in S-rich conditions, reaches a plateau around the minimum total pressure and then starts to increase again with increasing P_{Cd} under Cd-rich conditions. The extent of the plateau shortens as the temperature rises. On the Cd-rich side $D^*(\mathrm{Cd}) \propto P_{\mathrm{Cd}}^{\gamma}$ where $\gamma \approx 3$ at 675 °C and decreases to < 1 at and above 800 °C. An Arrhenius plot of $D^*(\mathrm{Cd})$ at $P_{\mathrm{Cd}}(\mathrm{sat})$ shows a break at ~ 700 °C with a larger activation energy below this temperature. Recent work confirms this feature [47]. The Arrhenius plot of $D^*(\mathrm{Cd})$ at the minimum total pressure also shows a break at ~ 650 °C but now $D^*(\mathrm{Cd})$ becomes virtually independent of temperature at least to 520 °C. Figure 10.3 shows Arrhenius plots of $D^*(\mathrm{Cd})$ at $P_{\mathrm{Cd}}(\mathrm{sat})$ and at the minimum total pressure. Donor doping enhances $D^*(\mathrm{Cd})$ under S-rich conditions but under Cd-rich conditions there is a conflict of evidence between doping producing no change, or a decrease, in $D^*(\mathrm{Cd})$. More recent results for $D^*(\mathrm{Cd})$ at $P_{\mathrm{Cd}}(\mathrm{sat})$ in highly In-doped (6×10^{19}–$3 \times 10^{20}\ \mathrm{cm}^{-3}$) CdS show $D^*(\mathrm{Cd})$ increasing with doping level [45]. An Arrhenius plot of these data is included in Fig. 10.3, which shows no break over the whole temperature span. It is somewhat surprising that the decrease in activation energy is so small compared with that in CdSe. As regards diffusion mechanisms in undoped CdS, the experimental evidence above 700 °C is consistent with a doubly ionized native donor ($\mathrm{Cd}_i^{\cdot\cdot}$ or $\mathrm{Cd}_i/\mathrm{V}_{\mathrm{Cd}}$ complex) over the whole existence region and, in the Cd-rich region, an ionized native defect also makes a contribution to $D^*(\mathrm{Cd})$ which is $\propto P_{\mathrm{Cd}}^{5/3}$ (e.g. $(2\mathrm{Cd}_i)^{\cdot}$ or a Cd_{S} diffusing via V_{Cd} or $(\mathrm{V}_{\mathrm{Cd}}\mathrm{V}_{\mathrm{S}})$). Below ~ 700 °C the situation is quite obscure. The cause of the breaks in the Arrhenius plots is unknown (Fig. 10.3). Although at 675 °C and high P_{Cd}, $D^*(\mathrm{Cd})$ is $\propto P_{\mathrm{Cd}}^3$, below ~ 620 °C $D^*(\mathrm{Cd})$ shows an inverse dependence on P_{Cd}, suggesting that V_{Cd} defects could now be dominant. The enhancement of $D^*(\mathrm{Cd})$ by donor doping again points to $\mathrm{V}'_{\mathrm{Cd}}$ and/or $\mathrm{V}''_{\mathrm{Cd}}$ being responsible.

In both ZnS and CdS, at high P_{M}, $D^*(\mathrm{M})$ has a superlinear dependence on P_{M}. Such a dependence implies diffusion by a double or triple M_i complex or by M_{X} diffusing via V_{M} or on both sublattices. A final point to make is that $D^*(\mathrm{M})$ varies with P_{M} in the würtzite but not in the zinc blende structures.

(b) Ternary alloys

The only self-diffusion results for $D^*(M)$ in widegap alloy systems are for $Cd_{0.74}Mn_{0.26}Te$ [48]. In the temperature range 450–800 °C at P_{Cd}(sat)

$$D^*(Cd) = 2.43 \exp(-2.11\,eV/kT)\,cm^2\,s^{-1}$$

$$D^*(Mn) = 3.75 \exp(-2.19\,eV/kT)\,cm^2\,s^{-1} \qquad (10.10)$$

and at P_{Te_2}(sat)

$$D^*(Cd) = 2.54 \exp(-2.16\,eV/kT)\,cm^2\,s^{-1}$$

$$D^*(Mn) = 2.89 \exp(-2.16\,eV/kT)\,cm^2\,s^{-1} \qquad (10.11)$$

It is easily seen that $D^*(Cd) \approx D^*(Mn)$ and that there is little difference between $D^*(M)$ at P_{Cd}(sat) and P_{Te_2}(sat). In addition these values of $D^*(M)$ are very close to those for $D^*(Cd)$ in CdTe. They have not been included in Fig. 10.2 in order to avoid confusing overlap.

The narrow banding of $D^*(M)$ already referred to in Fig. 10.2, and also evident for CdS and CdSe in Fig. 10.3 at P_{Cd}(sat), makes it very tempting to speculate that $D^*(M)$ in the appropriate ternary alloys can be assumed also to lie within the associated band.

10.6 CHEMICAL SELF-DIFFUSION

Measurements of the relaxation in the electrical conductivity of CdTe [49, 50] and CdS [51, 52], following a step change in P_{Cd} at the anneal temperature, have been used to obtain values of $D(\Delta)$. The results are given as Arrhenius plots in Fig. 10.4. Good agreement is evident for CdTe but the discrepancy for CdS is unexplained. Although $D(\Delta)$ generally should depend on Δ (i.e. on P_M) [3] no dependence was found in CdTe [50] or in CdS [51]. It has also been reported that $D(\Delta)$ in CdTe is reduced by Cu-[50] or In-doping [50, 54]. A further feature is the asymmetry of the relaxation in which $D(\Delta)$ is greater for a positive than for a negative step change in P_{Cd} [50, 54). The data in Fig. 10.4 represent positive step changes in P_{Cd}. An inconsistency in the interpretation of $D(\Delta)$ for CdTe in terms of native defects is discussed in ref. 2.

Arrhenius plots of $D(\Delta)$ for ZnS, ZnSe and ZnTe are also given in Fig. 10.4 using the results tabulated by Stevenson [53]. These $D(\Delta)$ were derived from the growth rates of reaction layers on Zn single-crystal faces immersed in the chalcogen vapour. Unknown factors in such measurements are the effects of the chemical reaction rate and the role of grain boundaries (the layers are virtually certain to be polycrystalline).

Our knowledge of $D(\Delta)$ and the defects that are involved clearly leaves much to be desired.

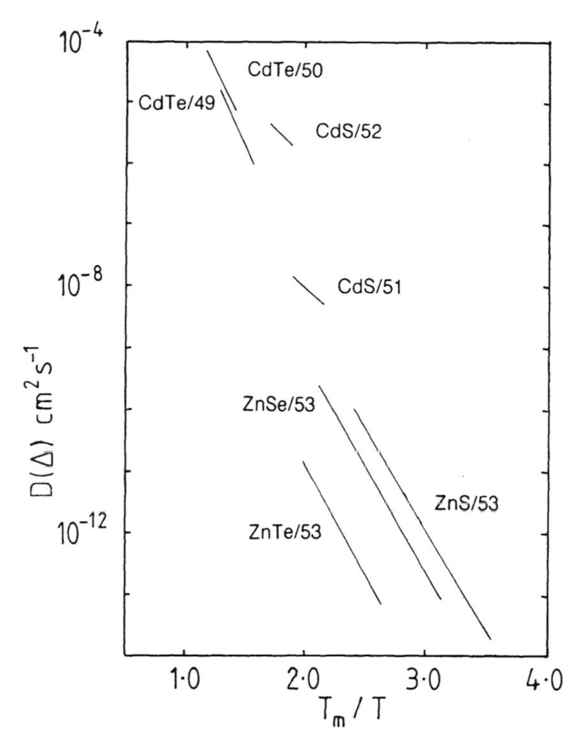

Fig. 10.4 Chemical self-diffusivities, $D(\Delta)$, versus T_m/T. The source reference is given against each plot.

10.7 IMPURITY DIFFUSION

From the discussion in section 10.3 it is not surprising to find that impurity diffusion profiles are often complex in the sense that, despite the imposition of the appropriate boundary conditions, they cannot be fitted to an erfc or gaussian variation. Such complex profiles frequently comprise two sections, as in Fig. 10.5, where it is possible to achieve approximate fits to both sections by either erfc or gaussian dependences. In this way the profile is arbitrarily divided into slow and fast components with each component characterized by a concentration-independent diffusivity. It must be recognized that this procedure is wholly empirical and that the division into fast and slow components is not proof of two separate diffusion mechanism. It is quite possible for a single diffusion mechanism to produce the type of profile shown in Fig. 10.5. The kink could reflect a change in the ENC so that the impurity no longer controls the concentrations of ionized native defects. The dissociative mechanism also can produce this type of profile [55]. In addition it needs to be remembered that a fast diffusion tail to a profile may arise from short-circuit paths [3].

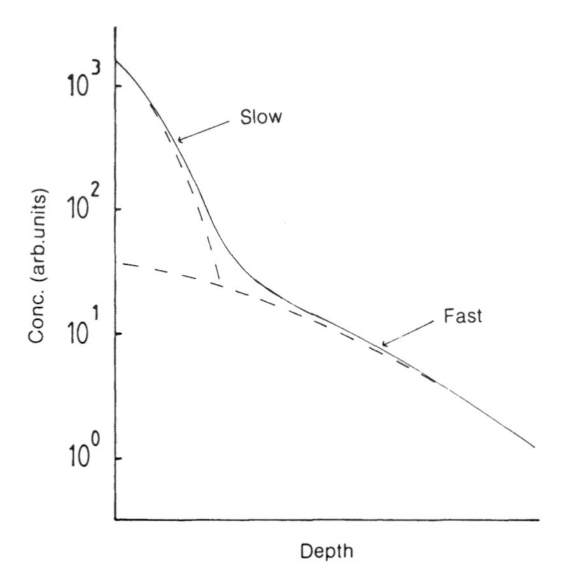

Fig. 10.5 Illustrative impurity diffusion profile which exhibits slow and fast diffusion sections: — — —, erfc or gaussian fits.

The format adopted is to consider diffusivities under specific impurity headings. Some selection of the available data has been made, e.g. limitation to single-crystal material and to experiments which give full details of the anneal conditions.

10.7.1 Group I (data are presented in Figs 10.6 and 10.7)

(*a*) *H*

Deuterium diffusion profiles in doped CdTe were obtained after anneals at 500 °C by secondary ion mass spectrometry (SIMS) under approximately minimum total pressure [58]. The dopants were P, As, Si and In. Two section profiles were found with acceptor doping (P, As) whereas only the fast section was present in the donor doped (Si, In) samples. The fast section, with a diffusivity $D^f(D) \simeq 2 \times 10^{-13}\,cm^2\,s^{-1}$, was independent of background doping and was attributed to diffusion by $D^\times$. It was proposed that the slow component, characterized by a diffusivity $D^s(D)$ of $\sim 10^{-15}$–$10^{-14}\,cm^2\,s^{-1}$, involved trapping of D by the acceptor impurities.

(*b*) *Li*

Complex profiles have been found in ZnTe [57] and CdTe [59] using nuclear reaction analysis and SIMS profiling respectively. In ZnTe the anneals were

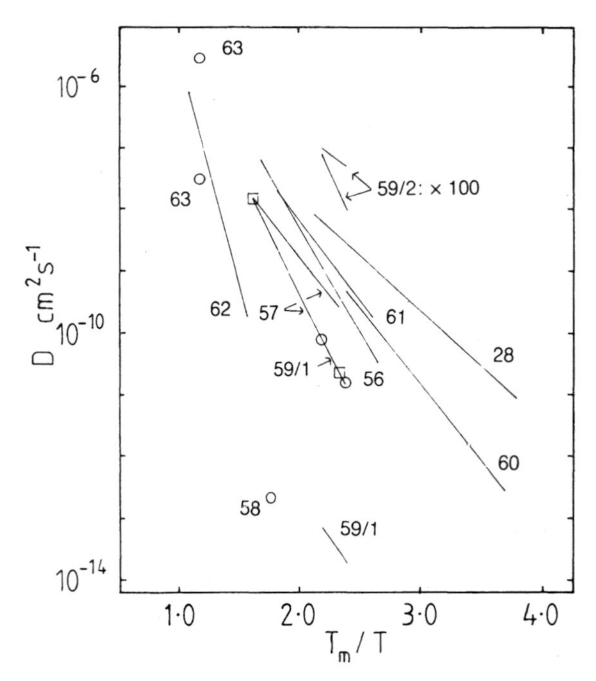

Fig. 10.6 Diffusivities, D, of group I impurities versus T_m/T in ZnSe (Ag [56], Cu [28]), ZnTe (Li [57]) and CdTe (D [58], Li [59], Cu [60, 61], Au [62, 63]). For D(Li) in CdTe, 59/1 refers to Li'_{Cd} and 59/2 refers to $Li_i^{\cdot}$. The upper and lower plots of 59/1 and 59/2 correspond to Te- and Cd-rich conditions respectively. Note that $D(Li_i^{\cdot})$ values are $\times 10^2$.

carried out under minimum total pressure and three section profiles were obtained. Arrhenius plots for the first two sections are given in Fig. 10.6. The results in CdTe were numerically fitted to simultaneous diffusion by Li'_{Cd} and $Li_i^{\cdot}$ to obtain $D(Li'_{Cd})$ and $D(Li_i^{\cdot})$ in Cd- and Te-rich samples. If Li'_{Cd} diffuses by a V_{Cd} mechanism then $D(Li'_{Cd})$ should be greater in the Te-rich material and this is observed (Fig. 10.6). In contrast, $Li_i^{\cdot}$ diffusion by an interstitial mechanism should be independent of the non-stoichiometry level and this appears to be approximately true (Fig. 10.6).

(*c*) *Cu*

Radiotracer profiles have been obtained in CdTe for Cu concentrations between 10^{15} and $10^{19}\,cm^{-3}$ [60, 61]. Whereas good erfc fits were found in ref. 60 over three orders of magnitude in concentration only the deeper-lying parts of the profiles in ref. 61 were fitted to an erfc. D(Cu) was independent of non-stoichiometry [61] and Fig. 10.6. shows that both sets of results are in quite satisfactory agreement. The evidence is consistent with Cu diffusing

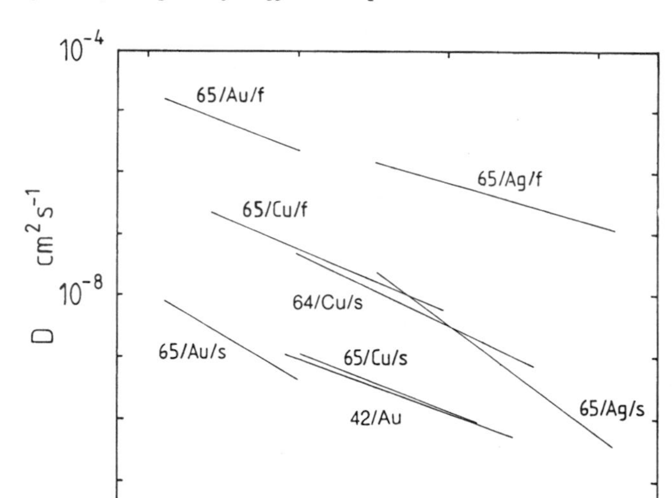

Fig. 10.7 Diffusivities, D, of group I impurities versus T_m/T in ZnS (Cu [64], Au [42]) and CdS (Cu, Ag, Au [65]). f and s refer to the fast and slow components respectively. In CdS the results for Cu and Ag were obtained under S-rich conditions, for Au f and s correspond to S-rich and minimum pressure conditions respectively. See Appendix 2 concerning ref. 42.

via the $(Cu_iV_{Cd})'$ defect [2]. The D(Cu) data reported by Mann *et al.* [66] are in error [2].

In ZnSe both radiotracer [28] and PL boundary methods [28, 56, 67] have given results for D(Cu) at P_{Zn}(sat) [28, 67] and P_{Se_2}(sat) [56] respectively. Erfc profiles were obtained in the radiotracer work with D(Cu) showing no change between undoped and Cl-doped samples. D(Cu) in Al-doped material, however, was substantially increased [28]. An Arrhenius plot of the radiotracer data [28] is given in Fig. 10.6. The PL results give Arrhenius plots lying a little above that of ref. 28, but for clarity are not shown in Fig. 10.6.

D(Cu) has been measured in ZnS and CdS by both optical boundary (ZnS [68], CdS [69]) and radiotracer methods (ZnS [64], CdS [65, 70]) under S-rich conditions [65, 68] and at the minimum total pressure condition [64, 65, 69, 70]. Two-section radiotracer profiles were found in both ZnS and CdS from which slow and fast diffusivities, D^s(Cu) and D^f(Cu), were derived. D(Cu) values based on PL boundary movement in ZnS [68] are no more than a factor 3 greater than D^s(Cu) [64]. The optical absorption measurements in CdS [69] are in close agreement with D^s(Cu) [65, 70] under the minimum pressure condition. Non-stoichiometry affects both D^s(Cu) and D^f(Cu) in CdS with each diffusivity showing a significant increase as P_{S_2}

decreases [65]. Representative Arrhenius plots for D(Cu) in ZnS and CdS are shown in Fig. 10.7.

(*d*) *Ag*

Measurements on Ag diffusion are limited to ZnSe [56] and CdS [65, 68, 71, 72] using the PL boundary method [56, 68] and radiotracer methods [65, 71, 72] under Se saturation in ref. 56 and all levels of non-stoichiometry in CdS. At S saturation both ref. 65 and ref. 72 found well-defined two-section profiles from which D^s(Ag) and D^f(Ag) values were obtained. As the S pressure is reduced, the radiotracer profile becomes a single erfc distribution [72]. It was also reported that D^s(Ag) decreased slightly with rising S pressure whereas D^f(Ag) was independent of P_{S_2} [72]. The profiles shown by Woodbury [71] seem more complex than those found in refs 65 and 72 and his D(Ag) results would seem to compare best with D^f(Ag) from refs 65 and 72. He also reported that D(Ag) was reduced by In-doping. The Arrhenius plot for D(Ag) in ZnSe [56] is given in Fig. 10.6 and similar plots for D^s(Ag) and D^f(Ag) from Zmija and Demianiuk's [65] results for CdS are shown in Fig. 10.7. The data of refs 71 and 72 lie between the D^s(Ag) and D^f(Ag) values from ref. 65. The PL result [68] mostly lies below D^s(Ag) [65] and is not shown in Fig. 10.7.

Changes occurring at room temperature in the electrical and PL properties of Ag-diffused CdTe imply a significant room temperature mobility of Ag [73].

(*e*) *Au*

Radiotracer [62] and Rutherford backscatter [63, 74] profiling techniques have been used to obtain D(Au) in CdTe under minimum total pressure conditions [62, 74] and under Cd and Te saturation [63]. The radiotracer profiles gave good fits to single gaussian distributions over two orders of magnitude in the Au concentration. The results, however, of ref. 74 are incorrect [2]. The spot measurements of D(Au) at 900 °C by Akutagawa *et al.* [63] show a strong effect of non-stoichiometry with D(Au) decreasing by a factor of 100 as the CdTe existence region is traversed from the Te to the Cd side. Arrhenius plots of the results from refs 62 and 63 are given in Fig. 10.6 where it can be seen that the data are consistent despite the limited information under Cd and Te saturation. The observed features can be explained if Au forms the complexes $(Au_{Cd}Au_{Te})^{\cdot}$ or $(Au_{Cd}Au_iV_{Te})^{\cdot}$ which diffuse via $V_{Cd}^{\times}$ [2].

Musa *et al.* [75] using Rutherford backscattering have observed diffusion lengths for Au in CdeTe at room temperature in the range 2–10 nm corresponding to $D(\mathrm{Au}) \approx 1.4 \times 10^{-21}\,\mathrm{cm^2\,s^{-1}}$.

Measurements of D(Au) in ZnS [42] and CdS [65, 76] have been made by radiotracer profiling at the minimum total pressure [42] and between the minimum total pressure and S saturation [65, 76]. In the case of ZnS, profiling was limited to approximately the first decade drop in Au concentration to which good gaussian fits were obtained. Whether or not a fast section to a profile existed was not determined. An Arrhenius plot of these data is included in Fig. 10.7. The situation in CdS is complicated. Zmija and Demianiuk [65] found two-section profiles which yielded D^s(Au) and D^f(Au): both diffusivities increased by less than approximately a factor of 10 as the S pressure changed from minimum pressure to saturation values. Although Nebauer [76] also found two-section profiles he attributed the fast section to dislocation traffic. The slow sections gave good gaussian fits at lower S pressures but as the pressure increased D^s(Au) became dependent on [Au]. It was further noticed that D^s(Au) showed a weak increase with S pressure to ~ 1 Torr, thereafter becoming independent. D^s(Au) results versus temperature were given for $P_{S_2} = 500$ Torr. Nebauer's values for D^s(Au) are at least a decade greater than those of ref. 65 for S saturation, i.e. substantial disagreement. Nebauer concluded that Au exists as Au'_{Cd} and diffused via V_{Cd}. Arrhenius plots of the largest D^f(Au) and smallest D^s(Au) from ref. 65 are shown in Fig. 10.7.

Nebauer and Quednau [77] have described the results of experiments on the electromigration of Au in CdS in which radiotracer profiling was used.

(*f*) *Diffusivities derived from the electrical properties of a diffused region*

The diffusion of Cu, Ag or Au into low resistivity material can lead to the formation of a high resistivity compensated surface layer. The thickness of such a layer, which can be determined by capacitance [78] or ultrasonic [79] techniques, is obviously related to the diffusivity of the impurity. Diffusivities for Cu [78, 79], Ag [79] and Au [79] in CdS and of Ag [80] in CdSe under approximately minimum total pressure conditions have been obtained in this way. Comparison of these data with the radiotracer results in CdS show that for Cu and Ag the radiotracer diffusivities exceed those from refs 78 and 79 by at least an order of magnitude. In the case of Au there is a cross-over at $\sim 720\,°C$ between the Arrhenius plots for the radiotracer D^s(Au) [65] and D(Au) [79] with the latter decreasing rapidly below D^s(Au) at lower temperatures. Overall it appears that diffusivities based on the thickness of the compensated region considerably underestimate the true diffusivity of the impurity. This needs to be borne in mind in the case of ref. 80 which is the only diffusion measurement that we have for Ag in CdSe.

10.7.2 Group II (data are presented in Figs 10.8 and 10.9)

(*a*) *Al*

Results for D(Al) are limited to ZnSe and ZnS. Arrhenius plots of D(Al) in ZnSe [56, 82] and in ZnS [68], based on PL [56, 68] and cathodoluminesence

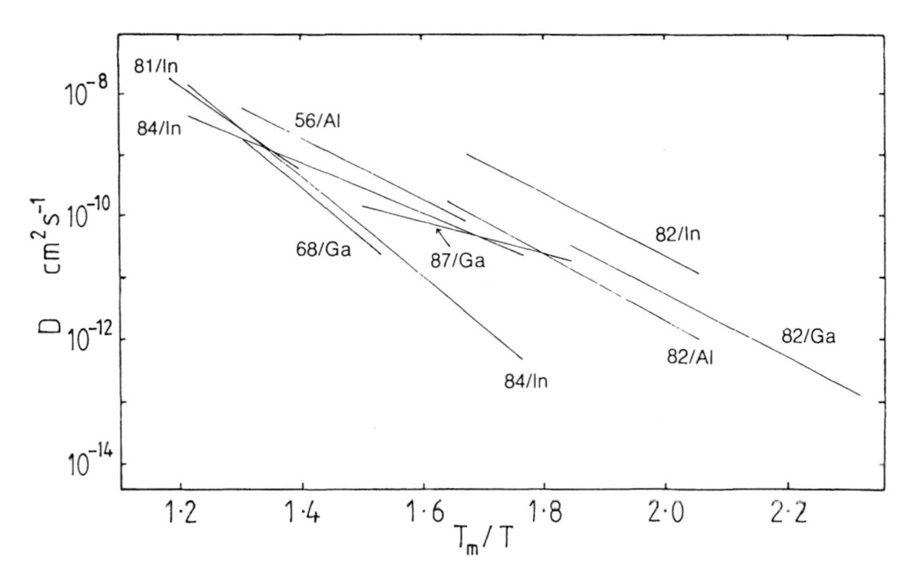

Fig. 10.8 Diffusivities, D, of group III impurities versus T_m/T in ZnTe (In [81]), ZnSe (Al [56, 82], Ga [68, 82, 87], In [81, 82]) and CdTe (In [84]). For CdTe the steeper line corresponds to Cd saturation, the other to Te saturation.

(CL) [82] boundaries, are given in Figs 10.8 and 10.9 respectively. All the diffusion anneals were performed under Zn-rich conditions and good agreement is evident in ZnSe between ref. 56 and ref. 82. Earlier measurements of D(Al) by Aven and Kreiger [85] using the PL technique and the same annealing conditions give values smaller than those of refs 56 and 82 by up to an order of magnitude. Bjerkeland and Holwech [86] obtained Al diffusion profiles in ZnSe at 950 °C under Zn-rich conditions using an optical absorption method, from which D(Al) was clearly concentration dependent. Matano analysis of the profiles gave $1 \times 10^{-11}\ \mathrm{cm^2\,s^{-1}} \leqslant D(\mathrm{Al}) \leqslant 7 \times 10^{-10}\ \mathrm{cm^2\,s^{-1}}$ for $6 \times 10^{18}\ \mathrm{cm^{-3}} \leqslant [\mathrm{Al}] \leqslant 3 \times 10^{20}\ \mathrm{cm^{-3}}$. The upper range of these D(Al) results is close to the PL values from ref. 56 (Fig. 10.8).

Bryant *et al.* [83], using an electron microprobe, obtained an Al diffusion profile in ZnS following an anneal at 1000 °C near Zn saturation. Their profile is a good fit to an erfc with $D(\mathrm{Al}) = 4.6 \times 10^{-10}\ \mathrm{cm^2\,s^{-1}}$. This diffusivity is included in Fig. 10.9. and it clearly differs substantially from the measurements of ref. 68.

(*b*) *Ga*

Measurements of D(Ga) have been reported only for ZnSe [68, 82, 86, 87] and CdS [39] based on PL [68] or CL [82] boundaries and profiling using electron microprobe analysis [39, 87] or optical absorption [86]. Figure 10.8 shows Arrhenius plots of D(Ga) in ZnSe from refs 68, 82 and 87. Although the anneal conditions for refs 68 and 82 are both the same (Zn-rich) there is

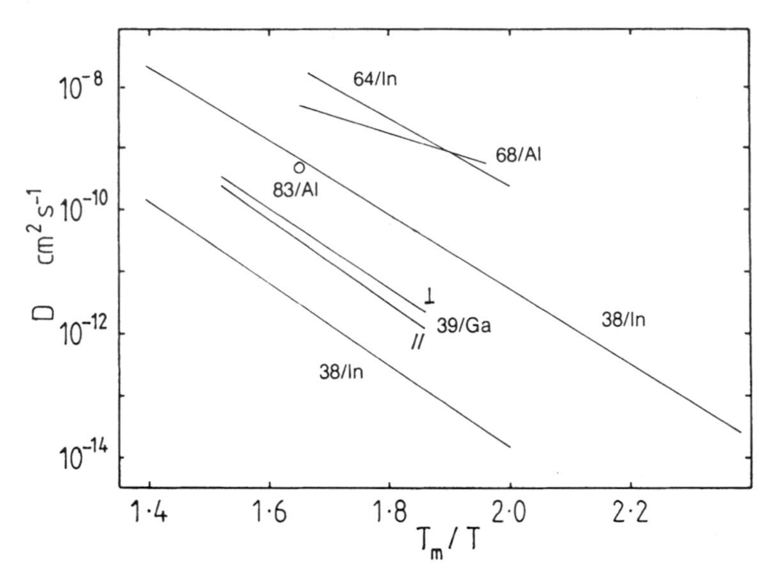

Fig. 10.9 Diffusivities, D, of group III impurities versus T_m/T in ZnS (Al [68, 83], In [64]) and CdS (Ga [39], In [38]). For In in CdS the upper line represents D(In) at an In concentration of $2 \times 10^{20}\,cm^{-3}$, i.e. concentration-dependent D(In); the lower line corresponds to concentration-independent diffusion of In.

a very marked discrepancy between the two sets of results. Donor doping caused D(Ga) to decrease [82]. The profiles obtained by Muranoi and Furukoshi [87] under minimum total pressure conditions gave erfc fits over two to three orders of magnitude in [Ga]. On the other hand, optical absorption profiling at 750 °C after annealing in Se-rich conditions indicated that D(Ga) increased with [Ga] [86] with D(Ga) ranging from $4 \times 10^{-12}\,cm^2\,s^{-1}$ at 5×10^{18} Ga cm^{-3} to $5 \times 10^{-11}\,cm^2\,s^{-1}$ at 7×10^{20} Ga cm^{-3}. These results bracket those of ref. 87. Obvious questions arising from these results in ZnSe are: is D(Ga) concentration dependent and what is the effect of non-stoichiometry?

In CdS, Jones and Mykura [39] found that D(Ga) increased with [Ga] under approximately minimum total pressure conditions. Their results showed that $D(Ga) \propto [Ga]$ for $1 \times 10^{19}\,cm^{-3} < [Ga] < 1 \times 10^{21}\,cm^{-3}$ at the higher temperatures whereas in the lower temperature range $D(Ga) \propto [Ga]$ up to higher values of [Ga] when D(Ga) started to decrease. D(Ga) was measured both perpendicular and parallel to the c axis with the former diffusivity proving to be greater by a factor $\leqslant 2$. Arrhenius plots of D(Ga) for $[Ga] = 2 \times 10^{20}\,cm^{-3}$ from the data of ref. 39 are given in Fig. 10.9. The linearity between D(Ga) was attributed to diffusion by the defect pair $(Ga_{Cd}V_{Cd})'$ with the ENC expressed by $[Ga_{Cd}^{\cdot}] = 2[V_{Cd}'']$ [39]. For this situation, linearity requires $[(Ga_{Cd}V_{Cd})'] \ll [Ga_{Cd}^{\cdot}]$ which seems an unlikely condition at the levels of [Ga] involved. Alternatively, if Ga is present predominantly as $(Ga_{Cd}V_{Cd})'$, with the ENC $[(Ga_{Cd}V_{Cd})'] = 2[Cd_i^{\cdot\cdot}] + 2[V_S^{\cdot\cdot}]$, linearity between D(Ga) and [Ga] will follow if the $(Ga_{Cd}V_{Cd})'$ pair diffuses by the $(Ga_{Cd}V_{Cd})^{\cdot}$ defect (assuming that it exists) or by the mediation of a Cd_i or V_S defect.

(c) *In*

Much greater attention has been given to characterizing the diffusion of In, compared with Al or Ga, and results are available for ZnSe, ZnTe, CdTe, ZnS and CdS.

D(In) has been measured in ZnSe across the range of non-stoichiometry using CL [82] and profiling by optical absorption [86] and by ion microprobe analysis [88]. An Arrhenius plot of the results of ref. 82, which were obtained under Zn-rich conditions, is given in Fig. 10.8. Spot measurements of D(In) at 560 °C [88] and 750 °C [86] have also been made following anneals under Zn- and Se-rich conditions respectively. At 560 °C under low P_{Zn} a concentration-independent D(In) value of $7.4 \times 10^{-13}\,cm^2\,s^{-1}$ was obtained whereas at a higher P_{Zn} D(In) increased but also became dependent on [In] with D(In) $\sim 1 \times 10^{-11}\,cm^2\,s^{-1}$ in the near-surface region. Under Se-rich conditions at 750 °C D(In) was concentration dependent and decreased from a value of $1.8 \times 10^{-9}\,cm^2\,s^{-1}$ for $2 \times 10^{20}\,In\,cm^{-3}$ apparently to level off at $\sim 4 \times 10^{-11}\,cm^2\,s^{-1}$ below $\sim 4 \times 10^{18}\,In\,cm^{-3}$. The results of refs 86 and 88 bracket those of ref. 82 which implies that D(In) is dependent on [In] at high P_{Zn} and P_{Se_2} and also has similar values at these two extremes.

Radiotracer experiments in ZnTe revealed erfc profiles following anneals under approximately minimum total pressure conditions [81]. An Arrhenius plot of these data is included in Fig. 10.8.

Several workers have investigated In diffusion in CdTe. These results have been reviewed by Watson and Shaw [84] who also presented their own data from the most detailed study of D(In) in CdTe so far available. There is a quite satisfactory agreement between the various sets of data, most of which were based on radiotracer profiling. The principal features are (i) D(In) is independent of [In], (ii) D(In) does not vary greatly between Cd and Te saturation above ~ 600 °C, (iii) below this temperature D(In) at Te saturation becomes increasingly larger than the diffusivity at Cd saturation as the temperature decreased, and (iv) below 400 °C and between 200 and 300 °C, under Te-rich conditions, D(In) becomes independent of temperature with a value of $\sim 2 \times 10^{-13}\,cm^2\,s^{-1}$. Arrhenius plots of D(In) at Cd and Te saturation are shown in Fig. 10.8. It was concluded in ref. 84 that several competing diffusion mechanisms contributed to D(In) which included (i) $In_{Cd}^{\cdot}$ diffusing via a neutral native defect for the ENC $n = [In_{Cd}^{\cdot}]$ and (ii) diffusion by the complex $(In_{Cd}V_{Cd})'$ with the ENC $[(In_{Cd}V_{Cd})'] = [In_{Cd}^{\cdot}]$.

Limited radiotracer profiling of In diffusion in ZnS under S-rich conditions indicates that the profiles consist of a fast and a slow section [64]. Figure 10.9 shows an Arrhenius plot of the D^s(In) results.

More comprehensive radiotracer measurements of D(In) in CdS have been made by Jones and Mykura [38]. They found at high P_{Cd} that D(In) was independent of [In] whereas at low P_{Cd}, $D(\mathrm{In}) \propto [\mathrm{In}]$. Arrhenius plots of their data are given in Fig. 10.9: at low P_{Cd} the values of D(In) refer to $[\mathrm{In}] = 2 \times 10^{20}\,\mathrm{cm}^{-3}$ and, as can be seen from Fig. 10.9, they are a factor of 200 or so greater than the diffusivities at high P_{Cd}. Although measurements were made for diffusion parallel and perpendicular to the c axis any anisotropy was within the precision of the radiotracer technique. Measurements of the optical absorption boundary did reveal, however, a small anistropy in D(In) with $D(\perp) > D(\parallel)$. Radiotracer profiles obtained under S-rich conditions at 810 and 900 °C by Chern and Kröger [89] show D(In) to vary with [In]. At 900 °C $D(\mathrm{In}) \approx 7.3 \times 10^{-16}[\mathrm{In}]^{1/3}\,\mathrm{cm}^2\,\mathrm{s}^{-1}$ for $2 \times 10^{17}\,\mathrm{cm}^{-3} \leqslant [\mathrm{In}] \leqslant 2.5 \times 10^{19}\,\mathrm{cm}^{-3}$ whereas at 810 °C a two-section profile was found with $D^s(\mathrm{In}) = 3.4 \times 10^{-11}\,\mathrm{cm}^2\,\mathrm{s}^{-1}$ and $D^f(\mathrm{In}) = 1.8 \times 10^{-10}\,\mathrm{cm}^2\,\mathrm{s}^{-1}$ over the range from 2.5×10^{17} to $2.5 \times 10^{18}\,\mathrm{In\,cm}^{-3}$. These values of D(In) from ref. 89 lie between the Arrhenius plots shown in Fig. 10.9 for results from ref. 38. Isoconcentration experiments, as expected, gave erfc profiles between 300 and 960 °C [90]. Below 450 °C, however, the diffusivity levelled off (i.e. became temperature independent) at $\sim 3 \times 10^{-14}\,\mathrm{cm}^2\,\mathrm{s}^{-1}$. This is a similar feature to that observed for D(In) in CdTe at low temperature [84]. Jones and Mykura [38] suggested that diffusion by $(\mathrm{In_{Cd}V_{Cd}})'$ with the ENC $[\mathrm{In}^{\cdot}_{Cd}] = 2[\mathrm{V}''_{Cd}]$ was the basis for the $D(\mathrm{In}) \propto [\mathrm{In}]$ relation. This is the same model introduced for D(Ga) in CdS and the discussion thereof, given above, is equally applicable.

10.7.3 Groups IV, V and VII (Data are presented in Fig. 10.10)

(*a*) *Sn*

Panchuk *et al.* [93] measured D(Sn) in CdTe by a radiotracer method at P_{Cd}(sat) and at approximately $0.1P_{Cd}$(sat). D(Sn) seemingly was independent of concentration. At 700 °C D(Sn) was also independent of P_{Cd} but with increasing temperature a strengthening dependence developed such that $D(\mathrm{Sn}) \propto P_{Cd}^{-2/3}$ at 925 °C. Their results indicate that below 785 °C D(Sn) is largely independent of P_{Cd}. Arrhenius plots of their data are given in Fig. 10.10 where it can be seen that at P_{Cd}(sat) the activation energy for diffusion (0.38 eV) is unusually small for a substitutional impurity. It was suggested that diffusion via V'_{Cd}, V''_{Cd} and $\mathrm{V}^{\cdot\cdot}_{Te}$ was involved but it is not clear how D(Sn) could be independent of [Sn] when account is taken of Sn occupying both Cd and Te lattice sites.

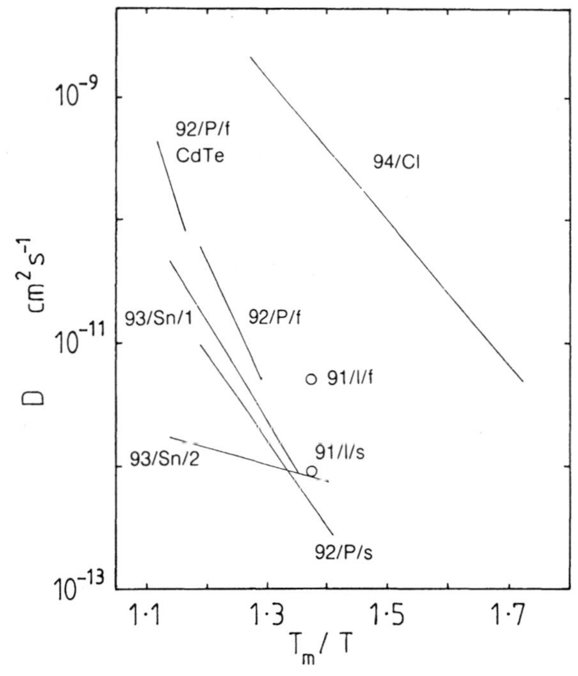

Fig. 10.10 Impurity diffusivities, D, versus T_m/T in CdS (I [91], CdSe (P [92]) and CdTe [Sn [93], P [92], Cl [94]). f and s refer to the fast and slow components respectively. For Sn [93] 1 and 2 denote the partial pressures 0.1 P_{Cd}(sat) and P_{Cd}(sat) respectively.

(*b*) *P*

Radiotracer profiles have been obtained in CdTe and CdSe [92]. In CdTe measurements were limited to 900 and 950 °C and two-section profiles were found. $D^f(P)$ was independent of P_{Cd} near P_{Cd}(sat) but with reducing pressure $D^f(P) \propto P_{Cd}^{-2/3}$. Although quantitative data were presented only for $D^f(P)$ the authors state $D^s(P)$ was also independent of P_{Cd} at high P_{Cd} but at low P_{Cd} there was too much scatter to define a relationship. An Arrhenius plot of their $D^f(P)$ data at high P_{Cd} is shown in Fig. 10.10: the highest $D^f(P)$ values measured at lower P_{Cd} are about a factor of 10 greater than the values in Fig. 10.10. Hall and Woodbury [92] suggested that $D^f(P)$ could be due to a P_i mechanism. The solubility analysis for P in CdTe by Selim and Kröger [14] indicates the presence also of $P_{Cd}^{\cdot\cdot\cdot}$, P'_{Te} and the complexes $(P_{Cd}P_i)^\times$, $(P_{Cd}2P_i)^\times$. The diffusion of P in CdTe appears to be highly complex and much more experimental diffusion data are required before a serious attempt at interpretation can be made. Despite the complexity of P-doping in CdTe, P diffusion has been used to make p–n junctions in n-type CdTe at 850 °C [95]. A junction depth of 30 μm was obtained after a two week anneal in Cd-rich conditions. By equating this depth to one or two diffusion lengths values of $2 \times 10^{-12}\,cm^2\,s^{-1}$ and $5 \times 10^{-13}\,cm^2\,s^{-1}$ are found for the effective $D(P)$. These crude estimates lie a factor of 6 or so below $D^f(P)$ (extrapolated) from ref. 92.

P diffusion profiles in CdSe between 800 and 1000 °C are similar in form to those found in CdTe [92]. $D^f(P)$, however, varied as $P_{Cd}^{-2/3}$ right up to P_{Cd}(sat) with P_{Cd} spanning a range of $\sim 4 \times 10^4$. $D^s(P)$ showed no clear dependence on P_{Cd}. It was also observed that P diffusion was unaffected by donor (In) or acceptor (Ag) doping. Figure 10.10 gives Arrhenius plots of $D^s(P)$ and $D^f(P)$ at P_{Cd}(sat). As regards the diffusion mechanism, similar remarks apply here as for CdTe.

(*c*) *Cl*

Radiotracer erfc profiles for Cl diffusion in CdTe were obtained in anneals near to P_{Cd}(sat) and to the minimum total pressure [94]. There was no significant difference in D(Cl) at these two partial pressures. An Arrhenius plot of D(Cl) for these data is given in Fig. 10.10. It was concluded that Cl diffused by the neutral vacancy pair $(V_{Cd}V_{Te})^{\times}$. Woodbury has noted also that Cl (as well as F and Br) is readily incorporated into CdS by diffusion but no diffusivity data were given [91].

(*d*) *I*

Woodbury [91] using radiotracer profiling found well-defined two-section profiles for I diffusion in CdS at 1000 °C under Cd-rich conditions. His data are included in Fig. 10.10.

10.7.4 Other impurities: O, Fe, Ni, Mn and Yb

(*a*) *O*

O is an ubiquitous contaminant and it is perhaps surprising that it has received so little attention. The very recent work of Yokota *et al.* [96] in CdTe demonstrates the importance of the vacuum quality and its relation to O contamination in material processing. The only measurements of O diffusivity appear to be those of Vodovatov *et al.* [97] in which a laser beam was used to vaporize material prior to mass spectrometric analysis. Their paper, however, has several quantitative inconsistencies: the Arrhenius equations do not fit the graphical data in their Fig. 2; the break in the Arrhenius plots in their Fig. 2 occurs at ~480 °C and not at 650 °C as given in the text. Using their Fig. 2 data the following Arrhenius equations were derived: above 490 °C in n- and p-type crystals $D(O) = 6 \times 10^{-10} \exp(-0.10\,eV/kT)\,cm^2\,s^{-1}$, below 480 °C in n-type material $D(O) = 5 \times 10^{-9} \exp(-0.25\,eV/kT)\,cm^2\,s^{-1}$

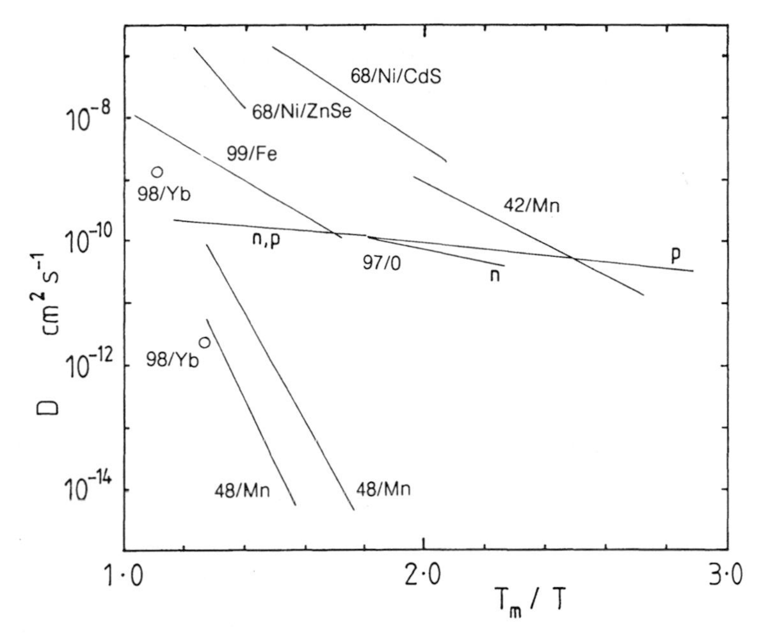

Fig. 10.11 Impurity diffusivities, D, versus T_m/T in ZnS (Mn [42]), ZnSe (Ni [68]), CdS (Ni [68], Yb [98]) and CdTe (Mn [48], O [97], Fe [99]). For results from ref. 97 n and p denote conductivity type. The right-hand plot of D(Mn) in CdTe [48] corresponds to Te(sat) and the left-hand one to Cd(sat). Note that in ref. 99 the pre-exponential factor for D(Fe) should be $1.16 \times 10^{-5}\,cm^2\,s^{-1}$ in order to fit the experimental data. See Appendix 2 concerning ref. 42.

and in p-type material $D(O) = 1 \times 10^{-9} \exp(-0.14\,eV/kT)\,cm^2\,s^{-1}$. The measurements of D(O) were made under approximately minimum total pressure conditions. It was also reported that D(O) was concentration-dependent [97], but not strongly so, and it is not clear for which [O] the Arrhenius plots in their Fig. 2 refer to. Arrhenius plots of the above expressions for D(O) are given in Fig. 10.11.

(*b*) *Fe*

Radiotracer profiling of Fe diffusion in CdTe at P_{Cd}(sat) gave erfc profiles and at 830 °C D(Fe) was independent of P_{Cd} on the Cd-rich side of the existence region [99]. It should be noted that in Panchuk *et al.*'s Arrhenius expression for D(Fe) the pre-exponential term should be $1.16 \times 10^{-5}\,cm^2\,s^{-1}$ and not $1.16 \times 10^{-6}\,cm^2\,s^{-1}$ as cited in order to agree with the graphical data in their Figs 1 and 2. An amended Arrhenius plot for D(Fe) is given in Fig. 10.11. The Fe solubility varied as $P_{Cd}^{-\gamma}$ with $0 < \gamma < 1$ and γ decreased with increasing temperature [99]. On the basis of their solubility evidence

Panchuk *et al.* suggested that Fe diffused by the dissociative mechanism involving $Fe_{Cd}^{\times}$ and $Fe_i^{\times}$ as the principal defects. It is easy to see, however, that such a model requires $D(\mathrm{Fe}) \propto P_{\mathrm{Cd}}^{-1}$ which is contrary to the observed independence.

(*c*) *Ni*

The PL boundary technique has been used to obtain *D*(Ni) in ZnSe and in CdS under Se- and S-rich conditions respectively [68]. Figure 10.11 shows Arrhenius plots of the results.

(*d*) *Mn*

D(Mn) has been measured by radiotracer profiling in CdTe [48] and in ZnS [42] at P_{Cd}(sat) and $P_{\mathrm{Te_2}}$(sat) for CdTe and near to the minimum total pressure for ZnS. Concentration-independent diffusivities were found and Arrhenius plots of *D*(Mn) for the two compounds are given in Fig. 10.11.

(*e*) *Yb*

Girton and Anderson [98] measured *D*(Yb) in CdS under S-rich conditions at 800 and 960 °C from profiles determined from the variation of PL intensity across a bevelled section through the diffusion zone. It was assumed that the PL intensity was $\propto$ [Yb]. The profiles exhibited broad peaks in the near-surface region so that *D*(Yb) values were derived from erfc fits to the part of the profiles lying deeper than the peak. *D*(Yb) increased with increasing S pressure. Figure 10.11 shows the highest values of *D*(Yb) measured at 800 and 960 °C.

10.8 INTERDIFFUSION

Ternary II–VI alloys provide the simplest interdiffusion system compared with possible quaternary II–VI alloys. A further aspect arises in the growth of epitaxial layers on non II–VI substrates. In the common case of a III–V substrate, interdiffusion across the interface gives rise to a quaternary system. This situation has been referred to in the literature as autodiffusion and references can be found in ref. 2. The topic is complicated by the possibility of chemical reactions at the interface in addition to component diffusion. More recent work in this area has been described by Golding *et al.* [100] and by Kassel *et al.* [101] for the CdTe/InSb and ZnSe/GaAs, ZnSe/AlAs systems respectively. Autodoping is beyond the scope of this chapter and will not be considered further.

Measurements of $\tilde{D}(x)$ so far have been confined to the ternary II–VI alloys for which equations (10.4) and (10.5) are applicable. We can expect

$\tilde{D}(x)$ to be a function of x, non-stoichiometry, doping and temperature. It turns out that for given levels of doping and non-stoichiometry $\tilde{D}(x)$ can be expressed in the usual Arrhenius form

$$\tilde{D}(x) = D_0(x)\exp[-Q(x)/kT] \tag{10.12}$$

where over particular ranges in x, $D_0(x)$ and $Q(x)$ may either be independent of x or vary exponentially ($D_0(x)$) or linearly ($Q(x)$). Interdiffusion profiles have been determined by electron microprobe, radiotracer or PL (bandgap) techniques in what may be termed bulk material experiments (i.e. the diffusion lengths, $2\sqrt{\tilde{D}t}$, are $\geqslant 1\ \mu$m) and which only involve a single interface. This distinction is made so as to separate such measurements from interdiffusion experiments based on (MQW) or superlattice (SL) structures in which $2\sqrt{\tilde{D}t} \ll 1\ \mu$m. With one exception [32] all the available data for $\tilde{D}(x)$ are derived from bulk-type experiments. The results from these latter experiments for the different alloys are presented in Tables 10.1 and 10.2. Data for the alloy $Cd_{1-x}Mn_xTe$ are also included in Table 10.2.

In those cases where the authors have not expressed $\ln D_0(x)$ and $Q(x)$ in terms of x their results have been fitted to a linear dependence. Generally the fits lie within a factor of 2 of the data. Further comments concerning experimental details and analysis can be found in ref. 2.

Imai *et al.* [32] reported $3.6 \times 10^{-21}\ \text{cm}^2\,\text{s}^{-1} \leqslant \tilde{D} \leqslant 2.2 \times 10^{-19}\ \text{cm}^2\,\text{s}^{-1}$ at 500 °C in MBE-grown ZnSe/ZnTe strained layer SLs comprising 10 Å ZnTe layers and 10 or 20 Å ZnTe layers. The anneal conditions were close to the minimum total pressure condition. The two orders of magnitude variation in $\tilde{D}$ was attributed without further explanation to differences in the growth conditions of the layers. The diffusion lengths involved were between 2 and 3 Å and for such small values consideration must be given to gradient energy terms and a non-continuum description of the diffusion process (section 10.9). In addition Table 10.1 indicates that $\tilde{D}(x)$ varies significantly with x in $ZnSe_{1-x}Te_x$ whereas Imai *et al.* assumed in their analysis of the first-order rocking curve satellites that $\tilde{D}$ was independent of x.

The results for $\tilde{D}(x)$ in $CdS_{1-x}Se_x$ [105] given in Table 10.1 agree with those of ref. 27 from Table 10.2. In the case of $CdSe_{1-x}Te_x$ there is reasonable agreement between ref. 104 (Table 10.1) and ref. 105 (Table 10.2) on the Cd-rich side whereas on the chalcogen-rich side $\tilde{D}$ from ref. 105 is a factor of ~ 40 greater than the value from ref. 104. The data from ref. 105 are preferred.

The radiotracer interdiffusion profiles at 900 °C for $Zn_{1-x}Cd_xS$ obtained by Nebauer [107] (Table 10.2) indicated that $\tilde{D}(x)$ increased with x in those profiles for which x varied between unity and $0.2 \leqslant x \leqslant 0.8$. An erfc fit, however, was obtained for a profile over which $x \geqslant 0.98$: the value of $\tilde{D}$ is given in Table 10.2. Both Nebauer [107] and Biter and Williams [108] agree that $\tilde{D}(x)$ increases with x but the magnitudes of $\tilde{D}(x)$ from refs 107 and

Table 10.1 The Arrhenius parameters for interdiffusion in widegap II–VI ternary alloys

Alloy	*Reference*	*Temperature (°C)*	*Composition range*	*Deviation from stoichiometry*[a]	D_0 *(cm² s⁻¹)*	*Q (eV)*
$Zn_{1-x}Cd_xSe$	[102]	700–950	$0 \leqslant x \leqslant 0.36$	M, minimum	6.4×10^{-4}	1.87
$Zn_{1-x}Cd_xTe$	[103]	700–1010	$0.1 \leqslant x \leqslant 0.9$	M, X	$0.29 \exp(2.32x)$	2.14
$ZnSe_{1-x}Te_x$	[104]	850–950	$0.1 \leqslant x \leqslant 0.9$	X	$0.14 \exp(10.5x)$	$2.22 + 0.56x$
				M	$2.5 \times 10^3 \exp(11.6x)$	$3.72 + 0.91x$
$CdS_{1-x}Se$	[105]	650–1000	$x \leqslant 0.01$	P_{S_2}(sat)	1.6×10^{-4}	1.5
	[106]	790–1010	$0 \leqslant x \leqslant 0.5$	minimum	0.17	2.60
$CdSe_{1-x}Te_x$	[104]	750–950	$0.1 \leqslant x \leqslant 0.5$	X	$117 \exp(-11.8x)$	$2.85–1.48x$
			$0.1 \leqslant x \leqslant 0.9$	M	$1.5 \times 10^{-2} \exp(-2.62x)$	$2.14–0.41x$

[a] M, X, minimum refer to metal-rich, chalcogen-rich and minimum total pressure conditions respectively during the interdiffusion anneal.

Table 10.2 Interdiffusivities in widegap II–VI ternary alloys at specific temperatures

Alloy	*Reference*	*Temperature (°C)*	*Composition range*	*Deviation from stoichiometry*[a]	$\tilde{D}$ *(cm² s⁻¹)*
$Zn_{1-x}Cd_xS$	[107]	900	$x \geqslant 0.98$	X	2.4×10^{-11}
	[108]	1100	$0.1 \leqslant x \leqslant 0.9$	X (Cl doped)	$5.0 \times 10^{-12} \exp(3.19x)$
$Cd_{1-x}Mn_xTe$	[48]	700	$0 \leqslant x \leqslant 0.26$	P_{Cd}(sat)	6.4×10^{-11}
				minimum	7.0×10^{-12}
				P_{Te_2}(sat)	2.6×10^{-11}
$ZnS_{1-x}Se_x$	[109]	1060	$0.4 \leqslant x \leqslant 1.0$	X	8.0×10^{-12}
		1070	$0 \leqslant x \leqslant 0.31$	X	5.0×10^{-13}
$CdS_{1-x}Se_x$	[27]	600	$0.19 \leqslant x \leqslant 0.91$	X	7.0×10^{-13}
		650	$0.19 \leqslant x \leqslant 0.91$	X	1.6×10^{-12}
$CdSe_{1-x}Te_x$	[105]	800	$x \leqslant 0.01$	P_{Te_2}(sat)	1.9×10^{-10}
				P_{Cd}(sat)	5.4×10^{-13}

[a] M, X, minimum refer to metal-rich, chalcogen-rich and minimum total pressure conditions respectively during the interdiffusion anneal.

108 are surprisingly close considering the 200 °C difference in the anneal temperatures.

In the common metal ternaries $CdS_{1-x}Se_x$ and $CdSe_{1-x}Te_x$ it has been found that $D(x) \propto P_{S_2}^{1/2}$ [105, 106] and $\tilde{D}(x) \propto P_{Te_2}^{1/2}$ [105] respectively. These are the same variations established for chalcogen self-diffusion in the binary compounds (section 10.5.1) so that it is reasonable to assume that in the $MX_{1-x}Y_x$ systems chalcogen interdiffusion proceeds by the same mechanism as in the binary material, at least in the chalcogen-rich domain.

For common chalcogen systems involving Se and Te the evidence in Tables 10.1 and 10.2 suggests that $\tilde{D}(x)$ is little affected by the level of non-stoichiometry. This again agrees with what has been found for metal self-diffusion in ZnSe, ZnTe, CdTe and $Cd_{0.74}Mn_{0.26}Te$ (section 10.5.2), so that similar diffusion mechanisms can again be inferred. Although there are no results for $\tilde{D}(x)$ versus non-stoichiometry in $Zn_{1-x}Cd_xS$ the binary self-diffusion data would lead to the expectation that $\tilde{D}(x)$ could vary significantly with P_M.

Given that the same diffusion mechanisms are operating in interdiffusion and in self-diffusion we can anticipate that the effects of doping will be similar in both regimes. In short, $\tilde{D}(x)$ in common metal ternaries should be independent of doping whereas donor, and possibly acceptor, doping will enhance $\tilde{D}(x)$ in common chalcogen systems. Such an enhancement could have important consequences for the stability of doped MQW and SL structures.

Finally, Al-Dallal [110] has described some observations of interdiffusion in the quaternary system CdSe–ZnTe at 545 and 580 °C. SIMS profiling revealed complex profiles. The widths of the interdiffusion zones were $\sim 2\,\mu m$ after 110 min at the lower temperature and $\sim 4\,\mu m$ after 86 min at the higher temperature, which correspond to interdiffusivities in the 10^{-12}–10^{-11} $cm^2\,s^-$ range.

10.9 DIFFUSION IN QUANTUM WELL STRUCTURES

Some of the general features that can affect diffusion in a quantum well have been recently noted [111]. These aspects all stem from the small spatial scale involved. As yet, however, there is a dearth of experimental measurements directly related to diffusion in quantum wells comprising widegap II–VI materials. It is an extremely important area, not least because of the stability of such structures. This section will therefore address those issues which have emerged from diffusion experiments on bulk materials, discussed in previous sections, and which will have a probable impact on the stability of QLSs.

Firstly, there is the stability of the interface between barrier and well in a heterostructure during growth. Adopting the criterion given in section 10.1, that the interdiffusivity $\leqslant 10^{-18}\,cm^2\,s^{-1}$, it is easily seen by extrapolation, using the results in Table 10.1, that at 300 °C this criterion is satisfied in all the systems listed with the sole exception of $CdS_{1-x}Se_x$ at S saturation where $\tilde{D} = 1 \times 10^{-17}\,cm^2\,s^{-1}$. Although such extrapolation always requires caution it seems certain, because of the results of Imai *et al.* [32], that the criterion is met in $ZnSe_{1-x}Te_x$. These conclusions, however, are for undoped material. In the binary materials there is ample evidence that donor doping (and, to a lesser degree, acceptor doping) enhances metal self-diffusion but not chalcogen self-diffusion. This means that doped common metal heterostructures are inherently more stable than common chalcogen heterostructures. It is quite possible that the above criterion would not be met in doped common chalcogen structures. The situation is basically the same as arises in the well-known compositional disordering of III–V MQWs due to impurity diffusion [111].

Secondly, there is the stability of dopant distributions (assuming stable interfaces between barrier and well) wherein two distinct cases can be recognized: out-diffusion of a dopant from one layer into adjacent layers and diffusional spread within a layer from an initially localised dopant distribution, e.g. δ doping. In the former case the confinement of the dopant will be determined by the dopant diffusivity in both the parent and adjacent layers whereas for the latter case it is the diffusivity within the parent layer that is relevant. Extrapolation of the impurity diffusivities described in section 10.7 shows that many exceed $10^{-18}\,cm^2\,s^{-1}$ at 300 °C. The validity of course of this extrapolation can be questioned because of (a) accuracy, (b) the assumption that the bulk diffusivity is the same as in the layer, and (c) the assumption that the diffusion mechanism is unchanged at lower temperatures. Item (c) is known to be invalid for In diffusion in CdTe [84] and CdS [90] where below ~450 °C D(In) becomes largely independent of temperature with values of 2×10^{-13} and $3 \times 10^{-14}\,cm^2\,s^{-1}$ respectively.

10.10 CONCLUSIONS

Over the last 30 years a wide range of diffusivity measurements of all descriptions have been made in the widegap II–VI materials. These have revealed patterns of behaviour for self-diffusion in the binary compounds where most progress in the identification of the diffusion mechanisms has been made. Even so, the progress has consisted largely of whittling down the number of possible mechanisms rather than achieving a positive and unambiguous identification. Impurity diffusion generally shows itself to be complex with the diffusivity varying with both concentration and deviation from stoichiometry. Any unravelling of the mechanisms responsible is made more difficult than in the self-diffusion case by the need for many more measurements, e.g. chemical and isoconcentration diffusion experiments as functions of concentration and deviation from stoichiometry. Even if all this could be achieved at a given temperature one might face the near certainty that any substantial change in temperature will create a quite different situation. Nonetheless, the data currently available have considerable empirical value. Recent theoretical calculations of defect energies [40, 112] offer the real prospect that the ambiguities, which can arise in the interpretation of diffusivity results in terms of specific diffusion mechanisms, may soon be removed or at least reduced.

Virtually all of the currently available diffusivity results refer to relatively high-temperature experiments in bulk material. With the increasing use and interest in epitaxial material it is necessary and inevitable that diffusivity measurements will focus on this type of material and will be performed at much lower temperatures. It appears likely that we are at the beginning of a new era in diffusivity measurements in II–VI materials.

APPENDIX 1

The following values of T_m(K), the binary compound melting point [113] were used to calculate T_m/T:

	S	Se	Te
Zn	2103	1793	1568
Cd	1748	1512	1365

APPENDIX 2

The Arrhenius parameters for D^*(S), D(Au) and D(Mn) in ZnS given by Williams [42] do not match his Arrhenius plots. Using the appropriate data from these plots the following Arrhenius parameters were derived:

	S	Au	Mn
$D_0(\mathrm{cm^2\,s^{-1}})$	2.1×10^{-4}	3.13×10^{-6}	8.0×10^{-5}
Q(eV)	1.44	0.73	1.04

The data for 42/Au and 42/Mn in Figs 10.7 and 10.11 are based on these new values.

REFERENCES

1. Sharma, B.L. (1989) *Defect Diffus. Forum*, **64–65**, 77.
2. Shaw, D. (1988) *J. Cryst. Growth*, **86**, 778.
3. Shaw, D. (1973) In *Atomic Diffusion in Semiconductors* (ed. D. Shaw), Plenum, London, Chapter 1.
4. Dzhafarov, T.D. (1989) *Phys. Status Solidi B*, **155**, 11.
5. Davis, G.D., Beck, W.A., Kilday, D.G., McKinley, J.T. and Margaritondo, G. (1989) *J. Vac. Sci. Technol*, *A*, **7**, 870.
6. Raisanen, A., Peterman, D.J., Wall, A., Chang, S., Haugstad, G., Yu, X. and Franciosi, A. (1989) *Solid State Commun.*, **71**, 585.
7. Pandey, K.C. (1986) *Phys. Rev. Lett.*, **57**, 2287.

8. Pantelides, S.T. (1990) In *Diffusion in Materials*, NATO ASI Series Vol. 179 (eds A.L. Laskar, J.L. Boquet, G. Brébec and C. Monty), Kluwer, Dordrecht, p. 523.
9. Kröger, F.A. (1974) *The Chemistry of Imperfect Crystals*, 2nd edn, North-Holland, Amsterdam, p. 302.
10. Brouwer, G. (1954) *Philips Res. Rep.*, **9**, 366.
11. Shaw, D. (1984) *J. Phys. C*, **17**, 4759.
12. Kukk, P.L. and Aarna, H.A. (1982) *Phys. Status Solidi A*, **69**, K109.
13. Kukk, P.L., Aarna, H.A. and Voogne, M.P. (1981) *Phys. Status Solidi A*, **63**, 389.
14. Selim, F.A. and Kröger, F.A. (1977) *J. Electrochem. Soc.*, **124**, 401.
15. Ido, T., Heurtel, A., Triboulet, R. and Marfaing, Y. (1987) *J. Phys. Chem. Solids*, **48**, 781.
16. Boquet, J.L., Brébec, G. and Limoge, Y. (1983) In *Physical Metallurgy*, 3rd edn (eds R.W. Cahn and P. Haasen), North-Holland, Amsterdam, Chapter 8.
16a. Tang, M.F.S. and Stevenson, D.A. (1990) *J. Phys. Chem. Solids*, **51**, 563.
17. Dzhafarov, T.D. (1977) *Phys. Stat. Sol. A*, **42**, 11.
Abdullaev, G.B. and Dzhafarov, T.D. (1987) *Atomic Diffusion in Semiconductor Structures*, Harwood, Chur, Switzerland. Chapter 2.
18. Queisser, H.J. (1961) *J. Appl. Phys.*, **32**, 1776.
19. Purdy, G.R. (1990) In *Diffusion in Materials*, NATO ASI Series Vol. 179 (eds A.L. Laskar, J.L. Boquet, G. Brébec and C. Monty), Kluwer, Dordrecht, p. 309.
20. Buonomo, A. and Di Bello, C. (1988) *Solid State Electron.* **31**, 1555.
21. Meijer, P.H.E., Keskin, M. and Napiorkowski, M. (1988) *J. Appl. Phys.*, **63**, 1608.
22. Vasilevskii M.I., and Pantaleev, V.A. (1984) *Sov. Phys. Solid State*, **26**, 33.
23. Hildebrand, O. (1982) *Phys. Status. Solidi A*, **72**, 575.
24. Rothman, S.J. (1990) In *Diffusion in Materials*, NATO ASI Series Vol. 179 (eds A.L. Laskar, J.L. Boquet, G. Brébec and C. Monty), Kluwer, Dordrecht, p. 269.
25. Yeh, T.H. (1973) In *Atomic Diffusion in Semiconductors* (ed. D. Shaw), Plenum, London, Chapter 4.
26. Tuck, B. (1988) *Atomic Diffusion in III–VI Semiconductors*, Hilger, Bristol, Chapter 2.
27. Taylor, H.F., Smiley, V.N., Martin, W.E. and Pawka, S.S. (1972) *Phys. Rev. B*, **5**, 1467.
28. Aven, M. and Halstead, R.E. (1965) *Phys. Rev.*, **137**, A228.
29. Zanio, K. (1970) *J. Appl. Phys.*, **41**, 1935.
30. Fleming, R.M., McWhan, D.B., Gossard, A.C., Wiegmann, W. and Logan, R.A. (1980) *J. Appl. Phys.*, **51**, 357.
31. Staudenmann, J.L., Horning, R.D., Knox, R.D., Reno, J., Sou, I.K., Faurie, J.P. and Arch, D.K. (1986) In *Semiconductor Based Heterostructures* (eds M.L. Green, J.E.E. Baglin, G.Y. Chin, H.W. Deckman, W. Mayo and D. Narasinham) Metallurgical Society, of AIME, Warrendale, PA, p. 41.
32. Imai, A., Kobayashi, M., Dosho, S., Kongai, M. and Takahashi, K. (1988) *J. Appl. Phys.*, **64**, 647.
33. van de Walle, G.F.A., van IJzendoorn, L.J., van Gorkum, A.A., van den Heuvel, R.J. and Theunissen, A.M.L. (1990) *Semicond. Sci. Technol.*, **5**, 345.
34. Kim, Y., Ourmazd, A. and Feldman, R.D. (1990) *J. Vac. Sci. Technol. A*, **8**, 1116.
35. Schubert, E.F., Tu, C.W., Kopf, R.F., Kuo, J.M. and Lunardi, L.M. (1989) *Appl. Phys. Lett.*, **54**, 2592.
36. Jones, E.D. (1972) *J. Phys. Chem. Solids*. **33**, 2063.
37. Sysoev, L.A., Fel'dman, A.Ya., Koraleva, A.D. and Kravchenko, N.G. (1969) *Inorg. Mater., USA*, **5**, 1889.
38. Jones, E.D. and Mykura, H. (1978) *J. Phys. Chem. Solids*, **39**, 11.

39. Jones, E.D. and Mykura, H. (1980) *J. Phys. Chem. Solids*, **41**, 1261.
40. Berding, M.A., van Schilfgaarde, M., Paxton, A.T. and Sher, A. (1990) *J. Vac. Sci. Technol. A*, **8**, 1103.
41. Zmija, J. (1973) *Acta Physi. Pol. A*, **43**, 345.
42. Williams, V.A. (1972) *J. Mater. Sci.*, **7**, 807.
43. Astles, M.G. and Blackmore, G. (1986) *J. Electron. Mater.*, **15**, 287.
44. Secco, E.A. (1958) *J. Chem. Phys.*, **29**, 406.
45. Jones, E.D. and Stewart, N.M. (1989) *J. Cryst. Growth*, **96**, 40, and private communication.
46. Borsenberger, P.M., Stevenson, D.A. and Burmeister, R.A. (1967) In *II–VI* Semiconducting Compounds (ed. D.G. Thomas), Benjamin, New York, p. 439.
47. Jones, E.D., Stewart, N.M. and Thambipillai, V. (1989) *J. Cryst. Growth*, **96**, 453.
48. Jamil, N.Y., Shaw, D. and Lunn, B., to be published.
49. Zanio, K. (1970) *J. Appl. Phys.*, **41**, 1935.
50. Rud', Yu.V. and Sanin, K.V. (1972) *Sov. Phys. Semicond.*, **6**, 764.
    Rud', Yu.V. and Sanin, K.V. (1974) *Inorg. Mater.*, **10**, 839.
51. Boyn, R., Goede, O. and Kushnerus, S. (1965) *Phys. Status Solidi*, **12**, 57.
52. Kumar, V. and Kröger, F.A. (1971) *J. Solid State Chem.*, **3**, 406.
53. Stevenson, D.A. (1973) In *Atomic Diffusion in Semiconductors* (ed. D. Shaw), Plenum, London, Chapter 7.
54. Chern, S.S. and Kröger, F.A. (1975) *J. Solid. State Chem.*, **14**, 299.
55. Tuck, B. (1988) *Atomic Diffusion in III–VI Semiconductors*, Hilger, Bristol.
56. Lukaszewicz, T. and Zmija, J. (1980) *Phys. Status Solidi A*, **62**, 695.
57. Martin, P. and Bontemps, A. (1980) *J. Phys. Chem. Solids*, **41**, 1171.
58. Svob, L., Heurtel, A. and Marfaing, Y. (1988) *J. Cryst. Growth*, **86**, 815.
59. Svob, L. and Marfaing, Y. (1982) *J. Cryst. Growth*, **59**, 276.
60. Woodbury, H.H. and Aven, M. (1968) *J. Appl. Phys.*, **39**, 5485.
61. Panchuk, O.E., Grytsiv, V.I. and Belotskii, D.P. (1975) *Inorg. Mater., USA*, **11**, 1510.
62. Teramoto, I. and Takayanagi, S. (1962) *J. Jpn. Phys. Soc.*, **17**, 1137.
63. Akutagawa, W., Turnbull, D., Chu, W.K. and Mayer, J.W. (1975) *J. Phys. Chem. Solids*, **36**, 521.
64. Nelkowski, H. and Bollman, G. (1969) *Z. Naturforsch. A*, **24**, 1302.
65. Zmija, J. and Demianiuk, M. (1971) *Acta Phys. Pol. A*, **39**, 539.
66. Mann, H., Linker, G. and Meyer, O. (1972) *Solid State Commun.*, **11**, 475.
67. Yamaguchi, M. and Shigematsu, T. (1978) *Jpn. J. Appl. Phys.*, **17**, 335. (1978).
68. Lukaszewicz, T. (1982) *Phys. Status Solidi A*, **73**, 611.
69. Szeto, W. and Somorjai, G.A. (1966) *J. Chem. Phys.*, **44**, 3490.
70. Clarke, R.L. (1959) *J. Appl. Phys.*, **30**, 957.
71. Woodbury, H.H. (1965) *J. Appl. Phys.*, **36**, 2287.
72. Slinkina, M.V., Zhakovskii, V.M. and Shukovskaya, A.S. (1984) *Sov. Phys. Solid State*, **26**, 1361.
73. Chamonal, J.P., Molva, E., Pautrat, J.L. and Revoil, L. (1982) *J. Cryst. Growth*, **59**, 297.
74. Hage-Ali, M., Mitchell, I.V., Grob, J.J. and Siffert, P. (1973) *Thin Solid Films*, **19**, 409.
75. Musa, A., Ponpon, J.P., Grob, J.J., Hage-Ali, M., Stuck, R. and Siffert, P. (1983) *J. Appl. Phys.*, **54**, 3260.
76. Nebauer, E. (1968) *Phys. Status Solidi*, **29**, 269.
77. Nebauer, E. and Quednau, F. (1984) *Phys. Status. Solidi A*, **26**, 225.
78. Sullivan, G.A. (1969) *Phys. Rev.*, **184**, 796.
79. Sullivan, J.L. (1973) *J. Phys. D*, **6**, 552.
80. Sullivan, J.L. (1975) *Thin Solid Films*, **25**, 245.

81. Yokozawa, M., Kato, H. and Takayanagi, S. (1968) *Denki Kakagu*, **36**, 282.
82. Takenoshita, H., Kido, K. and Sawai, K. (1986) *Jpn. J. Appl. Phys.*, **25**, 1610.
83. Bryant, F.J., Krier, A. and Zhong, G.Z. (1985) *Solid State Electron.*, **28**, 847.
84. Watson, E. and Shaw, D. (1983) *J. Phys. C*, **16**, 515.
85. Aven, M. and Kreiger, E.L. (1970) *J. Appl. Phys.*, **41**, 1930.
86. Bjerkeland, H. and Holwech, I. (1972) *Phys. Norv.*, **6**, 139.
87. Muranoi, T. and Furukoshi, M. (1981) *Thin Solid Films*, **86**, 307.
88. Kun, Z.K. and Robinson, R.J. (1976) *J. Electron. Mater.*, **5**, 23.
89. Chern, S.S. and Kröger, F.A. (1974) *Phys. Status Solidi A*, **25**, 215.
90. Jones, E.D. and Vere, D.M. (1985) *J. Cryst. Growth*, **72**, 184.
91. Woodbury, H.H., (1967) In *II–VI Semiconducting Compounds* (ed. D.G. Thomas), Benjamin, New York, p. 244.
92. Hall, R.B. and Woodbury, H.H. (1968) *J. Appl. Phys.*, **39**, 5361.
93. Panchuk, O.E., Shcherbak, L.P., Feichuk, P.I. and Savitskii, A.V. (1978) *Inorg. Mater., USA*, **14**, 41,
94. Shaw, D. and Watson, E. (1984) *J. Phys. C*, **17**, 4945.
95. Mandel, G. and Morehead, F.F. (1964) *Appl. Phys. Lett.*, **4**, 143.
96. Yokota, K., Yoshikawa, T., Inano, S., Morioka, T. and Katayama, S. (1990) *J. Appl. Phys.*, **56**, 866.
97. Vodovatov, F.F., Indenbaum, C.V. and Vanyukov, A.V. (1970) *Sov. Phys. Solid State*, **12**, 17.
98. Girton, D.G. and Anderson, W.W. (1969) *Trans. Metall. Soc. AIME*, **245**, 465.
99. Panchuk, O.E., Fesh, R.N., Savitskii, A.V. and Shcherbak, L.P. (1981) *Inorg. Mater. USA*, **17**, 1004.
100. Golding, T.D., Martinka, M. and Dinan, J.H. (1988) *J. Appl. Phys.*, **64**, 1873.
101. Kassel, L., Abad, H., Garland, J.W., Raccah, P.M., Potts, J.E., Haase, M.A. and Cheng, H. (1990) *Appl. Phys. Lett.*, **56**, 42.
102. Martin, W.E. (1973) *J. Appl. Phys.*, **44**, 5639.
103. Blömer, F. and Leute, V. (1973) *Z. Phys. Chem., N.F.*, **85**, 47.
104. Leute, V. and Blömer, F. (1974) *Z. Phys. Chem. N.F.*, **89**, 15.
105. Woodbury, H.H. and Hall, R.B. (1967) *Phys. Rev.*, **157**, 641.
106. Nakano, M. and Igaki, K. (1982) *Trans. Jpn. Inst. Met.*, **23**, 103.
107. Nebauer, E. (1973) *Phys. Status Solidi A*, **19**, K183.
108. Biter, W.F. and Williams, F. (1971) *J. Lumin.*, **3**, 395.
109. Asami, S., Ebina, A. and Takahashi, P. (1978) *Jpn. J. Appl. Phys.*, **17**, 779.
110. Al-Dallal, S. (1977) *Phys. Status Solidi A*, **44**, 183.
111. Shaw, D. (1990) in *Diffusion in Materials*, NATO ASI Series Vol. 179 (eds A.L. Laskar, J.L. Bouquet, G. Brébec and C. Monty, Kluwer, Dordrecht, p. 557.
112. Morgan-Pond, C.G., Schick, J.T. and Goettig, S. (1989) *J. Vac. Sci Tchnol. A7*, 354.
Schick, J.T. and Morgan-Pond, C.G. J. (1990) *J. Vac. Sci. Technol A*, **8**, 1108.
113. Lorenz, M.R. (1967) *Physics and Chemistry of II–VI Compounds* (eds M. Aven and J.S. Prener), North-Holland, Amsterdam, Table 2.3.

11

Doping and conductivity in widegap II–VI compounds

G.F. Neumark

11.1 INTRODUCTION

The achievement of good bipolar conductivity in widegap semiconductors has been an elusive aim of research for many years. An extensive review on II–VI compounds and their general properties has been presented by Hartmann *et al.* in 1982 [1]. Additional reviews, focusing more on the conductivity problem and on available dopants, are for instance those by Bhargava [2, 3], Dean [4], Marfaing [5], Neumark [6], Park and Shin [7] and Pautrat *et al.* [8]. In this presentation, I shall primarily emphasize work since 1981, i.e. that not covered by Hartmann *et al.* [1], as well as including some new concepts.

A main aim of the present article will be to present concepts which are common to the problem of obtaining good conductivity in all widegap materials; of primary relevance in this regard are solubility constraints on the dopants and some resultant consequences (section 11.2), as well as the role and types of compensation (section 11.3). However, it is helpful to illustrate these concepts in terms of specific materials. For this, one aspect is to decide which II–VI semiconductors to classify as 'widegap'; I shall, arbitrarily, select the 'wider' gap semiconductors, those with a bandgap (E_g) greater than 2 eV. I shall also omit the oxides, leaving ZnS ($E_g = 3.66$ eV at 300 K), ZnSe ($E_g = 2.67$ eV), ZnTe ($E_g = 2.25$ eV), and CdS ($E_g = 2.42$ eV), where the listed bandgaps are those given by Hartmann *et al.* [1]. It must also be mentioned in this connection that there is far more recent literature available on ZnSe than on the others; consequently the present review will focus primarily on this material. Regarding (recent) work on the others, ZnTe is the next most studied material, with still less on ZnS and CdS. A further point which should be made is that it has been possible to obtain reasonably well-conducting n-type ZnSe, ZnS, and CdS, but not ZnTe; as regards well-conducting p-type, this has been achieved relatively readily for ZnTe but, at best, only with difficulty for the others. The results up to 1982 have been summarized by Hartmann *et al.* [1]. For p-ZnSe the situation has since been reviewed by Neumark [6], emphasizing reproducibility problems and a likely role of twinning and resultant gettering. Since then, there have, however, been further reports of p-ZnSe, by Ohki *et al.* [9], Suemune *et al.* [10], Haase *et al.* [11] and Taike *et al.* [12]; the reproducibility as well as the role of twinning in these reports remain to be investigated. In addition, there has also been a recent report of good p-conductivity in ZnS [13]. In any case, it still seems far easier to obtain p-type in ZnTe and n-type in the others, and this difference at present is one of the more puzzling aspects of II–VI behaviour. There have recently been a number of suggestions for explaining these results: (1) Jansen and Sankey [14], who suggest energetics of formation of native defects (where a similar suggestion, but involving partly different defects, was made by Van Vechten in 1975 [15], based on more empirical calculations), (2) Dow *et al.* [16], who suggest tendencies of formation of shallow versus deep levels, and (3) Chadi [17], who suggests energetics of formation of DX-type centres (see further below and section 11.3 for a description of DX centres); at this point these explanations are all still tentative, and I shall therefore either not comment on them here, or comment only minimally.

In order to place the relation between doping and conductivity in focus, it seems worthwhile to review the requirements for good conductivity, i.e. the requirements for obtaining an adequate number of electrons in the conduction band (and correspondingly holes in the valence band). It is reasonably obvious that this requires first, a sufficient concentration of the donors, and, second, a sufficiently small energy separation between the donors and the band, so that the donors are readily ionized (i.e. the donors must be shallow). However, there is also a third requirement, which is that there be relatively low compensation. It is well known (e.g. refs 18–24) that wide bandgaps give a strong energy incentive for compensation, and that it is thus difficult to avoid this problem. However, it can readily be shown that even 90% compensation could, in principle, still lead to a carrier concentration which would be usable at least for some devices; as a specific example, such compensation would still give, for p-ZnSe with a substitutional Li concentration of $10^{17}\,\mathrm{cm}^{-3}$, a carrier concentration of $\approx 4 \times 10^{15}\,\mathrm{cm}^{-3}$ [25]. Given shallow dopant levels, which are known to exist in most cases (e.g. refs 1 and 2), it follows that lack of good conductivity must be caused either by very strong compensation (probably together with the presence of deep levels [26]) and/or an inadequate concentration of the shallow dopants. It can be noted that an adequate concentration of dopants should be obtainable if their solubility is high enough. However, recent work [27] has shown that there are severe constraints on dopant solubility in widegap materials. Under this circumstance, the only way of obtaining an adequate concentration is by non-equilibrium incorporation. It must here also be emphasized that well-conducting material (e.g. n-ZnSe, p-ZnTe etc.) *has* been obtained. A

Table 11.1 Reports of well-conducting n-ZnSe, 1988–

Growth method	*Best substrate temperature (°C)*	*Dopant*	*Best carrier concentration (cm^{-3})*	*Information on dopant 'Activity'*[a]	*Reference*
MBE	270	Ga	3×10^{17}	Deep luminescence increases with dopant flux	[28]
MOCVD	310	I	1.5×10^{19}	Deep luminescence increases with n	[29]
MOCVD	300–400	I	3×10^{19}	Deep luminescence increases with n; from SIMS, I concentration $\sim 10^{20}$ cm^{-3}	[30]
MOCVD	500	Al	1×10^{18}	Strong deep luminescence	[31]
		Cl	1×10^{17}	Deep luminescence increases with n	
MOCVD	—	Cl	2×10^{17}	Deep luminescence, for given ratio of partial pressures, increases with n	[32]
MOCVD	260–330	I	8×10^{18}	Deep luminescence increases with n; compensation $\sim$0.15–0.4, but is estimated only for $n \leqslant 4 \times 10^{17}$ cm^{-3}	[33]
MBE	300	Cl	$> 10^{19}$	Deep luminescence increases with n, but based on SIMS, compensation is not a serious problem	[34]

[a]Widegap materials tend to be compensated, with the result that frequently part of the dopant will not be on the donor site, i.e. it will not be "active". An example relevant to n-ZnSe is the complex [V_{Zn}–D] (the so-called "A" center), which gives deep luminescence.

listing for recent results, since 1988 (for specificity, for material with a carrier concentration greater than $\approx 10^{17}\,\text{cm}^{-3}$) for n-ZnSe is given in Table 11.1; similar results for n-ZnS have been reported by Yoshikawa *et al.* [35] and by Yasuda *et al.* [36]. However, it seems very likely [27] that cases of good conductivity have mostly (always?) indeed been obtained by appropriate non-equilibrium incorporation; this view is also corroborated by the fact that all the recent work (Table 11.1) has used molecular beam epitaxy (MBE) or metal–organic chemical vapour deposition (MOCVD).

11.2 SOLUBILITY CONSIDERATIONS

Recent work [27] has shown that the problem of obtaining good conductivity in widegap materials is caused, fundamentally, by solubility limits for shallow dopants rather than, as previously thought (e.g. refs 5, 18–25), merely from strong tendencies toward compensation. I shall therefore largely ignore the compensation aspect in this section, but—since compensation is nevertheless important, both in terms of the physics and also historically—I shall return to it in section 11.3. Regarding the solubility aspects, I shall first give the qualitative reasoning, then the quantitative development. Subsequently, I shall discuss the parameters in the theory, with the aim of showing in which respects one can—and cannot—manipulate conditions such as to obtain the best conductivity values. As the last part of this section, I shall then give evidence—using results in prior literature—that solubilities are, indeed, relatively limited.

A qualitative view of the importance of dopant solubility can readily be obtained by separating the incorporation of dopants into several steps. I will present this argument for n-type material, where a comparable argument can be made, *mutatis mutandis*, for p-type: (1) the dopant atom (in an external phase) is ionized; (2) the dopant ion is inserted into the host (3) the electron is inserted into the host, where it will effectively be incorporated at the Fermi level (basically, the electron gains the energy of the 'work function', i.e. the energy difference between the host vacuum level and the Fermi level—see for example Weiser [37] or Neumark [24]). It is apparent that steps (1) and (2) are independent of the Fermi level (of course, assuming, for step (2), that the dopant remains on the same site regardless of the value of the Fermi level). It is equally apparent that for step (3) the energy gain will decrease as the Fermi level increases; as a result, inserting a very low dopant concentration in a host (such as to leave the Fermi level essentially at the intrinsic level) will be far more favourable—in terms of energy—than putting in a high concentration, with a correspondingly high Fermi level (where a high Fermi level is of course required for a good conductivity). It is also obvious that this Fermi level problem becomes increasingly severe as the bandgap increases. Moreover, using estimates for the energies of inserting the bare ion obtained from calculations of Harrison and Kraut [38], it can be shown that, for example for ZnSe, these insertion energies are only about one-half of those resulting from the Fermi level considerations [27].

The argument about the role of dopant solubility can also, relatively easily, be made quantitative. Consider the free energy of incorporating dopants (of concentration N_d) in a host (with a concentration N_s of lattice sites). This free energy (μ_d) can be expressed as [19, 39]

$$\mu_d = B(T, P) + kT\ln(N_d/N_s) + E_F - E_D + kT\ln 2 - kT\ln\{1 + 2\exp[(E_F - E_D)/kT]\} \qquad (11.1)$$

where E_F is the Fermi energy, E_D is the donor energy (and where both are referred to the valence band), T is the temperature, P is the effective pressure of the host constituents (and is the actual pressure if the external phase is gaseous), k is Boltzmann's constant, $B(T, P)$ does not depend on the donor properties, and it is assumed that $N_s \gg N_d$. Assume now that this material is in equilibrium with the same external phase as an intrinsic host (Fermi level E_I) with a donor concentration N_i; it can be noted that although such a situation (intrinsic host with the same external phase) is, to a certain extent, hypothetical, it is still a valid concept, and moreover one can regard this intrinsic case as corresponding to fully compensated material. The chemical potentials will then be the same, and can be equated, with the result that [27]

$$kT\ln(N_d/N_i) = E_I - E_F - kT\ln\{1 + 2\exp[(E_F - E_D)/kT]\} + kT\ln\{1 + 2\exp[(E_I - E_D)/kT]\} \qquad (11.2a)$$

Considering only a single dopant, the expressions for E_I and for E_F are well known (e.g. ref. 40):

$$E_I = (1/2)E_G + (3/4)kT\ln(m_h^*/m_e^*) \qquad (11.3a)$$

where E_G is the bandgap, and where m_h^* and m_e^* are the density-of-states mass ratios for holes and electrons respectively. The Fermi level at relatively high temperatures (as generally used for growth or diffusion) is well approximated up to low degeneracy (up to $(E_F - E_G)/kT < 1$—see for example Appendix C3 of ref. 40) by

$$E_F \approx E_G - kT\ln[(N_c/N_d) - 0.27], \qquad (11.3b)$$

for $E_F > E_I$, with $E_F \approx E_I$ subsequently. Here, N_c is the conduction band (effective) density of states. A useful simplification can be obtained under conditions sufficiently non-degenerate such that $E_F \ll E_D$ and $E_I \ll E_D$, so that equation (11.2a) reduces to

$$kT\ln(N_d/N_i) \approx E_I - E_F \qquad (11.2b)$$

Since the 0.27 term in equation (11.3b) is now negligible, substitution of equations (11.3) into (11.2b) gives

$$2\ln N_d \approx (3/4)\ln(m_h^*/m_e^*) + \ln(N_i N_c) - E_G/2kT \qquad (11.4)$$

The role of the bandgap is now readily apparent!

Given a solubility in intrinsic material, a bandgap, a donor energy, the effective masses, and temperature, equations (11.2) and (11.3) can be solved self-consistently (with alternate use of equation (11.4) when it is applicable) to give a corresponding donor solubility. Of interest here are the maximum (i.e. equilibrium) solubilities, N_i^e and N_d^e respectively. One problem with this calculation is that intrinsic equilibrium solubilities are not known for widegap materials (in narrowgap materials, at high temperatures such as those used during processing for impurity incorporation, one would expect $E_F \approx E_I$, and thus that measured solubilities correspond to the intrinsic ones); however, results obtained with use of reasonable estimates are instructive. For specificity, one can compare the isocoric compounds ZnSe ($E_G = 2.8$ eV at 0 K), GaAs ($E_G = 1.5$ eV at 0 K), and Ge ($E_G = 0.74$ eV at 0 K), and assume that these have comparable equilibrium intrinsic solubilities [27]. Solubilities in Ge have been reported [41] up to $\approx 4 \times 10^{20}\,\mathrm{cm}^{-3}$ (from $\approx$ 700–1100 K), and the band gap of Ge is sufficiently low that these should correspond, approximately, to the intrinsic value. Assuming shallow dopants, and thus that $E_D \approx E_G$, and with reasonable values for the bandgaps as F(T) and effective masses [27], one obtains the results of Table 11.2.

It can be seen from Table 11.2 that, for example at 800 K, the solubility in ZnSe is a factor of 4×10^4 (!) lower than the selected intrinsic solubility. The actual value of $10^{16}\,\mathrm{cm}^{-3}$ is relatively quite low, particularly considering that we have used the highest solubility value in Ge (that of Ga and Al), where other values can be appreciably lower (down to $10^{15}\,\mathrm{cm}^{-3}$ for Ag and Fe, with even that for In being only 7×10^{18}—see ref. 41). In any case it is apparent from Table 11.2 that the decrease in solubility is pronounced for wide gaps, and moreover is a strong function of bandgap.

It is also instructive to consider, in more detail, the parameters of the theory. The first one I shall consider is the temperature, since the dependence of the solubility on this parameter is very apparent from the results given in Table 11.2. The reasons for this result are not hard to understand: the increase with T is partly due to the decrease of the Fermi level (for a given dopant concentration) with increasing temperature, and partly to the decrease in bandgap. I would now like to emphasize that the temperature value to be used is not arbitrary (see next paragraph). I would also like to recall that the theory applies to equilibrium. Dopant incorporation under

Table 11.2 Dopant solubilities (cm^{-3}) for an intrinsic solubility of $4 \times 10^{20}\,\mathrm{cm}^{-3}$

	T(K)		
	800	*1000*	*1200*
GaAs	8×10^{17}	2×10^{18}	5×10^{18}
ZnSe	1×10^{16}	1×10^{17}	6×10^{17}

equilibrium conditions will thus be constrained by the theory. (I here exclude doping under initially non-equilibrium conditions such as doping during low-temperature growth—it is apparent that for such conditions one is not dealing with solubility considerations.)

It is reasonably apparent that the appropriate temperature value to be used to obtain realistic dopant concentrations is either the treatment temperature, provided that this is applied long enough for equilibrium, or the temperature at which atomic mobility effectively ceases ($= T_{eff}$), whichever is lower. It can be noted that the solubility at the temperature used in actual device operation would in general be a non-equilibrium value, but the present concern is with dopant incorporation under equilibrium conditions. Moreover, assuming that the intrinsic solubility does not increase with decreasing temperature, it is apparent that it will be T_{eff} which is the temperature of primary relevance: since one wants the solubility to be as high as possible, one wants the incorporation temperature to be as high as possible, and, by definition, this is limited by T_{eff}. An immediate consequence is that not only should good dopants have a high solubility, but also they should have a relatively low atomic mobility (i.e. a low diffusion). (It is of interest that this temperature dependence, together with differences in diffusion coefficients, can explain the empirical observation that it is more difficult to obtain good conductivity in II–VI semiconductors than in the III–Vs, even for comparable bandgap values—see ref. 27.) In this connection it can also be noted that a good T_{eff} can be helped by a fast quench rate. A further consideration is that T_{eff} is also lower the further a dopant has to move before it is removed from the system; this means having as few internal 'sinks' (dislocations, precipitates, etc.) as possible.

A second parameter of the theory of course is the intrinsic solubility, which can be adjusted, in principle, by use of different dopants. However, in practice, this option is limited. First, one requires shallow dopants. Second, a number of dopants which give shallow levels on one lattice site behave differently on other sites. (Recall, as stated above, that the theory assumes dopants to remain on the same lattice site; this is implicit in the derivation of equation (11.2).) Included in this class are amphoteric dopants, which give donors (acceptors) on one site, and acceptors (donors) on another; a well-known case in the II–VI materials are the group I metals, which are acceptors on the cation site, and donors on the interstitial site (e.g. refs 42 and 43). A second type, probably, are the so-called DX centres. These have been extensively studied in the wider-gap III–V semiconductors, and the present theory [44, 45] is that the donors here provide shallow levels when they are on a substitutional lattice site, but that either they or lattice ions can switch to interstitial sites, where they give deep levels. It has recently been suggested by Chadi and Chang [46] that at least several of the acceptors in ZnSe behave similarly. It is indeed known that in widegap materials the 'standard' dopants often introduce deep levels (e.g. P and As in ZnSe—see, for example, ref. 2). Overall, the number of appropriate shallow dopants is thus limited, resulting in little or no choice in terms of finding ones with high intrinsic solubility (never mind a preference for those having, in addition, a relatively low diffusivity). Still a further constraint is provided by the drastic decrease of the solubility with Fermi level. To illustrate this, assume that our intrinsic value of $4 \times 10^{20}\,\mathrm{cm}^{-3}$ is low, and that higher values are available—however, how high? Suppose one can obtain a 10% solubility; for ZnSe, this would give an intrinsic value of $2 \times 10^{21}\,\mathrm{cm}^{-3}$, which in turn would lead to a donor solubility of $2 \times 10^{18}\,\mathrm{cm}^{-3}$ at 1200 K. This, of course, is not bad, but it nevertheless is still low compared with values obtainable in Ge and Si. The net result is that to achieve values such as those in Ge and Si, one would require extremely high intrinsic solubilities (up to, or greater than (??), 100%).

An additional parameter in equation (11.2) is the Fermi level, E_F. As mentioned above, the values given in Table 11.2 are based on having only a single dopant, with the Fermi level then given by equation (11.3b). For additional dopants (or contaminants), the Fermi level of course becomes a parameter. A suggestion [27, 47] for non-equilibrium dopant incorporation thus is the following. First, incorporate both the primary (desired) dopant and a relatively mobile secondary (compensating) species. Second, remove (via gettering or alternate means) the secondary dopant. Under these conditions, the concentration of incorporated primary dopant should correspond, at least approximately, to the intrinsic solubility, and removal of the secondary dopant would then lead to excess primary dopant, and thus to good conductivity.

A further parameter, which has been only implicit so far, is the pressure of the constituents. This comes about from the $B(T, P)$ term in equation (11.1). In deriving equation (11.2), it is of course assumed that this pressure is the same in the intrinsic and in the doped cases. Nevertheless, in any doping experiment, the pressure is an adjustable parameter. The aim, for good doping, of course is to adjust the pressure so as to obtain the maximum incorporation of the desired dopant. The mass action considerations which show how to best achieve this aim are exhaustively discussed by Kröger [48, 49]; specific application to ZnSe is for instance given by Ray and Kröger [50, 51].

It remains to show that there indeed is evidence of poor dopant solubility in the II–VI materials. This evidence consists of (at least) three types of results. First, there are reports showing low carrier concentrations (or relatively low, compared with Ge and Si) under conditions such that one would expect equilibrium; four examples, two each for ZnSe and for ZnTe, are given in Table 11.3. Second, there are reports of precipitation if one treats material at a temperature lower than the initial processing temperature. Six examples, three each for ZnSe and for ZnTe, are given in Table 11.4. Third, when good conductivity has been reported, this has in general been by non-equilibrium methods. I here focus on n-ZnSe. All recent reports of good

Table 11.3 Observation of low carrier concentrations under equilibrium processing

	Temperature	Concentration (cm^{-3})	Reference
ZnSe (Al)[a]	1200–1300 K	$n < 2 \times 10^{18}$	[50]
ZnSe (Ga)[b]	1200–1300 K	$n < 10^{18}$	[51]
ZnTe (excess Te)	Close to room temperature (travelling heater method, very slow cooling)	$p < 10^{16}$	[52]
ZnTe (In)	Close to room temperature (travelling heater method, very slow cooling)	$n < 10^{11}$	[53]

[a]For highest doping, spectrographic analysis gave [Al] = 7[n].
[b]Precipitation on cooling; room temperature concentration lower.

Table 11.4 Observation of precipitation in ZnSe and ZnTe with high doping

	Conditions	Reference
ZnSe (Ga)	Cooling from 1100 K–1300 K	[51]
ZnSe (Ga, In)	Heating at 1100 K after growth at 1400 K	[54]
ZnSe (Li)	Heating at 600 K–700 K after growth at 1100 K	[25]
ZnTe (Li)	Heating at 1000 K after growth at 1400 K	[55]
ZnTe (Cu)	Heating at 1000 K after growth at 1400 K	[56]
ZnTe (In)	Slow cooling after growth at 1100 K	[53]

conductivity, as presented in Table 11.1, use such non-equilibrium methods; specifically, all use dopant incorporation during low-temperature growth (MBE or MOCVD). Moreover, earlier results with good conductivity in bulk material have been obtained via heat treatment in excess Zn. This method was first reported by Aven and Woodbury [57], who also showed that this approach resulted in the removal of Cu (an acceptor on the Zn site). In general this method thus extracts accidental acceptor impurities, i.e. effectively the desired (donor) dopant has now been introduced together with a compensating—and relatively mobile—species, with the subsequent removal of the compensating species. As shown above, such compensation increases the donor solubility, and removal of the compensating acceptor then gives non-equilibrium donor incorporation.

A further interesting result on solubility can be obtained by re-analysis of data of Ray and Kröger [51]. In this work they determined the electron concentration, at high temperatures (800, 900, and 1000 °C), due to Ga doping, as a function of Zn pressure and of Ga concentration. The electron concentration, at low Zn pressures, increased with Zn pressure. However, for the two highest-doped Ga samples (nominal concentrations of $5 \times 10^{19}\,\text{cm}^{-3}$ and $6 \times 10^{20}\,\text{cm}^{-3}$) two observations are noteworthy: first, the electron concentration saturated at the higher Zn pressures; second, this saturated value was the same for both samples. In view of these two results, as well as of the high temperatures, it seems very likely that this carrier concentration corresponds to the equilibrium dopant solubility. The electron concentrations were $3 \times 10^{17}\,\text{cm}^{-3}$ at 800 °C, $6 \times 10^{17}\,\text{cm}^{-3}$ at 900 °C, and $1 \times 10^{18}\,\text{cm}^{-3}$ at 1000 °C. Carrying out calculations analogous to those for Table 11.2, but using effective mass values more specific to ZnSe ($m_e^* = 0.16$ and $m_h^* = 0.8$), and with an intrinsic solubility of $2 \times 10^{20}\,\text{cm}^{-3}$ (independent of temperature), one obtains the very same values as the electron concentrations of Ray and Kröger! Unless one is to regard this result as coincidence, this confirms that the present solubility considerations give values in agreement with experiment.

11.3 COMPENSATION ASPECTS

It was already realized in the 1950s [18, 19] that wide bandgaps provide an energy incentive for compensation. Thus, if an electron in a (shallow) level close to the conduction band can transfer to an acceptor level close to the valence band, the energy gain for the system will be approximately the bandgap energy. Of course, to get this gain, one has to provide the formation energy of the acceptor species. The usual assumption has been that this energy balance would be favourable, and this view tends to be confirmed by the frequent observation of high-resistivity material, despite extensive efforts at doping. The main argument in this area has not been the tendency towards compensation, but rather whether it was caused predominantly by native defects or by impurities. I shall review some aspects of this question here, but first include some new concepts, which have arisen only relatively recently, regarding types of 'compensation'.

The standard view of compensation is that, in addition to the desired dopant (e.g. a donor, of concentration N_D), there will be some species which gives a level of opposite type (i.e. an acceptor level, of concentration N_A). This means that there will be, at temperatures high enough for full ionization (but low enough so that band-to-band excitation can be neglected), only $N_D - N_A$ available carriers. However, recent work [46] has suggested an alternate possibility (specifically, applied to As- and P-doped ZnSe). Namely, lattice relaxation can lead to a different location—and thus level—for a dopant state; moreover, if the new level is either of opposite type [46] or merely deep even if the same type (usually, a large lattice relaxation would be assumed, so there can easily be a large change in energy), there will be loss of conductivity. Several factors are still worth noting in connection with this suggestion. First, the new levels, whether of opposite type or merely deep, will provide a higher solubility for the impurity (see section 11.2); recall that in section 11.2 it was pointed out that an implicit assumption of the theory was that the impurity would remain on the same lattice site regardless of the conditions. Second, as already pointed out by Chadi and Chang [46], the centres formed by this process are expected to be analogous to the well-known DX centres in the III–V compounds (e.g. refs 58 and 59). One well-known signature of DX centres is that the levels appear deep when the sample is cooled in the dark; another is the occurrence of persistent photoconductivity. It can be noted that deep levels have been reported in heavily doped II–VIs for many years (e.g. in n-CdTe already in 1959 by deNobel [60] and, for example, in n-ZnSe by Jones and Woods [54]); similarly, persistent photoconductivity has also been known since at least the early 1960s (Litton and Reynolds [61] for CdS, Lorenz and Woodbury [62] for CdS and CdTe, and Lorenz *et al.* [63] for ZnSe). (Note that Chadi and Chang [46] gave their suggestion for p-type ZnSe but, based on the experimental data, DX-type centres apparently also occur in n-type.) Overall, this effect will very probably have to be considered in future work on the II–VIs; one specific aspect would be to find dopants which are least likely to give DX-type centres.

Reverting now to a discussion of the more standard type of compensation, one question here is whether opposing types of defects are more likely to be caused by native defects ('self') or by impurities ('impurity' compensation). In the early work (e.g. by Kröger and Vink [18], Longini and Greene [19], Mandel [20], Mandel *et al.* [21]), the theory was developed considering only native defects. It was subsequently suggested (e.g. by Dean *et al.* [64], Dean [4] and Neumark [24]) that compensation by impurities was more likely. Nevertheless, intrinsic defects (including anti-sites) are also still being considered (e.g. by Van Vechten [22, 23], Jansen and Sankey [14, 65] and Sankey and Jansen [66]).

Some further comments can be made regarding this question of impurity versus native defects compensation. First, in principle, one could argue that, in the 'high-purity limit', it could be only native defects which play a role. However, doped material is, by definition, not 'high-purity'. In practice, there are two ways in which impurities can play a major role. One is if one is using amphoteric impurities, i.e. impurities which act as acceptors (donors) on one lattice site, and as donors (acceptors) on a different lattice site; well-known examples for the II–VIs are the alkali metals, which are acceptors on the metal site and donors on the interstitial site, and the group IVs, which are expected to be (double) acceptors on the anion site and donors on the metal site. Of course, it can be argued that amphoteric impurities are not expected to be good choices and should not be used. However (as already mentioned in section 11.2), the choice of dopants is, in fact, limited (one 'wants', at least, shallow levels, as good a solubility as possible, a low tendency for formation of DX-type centres, and a low ionic mobility); moreover, Si has been used very successfully as an n-type dopant in GaAs and related compounds, despite its being amphoteric. With amphoteric dopants there will always be a limit on obtainable carrier concentrations [24], depending on the incorporation energies at the two sites, on temperature, and on constituent partial pressures, but if this limit is high enough, they can be used. A nice experimental determination of this limit for Li in ZnSe ($< 10^{17}\,cm^{-3}$ in Se-stabilized MBE growth) has recently been obtained by Haase *et al.* [11]. A second way in which impurities play a role is by complexing with native defects; in a sense, this is of course 'mixed' compensation, but it must be recalled that, in terms of any arguments for 'self' compensation based on the energetics of pure native defects, one is now dealing with different species. In any case, a Zn-vacancy donor complex, the so-called 'A' centre, has been known for many years in both ZnS (e.g. refs 67 and 68) and ZnSe (e.g. refs 69 and 70); moreover, Watkins and coworkers (e.g. ref. 70) have shown that, when Zn vacancies are created in ZnSe by electron irradiation, these then disappear on annealing (at quite low temperatures, $\approx 450\,K$), with the simultaneous formation of A-type centres. A further indication of the importance of these complexes is also given by Table 11.1, which gives a list of well-conducting n-ZnSe; even here, when samples were checked for deep luminescence (presumably due to A centres) and/or total dopant concentration, it was found that (at least at high doping) deep luminescence was present and/or the concentration of the dopant species was far higher than the electron concentration (indicating that a good part of the dopant was on a different site from the one giving donors). It can also still be mentioned that one reason why impurity compensation was considered likely was that in carefully prepared II–VI materials no signatures of pure intrinsic vacancies or interstitials were found in the luminescence spectra [4, 64, 71]. Indeed, for the Zn defect species in ZnSe—both interstitials and vacancies—Watkins and coworkers (e.g. refs 70 and 72) have determined the luminescence characteristics via non-equilibrium incorporation; to my knowledge, these characteristics have not been observed in other ZnSe samples. Nevertheless, it is of course difficult to exclude a role for 'pure' native defects, since lack of observation (as in luminescence) cannot prove their absence.

In terms of conductivity in compensated material, it should also be recalled that this will be particularly badly degraded (by many orders of magnitude) in the presence of deep levels together with 'strong' compensation [26]. With a concentration of shallow majority dopants ($N_{maj}{}^{S}$), deep majority dopants ($N_{maj}{}^{D}$), and of minority species (N_{min}), such 'strong' compensation is defined as the case where $N_{maj}{}^{S} < N_{min}$.

11.4 SUMMARY

In the present review I have emphasized two aspects of the conductivity problem of widegap materials. One is that of the limited solubility, the other is that of the tendency towards compensation. It is here also hopefully apparent that both these aspects are basically different manifestations of the same problem, namely that excess carriers in such materials result in a high-energy state for the system.

Regarding the solubility aspect, the resultant constraints have been discussed, and means of circumventing the resultant problems have been given. Basically, to obtain conductivity, one requires non-equilibrium incorporation of 'good' dopants. As brought out, there are various requirements on 'good' dopants, including that they provide shallow levels, have as good a solubility as possible, have a low tendency for formation of DX-type centres, and have a low ionic mobility.

As to compensation, one aspect of high interest is the (old) question of 'self' versus 'impurity' compensation. This has not yet been resolved. However, improved first-principles state-of-the-art defect calculations are at present being carried out [14, 73], so that an answer may soon be forthcoming.

REFERENCES

1. Hartmann, H., Mach, R. and Selle, B. (1982) *Curr. Top. Mater. Sci.*, **9**, 1.
2. Bhargava, R. (1982) *J. Cryst. Growth*, **59**, 15.
3. Bhargava, R. (1988) *J. Cryst. Growth*, **86**, 873.
4. Dean, P.J. (1979) In *Defects and Radiation Effects in Semiconductors, Inst. Phys. Conf. Ser.*, **46**, 100.
5. Marfaing, Y. (1981) *Prog. Cryst. Growth Charact.*, **4**, 317.
6. Neumark, G.F. (1989) *J. Appl. Phys.*, **65**, 4859.
7. Park, Y.S. and Shin, B.K. (1977) In *Topics in Applied Physics*, Vol. 17, Electroluminescence (ed. J.I. Pankove), Springer, Berlin, p. 133.
8. Pautrat, J.L., Francou, J.M., Magnea, N., Molva, E. and Saminadayar, K. (1985) *J. Cryst. Growth*, **72**, 194.
9. Ohki, A., Shibata, N. and Zembutsu, S. (1988) *Jpn. J. Appl. Phys.*, **27**, L909.
10. Suemune, I., Yamada, K., Masato, H., Kanda, T., Kan, Y. and Yamanishi, M. (1988) *Jpn. J. Appl. Phys.*, **27**, L2195.
11. Haase, M.A., Cheng, H., DePuydt, J.M. and Potts, J.E. (1990) *J. Appl. Phys.*, **67**, 448.
12. Taike, A., Migita, M. and Yamamoto, H. (1990) *Appl. Phys. Lett.*, **56**, 1989.
13. Iida, S., Yatabe, T. and Kinto, H. (1989) *Jpn. J. Appl. Phys.*, **28**, L535.
14. Jansen, R.W. and Sankey, O.F. (1989) *Phys. Rev. B*, **39**, 3192.
15. Van Vechten, J.A. (1975) *J. Electrochem. Soc.*, **122**, 419.
16. Dow, J.D., Hong, R.-D., Klemm, S. *et al.* (1991) *Phys. Rev.*, B**43**, 4396.
17. Chadi, D.J. (1990) private communication.
18. Kröger, F.A. and Vink, H.J. (1956) *Solid State Phys.*, **3**, 307.
19. Longini, R.L. and Greene, R.F. (1956) *Phys. Rev.*, **102**, 992.
20. Mandel, G. (1964) *Phys. Rev.*, **134**, A1073.
21. Mandel, G., Morehead, F.F. and Wagner, P.R. (1964) *Phys. Rev.*, **136**, A826.

22. Van Vechten, J.A. (1980) In *Handbook on Semiconductors*, Vol. 3 (ed. S.P. Keller), North-Holland, Amsterdam, p. 1.
23. Van Vechten, J.A. (1985) *Mater. Res. Soc. Symp. Proc.*, **46**, 83.
24. Neumark, G.F. (1980) *J. Appl. Phys.*, **51**, 3383.
25. Neumark, G.F. and Herko, S.P. (1982) *J. Cryst. Growth*, **59**, 189.
26. Neumark, G.F. (1982) *Phys. Rev. B*, **26**, 2250.
27. Neumark, G.F. (1989) *Phys. Rev. Lett.*, **62**, 1800.
28. de Miguel, J.L., Shibli, S.M., Tamargo, M.C. *et al.* (1988) *Appl. Phys. Lett.*, **53**, 2065.
29. Shibata, N., Ohki, A. and Zembutsu, S. (1988) *Jpn. J. Appl. Phys.*, **27**, L251.
30. Shibata, N., Ohki, A., and Katsui, A. (1988) *J. Cryst. Growth*, **93**, 703.
31. Kamata, A., Uemoto, T., Okajima, M., Hirahara, K., Kawachi, M. and Beppu, T. (1988) *J. Cryst. Growth*, **86**, 285.
32. Kamata, A., Uemoto, T., Hirahara, K. and Beppu, T. (1989) *J. Appl. Phys.*, **65**, 2561.
33. Yoshikawa, A., Nomura, H., Yamaga, S. and Kasai, H. (1989) *J. Appl. Phys.*, **65**, 1223.
34. Cheng, H., DePuydt, J.M., Potts, J.E. *et al.* (1989) *J. Cryst. Growth*, **95**, 512.
35. Yoshikawa, A., Yamaga, S., Tanaka, K. and Kasai, H. (1985) *J. Cryst. Growth*, **72**, 13.
36. Yasuda, T., Hara, K. and Kukimoto, H. (1986) *J. Cryst. Growth*, **77**, 485.
37. Weiser, K. (1960) *J. Phys. Chem. Solids*, **17**, 149.
38. Harrison, W.A. and Kraut, E.A. (1988) *Phys. Rev. B*, **37**, 8244.
39. Reiss, H. (1953) *J. Chem. Phys.*, **21**, 1209.
40. Blakemore, J.S. (1987) *Semiconductor Statistics*, Dover Publications, New York.
41. Trumbore, F.A. (1960) *Bell Syst. Tech. J.*, **39**, 205.
42. Henry, C.H., Nassau, K. and Shiever, J.W. (1971) *Phys. Rev. B*, **4**, 2453.
43. Neumark, G.F., Herko, S.P., McGee, T.F., III and Fitzpatrick, B.J. (1984) *Phys. Rev. Lett.*, **53**, 604.
44. Chadi, D.J. and Chang, K.J. (1988) *Phys. Rev. Lett.*, **61**, 873.
45. Chadi, D.J. and Chang, K.J. (1989) *Phys. Rev. B*, **39**, 10063.
46. Chadi, D.J. and Chang, K.J. (1989) *Appl. Phys. Lett.*, **55**, 575.
47. Neumark, G.F. (1990) *US Patent 4,904,618.*
48. Kröger, F.A. (1964) *The Chemistry of Imperfect Crystals*, North-Holland, Amsterdam.
49. Kröger, F.A. (1973) *The Chemistry of Imperfect Crystals*, 2nd edn, North-Holland, Amsterdam.
50. Ray, A.K. and Kröger, F.A. (1978) *J. Electrochem Soc.*, **125**, 1348.
51. Ray, A.K. and Kröger, F.A. (1978) *J. Electrochem Soc.*, **125**, 1355.
52. Triboulet, R. and Didier, G. (1975) *J. Cryst. Growth*, **28**, 29.
53. Wald, F.V. (1976) *Phys. Status Solidi A*, **38**, 253.
54. Jones, G. and Woods, J. (1976) *J. Phys. D*, **9**, 799.
55. Bensahel, D., Dupuy, M. and Pfister, J.C. (1979) *Phys. Status Solidi A*, **55**, 211.
56. Bensahel, D. and Dupuy, M. (1979) *Phys. Status Solidi A*, **56**, 99.
57. Aven, M. and Woodbury, H.H. (1962) *Appl. Phys. Lett.*, **1**, 53.
58. Lang, D.V. (1986) In *Deep Centers in Semiconductors* (ed. S.T. Pantelides), Gordon and Breach, p. 489.
59. Mooney, P.M. (1990) *J. Appl. Phys.*, **67**, R1.
60. deNobel, D. (1959) *Philips Res. Rep.*, **14**, 361, 430.
61. Litton, C.W. and Reynolds, D.C. (1962) *Phys. Rev.*, **125**, 516.
62. Lorenz, M.R. and Woodbury, H.H. (1963) *Phys. Rev. Lett.*, **10**, 215.
63. Lorenz, M.R., Aven, M. and Woodbury, H.H. (1963) *Phys. Rev.*, **132**, 143.
64. Dean, P.J., Venghaus, H., Pfister, J.C., Schaub, B. and Marine, J. (1978) *J. Lumin.*, **16**, 363.
65. Jansen, R.W. and Sankey, O.F. (1987) *Solid State Commun.*, **64**, 197.

66. Sankey, O.F. and Jansen, R.W. (1988) *J. Vac. Sci. Technol. B*, **6**, 1240.
67. Schneider, J., Rauber, A., Dischler, B. and Holton, W.C. (1965) *J. Chem. Phys.*, **42**, 1839.
68. Shionoya, S., Koda, T., Era, K. and Fujiwara, H. (1963) *J. Phys. Soc. Jpn.*, **18**, Suppl. 2, 299.
69. Dunstan, D.J., Nicholls, J.E., Cavenett, B.C. and Davies, J.J. (1980) *J. Phys. C*, **13**, 6409.
70. Watkins, G.D. (1977) In *Radiation Effects in Semiconductors, Inst. Phys. Conf. Ser.*, **31**, 95.
71. Dean, P.J. (1979) *J. Lumin.*, **21**, 75.
72. Watkins, G.D. (1990) In *Proceeding International Conference on Science and Technology of Defect Control in Semiconductors*, North Holland, Amsterdam. p. 933.
73. Laks, D.B., Van de Walle, C.G., Neumark, G.F. and Pantelides, S.T. (1991) Proceedings 20th. International Conference on the Physics of Semiconductors, *Phys. Rev. Lett.*, **66**, 648.

Part Three: Devices

12

II–VI electroluminescent devices

J. Woods

12.1 INTRODUCTION

Electroluminescence (EL) is a process in which light is emitted from a substance as a direct result of the passage of an electrical current through it. The effect was first discovered in II–VI compounds in 1936 by Destriau [1] working with a mixture of powdered zinc sulphide and zinc oxide suspended in castor oil between two electrodes. On application of an alternating electric field a weak green luminescence was observed. After some initial disbelief, an extensive research effort in the 1950s and 1960s led to the manufacture of a variety of plastic and ceramic EL lamps. These were of low intensity, but found a range of applications as dark room safelights, exit signs in areas of reduced lighting, clock and dial faces etc.

The first matrix display was proposed by Piper [2] in 1953 and this stimulated research into a matrix-addressed TV display, the so-called TV on the wall. At the same time it was hoped that an improved performance of powder EL lamps would lead to luminescent walls and ceilings. This first major phase of EL research was severely reduced in the mid-1960s when the prospects of achieving these ambitious objectives seemed remote. In addition to problems with matrix addressing, the basic difficulty was that the luminance × life product was insufficient for practical applications. The luminance (brightness) of a panel could be increased by increasing the drive voltage or frequency, but this was more than offset by a reduced life to half-luminance. The early work has been described in considerable detail in a number of books and review articles [3–6].

In parallel with the work on powders, a smaller effort was devoted to a study of thin films of ZnS:Mn, but although Thornton [7] was able to obtain a luminance of 1000 foot-lambert in 1962, work proceeded slowly because his films degraded rapidly in a matter of hours. The first serious attempts to develop a thin film ACEL matrix display device were made by Soxman and his associates at Sigmatron [8] from 1964 to 1970, and although a life of 1000 h was claimed in 1972, the programme was abandoned in the early 1970s.

Meanwhile work had been in progress at the Sharp Corporation in Japan under Mito from about 1950, and in 1974 Inoguchi *et al.* [9] reported a thin film ACEL sandwich device with much improved stability. This event, coupled with the discovery of a hysteresis (memory) effect in the following year [10], provided the trigger for a world-wide resurgence of interest in EL display devices based on ZnS:Mn. Progress has been slow but steady. A 240 × 320 line panel was demonstrated in 1978, and by 1988 three manufacturers were offering monochrome ACEL display devices with up to 480 × 640 picture elements (pixels). Large-area, high-definition monochrome and colour displays are currently in development. One manufacturer employs films of ZnS:Mn put down by atomic layer epitaxy (ALE), a process pioneered by Suntola and his coworkers [11]. In other panels the active layers are deposited by evaporation.

In contrast, a totally different approach was adopted by Vecht and his group [12], who developed DCEL powder cells, where the grains of ZnS:Mn were coated with copper sulphide. The panel was subjected to a forming treatment to produce a thin insulating layer of ZnS, in which a high electric field would develop and generate the EL emission. Panels based on this process have been commercially available for some time.

Not surprisingly perhaps, work has also continued on DCEL in thin films of ZnS. Some of the recent work is described by Blackmore *et al.* [13] and Cattell *et al.* [14]. No practical device has yet emerged.

To summarize, large-area EL display devices have been produced in both thin film and powder form. They are either a.c. or d.c. driven. Each of the four basic types will be discussed in more detail in section 12.2. These large-area devices are all based on polycrystalline, insulating layers of ZnS. *A priori* considerations would suggest that higher efficiencies should be achieved in single-crystal devices, where grain boundaries are absent. Much effort has gone into the growth of bulk single crystals, and the deposition of single-crystal epitaxial layers of the II–VIs, as described in Chapters 1–4, but no light-emitting diode (LED) capable of displacing the III–V compound market leaders has been produced.

The main reason for this is that it has proved exceedingly difficult to dope the widegap II–VIs to display amphoteric semiconduction. ZnS, ZnSe, CdS and CdSe can all be readily prepared as low-resistivity, n-type semiconductors, but the production of low-resistivity, p-type behaviour is more difficult. Success would appear to be closer at hand with ZnSe where ion implantation and the doping of epitaxial layers with such impurities as Li or N has led to various reports of success. However, even when satisfactory p-type semiconduction is achieved in ZnSe the preparation of a good ohmic contact remains a serious problem [15]. As a result of these difficulties the production of a p–n homojunction remains a formidable task. These topics form the subject matter of Chapters 11 and 13, and will be touched on here briefly in section 12.3.

A p–n homojunction of a II–VI compound in forward bias passing $100\,\mathrm{mA\,cm^{-2}}$ and emitting in the green ($h\nu = 2.5\,\mathrm{eV}$) would have the surface brightness of a tungsten lamp operating at 3000 K. Blue LEDs and lasers would be possible, but no commercially viable device has been produced and research continues. Since homojunctions proved so elusive, attention turned naturally to heterojunctions and to Schottky (MS) and metal–insulator– semiconductor (MIS) diodes as potential LEDs. Reasonable success has been achieved with red, yellow and green emitting devices [16, 17] but none of these has been able to replace GaAsP and GaP LEDs.

12.2 FLAT PANEL DEVICES

12.2.1 A.C. powder lamps

The first type of EL panel to be developed was the a.c. powder device derived from the pioneer work of Destriau [1]. The active ingredient is a layer of ZnS or (ZnCd)S suitably doped with Cu and Cl (alternatively Br) to give green or blue emission depending on the impurity concentrations. Mn is added if a yellow emission is required. The doped ZnS powder with a grain size in the range 1–20 μm is suspended in a dielectric resin such as cyanoethyl cellulose ($\varepsilon \sim 20$), and then spread, sprayed or screen printed on to glass coated with a transparent conducting layer of tin oxide (TO) or indium tin oxide (ITO). After this, a layer of barium titanate is usually applied. This has a high dielectric constant and prevents localized dielectric breakdown in weak spots in the ZnS. The final back electrode is formed by spraying aluminium powder. In addition to acting as an electrical contact this layer reflects the emitted light forward. The whole device is then hermetically sealed in a polymer.

All ACEL powder panels depend on the use of copper as an added impurity to the ZnS. It must be present in quantities which exceed the solubility limit. The reason for this is that some of the copper is incorporated substitutionally into the ZnS lattice as a luminescence activator, while yet more is precipitated in the ZnS as Cu_xS. The cooling schedule during powder processing is rather critical, since it controls the distribution of the copper. It is standard procedure to wash the powder in KCN to remove the surface Cu_xS with which most grains are coated after firing. Panels are also prepared in ceramic or flexible form.

On application of an alternating electric field of the order of $10^4\,\mathrm{V\,cm^{-1}}$ uniform EL is emitted with an initial luminescence of about 20 foot-lambert, at the usual operating frequency of 400 Hz. Applied voltages are around 115 V a.c. The luminance decreases steadily with continuous operation, and nowadays the time to half the initial luminance is 2000–3000 h. The light sum is limited so that there is a shorter life at higher brightnesses. The degradation is associated with the migration of copper in the high electric field. Panels also have low discrimination and contrast ratios owing to the high reflectivity of the whitish phosphor powder. An ultrathin, black absorbing layer in front of the phosphor is provided by some manufacturers to improve the contrast in high-light ambient conditions.

The mechanism of the ACEL process in powders has been the subject of much discussion and speculation, but the most convincing explanation is that due to Fischer [18, 19], who made extensive microscopic observations of special experimental cells in operation. The Cu_xS is precipitated as needles in the bulk of the ZnS grains. These form at cracks, voids, dislocations and stacking faults. Since ZnS can occur in two crystallographic modifications with either the wurtzite or zinc blende structures, cooling from high temperatures leads to crystallites with a mixed hexagonal–cubic structure and with large numbers of stacking faults, providing sites for the precipitation of the excess copper in the form of Cu_xS. This may be why ZnS is uniquely successful as a medium for this type of EL panel.

Cu_xS is a good p-type semiconductor, while the ZnS is an insulator. The applied electric field is deformed in the vicinity of a Cu_xS needle, and field concentration by as much as 1000 times can occur. Fischer suggests that electrons and holes are injected into the ZnS from opposite ends of a needle. With an applied alternating field, electrons and holes are injected alternately from the same end. The holes are trapped at the luminescence centres, while the electrons remain mobile or are trapped at shallow electron traps. On field reversal the electrons recombine with the holes and characteristic light is emitted.

The justification for this argument is that light was observed to be emitted locally in the form of a comet. The light from a single particle appears in double comet lines, each half of which lights up alternately, as the field is reversed and the nearer electrode turns positive. There might be more than 20 comet lines in any one ZnS grain lying along particular crystallographic directions.

The intensity, B, of the time-integrated emission from a comet is found to vary with applied voltage, V, as

$$B = A\exp(-C/V) \qquad (12.1)$$

The constants A and C are different for comets of different length.
According to Fischer, the integrated light intensity of a large number of comets of different lengths is accurately given by the expression

$$B = A\exp(-C/V^{1/2}) \qquad (12.2)$$

over many orders of magnitude of luminance. Numerous investigators from Destriau onwards have reported that the luminance of their panels is described by equation (12.2).

The comet theory is attractive because it can explain the importance of the unique combination of ZnS and copper. It is a happy feature of the combination that the solubility limit of Cu in the ZnS lattice is reached before the emission associated with the substitutional copper is subjected to concentration quenching. The theory can also explain why yellow EL is obtained from layers of ZnS containing Mn and Cu. Lastly the degradation with ageing is accompanied by a decrease in the volume of comets. Comets also break in two.

ACEL powder devices find numerous applications in areas of low lighting. They are used for the illumination of instruments, keyboards, aircraft control panels, maps and as back lighting for liquid crystal and even plasma displays. Their possible use in car and boat instrumentation has also been considered. ACEL powder devices were used in the Apollo and LM modules for the manned lunar programme as event timers and navigational displays. Solid state inverters to convert 12 V d.c. to 170 V a.c. at 750 Hz are readily available.

12.2.2 A.C. thin film display devices

(a) Matrix displays

The first phase of enthusiasm for EL devices lasted until about 1964. Activity then went into steep decline when no solution to the problem of combining high brightness with long life appeared to be forthcoming. The difficulty of effective matrix addressing also remained to be solved. It was obviously relatively easy to produce an alphanumeric 5- or 7-bar display by etching appropriate patterns into the tin-oxide-coated glass before the phosphor layer was applied. However, TV on the wall requires an adequate gray scale, a large array of pixels and some method of addressing them.

A matrix display consists of two sets of mutually orthogonal parallel lines of electrodes. One set provides the horizontal electrodes on one side of the active layer, while the orthogonal set provides the vertical electrodes on the opposite side. These are known as the row and column electrodes respectively. The points of intersection of the rows and columns constitute the pixels, and when a voltage is applied to a particular row and column the pixel at their point of intersection lights up. Matrix addressing is illustrated in Fig. 12.1.

Work on matrix displays intensified in the mid-1970s following the report by Inoguchi *et al.* [9, 20] of a thin film ACEL panel with a picture size of 48 mm × 36 mm and 120 × 90 pixels. This panel consisted of an active layer of ZnS:Mn sandwiched between two insulating layers. A schematic arrangement of the device is shown in Fig. 12.2. The active layer contained up to 5% by weight of Mn and was put down by electron beam deposition in vacuum. A similar technique was used to put down the insulating layers of Y_2O_3. Other insulating layers were also investigated including Si_3N_4 reactively sputtered from a Si cathode, or a composite of Si_3N_4–Al_2O_3 successively reactively sputtered from Si and Al cathodes. The sandwich structure was fabricated on tin-oxide-coated glass, and aluminium was used

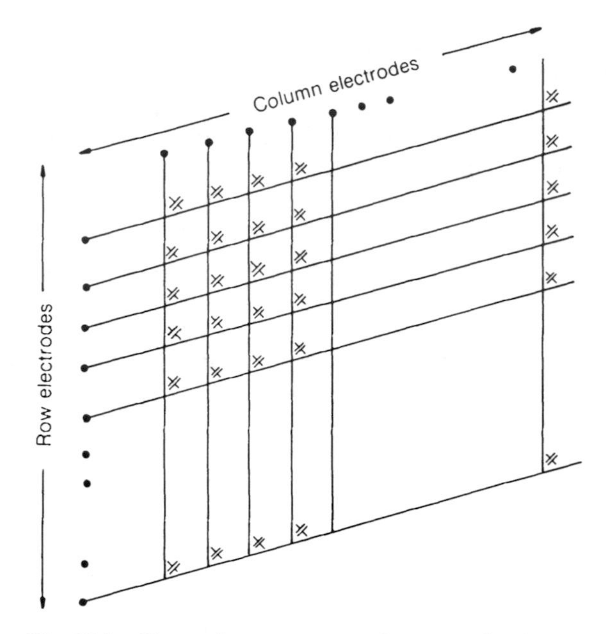

Fig. 12.1 Electrode arrangement for a matrix display.

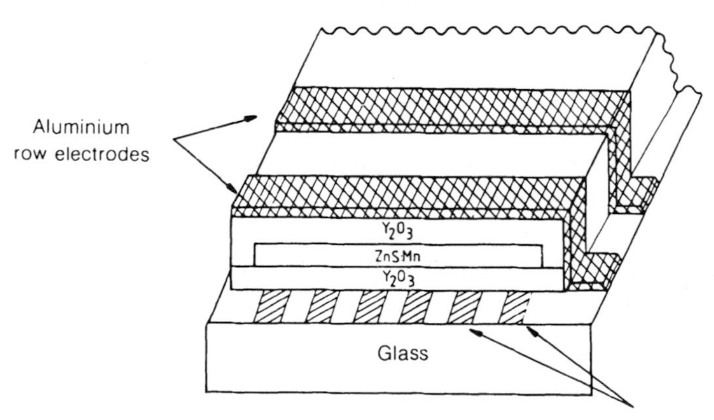

Fig. 12.2 Schematic diagram of a thin film ACEL display panel.

as the back contact. The device had to be protected from humidity since any water hydrolyses under the electric field and delaminates the film. Silicone oil and solid polymer with metal cladding have been used as effective sealants by different manufacturers.

The device emits a bright yellowish-orange luminescence when an alternating electric field is applied. This is the characteristic emission of Mn^{2+} in ZnS and occurs in a broad band centred at 585 nm (Fig. 12.3). Typical

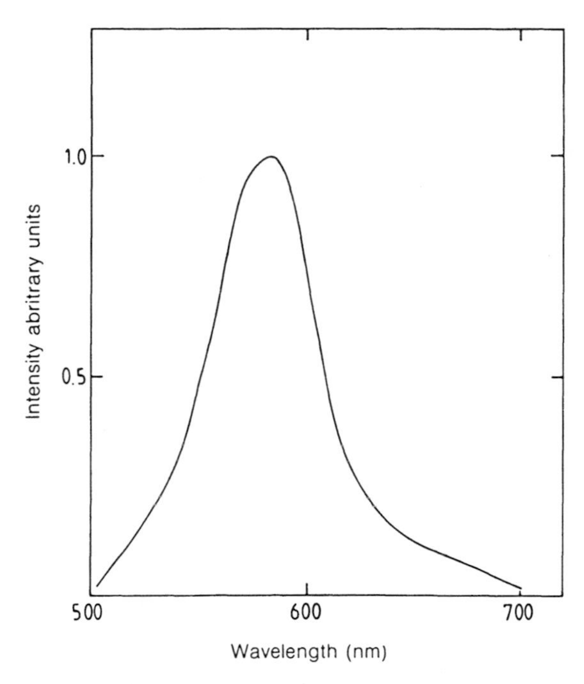

Fig. 12.3 Spectral distribution of the Mn^{2+} emission from a thin film ACEL panel.

brightness–voltage characteristics of the device are shown in Fig. 12.4. Clearly the brightness is strongly dependent on the applied voltage in the low-voltage region, and tends to saturate at high voltages. The saturation brightness is about 1500 foot-lambert at 5 kHz. The curves illustrate an important feature of this type of sandwich device, i.e. that during the first several hours of operation the B–V characteristic moves to higher voltages although the saturation brightness does not change. This shift of the characteristic slows and stops after a certain time whereupon a much more stable performance is obtained. This stabilization process can be accelerated by raising the temperature to about 200 °C while operating the device. In this way the performance can be stabilized in about 1 h. Inoguchi *et al.* [9] demonstrated that, following this initial burn-in procedure, a device could be operated continuously without loss of brightness for more than 20 000 h. Some manufacturers now claim 40 000 h minimum life. The great increase in life compared with the earlier ACEL powder panels is attributed to the absence of copper in the thin film device.

Another important feature of the double insulating sandwich device is the polarity or memory effect [20, 21]. When a device is driven by a series of pulses, the brightness at each pulse is strongly dependent on the polarity of the preceding pulse. This is illustrated in Fig. 12.5. Here, the first three voltage

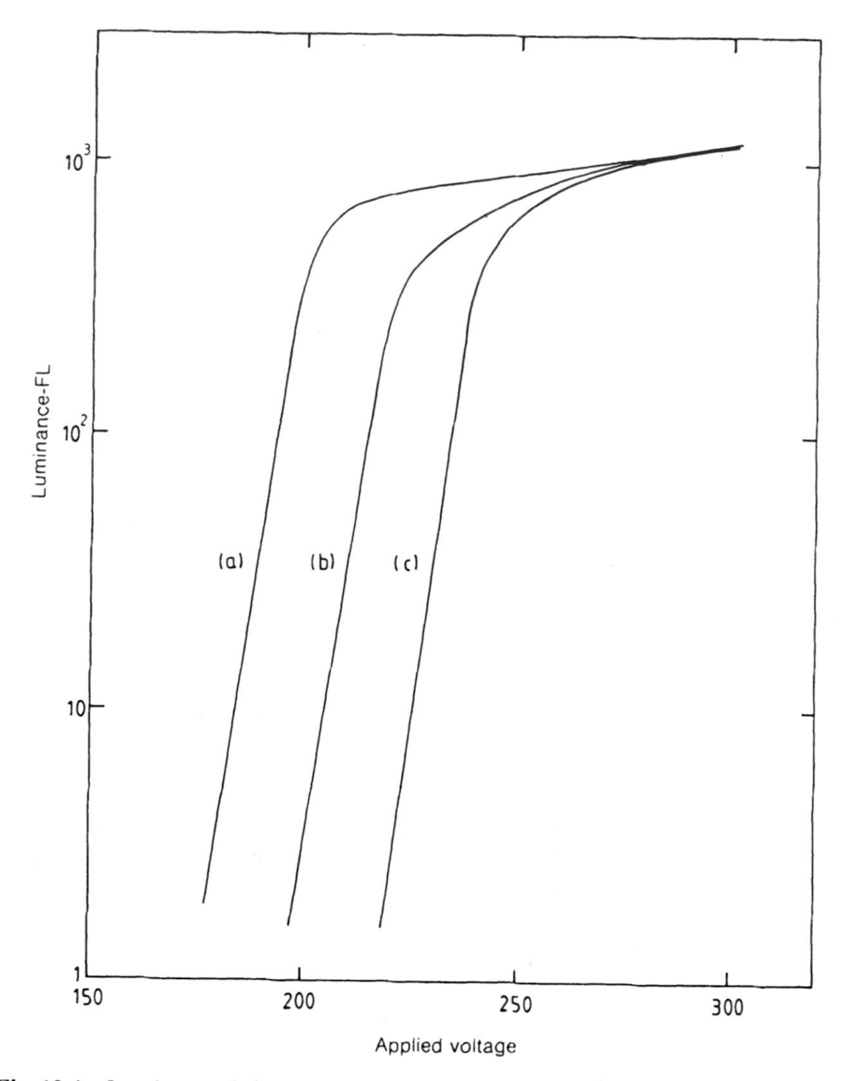

Fig. 12.4 Luminance (brightness) of a TFACEL panel as function of applied voltage (a) as prepared, (b) after 1 h of operation and (c) after 80 h of operation.

pulses are all of the same size. As a result the light output at each succeeding pulse is progressively reduced. When the polarity of the voltage is reversed, the light output at the first reversal is much larger, but if the subsequent pulses are all of the same size the light output is reduced once more. This is very convenient for display devices where the maximum light output is achieved by using pulses of alternating polarity. Pulse width modulation can be used to achieve a gray scale.

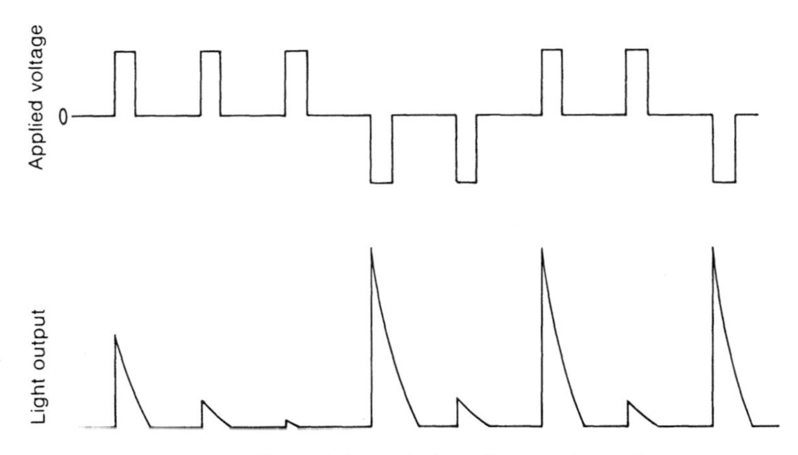

Fig. 12.5 Illustrating the effect of the polarity of successive voltage pulses on the light output from a TFACEL panel.

(*b*) *Theory of operation*

Although the electronic processes involved in the ACEL device are not fully understood, there is considerable agreement about many features [22–25]. Thus it is suggested that, when a pulse is applied, electrons tunnel from electron traps at the negative insulator–ZnS interface into the conduction band of the phosphor until they acquire a kinetic energy in excess of 2.5 eV when they are able to excite the Mn^{2+} $3d^5$ ions by impact from the $^6A_1$ ground state to the $^4T_1$ first excited state. The characteristic yellow–orange luminescence is emitted when the ions return to the ground state. The transition is spin forbidden and has a long lifetime of 1.8 ms in cubic and 1.2 ms in hexagonal ZnS. The emission peaks at 2.1 eV, i.e. 585 nm. Concentration quenching of the emission sets in at a Mn^{2+} content of about 0.5 at.%. Thermal quenching does not occur until well above room temperature. Electrons arriving at the positive insulator–ZnS interface are trapped, and await the reversal of the applied voltage before restarting the process. If the next pulse is of the same polarity as the preceding one, the system is already partially polarized, so that the light output is necessarily reduced, since not only is the internal electric field lower, but there are now fewer electrons trapped at the cathode insulator–ZnS interface.

(*c*) *Hysteresis and switching effects*

When an alternating voltage is applied to a double insulator sandwich an important hysteresis effect occurs in some devices. This is illustrated in Fig. 12.6. When a voltage is first applied electrons thermally excited into the conduction band of the ZnS are accelerated there and the weak luminescence

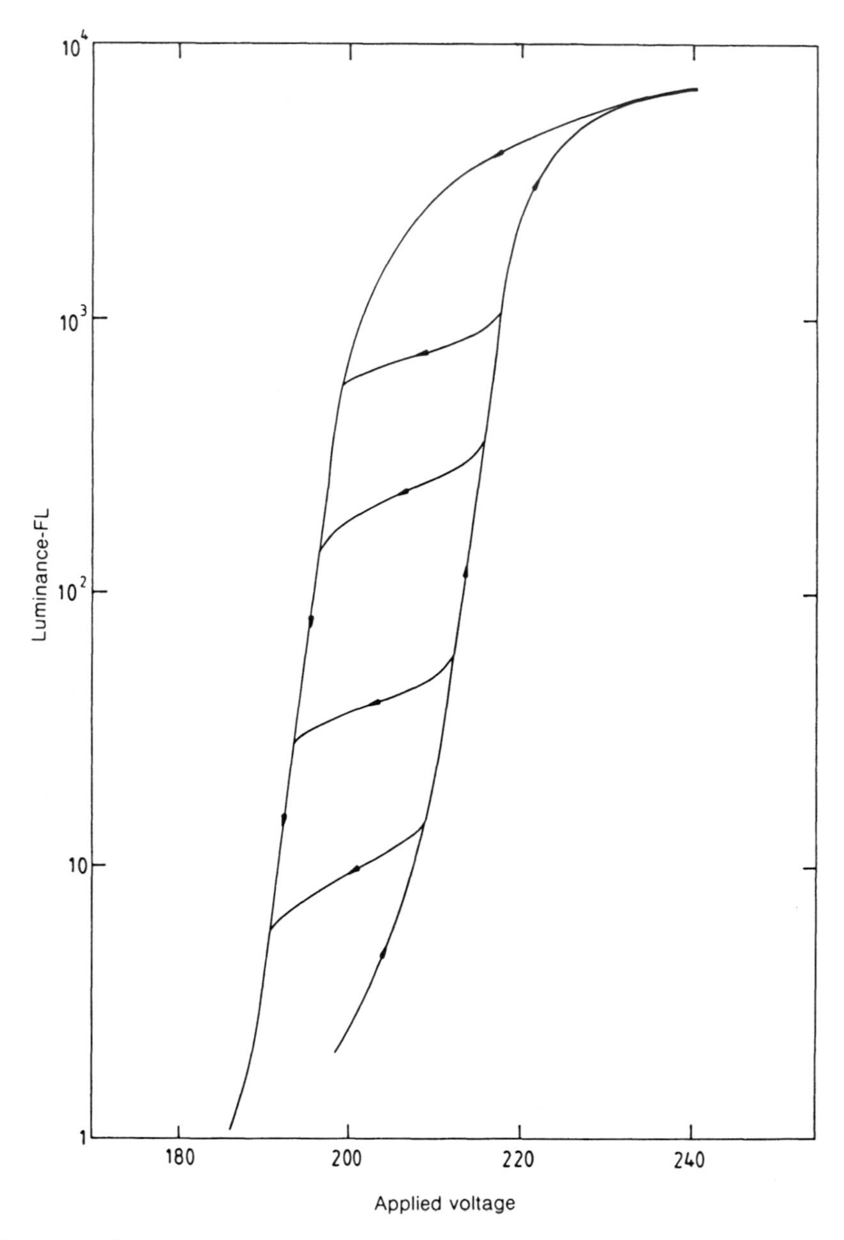

Fig. 12.6 Illustrating the hysteresis in the EL emission with rising and falling applied voltage.

increases slowly with voltage. At the threshold for release of electrons from the interface traps a steep increase in the number of accelerated electrons takes place and a multiplication and avalanche process probably occurs. The light output saturates when all the electrons originally trapped at the interface have been released.

When the voltage is decreased from the saturation region, the number of electrons being accelerated in the conduction band remains close to the value at saturation until the voltage is reduced below a second, falling threshold value. This ensures that the light output is maintained at a higher level than on the increasing voltage branch. Below the second threshold, the free electrons are rapidly trapped, and the luminance falls steeply. The electronic processes involved have been revealed by the elegant work of Yang and Owen [26] and Howard *et al.* [27]. It is not clear which traps are involved in this phenomenon, but it is interesting to note that hysteresis occurs only with active layers of ZnS:Mn when the Mn content exceeds 1 at.%, and does not occur with Al_2O_3 insulating layers.

The hysteresis function can be exploited very successfully in matrix displays. A voltage, V_s, roughly in the middle of the hysteresis loop will sustain a luminescence output on the upper falling limb of the hysteresis curve. If a train of alternating pulses of amplitude V_s is applied initially, a light output on the lower, rising limb of the hysteresis curve will be obtained. This will be very weak. Now when a synchronized pulse of the same polarity is superimposed on the train so that the resultant pulse is much larger, the light output is swept towards or into saturation depending on the magnitude of the net voltage. When the voltage returns to V_s, a high light output on the falling upper limb of the hysteresis curve results. This will be sustained until an erase pulse of the opposite polarity reduces the net applied pulse to zero. After a few erase pulses the pixel is reset to the sustaining voltage. A gray scale can be obtained utilizing this principle because of the minor hysteresis loops which can be traced as shown in Fig. 12.6. This switching facility obviously introduces the principle of storage into a TV display between successive frames. Another advantage of the memory effect is that a pixel can also be switched from a low to a high brightness state by optical irradiation or by electron bombardment. This latter possibility obviously introduces the possibility of manufacturing a storage CRT [28, 29] in which the screen is an EL sandwich device with memory. Much remains to be explained before the detailed electronic and trapping processes are understood.

(*d*) *Physical processes*

There are now a variety of papers available in the literature discussing the physics of EL devices, of which those by Allen [23], Müller [24] and Müller and Mach [25] give a good insight into the processes involved. Müller and

Mach show that a pixel in an ACEL device has an efficiency of about $3\,\mathrm{lm\,W^{-1}}$ at the optimum Mn doping of $3 \times 10^{20}\,\mathrm{cm^{-3}}$. This, however, is only some 10% of the internal efficiency which at $30\,\mathrm{lm\,W^{-1}}$ is about double that of a tungsten lamp. The loss of light output is due to total internal reflection because of the high refractive index (2.4) of ZnS. They calculate by rule of thumb that 12.5 Mn^{2+} centres are excited in 1 μm of electron path length in ZnS doped with the optimum Mn^{2+} concentration. They show that the maximum possible applied voltage across 1 μm of ZnS is 166 V, and therefore that the maximum possible number of excitations is $166/2.5 = 66.4$, so that only 1 in 5 of the Mn^{2+} ions are excited successfully. Other authors have also concluded that only a fraction of the Mn^{2+} centres are actually excited.

(*e*) *Commercial displays*

As stated earlier the first flat panel ACELTV display was demonstrated in 1978. The contrast ratio with a thin film is much better than with a powder panel, since a film is basically non-reflecting, unlike the white particles of the powder. The new display dispensed with the black absorber layer which had been used with a.c. powder panels. Instead, a circularly-polarizing neutral filter was placed in front of the display to improve the contrast in high ambient illumination.

Today there are three major manufacturers of a.c. thin film displays, all of which are monochrome based on ZnS:Mn. Pixel luminance ranges from 30 to 38 foot-lambert. Various display areas and dot formats are available, depending on the application. The largest format currently available is 640 columns and 480 rows although 1024×800 devices are in development. The panels have wide viewing angle and can display information as rapidly as a conventional CRT. Lives in excess of 20 000 or even 40 000 h are claimed. There are 16 levels of gray scale and operating temperatures lie between −20 °C and 165 °C.

The row electrodes are generally at the back of the panel and are therefore of aluminium, while the columns are etched into the transparent conducting oxide on the glass. A PCB on which components are mounted is attached to the glass substrate to make a completely flat package. A d.c.–a.c. converter is supplied so that battery operation is possible. Although all the early work on the electronic drive system was concentrated on TFTs, drive systems are now based on MOS ICs with CMOS logic and DMSO high voltage drivers on the same chip. In addition to a power supply, a display usually requires four basic signals to operate, namely video data, video clock, and horizontal and vertical synchronization.

Although a full-colour TV display has not yet been developed, flat panel monochrome displays are finding a wide range of applications in office and home automation via desk top and lap top PCs etc., test and measurement instrumentation, military hardware, traffic and transportation equipment and systems, industrial process control and medical instrumentation.

(*f*) *Colour*

Current activity is directed towards improved film quality using chemical vapour deposition (CVD) techniques [30] as well as the ALE technology developed in Finland. Large-area [31] as well as full-colour displays [32] are well under way. Although the luminance and efficiency of primary red, green and blue phosphors for use in ACTFEL displays do not approach those of ZnS:Mn, considerable progress has been made in recent years [33–36] so that a full-colour display is now quite feasible. A wide variety of phosphor systems such as ZnS with rare earth activators and alkaline earth hosts such as SrS and CaS have been investigated with varying success. The first large-area matrix-addressed multicolour (red, yellow, green) display was reported in 1986. This employed stacked transparent, red and green planes. More recently phosphor patterning techniques have allowed three primary colour phosphors to be patterned on a single substrate. One such device [32] uses red ZnS:Sm, green ZnS:Tb and blue SrS:Ce phosphors in parallel stripes aligned with the column electrodes. The chromaticities of the red and green phosphors are adequate for most applications, but that of the blue needs to be deeper for TV display; however, it is quite acceptable for many data display purposes. Research is still concentrating on finding a better blue phosphor. As for brightness all other phosphors are distinctly inferior to the yellow–orange ZnS:Mn. The pixel luminances at a 60 Hz frame rate of the red, green and blue phosphors mentioned here are 0.96 foot-lambert for the red, 6.5 foot-lambert for the green and 0.96 foot-lambert for the blue. In principle, the three-colour display could be addressed at 240 Hz when the luminances would extrapolate to 3.8 foot-lambert (red), 26 foot-lambert (green) and 3.8 foot-lambert (blue). This corresponds to an average white luminance of 1.72 foot-lambert. Although the black–white contrast in an ambient illumination of 500 lx would be useable it has to be conceded that higher luminance is very desirable.

(*g*) *Recent developments*

A variety of deposition techniques have been investigated in recent years in attempts to improve the EL characteristics of ZnS:Mn thin films. Metal organic chemical vapour deposition (MOCVD) techniques have been used [37] and offer the advantage of lower growth temperatures and better control of impurity incorporation. A different CVD procedure based on halide or hydride transport has also been used with some success. In a very recent development Mikami *et al.* [30] describe a process in which ZnS:Mn films are prepared in a low-pressure CVD reactor in which a stream of ZnS is provided by passing hydrogen over heated ZnS powder, and the manganese is introduced as the gaseous chloride by passing HCl over manganese. The growth process of the ZnS is greatly influenced by the presence of the manganese chloride, and the manganese transport can be strictly controlled by adjusting the HCl flow rate. Films were put down on Si_3N_4 or NiO_x and made up into ACEL cells. Undoped ZnS has the (111)-oriented cubic structure, while Mn-doped films are hexagonal with (0001) preferred orientation. Epitaxial layers can be grown on (111)Si at 500 °C. The PL of the CVD films exhibits a strong blue self-activated (SA) emission in addition to the characteristic managanese yellow luminescence. The SA emission is quenched as the Mn concentration is increased. The luminance–voltage characteristic of the ACEL cells is remarkably steep with pronounced saturation effects at higher voltage. The light output is yellow (the SA emission is weak in EL), with a luminance exceeding 1000 foot-lambert at 1 kHz. This technology is now at a pre-production stage.

The ALE process introduced in Finland is also in the course of continuous development, since high quality films can be produced by this method. The essential feature of ALE is that ZnS is grown from Zn and S (or H_2S) vapours. When zinc vapour is introduced, cation–anion bonds will allow the Zn to adhere as a monatomic layer on an anion surface such as sulphur or oxygen. However, the temperature is maintained sufficiently high to prevent zinc–zinc bonds forming, so that no more than one monolayer of zinc is produced. Excess zinc is flushed away and the zinc vapour replaced with sulphur (or H_2S). ZnS forms as the sulphur bonds to the zinc layer. Once a monolayer of ZnS is obtained the sulphur bonds very weakly and in its turn the excess sulphur is swept away. The whole cycle is repeated many times as a film of Zn is built up monomolecular layer at a time. ALE active layers are used in the Finlux ACEL display devices.

12.2.3 D.C. powder panels

Systematic studies of DCEL powder displays did not begin until the late 1960s. At first panels were operated under continuous d.c. conditions but, since this led to fairly rapid degradation, pulsed operation became the norm. This of course is particularly appropriate for multiplexed matrix displays.

The DCEL powder panel [12, 38–41] is a sandwich device in which a copper-coated ZnS:Mn, Cu powder, suspended in an organic binder, is spread or sprayed on conducting glass. The sandwich is completed with an evaporated aluminium back electrode.

Cell panels have to be subjected to a forming process before a useful electroluminescence output is obtained. During the preparation of the phosphor powder, the individual grains are coated with copper sulphide. Since copper sulphide is a good semiconductor, a layer of ZnS:Mn, Cu in a DCEL cell has a relatively low resistivity of 100–200 Ω cm. To form the device a ramped d.c. voltage is applied with the aluminium electrode negative.

At first the current increases with increasing voltage and no light is emitted, until the current begins to decrease at a particular voltage when the emission of visible light begins. This initial phase of forming is accompanied by a reduction in current by between 3 and 100 times depending on the pre-treatment of the powder. Continued increase of the voltage leads to a further increase in resistance of the panel. Standard forming procedures to obtain optimum electroluminescence have been devised. When completed, the panels are hermetically sealed in an argon atmosphere to ensure longer life.

Typical phosphor layer thicknesses are about 15–20 μm and individual ZnS grains are in the range 0.5–2.0 μm. What appears to happen on forming is that the copper migrates away from the surfaces of the particles of ZnS immediately adjoining the transparent front electrode so that a high resistance layer, ideally one particle ($\sim 1\ \mu$m) thick, is left on top of a base of highly conducting unformed phosphor. Practically all the applied voltage is then dropped across the formed region, so that a voltage of 100 V will lead to a very high field of the order $10^6\ \mathrm{V\,cm^{-1}}$. In this condition EL emission is obtained from the formed layer.

Developments in phosphor preparation with strict control of particle size to ensure a uniform active layer thickness have shown that optimum DCEL performance in ZnS powder is obtained with concentrations of 0.32 wt% Mn and 0.048 wt% Cu. Forming is improved if a layer of Al_2O_3, 40–60 Å thick, is evaporated onto the ITO-coated glass before the active layer is put down. The function of the Al_2O_3 is to reduce the current necessary for forming and to allow the whole display area to be formed simultaneously. Alphanumeric DCEL displays are normally operated at 30–50 V above the final forming voltage using 120 V pulses of 17 μs duration at 240 Hz. The emission is the characteristic yellow–orange 585 nm luminescence of Mn^{2+}. Response to the leading edge of the pulse is rapid, $\sim 3\ \mu$s, but light continues to be emitted well after the voltage returns to zero. The luminance is a highly superlinear function of applied voltage once a threshold of 30–40 V is exceeded (Fig. 12.7). As with the ACEL thin film devices saturation of the light emission occurs, and with the d.c. powders this is usually at about 300 foot-lambert.

Although panels with a brightness of 1000 foot-lambert can be produced, they have a very poor maintenance especially if they are operated at constant voltage. It was soon discovered that much better life is obtained if panels are operated under conditions of constant power dissipation. The degradation under constant voltage is attributed to three basic processes:

1. continued forming, i.e. the high-resistance active layer slowly broadens as the copper migration continues;
2. the resistance of the unformed supporting layer of ZnS coated with copper sulphide slowly increases, leading to an increased series resistance;
3. the active high-field region becomes more conducting with time as copper diffuses out.

The overall efficiency of a device which is of the order of $2\ \mathrm{lm\,W^{-1}}$ does not change greatly during a few thousand hours operation even while the brightness is falling substantially during constant-voltage operation. By operating devices under pulsed conditions at constant power, by improving

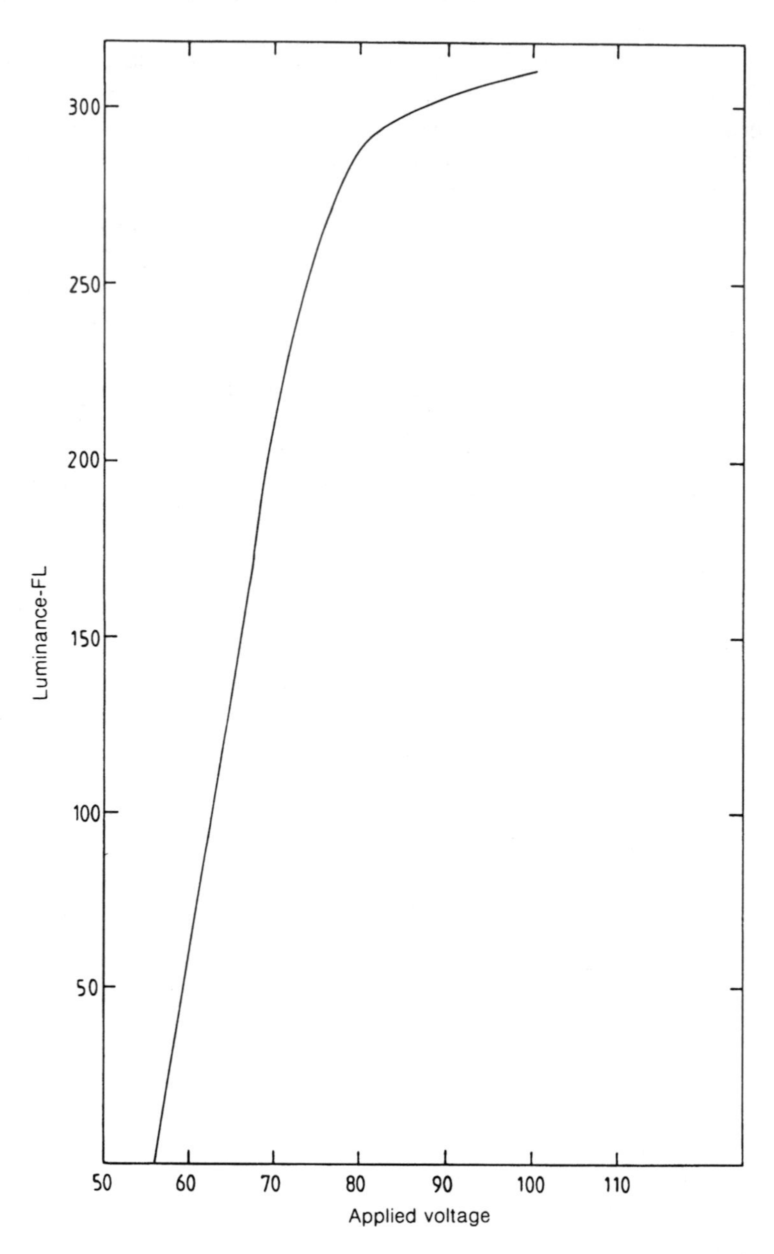

Fig. 12.7 Typical brightness versus voltage curve for a formed DCEL powder device.

the final encapsulation processing and by incorporating feedback into the drive circuitry so that the light output can be monitored and the voltage increased as necessary, a satisfactory life up to 10 000 h can be maintained, and it is claimed that 20 000 h should be possible.

The first commercial DCEL powder panels were produced in the early

1980s. These were limited to fixed legend and mimic displays. Today the technology has advanced and displays suitable for desk top computers are available [42]. One such panel [43] has an active display area of 117 mm × 195 mm with 640 × 256 pixels capable of displaying 25 lines of 80 characters. Pixel size is 0.406 mm × 0.229 mm and the luminance is 25 foot-lambert. The power consumption is 15 W. This display can be used with a resistive touch switch overlay. DCEL displays have the advantage over ACEL that they are resistive rather than capacitive. This means that they are more power efficient at high-frequency refresh rates. This has been exploited in the design of a column driver to supply video data to the columns at a high data rate, which increases the luminance and the power efficiency. A gray scale can be achieved by frame-rate modulation. This is an alternative to pulse-width modulation which has also been exploited in DCEL devices.

So far only the monochrome yellow–amber Mn^{2+} displays are available. Red, green and blue phosphors based on CaS and SrS are under development. Interesting features of the alkali earth materials, particularly those doped with cerium, are that they do not saturate as readily as ZnS:Mn phosphors. The response times of the various rare earth activators are usually different so that a two-colour display may be obtained by mixing two individual coloured phosphors in the same layer. However, much remains to be done before multicolour displays become commonplace.

12.2.4 Thin film DCEL devices

DCEL thin film displays might be expected to have a number of advantages over ACEL thin film devices, such as lower voltage operation and easier drive circuitry. As it is, however, a.c. driven displays are far more developed. The reasons for this are that problems of maintenance and localized dielectric breakdown at weak spots and pinholes in TFDCEL displays, have not been adequately solved, and no large-area devices have been produced in quantity.

(*a*) *Structure and characteristics*

The first DCEL thin film devices [7, 44] were simple structures with evaporated layers of ZnS:Mn,Cu,Cl sandwiched between tin-oxide-coated glass and aluminium electrodes. Although high brightness up to 1000 foot-lambert could be achieved, life was short since the devices were operated in the avalanche region where localized dielectric breakdown occurred with catastrophic results. This difficulty was effectively overcome in ACEL thin film devices by utilizing insulating layers of high dielectric strength and structural perfection so that breakdown at pinholes or weak spots in the ZnS did not occur.

Okamoto *et al.* [45] attempted to overcome the breakdown problem by fabricating a device with a control layer, the function of which was to limit the current flow at weak spots and thus to prevent destructive breakdown. Their device consisted of a 300 nm thick layer of ZnS:Mn deposited by electron beam evaporation onto conducting glass. This active layer was covered by a 400 nm thick resistive layer of ZnSe. The device was operated by unipolar pulses 2 ms wide at 200 Hz. Characteristic yellow–orange Mn luminescence was emitted at a threshold of 80 V. A luminance of about 50 foot-lambert was reached at 100 V. Current measurements suggested that the ZnS layer was in the avalanche regime when the light emission was observed, and that the differential resistance was zero.

The next development of note was that described at some length in the papers by Blackmore *et al.* [13] and Cattell *et al.* [14]. Blackmore *et al.* draw attention to the fact that in insulating ZnS ($\rho \sim 10^{12}\,\Omega\,\mathrm{cm}$) electroluminescence is observed following the onset of a dissipative conduction greater than $1\,\mathrm{A\,cm^{-2}}$ when a critical field of the order of $10^6\,\mathrm{V\,cm^{-1}}$ is exceeded. They point out the remarkable fact that despite a voluminous literature there is little mention of destructive effects accompanying the emission of electroluminescence as a large current passes through an insulator.

Blackmore *et al.* prepared thin film DCEL test structures using cadmium stannate as the transparent conducting coating on glass. Their first simple structures showed normal dielectric breakdown. ZnS:Mn films, 0.7 μm thick, were sputtered in argon onto the transparent conductor. The second electrode was of evaporated aluminium. When a ramped voltage was applied small bursts of current ($\sim 1\,\mathrm{nA\,cm^{-2}}$) occurred at low fields. If the ramp was stopped when these bursts began they rose rapidly before decaying to zero. As the voltage was increased still further the frequency of these bursts increased and they were accompanied by blue sparks at the aluminium electrode, and damage craters appeared. A sample might be destroyed by a non-self-healing breakdown event but, if it did not, most of the active film area would be covered with craters. This is entirely similar to self-healing localized destructive breakdown (LDB) in insulating films such as Ta_2O_5 [46]. Steady yellow Mn^{2+} emission was not observed. Similar effects were found with epitaxial layers of ZnS:Mn put down on silicon substrates by MOCVD techniques.

The frequency of LDB events could be reduced slightly by incorporating control layers of amorphous silicon or MnO_2 with resistivities of the order of $10^4\,\Omega\,\mathrm{cm}$ and then uniform yellow emission was obtained. A dramatic reduction in the incidence of LDBs was found in a hybrid device [47] in which a layer of ZnS:Mn was deposited on conducting glass by MOCVD techniques and then covered with a layer of undoped ZnS powder which had previously been treated in a copper chloride solution to coat each particle with Cu_xS. In pulsed operation with the copper sulphide layer negative, broadly similar results to those from DCEL powder panels described in the preceding section were obtained.

However, it was later found [13] that LDB rates could also be reduced

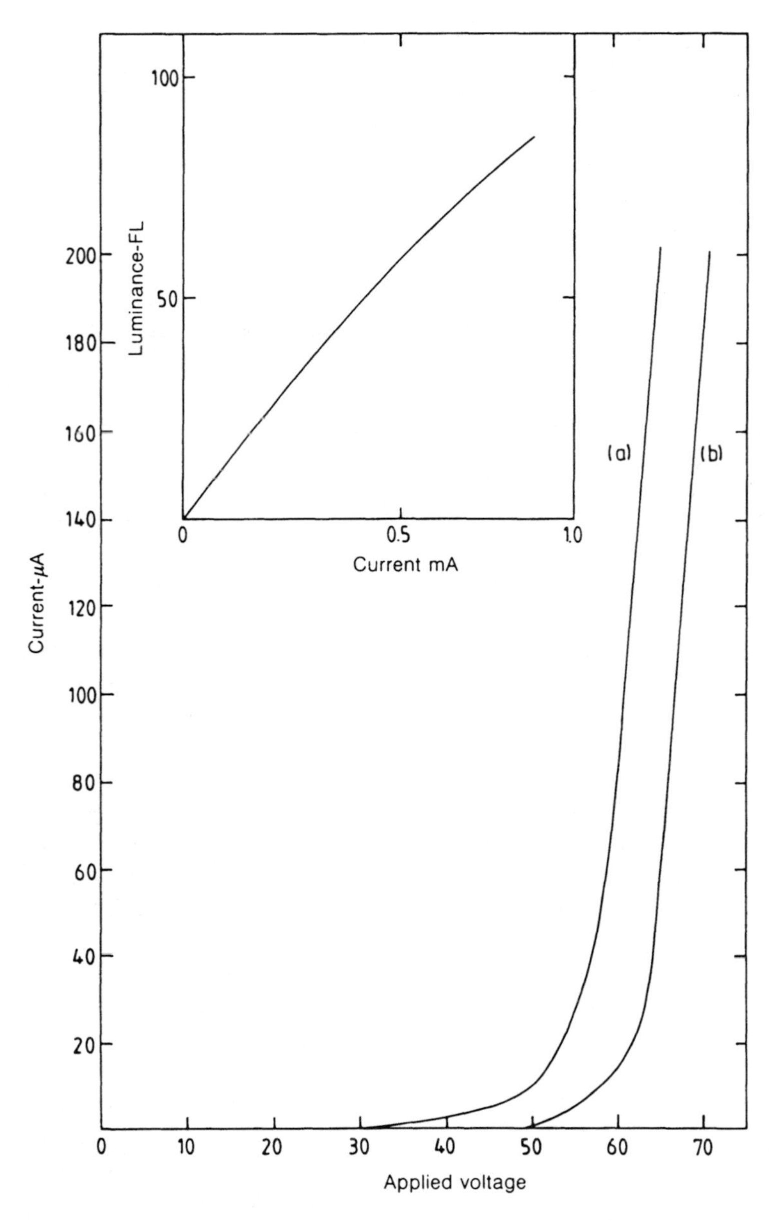

Fig. 12.8 Brightness as a function of voltage for a thin film DCEL device (a) without a control layer and (b) with a control layer. Inset shows the brightness as a function of current. Device area, 6 mm^2.

by sputtering ZnS:Mn in an argon–10% hydrogen gas mixture and administering a heat treatment to 450 °C in flowing argon. Test devices were prepared with and without additional control layers, which when used were of amorphous silicon ($\rho \sim 10^4\,\Omega$ cm) or 20% Ni–SiO_2 cermet (100 Ω cm). Some typical current–voltage–brightness data are shown in Fig. 12.8. The steep rise in current occurred at a voltage which varied linearly with ZnS film thickness. The time-dependent behaviour of the current and voltage when the device was driven from a constant-voltage source through a 1 MΩ load resistor showed glitches (spikes) on both the current and voltage traces displayed on an oscilloscope. These were more obvious on the voltage trace and were associated with LDB events which appeared as sudden bright yellow spots on a uniform yellow luminescent background. The scintillations appeared and disappeared in about 1 s, leaving behind small craters. The LDB rate increased rapidly as the applied voltage increased and became excessive when the current density exceeded 10 mA cm^{-2}. An interesting feature is that the LDB rate was dramatically reduced when the ZnS layer was illuminated with broad-band ultraviolet light through the conducting glass electrode. The authors conclude that in many structures uniform electroluminescence from ZnS:Mn occurs simultaneously with short-lived LDB events. A satisfactory device requires that the localized dielectric breakdown be reduced, if not eliminated. The procedure of sputtering in an argon–hydrogen mixture goes well on the way to achieving this.

(*b*) *Physical processes*

In the companion paper, Cattell *et al.* [14] offer a model to explain these observations. Here electrons injected into the ZnS:Mn from the transparent conducting electrode are accelerated in the high applied field ($\sim$ 1 MV cm^{-1}) until they excite the Mn^{2+} emission by impact. The field is high enough for the injected electrons to multiply by inducing free electron–hole pair production as well as impurity ionization. The current is carried by both electrons and holes. The holes are slow moving and provide a space charge in front of the cathode; indeed some of the holes are probably immobilized in hole traps. The positive space charge leads to an enhanced electron injection, resulting in increased ionization, more holes and a larger space charge and so on. In short a current-controlled negative differential resistance is obtained as breakdown occurs and the current increases very rapidly for a small change in applied voltage. This model is adapted from O'Dwyer's [48] theory of space-charge-enhanced injection, which has been used to explain dielectric breakdown in films of SiO_2 [49] and Ta_2O_5 [46]. The model requires slow-moving or trapped holes in front of the cathode. Hole mobilities should be low (quite feasible in polycrystalline material) or a hole trap density of about 10^{17} cm^{-3} is necessary. Hole trap densities of this magnitude have been invoked to explain the hysteretic memory effect in

ACEL thin films [26]. The dielectric breakdown model predicts an increased tendency to filamentary conduction with increased active layer thickness and cathode barrier height. For optimum performance and life, filamentary conduction should be eliminated.

Cattell *et al.* suggest that the model can also be used to explain the stabilizing effect of Cu_xS in DCEL powder and hybrid devices. In hybrid devices attempts are made to decouple the processes of electron injection and of acceleration in the ZnS. When p-type Cu_xS is used as a cathode for n-type ZnS:Mn, there may be little or no barrier to the injection of electrons from the Cu_xS. Although there are few electrons in the conduction band of the Cu_xS, a sufficiently high field would deplete the Cu_xS side of the junction, and electrons would be able to tunnel from the valence band of the Cu_xS to the conduction band of ZnS. There would be no barrier to injection into the ZnS. Breakdown would be less likely in the Cu_xS because of the thinness of the layers, and because the lower energy necessary for avalanching in a material with a bandgap of $\sim$ 1 eV would imply that electron injection would occur at lower fields than usual for ZnS. This they argue would lead to a much more stable device.

So far no large-area DCEL thin film displays have been produced. A UK consortium has achieved matrix displays with 128×128 lines in a $70\,mm \times 60\,mm$ area with a luminance of 10–20 foot-lambert at a drive voltage of 60 V. Displays were provided with control layers of amorphous silicon. Lives of 5000–6000 h have been achieved for individual pixels operated at 10 foot-lambert. Much more work is required if a viable display is to be produced.

12.3 LIGHT-EMITTING DIODES

The early work on ACEL powder panels in the 1950s and 1960s stimulated interest in EL from single-crystal devices, initially on ZnS and then on ZnSe, ZnTe and CdS, and a large volume of information was accumulated which is well summarized by Henisch [3] and Ivey [4]. ZnS, which was difficult to grow, and more difficult to produce in low-resistivity form, was studied in the hope that the physical processes involved in the complex powder devices would be better understood. When this hope was not realized, attention turned to other II–VI compounds in attempts to produce bright, stable, single-crystal LEDs. Much ingenuity was expended, as described by Fischer [5], but no commercial device was forthcoming.

More recently [50] a new series of attempts have been made to produce LEDs on single crystals of ZnS and ZnSe. ZnSe is a particularly suitable material. It has a bandgap at room temperature of 2.67 eV, and can readily be produced in single-crystal form with a high n-type conductivity (10^{-1}–$1.0\,\Omega^{-1}\,cm^{-1}$). Unfortunately, low-resistivity p-type ZnSe has not been satisfactorily prepared, with the result that no useful homojunction device has yet been produced. Numerous attempts have been, and are being, made to prepare p-type ZnSe and ZnS in the quest for a blue-light-emitting diode which would have a variety of applications. The problem of conductivity control is discussed in detail in Chapter 11, while the progress with homojunction devices is described in Chapter 13. The discussion in the present section is confined to Schottky (MS) and MIS diodes.

12.3.1 Zinc selenide

MS and MIS diodes prepared on ZnSe doped with manganese, and operated in reverse bias, are probably the brightest and most efficient II–VI LEDs yet produced. Such devices have been described by Allen *et al.* [16] and by Ozsan and Woods [17] who obtained a luminance of about 300 foot-lambert in the yellow–orange at 15 V and 10mA mm^{-2}.

The diodes were fabricated from bulk single-crystal dice with millimetre dimensions, doped with Mn and subjected to the so-called zinc extraction process to reduce the resistivity to 1–10 Ω cm. Opposite faces of a die were equipped with an indium and a gold contact to provide an ohmic and a rectifying contact respectively. When required, insulating layers were produced on the ZnSe surface before depositing the gold either by refluxing in boiling acetone or by depositing an insulating layer of ZnS.

In reverse bias both MS and MIS diodes on ZnS begin to emit light with a threshold of 4–6 V. Initially the light output increases linearly with the current but ultimately saturates and eventually begins to decrease. The maximum luminance obtainable is usually about 500 foot-lambert at a power efficiency of some $2 \times 10^{-3}\%$. In operation under constant current the luminance of life-tested diodes at first rose from 200 foot-lambert to a maximum of 350 foot-lambert, but after 1000 h it had fallen to 75% of its initial value.

The mechanism of the EL process is that the emission is due to impact excitation of Mn^{2+} ions by electrons accelerated in the high electric field of the depletion region of the MS or MIS devices [51, 52]. The electrons are injected into the ZnSe by tunnelling through the insulating layer and the top of the depletion region.

It is not necessary to dope ZnSe with Mn to obtain a yellow LED. Park *et al.* [53] obtained a luminance of 200 foot-lambert from an aluminium-doped ZnS diode. Variations in colour from yellow towards red can be achieved by adding copper as an impurity. Attempts have also been made to implant rare earth ions into the surface layers of semiconducting ZnSe. This has been achieved for example with erbium, thulium and neodymium [54–56]. The characteristic emission of the trivalent rare earth ion is excited in reverse bias electroluminescence.

There have been a number of reports of light emission from MIS diodes of ZnS operated in forward bias. With MS devices when the insulating layer is absent, little or no EL is observed. Jaklevic *et al.* [57] first suggested that minority carriers could be injected into n-type ZnSe by tunnelling through the insulating layer. Electron–hole recombination in the ZnSe would then lead to radiative emission, but the relative independence of the quantum efficiency of the thickness of the interfacial layer up to several thousand ångströms cannot be explained in terms of such a process. Lawther and Woods [58] have suggested, following Watanabe *et al.* [59], that the holes are excited from the metal over the hole barrier into the ZnSe by a two-step process involving the de-excitation of hot electrons which have entered the metal in forward bias. Theoretical calculations by Jones and Abram [60] demonstrate that this is a plausible process.

Lawther and Woods investigated MIS structures on ZnSe doped with Mn or Al. The EL emission was similar to that observed in photoluminescence, which is to be expected for a process of electron–hole recombination. All devices had thresholds for light emission in the range 1.2–1.4 V and the maximum luminance obtained was in the range 10–20 foot-lambert.

If strict precautions are taken to eliminate all accidental impurities from the ZnSe, it is possible to prepare MIS diodes in which the radiative energy in forward bias is concentrated in exciton emission. Fan and Woods [61, 62] have studied such exciton emission at temperatures down to 30 K, below which carrier freeze-out occurs. At 30 K the emission consists mainly of the 1LO and 2LO phonon-assisted replicas of the free-exciton recombination. The visual effect is blue. At room temperature the emission merges into a single broad band at 465 nm and is reduced in intensity. The visual effect is still blue but the quantum efficiency is low.

Fan and his coworkers [63, 64] have studied devices in which the insulating layers consist of ZnSe instead of ZnS. The emission is seen to be spotty under the microscope and it is suggested that microplasmas [65] in the I layers are the source of the holes which are injected into the ZnSe substrate.

In contrast Jones *et al.* [66] prepared a ZnSe LED with an epitaxial ZnS

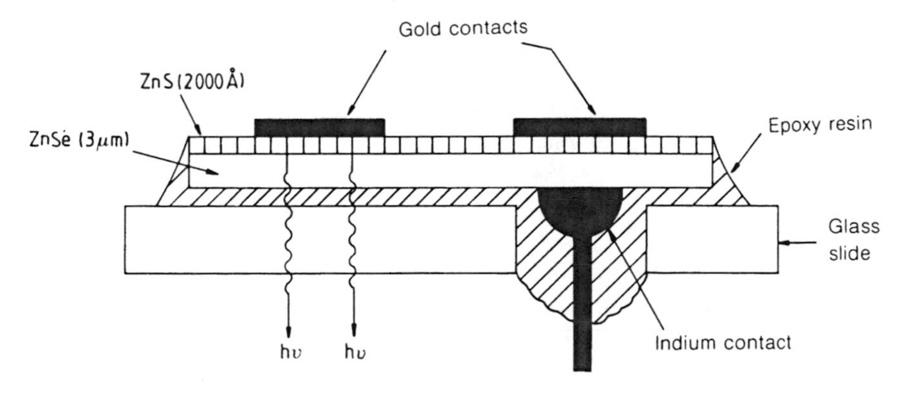

Fig. 12.9 ZnSe MIS device with epitaxial ZnS I layer.

I layer. A 200 nm thick ZnS insulating layer and a 3 μm thick insulating layer of semiconducting ZnSe were deposited sequentially by MOCVD onto (1 0 0)-oriented GaAs. An ohmic contact was made to the ZnSe with indium and the device was mounted on to a glass slide with epoxy resin; see Fig. 12.9. The complete assembly was then etched in a mixture of 95% H_2O_2 (100 volumes) and 5% NH_3 (35% solution) to remove the GaAs substrate. This etch does not attack ZnS appreciably. The device was completed by the deposition of evaporated gold contacts. In operation in forward bias the electroluminescence was found to be uniform over the whole electrode area, lending support to the suggestions of Lawther and Woods.

12.3.2 Zinc sulphide

One of the major problems in preparing LEDs in single-crystal ZnS has been the difficulty of reducing the resistivity to a value in the region of 10 Ω cm, although a number of authors claim to have done so. This topic is discussed in detail by Thomas *et al.* [67] who advocate a procedure for bulk single crystals of heating at 1000 °C in a molten mixture of Zn + 1% Ga + 0.5% Al. Suggested mechanisms for the success of this procedure are put forward in a later paper [68]. A second obstacle is that it is difficult to make ohmic contacts on semiconducting ZnS. In spite of this, quite good results have been obtained with manganese-doped ZnS where the characteristic orange–yellow Mn^{2+} emission is excited in MS diodes in reverse bias in exactly the same way as with ZnSe:Mn devices. However, there does not appear to be any advantage to be gained by using the sulphide rather than the selenide.

On the other hand, the wider bandgap of the ZnS (3.66 eV) offers the possibility of shifting the colour of the emission to the blue if the self-activated emission can be excited [69]. Indeed, Ozsan and Woods excited a blue emission at 4800 Å in reverse bias with a luminance of 90 foot-lambert for a power dissipation of 600 mW at 20 V.

Forward bias EL in ZnS MIS structures similar to that described for ZnSe devices has also been investigated fairly extensively. Thomas *et al.* [70] have demonstrated how the efficiency of the hole injection process depends on the work function of the positive (blocking) electrode and have shown that by using sulphur nitrogen polymer, $(SN)_x$, a material with a work function 1 eV greater than that of gold, the efficiency of forward bias blue EL can be increased by a factor of 100 compared with the best result obtained by Lawther and Woods [71] using gold. Thomas *et al.* estimated the luminance at 20 mA cm^{-2} as 10 foot-lambert.

None of these blue-emitting devices is sufficiently bright or long lived to satisfy practical demands. Hopes are now concentrated on preparing p-type layers of ZnSe or ZnS by CVD, MOVPE or MBE techniques so that efficient p–n homojunctions can be produced.

REFERENCES

1. Destriau, G. (1936) *J. Chim. Phys.*, **33**, 587.
2. Piper, W.W. (1953) *US Patent 2698915.*
3. Henisch, H.K. (1962) *Electroluminescence*, Pergamon, Oxford.
4. Ivey, H.F. (1963) *Adv. Electron.*, Suppl. I.
5. Fischer, A.G. (1966) In *Luminescence in Inorganic Solids* (ed. P. Goldberg), Academic Press, New York, pp. 541–602.
6. Pankove, J.I. (ed.) (1977) *Topics in Applied Physics*, Vol. 17, *Electroluminescence*, Springer, Berlin.
7. Thornton, W.A. (1962) *J. Appl. Phys.*, **33**, 123.
8. Soxman, E.J. and Ketchpel, R.D. (1972) *Electroluminescent Thin Film Research, JANAIR Rep. 720903.*
9. Inoguchi, T., Takeda, M., Kakihari, Y., Nakata, Y. and Yoshida, M. (1974) *SID Int. symp. Dig.*, 84.
10. Takeda, M., Kakihara, Y., Yoshida, M., Kawaguchi, M., Kishishita, H., Yamauchi, Y., Inoguchi, T. and Mito, S. (1975) *SID Int. Symp. Dig.*, S7.8.
11. Suntola, T., Antson, J., Pakkala, A. and Lindfors, S. (1980) *SID Int. Symp. Dig.* p. 108, Suntola, T. *US Patent 4058430.*
12. Vecht, A., Werring, N.J., Ellis, R. and Smith, P.J.F (1968) *J. Phys. D.* **1**, 134.
13. Blackmore, J.M., Cattell, A.F., Dexter, K.F., Kirton, J. and Lloyd, P. (1987) *J. Appl. Phys.*, **61**, 714.
14. Cattell, A.F., Inkson, J.C. and Kirton, J. (1987) *J. Appl. Phys.*, **61**, 722.
15. Haase, M.A., Cheng, H., De Puydt, J.M. and Potts, J.E. (1990) *J. Appl. Phys.*, **67**, 448.
16. Allen, J.W., Livingstone, A.W. and Turvey, K. (1972) *Solid State Electron.*, **15**, 1363.
17. Ozsan, M.E. and Woods, J. (1975) *Solid State Electron.*, **18**, 519.
18. Fischer, A.G. (1962) *J. Electrochem. Soc.*, **109**, 1043.
19. Fischer, A.G. (1963) *J. Electrochem. Soc.*, **110**, 733.
20. Inoguchi, T. and Mito, S. (1977) Vol. 17, In *Topics in Applied Physics, Electroluminescence*, (ed. J.I. Pankove), Springer, Berlin, p. 197.
21. Taniguchi, K., Tanaka, K., Ogura, T., Kakihara, Y., Nakajima, S. and Inoguchi, T. (1985) *Proc. SID*, **26**, 231.
22. Smith, D.H. (1981) *J. Lumin.*, **23**, 209.
23. Allen, J.W. (1984) *J. Lumin.*, **31–32**, 1984.
24. Müller, G.O. (1984) *Phys. Stat. Sol. A*, **81**, 597.
25. Müller, G.O. and Mach, R. (1988) *J. Lumin.*, **40–41** 92.
26. Yang, K.W. and Owen, S.J.T. (1983) *IEEE Trans. Electron Devices*, **30**, 452.
27. Howard, W.E., Sahni, O. and Alt, P.M. (1982) *J. Appl. Phys.*, **53**, 639.
28. Howard, W.E. and Alt, P.M. (1977) *Appl. Phys. Lett.*, **31**, 399.
29. Sahni, O., Alt, P.M., Dove, D.B., Howard, W.E. and McClure, D.J. (1981) *IEEE Trans. Electron Devices*, **28**, 708.
30. Mikami, A., Terada, K., Tanaka, K., Taniguchi, K., Yoshida, M. and Nakajima, S. (1989) *SID Int. Symp. Dig.*, 309.
31. Schmachtenberg, R., Jenness, T., Ziuchkovski, M. and Flegal, T. (1989) *SID Dig.*, 58.
32. Barrow, W.A., Coovert, K.E., King, C.N. and Ziuchkovski, M. (1989) *SID Dig.*, 284.
33. Barrow, W.A., Coovert, R.E., King, C.N. and Ziuchkovski, M. (1988) *SID Int. Symp. Dig.*, 284.
    King, C.N. (1989) *Jpn. Disp.*
34. Ogura, T., Mikami, A., Tanaka, K., Taniguchi, K., Yoshida, M. and Nakajima, S. (1986) *Appl. Phys. Lett.*, **61**, 1570.

35. Mikami, A., Ogura, T., Tanaka, K., Yoshida, M. and Nakajima, S. (1987) *Appl. Phys. Lett.*, **61**, 3028.
36. Tanaka, K., Mikami, A., Ogura, T., Taniguchi, K., Yoshida, M. and Nakajima, S. (1968) *Appl. Phys. Lett.*, **48**, 1730.
37. Jones, A.P.C., Brinkman, A.W., Russell, G.J., Woods, J., Wright, P.J. and Cockayne, B. (1987) *Semicond. Sci. Technol.*, **2**, 621.
38. Vecht, A., Werring, N.J., Ellis, R. and Smith, P.J.F. (1973) *Proc. IEEE*, **61**, 902.
39. Vecht, A. (1982) *J. Cryst. Growth*, **59**, 81.
40. Vecht, A. and Werring, N.J. (1970) *J. Phys. D*, **3**, 105.
41. Chadha, S.S., Haynes, C.V. and Vecht, A. (1988) *SID Int. Symp. Dig.*, 35.
42. Mayo, J. and Channing, D. (1990) *Inf. Disp.*, **6**, 6.
43. Miller, G.B. and Werring, N.J. (1989) *Displays*, **10**, 203.
44. Miyata, T., Nakagawa, S., Hirayama, S., Hasegawa, H. and Korenaga, M. *Jpn. J. Appl. Phys.*, **9**, 615.
45. Okamoto, K., Tanaka, S., Kobayashi, H. and Sasakura, H. (1978) *IEEE Trans. Electron Devices*, **25**, 1170.
46. Klein, N. (1969) *Advances in Electronics and Electron Physics*, Vol. 26, Academic Press, New York, p. 309.
47. Cattell, A.F., Cockayne, B., Dexter, K. and Kirton, J. (1983) *IEEE Trans. Electron Devices*, **30**, 471.
48. O'Dwyer, J.J. (1969) *J. Appl. Phys.*, **40**, 3887.
49. Schatzkes, M., Av-Ron, M. and Anderson, R.M. (1974) *J. Appl. Phys.*, **45**, 2065.
50. Woods, J. (1981) *Displays*, **2**, 251.
51. Mach, R., Gericke, W., Trepton, H. and Ludwig, W. (1978) *Phys. Status Solidi A*, **49**, 341.
52. Allen, J.W. (1973) *J. Lumin.*, **7**, 228.
53. Park, Y.S., Geesner, C.R. and Shin, B.K. (1972) *Appl. Phys. Lett.*, **21**, 567.
54. Yu, C.C. and Bryant, F.J. (1978) *Solid State Commun.*, **28**, 836.
55. Bryant, F.J., Hagston, W.E. and Krier, A. (1984) *J. Lumin.*, **31–32**, 948.
56. Zhang, G.Z. and Bryant, F.J. (1982) *J. Phys. D*, **15**, 705.
57. Jaklevic, R.C., Donald, D.K., Lambe, J. and Vassell, W.C. (1963) *Appl. Phys. Lett.*, **2**, 7.
58. Lawther, C. and Woods, J. (1977) *Phys. Status Solidi A*, **44**, 693.
59. Watanabe, H., Chikamara, T. and Wada, M. (1974) *Jpn. J. Appl. Phys.*, **13**, 357.
60. Jones, R.E. and Abram, R.A. (1988) *Solid State Electron.*, **31**, 989.
61. Fan, X.W. and Woods, J. (1981) *IEEE Trans. Electron Devices*, **28**, 428.
62. Fan, X.W. and Woods, J. (1981) *J. Phys. C*, **14**, 1863.
63. Fan, X.W., Zhang, J.Y., Zhang, Z.S., Wang, S.Y., Lu, A.D., Li, W.Z., Jiang, J.X. and Sun, Y.W. (1984) *J. Lumin.*, **31–32**, 957.
64. Zhang, J.Y. and Fan, X.W. (1988) *J. Lumin.*, **40–41**, 798.
65. Wang, S. and Fan, X.W. (1988) *J. Lumin.*, **40–41**, 800.
66. Jones, A.P.C., Brinkman, A.W., Russell, G.J., Woods, J., Wright, P.J. and Cockayne, B. (1986) *Semicond. Sci. Technol.*, **1**, 41.
67. Thomas, A.E., Russell, G.J. and Woods, J. (1983) *J. Cryst. Growth*, **63**, 265.
68. Thomas, A.E., Russell, G.J. and Woods, J. (1984) *J. Phys. C*, **17**, 6219.
69. Ozsan, M.E. and Woods, J. (1974) *Appl. Phys. Lett.*, **25**, 489.
70. Thomas, A.E., Woods, J. and Hauptman, Z.V. (1983) *J. Phys. D*, **16**, 1123.
71. Lawther, C. and Woods, J. (1978) *Phys. Status Solidi A*, **50**, 491.

13

Preparation of widegap II–VI homojunction devices by stoichiometry control

J.-i. Nishizawa and K. Suto

13.1 INTRODUCTION

Although many different crystal growth methods have been applied to the growth of widegap II–VI compounds, they failed to produce conductivity-type-controlled crystals. As shown in Table 13.1, ZnSe and most of the other widegap II–VI compounds showed n-type conductivity, but p-type crystals were not available even with heavy dopings of acceptor impurities. The exception was ZnTe, which showed only p-type conductivity. Because of these characteristics, injection light-emitting diodes (LED), laser diodes and other devices which needed p–n junctions were not realized in widegap II–VI compounds although they have direct bandgaps which have the potential to cause strong light emission.

On the other hand, one of the authors and coworkers have extensively studied the deviation from stoichiometry in GaAs and other III–V compound semiconductors, and developed a novel crystal growth method called the temperature difference method under controlled vapour pressure (TDM CVP) which can grow crystals with stoichiometry controlled by the applied vapour pressure [1–5]. In 1982, TDM CVP was extended to ZnSe and was successfully used in growing for the first time p-type ZnSe crystals and also ZnSe p–n homojunctions which emitted pure blue light with a wavelength of 470 nm at room temperature [6–9]. This result means that the difficulty in widegap II–VI compounds was due to natural deviation from stoichiometry which conventional growth methods were not able to control.

The TDM CVP, in which vapour pressures of material consisting elements are applied over the melt, can be widely applied for the growth of any of the widegap II–VI compounds. From Table 13.1, we can expect even violet or ultraviolet LEDs and laser diodes when the stoichiometry-controlled growth methods are established in these materials following the growth of p-type ZnSe.

Table 13.1 Properties of several II–VI compounds

Crystal	*Conductivity type*	*Energy gap (eV)*	*Lattice constant (Å)*	*Melting point (°C)*
ZnS	n	3.66	5.409	1830
ZnSe	n	2.67	5.669	1520
ZnTe	p	2.27	6.104	1295
CdS	n	2.41	5.802	1475
CdSe	n	1.67	6.05	1239

In this chapter, we first discuss the principle of the stoichiometry-controlled crystal growth by TDM CVP which was established in III–V compounds, particularly in GaAs. Then, the growth of p-type ZnSe and formation of blue-light-emitting ZnSe p–n junctions are described. The established method for p–n junction formation is the impurity diffusion technique under controlled vapour pressure. We finally describe the liquid phase epitaxial technique which is not yet well established, but is expected to be more promising from experience in III–V compounds.

13.2 FUNDAMENTALS OF STOICHIOMETRY-CONTROLLED CRYSTAL GROWTH BY THE TEMPERATURE DIFFERENCE METHOD UNDER CONTROLLED VAPOUR PRESSURE

Before mentioning TDM CVP crystal growth, we must describe our experiment on the heat treatment of GaAs under As vapour pressure [10]. GaAs wafers were heat treated using a two-zone furnace as shown in Fig. 13.1. We measured the lattice parameters and induced acceptor densities as well as deep-level luminescence (Fig. 13.2) and found that they showed minima at the same arsenic vapour pressure which we called P_{min}^{GaAs}. It would be most reasonable to consider that P_{min}^{GaAs} is the equilibrium vapour pressure which gives the exact stoichiometry of GaAs.

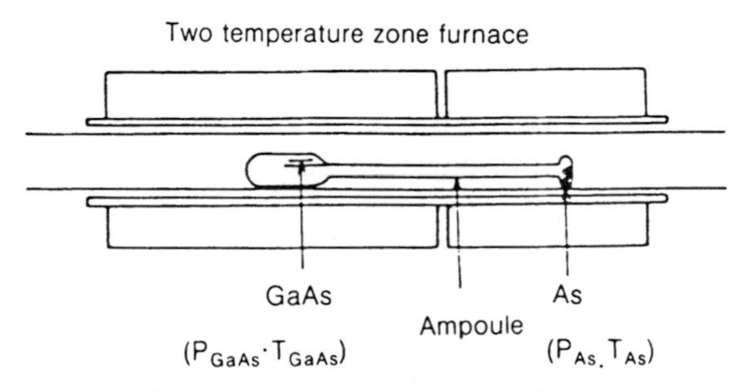

Fig. 13.1 Experimental set-up for heat-treatment under an applied vapour pressure.

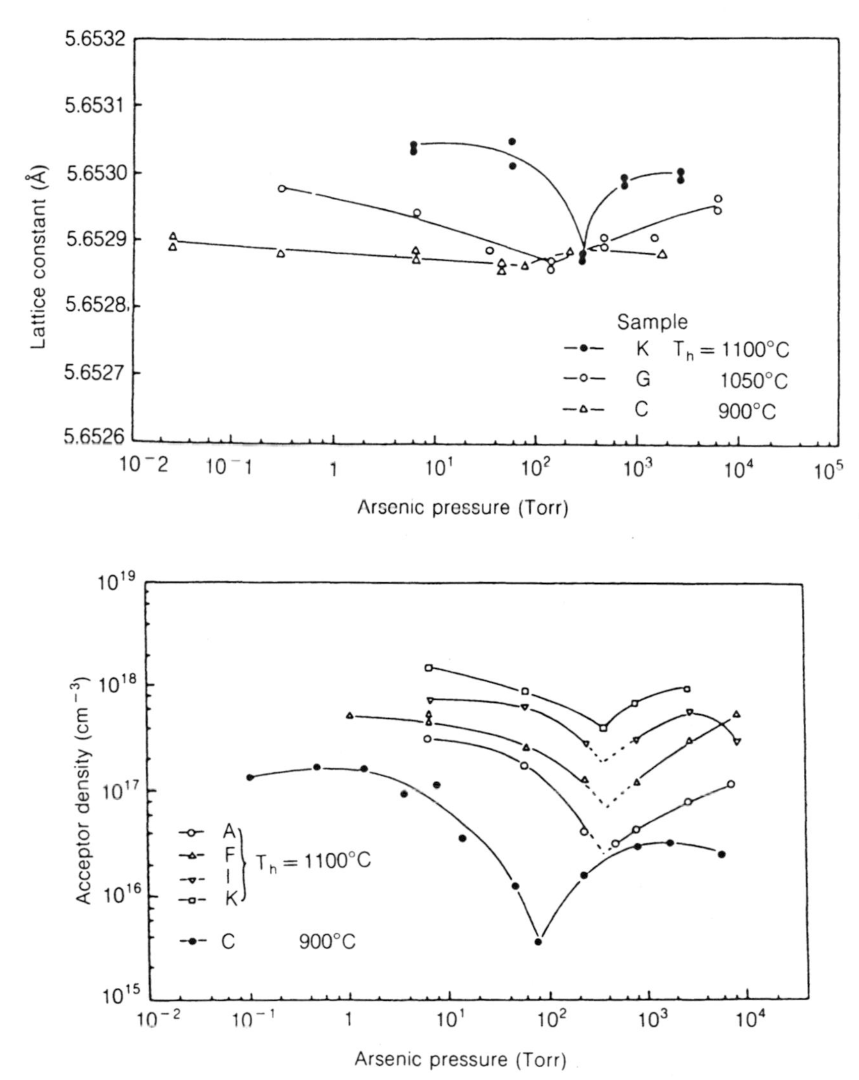

Fig. 13.2 (a) Lattice constants and (b) acceptor densities of heat-treated GaAs crystals as a function of applied arsenic vapour pressure.

In parallel to the heat treatment experiment, we carried out growth experiments with GaP and GaAs by the method schematically shown in Fig. 13.3 [1–5].

The vapour pressure of the group V element (As in case of GaAs) is applied over the molten phase, while the substrate is placed at the bottom of the melt. The temperature difference formed in the melt causes the diffusion of the source materials from the top of the melt to the substrate so that it

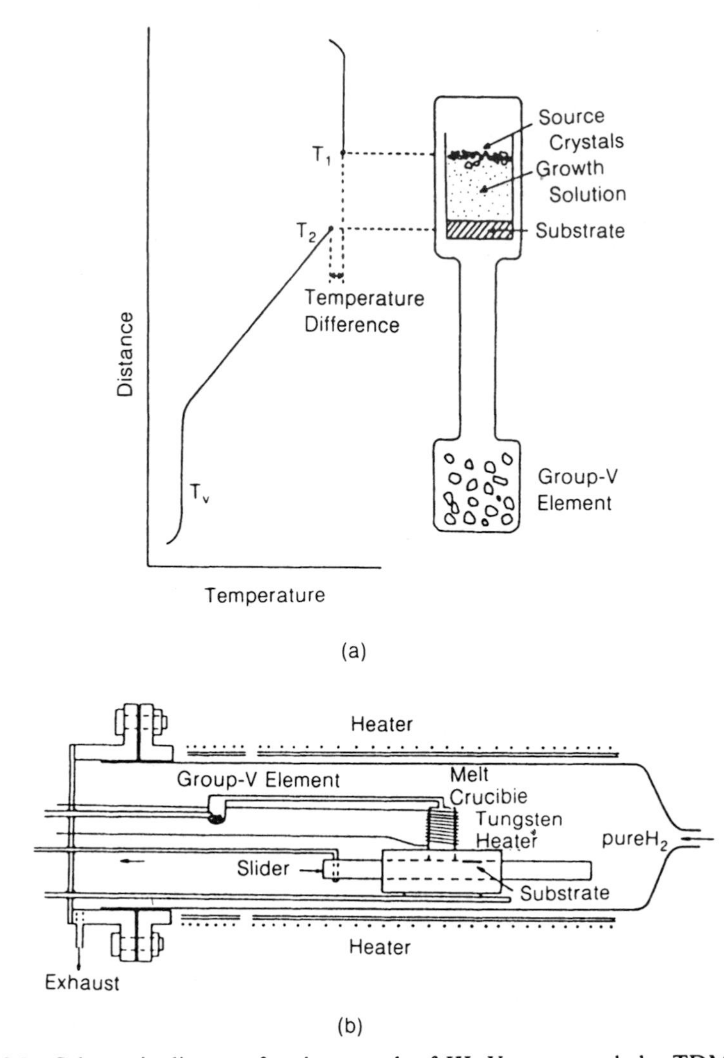

Fig. 13.3 Schematic diagram for the growth of III–V compounds by TDM CVP: (a) vertical closed system; (b) LPE apparatus in which pure hydrogen gas flows.

determines the growth speed, while the temperature of the zone containing the group V element determines the applied vapour pressure for stoichiometry control, independent of growth speed. In the case of III–V compounds, the vapour pressure of the group III element (Ga in the case of GaAs growth) is so low that we have only to consider the vapour pressure of the group V element. However, in the case of II–VI compounds, we must take into account the vapour pressures of both II and VI group elements as will be discussed later.

At first sight, no effects of the vapour pressure on the properties of the segregated crystals were expected because the saturation solubility in the melt was usually considered to be constant at a growth temperature. However, it was discovered that the crystal properties change as a function of the applied vapour pressure. Figure 13.4 shows the carrier density, lattice parameter and dislocation density in the grown GaAs epitaxial layers as a function of the applied arsenic vapour pressure. They showed the minima at the same arsenic vapour pressure, which we called the optimum vapour

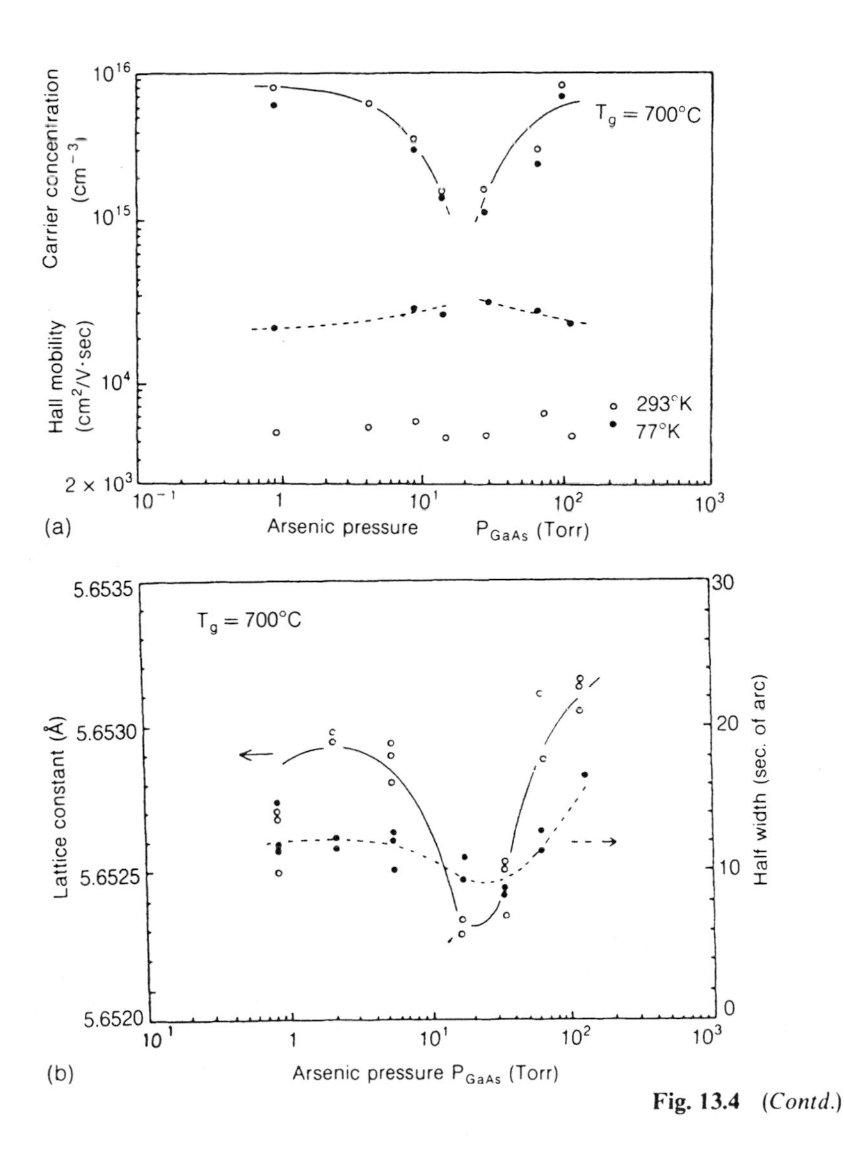

Fig. 13.4 (*Contd.*)

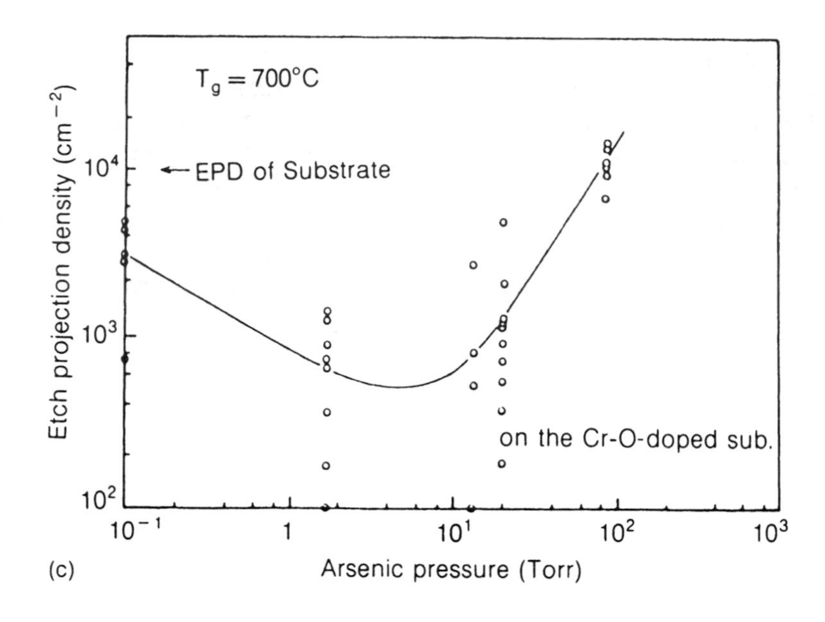

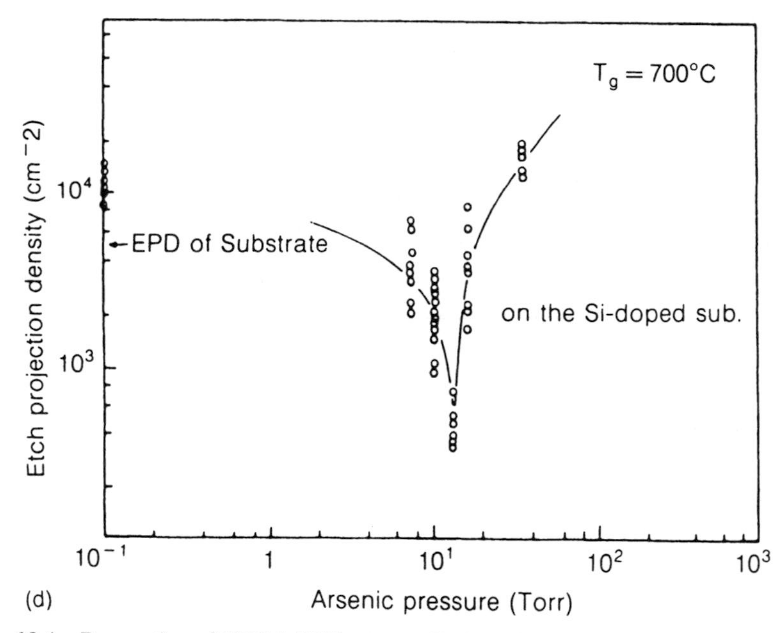

Fig. 13.4 Properties of TDM CVP grown GaAs epitaxial layers as a function of applied arsenic vapour pressure: (a) carrier concentrations and Hall mobilities, (b) lattice constants, (c) dislocation densities when Cr- and O-doped substrates are used, and (d) dislocation densities when Si-doped substrates are used.

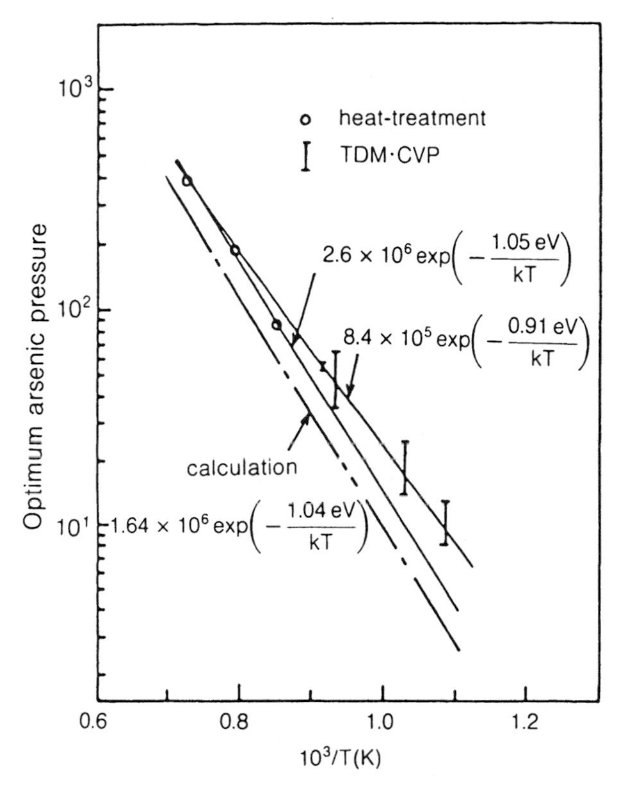

Fig. 13.5 Optimum arsenic vapour pressure for GaAs as a function of the inverse of the temperature.

pressure. Other than these parameters, deep-level densities found by photoluminescence and photocapacitance, densities of root-like faults as well as stacking faults showed minima at the optimum vapour pressure (only the lattice parameter of GaP showed a maximum at an optimum phosphorus vapour pressure).

The optimum vapour pressures for GaAs and GaP were measured as a function of temperature and found to be described as follows.

$$P_{\mathrm{GaAs,Opt}} = 2.6 \times 10^6 \exp(-1.05\,\mathrm{eV}/kT_g)\,\mathrm{Torr} \qquad (13.1)$$

$$P_{\mathrm{GaP,Opt}} = 4.67 \times 10^6 \exp(-1.01\,\mathrm{eV}/kT_g)\,\mathrm{Torr} \qquad (13.2)$$

The most important result was that the optimum vapour pressure coincided with the stoichiometric vapour pressure, P_{min}, found in the heat treatment experiment as shown in Fig. 13.5. Therefore, it was concluded that the vapour pressure applied over the solution affected the deviation from stoichiometry of the segregating crystals in such a way that they were in equilibrium with the applied vapour pressure. This effect was hard to understand if the saturation solubility of the solution was constant as usually assumed.

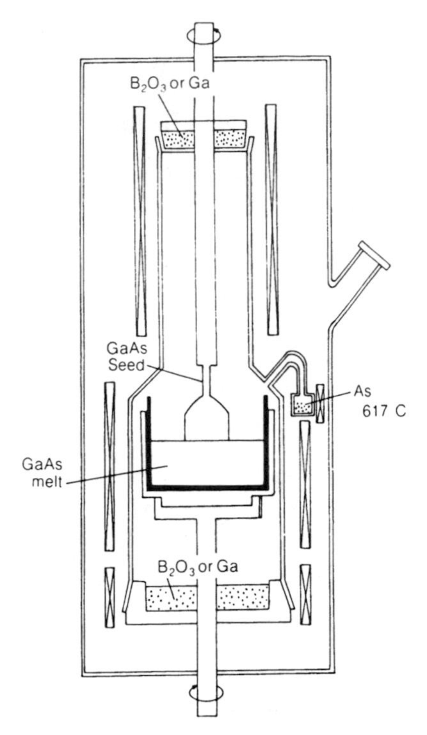

Fig. 13.6 Schematic diagram of the Czochralski method under controlled arsenic vapour pressure.

We have shown experimentally that the saturation solubility slightly changes as a function of the applied vapour pressure until it attains an equilibrium theoretically described as a three-phase equilibrium [4, 11].

$$\mu_{\mathrm{As}}^{\mathrm{gas}} = \mu_{\mathrm{As}}^{\mathrm{liquid}} = \mu_{\mathrm{As}}^{\mathrm{solid}} \tag{13.3}$$

$\mu_{\mathrm{As}}^{\mathrm{gas}}$ is the chemical potential in the gas phase and is described by only the applied vapour pressure and temperature as follows.

$$\mu_{\mathrm{As}}^{\mathrm{gas}} = \tfrac{1}{4} kT \ln P_{\mathrm{As_4,applied}} + f(T) \tag{13.4}$$

$\mu_{\mathrm{As}}^{\mathrm{liquid}}$ is the chemical potential of As in the liquid solution which depends on the change of the saturation solubility. $\mu_{\mathrm{As}}^{\mathrm{solid}}$ is the chemical potential of As in the solid phase which is a function of the deviation from stoichiometry of the segregating GaAs crystal. Subsequently, Ivaschenko [12] and we [5] actually calculated the optimum vapour pressure on the assumption of equation (13.3), and obtained a good agreement with the experimental optimum vapour pressure. The dashed line in Fig. 13.5 shows our calculation.

It was found that the dominant non-stoichiometric point defects were arsenic vacancies at low arsenic vapour pressures and arsenic interstitial atoms at high arsenic pressures [13, 14], the total amount of which reached a minimum at the optimum vapour pressure. The most striking result is that the dislocation density greatly decreased at the optimum vapour pressure (less than $10\,cm^{-2}$ [3] and $100\,cm^{-2}$ [14] for GaP and GaAs respectively) so that we obtained nearly perfect crystals of GaP and GaAs.

The TDM CVP crystal growth was extended to melt growth by Akai. He applied the optimum vapour pressure determined by equation (13.1) [15] to the three-zone horizontal Bridgman (HB) growth of GaAs [16]. The optimum vapour pressure at the melting point 1238 °C of GaAs becomes 813 Torr which corresponds to the temperature of arsenic chamber, 617 °C. Later, Parsey *et al.* more clearly showed that the dislocation density drastically decreased in the narrow temperature range $617\,°C \pm 2\,°C$ [17] in their Bridgeman growth experiment. The TDM CVP was also applied to the Czochralski method by Tomizawa *et al.* with the apparatus illustrated in Fig. 13.6 [18]. The arsenic pressure was controlled by the temperature of the As deposition zone T_{As}. The temperature gradient just above the melt surface was made to be about $10\,°C\,cm^{-1}$. This value was much lower than the case of the conventional liquid encapsulated Czochralski (LEC) method. As a result, the etch-pit densities (EPDs) were as low as $2 \times 10^3\,cm^{-2}$, which is lower than EPDs in the conventional LEC method by a factor of about 10. The lowest EPD was obtained at $T_{As} = 616\,°C$, the same as in HB growth.

From all these results, it can be said that the vapour pressure control of crystal growth—TDM CVP—is effective for growing very-high-quality compound crystals with exact stoichiometry and the least dislocation and other defect densities.

13.3 GROWTH OF CONDUCTIVITY-CONTROLLED II–VI COMPOUND CRYSTALS BY THE TEMPERATURE DIFFERENCE METHOD UNDER CONTROLLED VAPOUR PRESSURE

The fundamental procedure for the stoichiometry-controlled growth method in II–VI compounds is the same as in III–V compounds as shown in Fig. 13.7. However, we must take into account the fact that not only the vapour pressures of the group V elements Se or S but also the vapour pressures of the group II elements Zn or Cd are very high at growth temperatures as shown in Fig. 13.8. This is different from the situation in III–V compounds for which the vapour pressures of the group V elements are very high but those of the group III elements are so small that we can usually neglect their effects.

We must also take into account the solubilities of II–VI compounds in the solvents of the group II or VI elements. The solubility of ZnSe in the Zn solvent is much lower than that in the Se solvent. Similar situations also occur in the case of ZnTe, ZnS, CdS and CdSe. Therefore, p-type ZnSe crystal

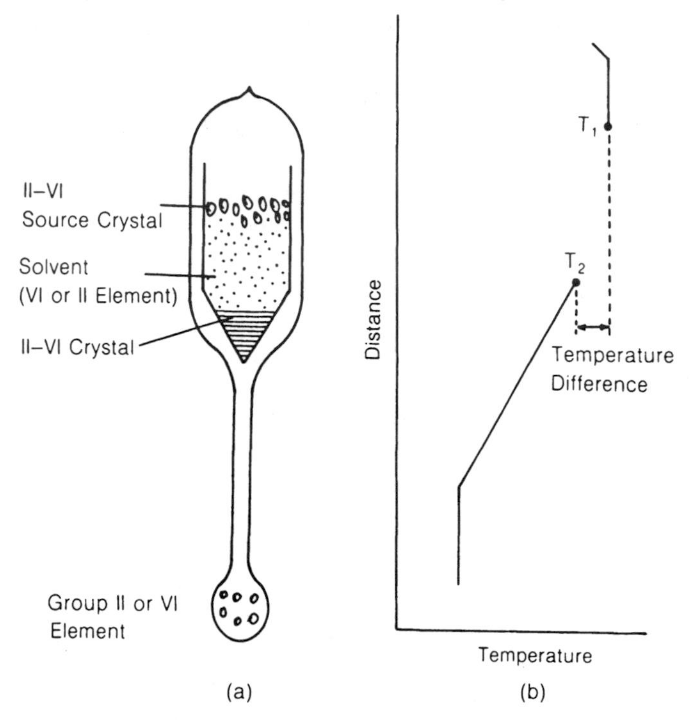

Fig. 13.7 (a) Schematic diagram of the growth system for II–VI compounds by TDM CVP and (b) the temperature distribution of the furnace.

growth by TDM CVP was first made by using Se solvent and the Zn vapour pressure was applied on the top surface of the solution from elemental Zn placed in the vapour pressure controlling zone [6, 7]. The growth temperature was 1050 °C, which was much lower than the maximum melting point of ZnSe, 1529 °C, while the Zn zone temperature was chosen to be various different values. Li was incorporated as an acceptor impurity in the range 5×10^{-3}–5×10^{-2} mol% into the selenium solution. Figure 13.9 shows the sliced p-type ZnSe crystals.

The conductivities and carrier concentrations of p-type ZnSe crystals depended on the applied Zn vapour pressure as shown in Fig. 13.10. Hole concentration showed a minimum at a Zn vapour pressure of 7.2 atm [7, 8]. Also, X-ray double-crystal rocking curve measurements were made and the half-width was found to show the lowest value (15 seconds of arc) at the same Zn vapour pressure (Fig. 13.11). This value is comparable with those of high-quality GaAs layers grown by TDM CVP.

ZnSe crystals thus grown usually show the near-edge emission around 460 nm and deep-level luminescence in the green or orange band. Their relative intensities are shown as a function of the Zn vapour pressure in

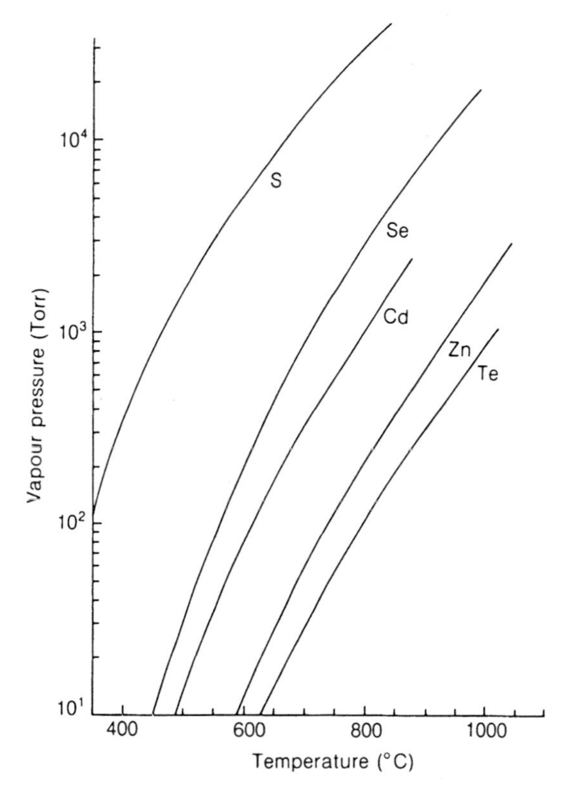

Fig. 13.8 Vapour pressure of several group II and VI elements as a function of temperature.

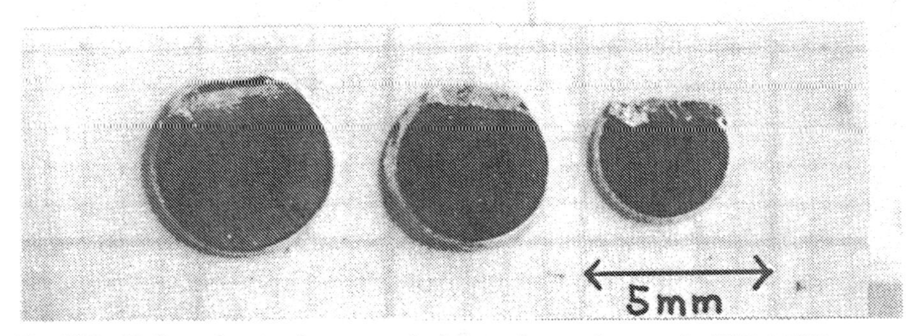

Fig. 13.9 ZnSe wafers cut from a conical-shaped crystal grown by TDM CVP.

Fig. 13.12. It is shown that the deep-level luminescences also reach minima at the same optimum vapour pressure, while the near-edge emission becomes strongest. Figure 13.13 shows an example of the near-edge emission. From these results it is considered that nearly stoichiometric ZnSe cyrstals with p-type conductivity were grown from selenium solution supplied with

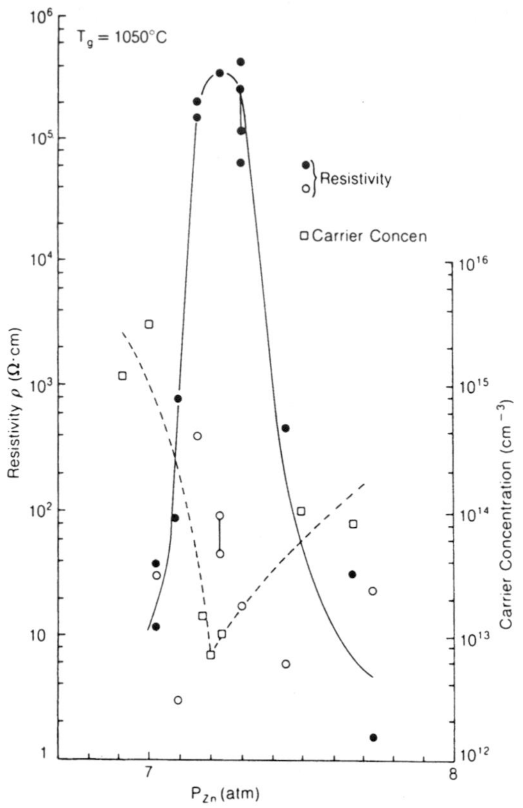

Fig. 13.10 Resistivities and carrier concentrations of p-type ZnSe crystals as a function of applied Zn vapour pressure.

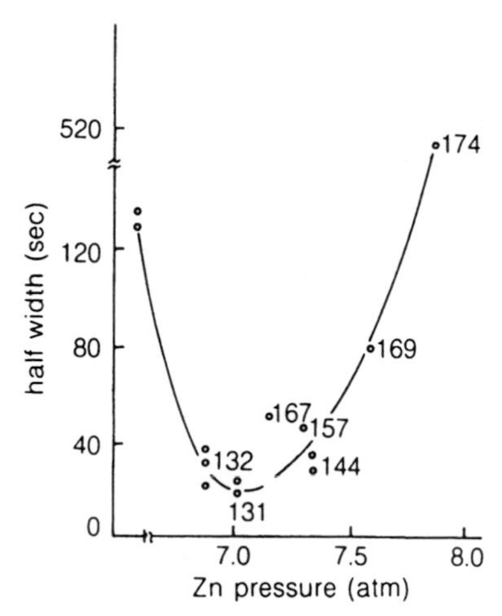

Fig. 13.11 Half-width of X-ray rocking curves as a function of applied Zn vapour pressure.

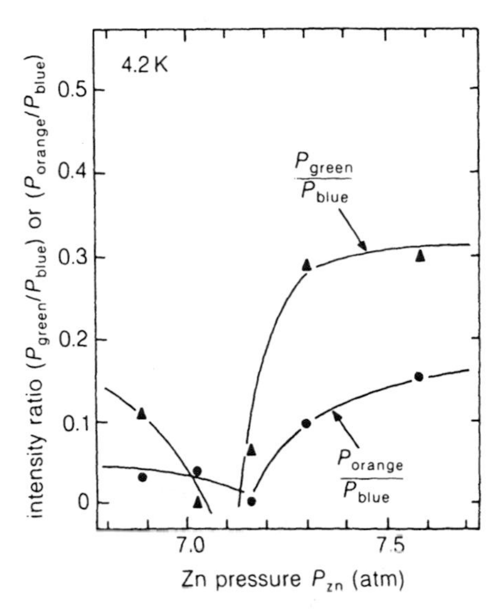

Fig. 13.12 Intensity ratios, P_{green}/P_{blue} and P_{orange}/P_{blue}, as a function of Zn vapour pressure.

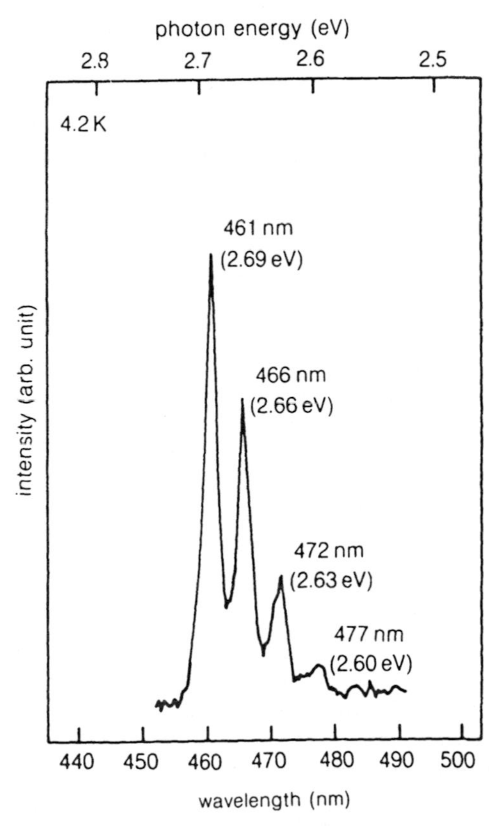

Fig. 13.13 Photoluminescence spectrum of TDM CVP-grown ZnSe at 4.2 K.

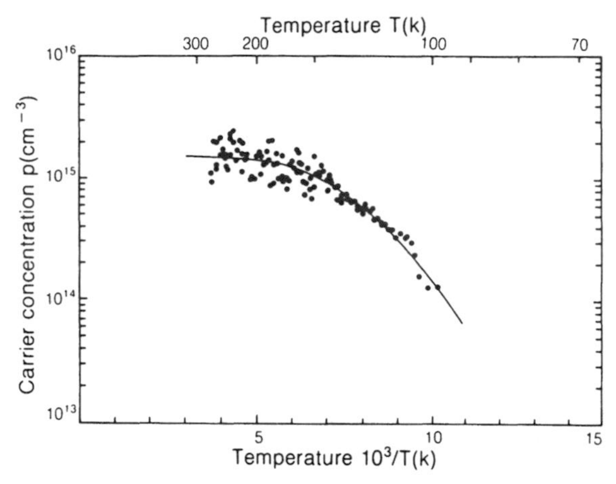

Fig. 13.14 Temperature dependence of the carrier concentration of p-type ZnSe.

the optimum Zn vapour pressure. In this case, however, the high vapour pressure of selenium from Se solution itself is more important for the successful growth of p-type ZnSe crystals.

As for the acceptor impurities in ZnSe crystals, shallow levels had not been known until the p-type crystals became available. Figure 13.14 shows the first experimental result on the shallow levels of Li acceptors obtained by Hall measurement in the TDM CVP-grown ZnSe. The ionization energy E_A was found to be 70–85 meV. Other impurities such as Na were also tried as later described.

13.4 ZnSe p–n JUNCTIONS EMITTING BLUE LIGHT

13.4.1 Fabrication of the p–n junction

The p–n junctions were produced by impurity diffusion under controlled Se vapour pressure on the TDM CVP-grown p-type ZnSe crystals [7]. Li-doped p-type ZnSe crystals grown under the optimum Zn vapour pressure had a hole concentration $p = 6.4 \times 10^{12}\,\mathrm{cm}^{-3}$ and mobility $\mu_p \sim 78\,\mathrm{cm}^2\,\mathrm{V}^{-1}\,\mathrm{s}^{-1}$. Although the hole concentration in p-type ZnSe grown at different Zn vapour pressures increases, for example, to $p \sim 3 \times 10^{15}\,\mathrm{cm}^{-3}$, the mobility decreases to $20\,\mathrm{cm}^2\,\mathrm{V}^{-1}\,\mathrm{s}^{-1}$, and deep level luminescences were found in their p–n junction luminescences as later described.

The As-grown crystal was sliced into sections of 1 mm thickness, mechanically polished, and etched by a Br–methanol solution. Then, impurity diffusion was carried out in an evacuated quartz tube having two zones. In the diffusion zone, the slices are placed in a Zn solution containing a small amount of Ga, while Se was put in the vapour pressure zone. The diffusion of Ga was carried out at 740 °C for 1 h at various different Se vapour pressures. As a result, n-type layers with thicknesses of 5–7 μm were obtained.

13.4.2 p–n junction characteristics

The p–n junction characteristics of the fabricated diodes were examined from their I–V chracteristics. The I–V characteristics can be described as

$$I = I_0\left[\exp\left(\frac{qV}{nkT}\right) - 1\right]$$

The n values of the fabricated diodes ranged from 1.4 to 1.8, which means that recombination in the vicinity of the junction interface is significant. In order to make clear whether a p–n junction or a metal–insulator–semiconductor structure is formed, the diffusion potential ϕ of the junction should be obtained. This was done by the following two methods. The I–V characteristics of the p–n junction can be described as

$$I = \frac{qD_n}{W} n \exp\left(\frac{\phi}{kT}\right)\left[\exp\left(\frac{qV_a}{kT}\right) - 1\right] \sim \frac{qD_n}{W} n \exp\left(-\frac{\phi}{kT}\right)\left(1 + \left|\frac{qV_a}{kT}\right| + \cdots - 1\right)$$

when the applied voltage V_a is small. Therefore, the resistance $R_0 = \mathrm{d}V/\mathrm{d}I$ at low applied voltage can be given by

$$R_0 = \frac{W}{AqD_n n} \exp\left(\frac{\phi}{kT}\right)$$

Figure 13.15 shows the measured R_0 as a function of temperature. Its slope gives $\phi = 2.59$ eV, which is close to the bandgap energy of ZnSe, 2.7 eV (room-temperature value).

Another method is to measure the capacitance–voltage (C–V) characteristics. The diffusion potential was found to be 2.5–2.7 eV from plotting $1/C^2$ versus V as shown in Fig. 13.16. Therefore, we can conclude that a p–n junction of ZnSe was really formed.

13.4.3 Emission spectra and their vapour pressure dependence

As described in section 13.3, the near-edge emission (blue light emission) from the bulk p-type crystal is strongest when the optimum vapour pressure (7.2 atm when $T_G = 1050$ °C) is applied. Figure 13.17 shows typical injection luminescence spectra at different temperatures for a diode fabricated from a crystal grown under a Zn vapour pressure deviating from the optimum pressure. The deep-level luminescences increase with increasing temperature, while the near-edge emission decreases. Therefore, the colour of the diode

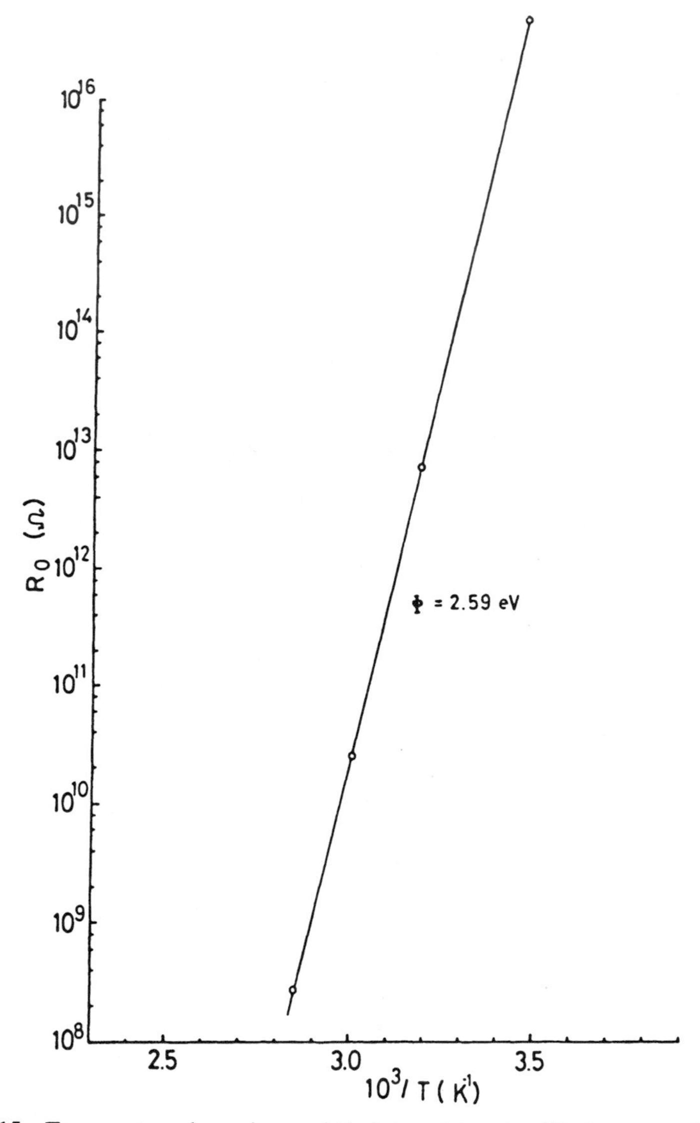

Fig. 13.15 Temperature dependence of R_0 determining the difusion potential, ϕ, of a ZnSe p–n junction.

changes from violet to orange as the temperature increases. Figure 13.18 shows the photon energies of near-edge emission and deep-level emission as a function of temperature. The temperature dependence for the near-edge emission nearly follows the temperature dependence of the bandgap energy, while three different deep-level luminescences show different behaviours. On the other hand, Figure 13.19 shows the intensity ratio $P_{\mathrm{Deep}}/P_{\mathrm{Edge}}$ at 77 K for

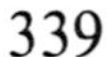

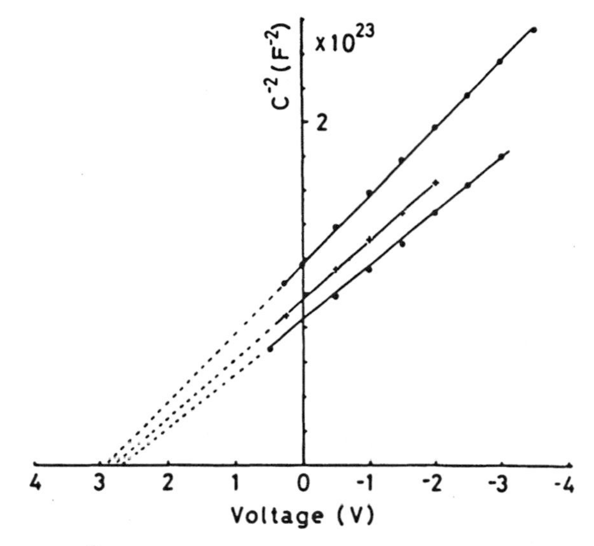

Fig. 13.16 C^{-2} as a function of applied voltage for ZnSe p–n junctions.

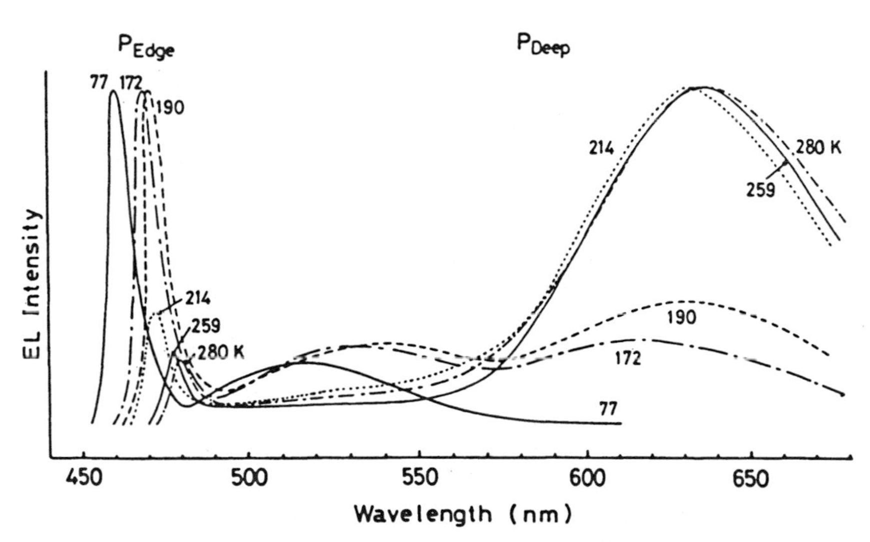

Fig. 13.17 Emission spectra of a ZnSe p–n junction at different temperatures.

various diodes as a function of Zn vapour pressure applied when the substrate crystal is grown. The result is similar to the photoluminescence measurement for the bulk crystals (cf. Fig. 13.12).

As shown in Fig. 13.19(b), the edge emission itself shows a maximum at the optimum vapour pressure. Therefore, it is clear that the applied Zn

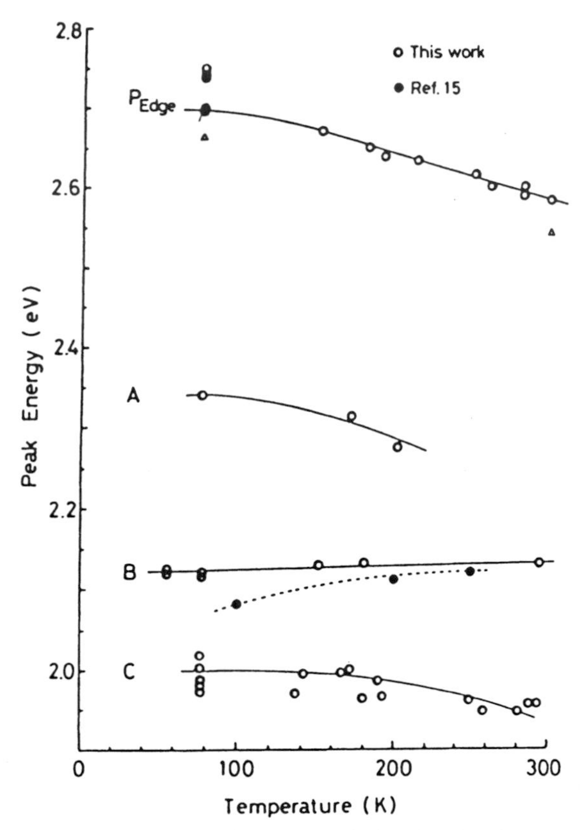

Fig. 13.18 Peak photon energies of several emission bands of ZnSe p–n junctions.

vapour pressure suppresses the generation of deep-level centres which should be related to non-stoichiometry, and as a result, enhances the intrinsic recombinations near the bandgap energy.

However, these considerations are not sufficient for obtaining room-temperature blue light emission. In the fabrication processes, defects are formed and deep-level luminescences are still observed at room temperature even for crystals grown under optimum Zn pressure. The temperature behaviours of the spectra are similar to that shown in Fig. 13.17. Therefore, an Se vapour pressure is applied during the diffusion process using the two-zone quartz tube described previously. Figure 13.20 shows P_{Deep}/P_{Edge} and P_{Edge}, measured at room temperature, as a function of applied Se vapour pressure for diodes fabricated from a crystal grown under the optimum Zn vapour pressure. In this diffusion process, the diffusion temperature was 740 °C and the optimum Se pressure was found to be about 200 Torr.

As a result, a pure blue light emission without any deep-level emission was obtained as shown in Fig. 13.21. The peak wavelength was 480 nm and its half-width was 7 nm at 300 K. These diodes had a brightness of 2 mcd at a current of 2 mA (Fig. 13.22).

13.5 LIQUID PHASE EPITAXY

In an earlier section, p–n junction fabrication on a bulk crystal grown at 1050 °C by the impurity diffusion technique was described. Liquid phase epitaxial growth of p–n junctions will be more promising, because it can lower the growth temperature so that the concentration of equilibrium

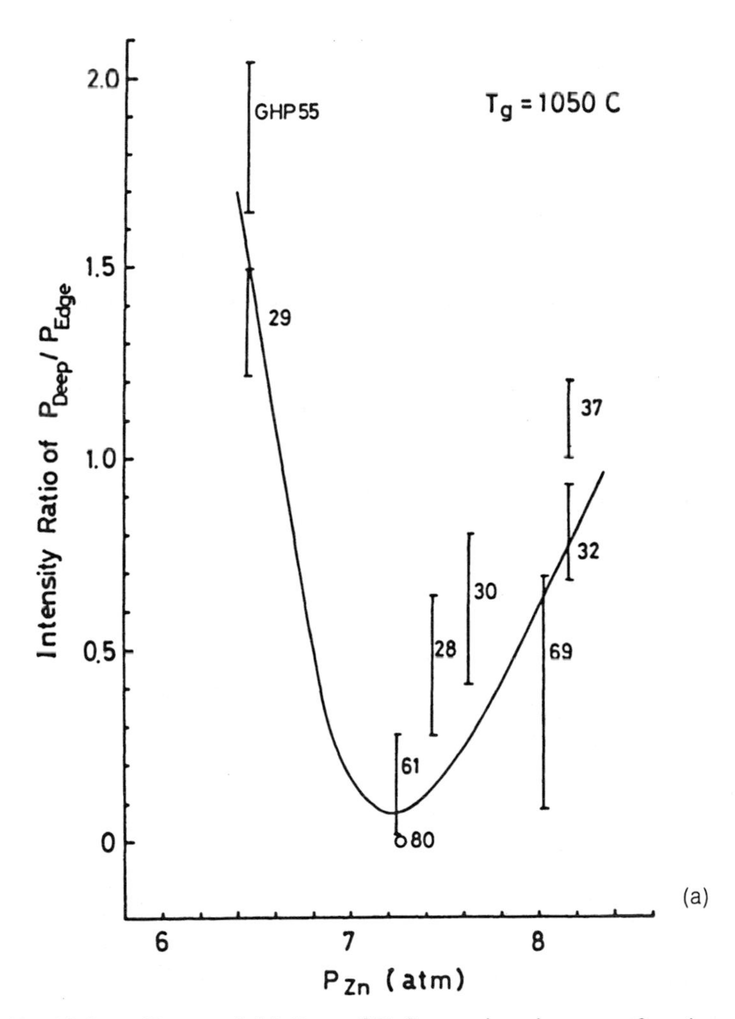

Fig. 13.19 (a) P_{Deep}/P_{Edge} and (b) P_{Edge} of ZnSe p–n junctions as a function of the Zn vapour pressure for substrate growth.

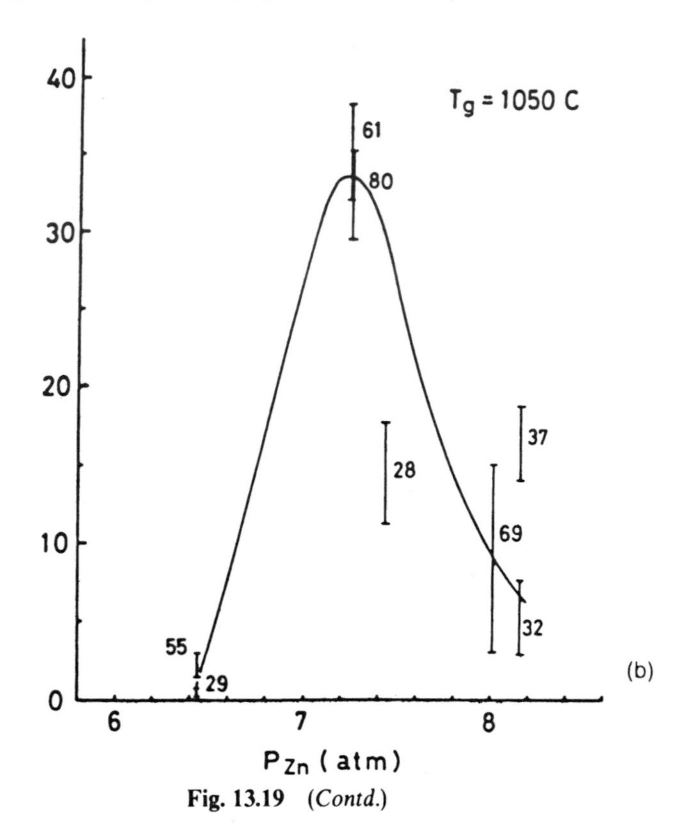

Fig. 13.19 (*Contd.*)

non-stoichiometric defects will be decreased. Also, the defect generation inevitable for diffusion processes can be avoided.

However, there are several problems for vapour pressure controlled liquid phase epitaxy in II–VI compounds. The Se solution has so high a vapour pressure that a closed-tube system must be used. Figure 13.23 illustrates the closed-tube liquid phase epitaxial growth system with and without Zn vapour pressure application. The specific gravity of ZnSe crystals is larger than that of elemental Se (5.28 g cm^{-3} and 4.80 g cm^{-3} respectively). Therefore, the quartz ampoule is provided with a narrow neck portion at a higher level of the Se solution in order to prevent the source ZnSe crystals dropping down to the bottom of the solution . In the case of III–V compound crystals, such a consideration is not necessary. In the present experiment, the ampoule has an 8 mm inner diameter and a 50–80 mm length. The growth temperature T_G is in the range from 950 °C to 650 °C, the growth times is 2 h and 10–20 μm single epitaxial layers are grown. ZnSe substrate crystals for epitaxial growth are those grown by TDM CVP as described in section 13.2.

When $T_G = 950$ °C, smooth single-crystal epitaxial layers are grown but, when T_G is in the range 850–650 °C, the grown layers have a worse surface

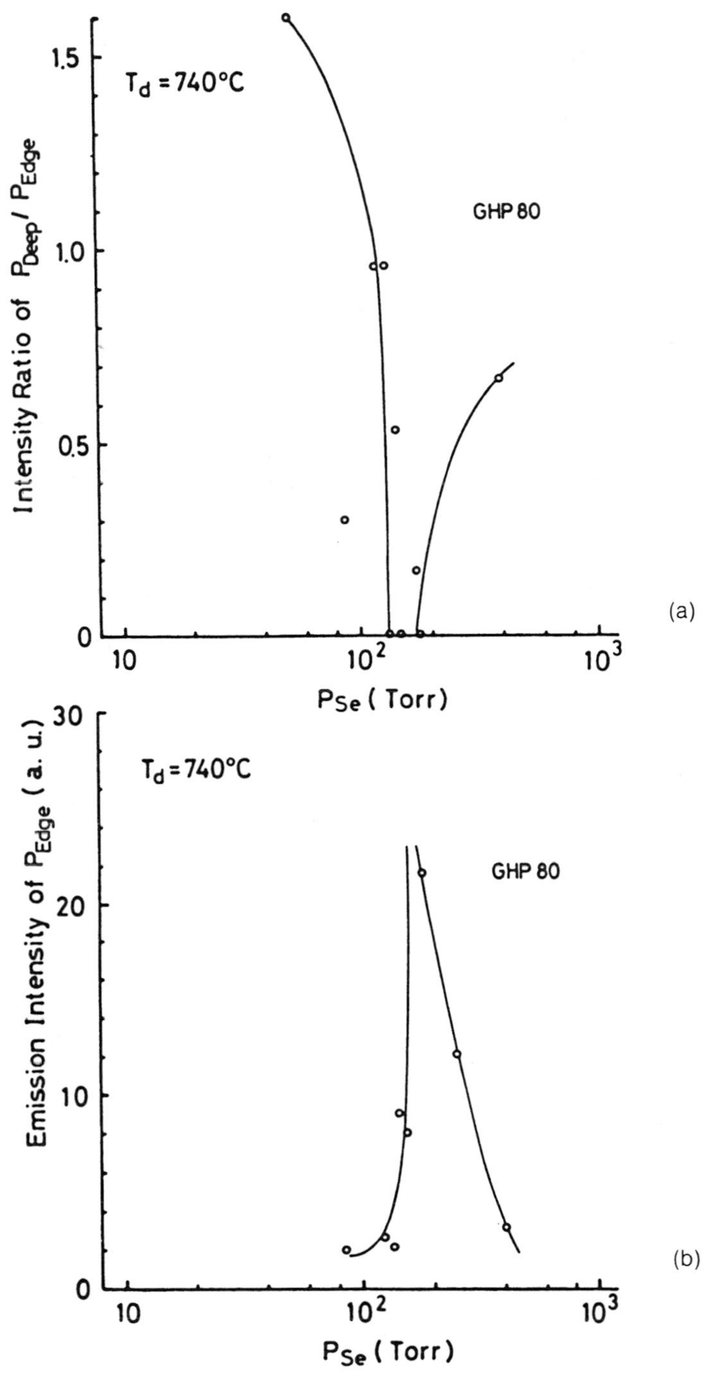

Fig. 13.20 (a) P_{Deep}/P_{Edge} and (b) P_{Edge} of ZnSe p–n junctions as a function of applied Se vapour pressure.

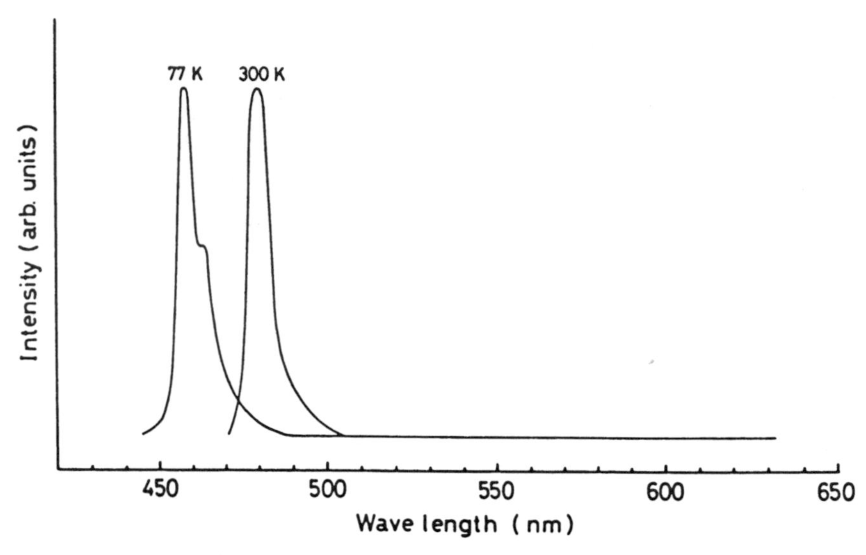

Fig. 13.21 Emission spectra of a ZnSe p–n junction fabricated under optimum vapour pressure conditions.

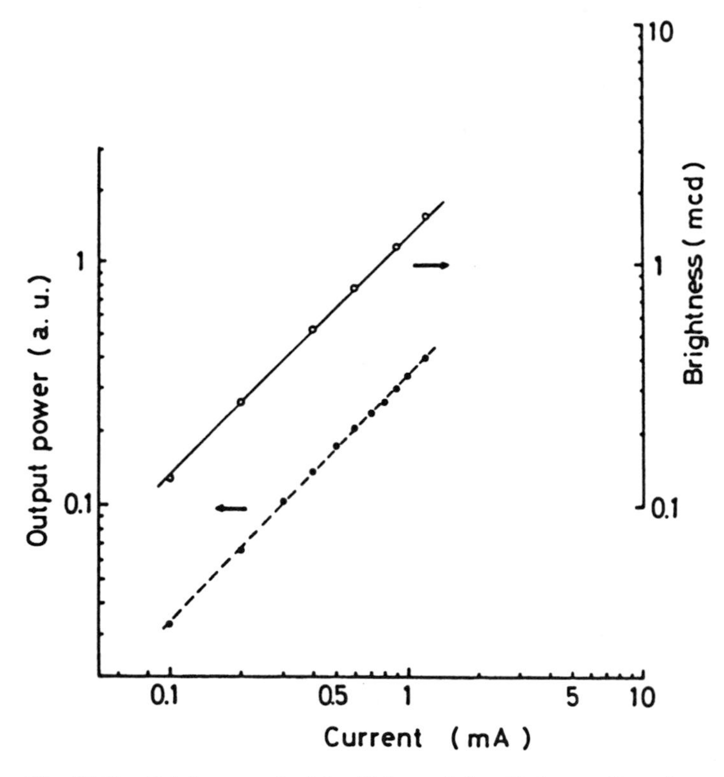

Fig. 13.22 Brightness of a blue-light-emitting ZnSe p–n junction.

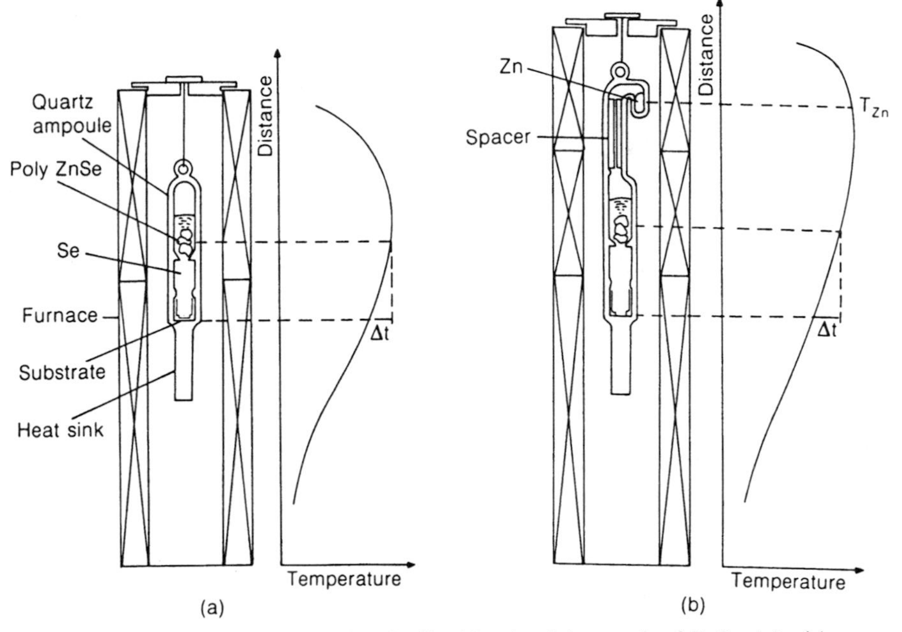

Fig. 13.23 Experimental set-up for the liquid epitaxial growth of ZnSe: (a) without and (b) with applied Zn vapour pressure.

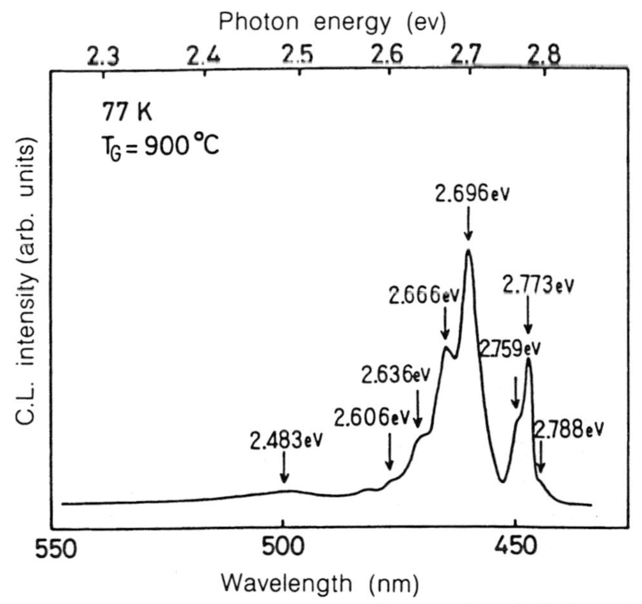

Fig. 13.24 Cathode luminescence spectrum of an undoped ZnSe epitaxial layer grown at 900 °C.

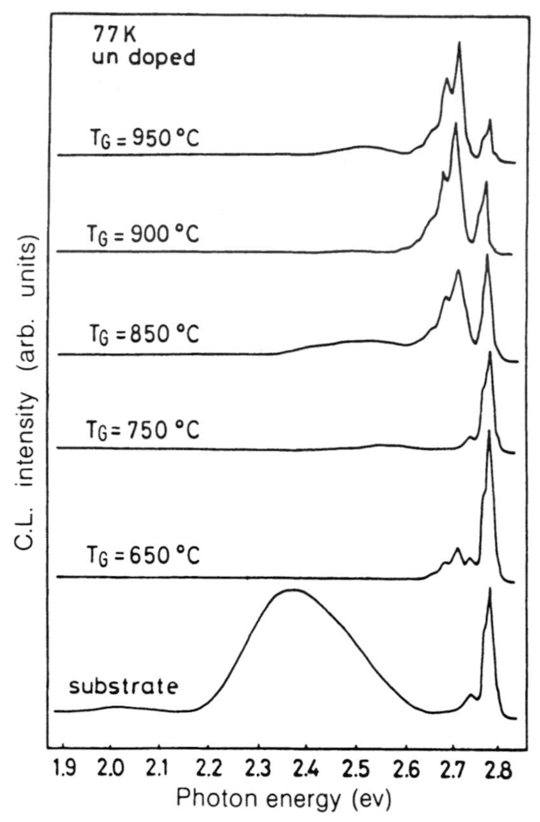

Fig. 13.25 Cathode luminescence spectra of undoped ZnSe epitaxial layers grown at various difference temperatures.

morphology, tending to consist of islands with selenium inclusions between them, although X-ray double-crystal rocking curve measurement has shown that they are single epitaxial layers. This result is due to the much smaller solubility of ZnSe in Se solution at lower temperatures together with a smaller surface migration distance.

Figure 13.24 shows the cathode luminescence (CL) spectrum at 77 K of an undoped epitaxial layer without Zn application grown at 900 °C. The edge emission consists of different kinds of exciton (Ex) recombinations around 2.773 eV and donor–acceptor (DA) pair recombination at 2.696 eV and its phonon replicas with a 30 meV interval. A deep-level emission peak at 2.48 eV is observed, although it is weak. With changing growth temperature, the spectral behaviour changes as shown in Fig. 13.25. At higher temperatures, 950 °C and 900 °C, the spectra are similar. On the other hand, at lower temperatures, Ex increases while DA decreases. Moreover, no deep-level emission is observed at $T_G = 650\,°C$. Therefore, it is considered that epitaxial layers grown at 950–900 °C contain more deep levels and impurities compared with those grown at lower temperatures although the surface morphologies are better, unless Zn vapour pressure is not applied. Therefore, epitaxial growth with application of Zn vapour pressure has been performed by the system shown in Fig. 13.23(b). Figure 13.26 shows the CL spectral behaviour as a function of applied Zn vapour pressure. It is seen that the Ex intensity and the relative intensity Ex/DA both show a maximum at $P_{Zn} = 3.0$ atm for an epitaxial growth temperature of 950 °C. This should be the stoichiometric Zn vapour pressure when Se solvent is used. Although

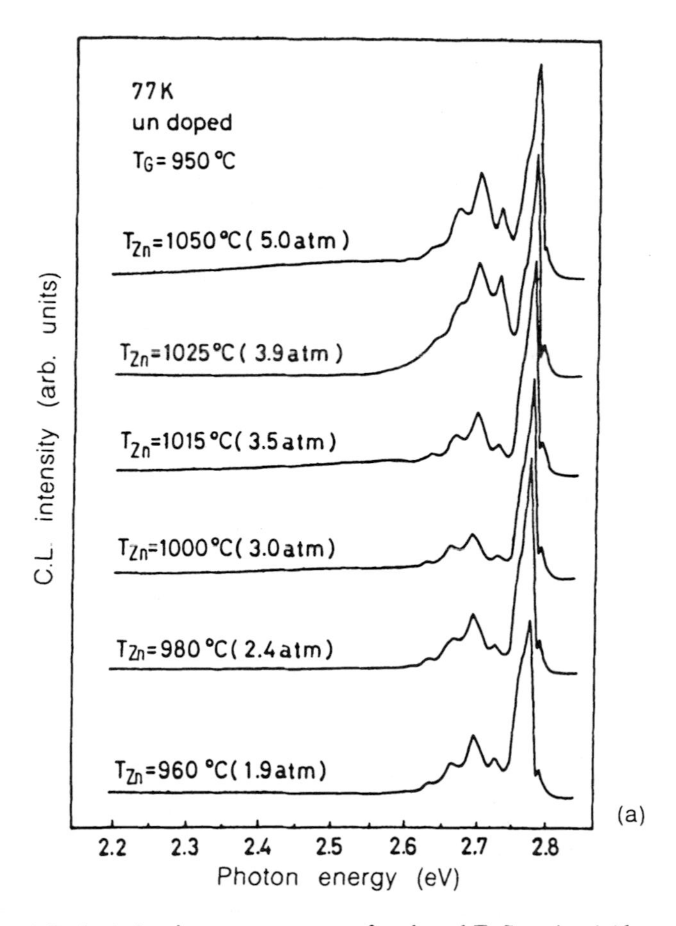

Fig. 13.26 (a) Cathode luminescence spectra of undoped ZnSe epitaxial layers grown under various different Zn vapour pressures. (b) Intensity ratio Ex/DA as a function of Zn vapour pressure.

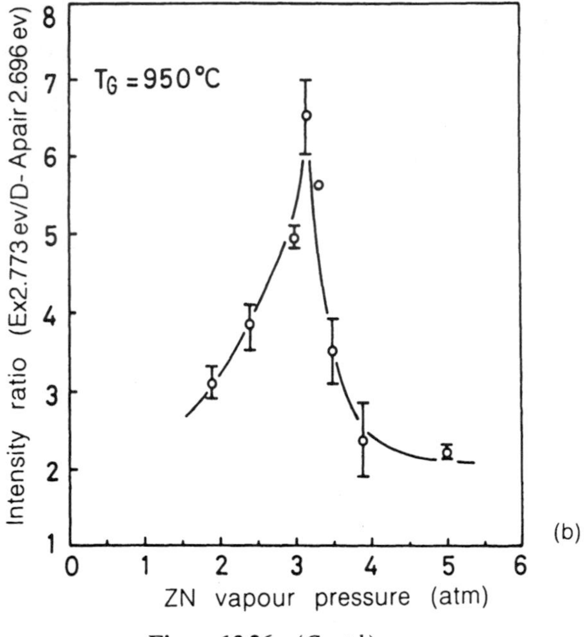

Figure 13.26 (*Contd.*)

the optimum Zn vapour pressure has not yet been obtained at temperatures other than $T_G = 1050\,°C$ ($P_{Zn} = 7.12$ atm) at 950 °C ($P_{Zn} = 3.0$ atm), we assume the following temprature dependence:

$$P_{Zn}^{Opt} = A\exp\left(-\frac{E}{kT}\right)\text{atm}$$

Then, we obtain $E = 1.2$ eV. This value is not so significantly different from the values in III–V compounds (see section 13.2).

As a next step, we have performed the epitaxial growth of p-type ZnSe layers by incorporating several different kinds of dopants, Au, Na_2Se, Li_2Se or $Li_2SO_4 \cdot H_2O$. Figure 13.27 compares the CL spectra both with and without application of the Zn optimum vapour pressure for doped layers. In the case of Na- and Li-doped epitaxial layers, the effect of Zn vapour pressure is to increase Ex and to decrease A.

While the layers grown without Zn vapour pressure with the doped acceptors have become semi-insulating, the layers with the applied Zn vapour pressures showed p-type conductivity. The hole concentration ranged from 8×10^{13} to $3 \times 10^{17}\,cm^{-3}$ as summarized in Table 13.2. The highest hole concentration ($3 \times 10^{17}\,cm^{-3}$) was obtained for epitaxial layers incorporated with Na_2Se at a level of 2.5 mol% or less in the Se solution. At present, the doping level is not yet well controlled. Also, we must develop double-layer

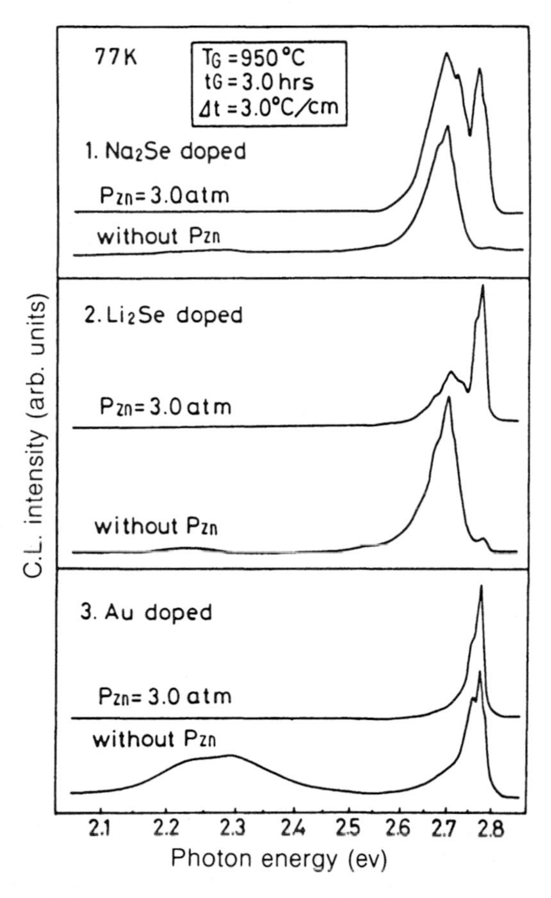

Fig. 13.27 Cathode luminescence spectra of ZnSe epitaxial layers doped with acceptor impurities.

Table 13.2 Carrier concentration and Hall mobilities of p-type ZnSe epitaxial layers

Dopant	*Dopant concentration in Se solution (mol.%)*	*Type*	*Carrier concentration (cm^{-3})*	*Hall mobility ($cm^2\,V^{-1}\,s^{-1}$)*	*Resistivity ($\Omega\,cm$)*
Na_2Se	$(1\text{–}3) \times 10^2$	p	$2 \times 10^{16}\text{–}3 \times 10^{17}$	4–90	2–90
$Li_2SO_4 \cdot H_2O$	-3×10^2	p	-6×10^{15}	–300	60
Li_2Se	-3×10^2	p	-8×10^{13}	450	3

$T_G = 950\,°C$, $P_{Zn} = 3.0$ atm, room temperature resistivities.

epitaxy for obtaining liquid phase epitaxial p–n junctions. However, we have shown that p-type ZnSe epitaxy became practical by TDM CVP.

13.6 CONCLUSION

After long efforts to obtain conductivity-controlled II–VI compounds and their p–n junctions by various conventional growth methods, the TDM CVP, which had shown drastic effects for stoichiometry control in III–V compounds, was first successfully applied to the growth of p-type ZnSe and blue-light-emitting diodes of ZnSe p–n junctions. Because the p–n junction is the key device for various applications of semiconductors, this result should be one of the great advances in II–VI compound semiconductor technology, and also for the fundamental understanding of the properties of II–VI compounds. A blue LED has long been required as the last of the three primary colours. Because ZnSe and other widegap II–VI compounds have a direct bandgap structure, a laser diode with blue light will also not be impractical. Not only a blue LED but also green to ultraviolet LEDs and laser diodes are targets worth developing in the future.

REFERENCES

1. Nishizawa, J., Shinozaki, S. and Ishida, K. (1973) *J. Appl. Phys.*, **14**, 1638.
2. Nishizawa, J. and Okuno, Y. (1975) *IEEE Trans. Electron Devices*, **22**, 716.
3. Nishizawa, J., Okuno, Y. and Tadano, H. (1975) *J. Cryst. Growth*, **31**, 215.
4. Nishizawa, J. and Okuno Y. (1980) *Optoelectronics* (ed. M. Herman), PWN-Polish Scientific Publishers, Chapter 5, pp. 101–130.
5. Nishizawa, J., Okuno, Y. and Suto, K. (1986) In *JARECT 19, Semiconductor Technologies* (ed. J. Nishizawa), OHM and North-Holland, pp. 17–80.
6. Nishizawa, J., Itoh, K. and Okuno, Y. (1983) In *Abstracts Conf. Berg- une Hüttenmannisher Tag, Freiberg, June* 1983.
7. Nishizawa, J., Itoh, K., Okuno, Y. and Sakurai, F. (1985) *J. Appl. Phys.*, **57** (6), 2210.
8. Nishizawa, J., Suzuki, R. and Okuno, Y. (1986) *J. Appl. Phys.*, **59** (6), 2256.
9. Nishizawa, J. and Okuno, Y. (1986) In *JARECT Semiconductor Technologies* (ed. J. Nishizawa), OHM and North-Holland, pp. 325–326.
10. Nishizawa, J., Otsuka, H., Yamakoshi, S., and Ishida, K. (1974) *Jpn. J. Appl. Phys.*, **13**, 46.
11. Nishizawa, J., Kobayashi, K. and Okuno, Y. (1980) *Jpn. J. Appl. Phys.*, **19**, 345.
12. Ivaschenko, A.I. (1983) In *Conf. Berg- und Hüttenmannisher Tag, Freiberg, 1983.*
13. Nishizawa, J., Toyama, N. and Oyama, Y. (1983) *Optoelectronic Materials and Devices* (ed. M. Herman), PWN–Polish Scientific Publishers, pp. 27–77.
14. Nishizawa, J., Shiota, I. and Oyama, Y. (1986) *J. Phys. D*, **19**, 1073.
15. Akai (1975) Private communication.
16. Suzuki, T. and Akai, S. (1971) *Bussei*, **12**, 144 (in Japanese).
17. Parsey, J.M., Jr., Nanishi, Y., Laogwski, J. and Gatos, H.C. (1981) *J. Electrochem. Soc.*, **128**, 937.
18. Tomizawa, K., Sassa, K., Shimauki, Y. and Nishizawa (1984) *J. Electrochem. Soc.*, **131**, 2394.

14

Optical non-linearities and bistability in II–VI materials

A. Miller

14.1 INTRODUCTION

This chapter reviews recent progress in the study of optical non-linearities in widegap II–VI semiconductors for applications in optical bistability and all-optical circuits. Semiconductors make appropriate materials for developing digital optical elements because of the existence of sensitive non-linear optical responses, the possibilities offered by novel structures and low dimensions, their state of advanced technological development, and compatibility with existing electronics technologies. The first observations of optical bistability in semiconductors in 1978 employing ZnS [1] and widegap II–VI materials have continued to play a major role in the development of this subject [2–6].

It is already common to superimpose information or data onto a laser beam for the communication, storage, scanning and display of information. These are now mature technologies with large commercial markets in components and systems. In these applications, data are converted back and forth between optical and electrical form to allow processing of the data to be carried out electronically. A challenge for optics is to extend its role to the processing stage so that some of the unique properties of laser light such as high bandwidth, coherence, and parallel interconnect capabilities can be fully exploited for higher-performance systems. Digital optical processors have been a primary motivation but light also lends itself to neural computing which depends on large numbers of interconnects and simple non-linear processing units. There is therefore a strong incentive to develop optical components which allow the control of one light beam by another and perform equivalent functions to transistors and other electronic devices.

The interaction of two optical beams requires a material with a non-linear optical response. In free space, photons do not interact with each other and, in most materials, the interaction is only very weak (i.e. the non-linear optical coefficients are small; however, the interaction can be greatly enhanced by exploiting resonances between the frequency of the light and electronic transitions in the medium. A non-linear optical component which has been studied extensively in recent years is the optically bistable device which offers such functions as logic, storage and gain at relatively low optical powers when resonant interactions are employed.

Optical bistability is a phenomenon which can be viewed from a range of perspectives from the fundamentals of quantum optics to the provision of elements for future optical computers. Stated in simple terms, an optically bistable system is one in which two stable transmission (or reflection) states are possible for a given input optical power. There are many subtle facets to optical bistability. When viewed as a discontinuous non-equilibrium phase transition, fundamental aspects include photon statistics, critical slowing down, instabilities and chaos, but the principal driving force behind research into optical bistability has been the provision of all-optical devices which exhibit memory and logic for digital optical processing. As a bonus, the prospect of optical computing using optically bistable switches has stimulated a great deal of non-linear optical materials research. Indeed, being a critical phenomenon, predicting the precise conditions for optical bistability can provide a severe test for theories of light–matter interaction.

There are two basic requirements for a system to yield optical bistability. One is an optical non-linearity and the other is some form of feedback. In this chapter, we first discuss some of the different forms of optical nonlinearities observed in widegap II–VI semiconductors in section 14.2, then describe two forms of optical bistability in section 14.3. Examples of optical bistability in II–VIs are given in sections 14.4 and 14.5 with an emphasis on II–VI non-linear optical interference filters which have proved to be useful devices for exploring architectures for digital optical computing. Section 14.6 gives some examples of the use of these devices in all-optical circuits. This chapter does not attempt to provide a comprehensive review of research into the rich and diverse non-linear optical phenomena observed in II–VI materials. Rather, the emphasis is on providing a perspective on progress towards understanding and harnessing optical non-linearities in II–VIs for optically bistable and associated devices.

14.2 OPTICAL NON-LINEARITIES

Many types of non-linear optical phenomena have been studied in semiconductors. Here we are interested only in processes which cause significant changes of absorption and refractive index by optical excitation. Three categories of non-linear absorption–refraction may be distinguished. These fall within different regimes of response time and sensitivity which therefore determine their potential application.

The fastest non-linear optical interactions occur via bound electrons. These processes are often adequately described classically whereby an intense light field produces anharmonic components in vibrations of the bound electrons. The alternative quantum mechanical approach is to describe the interaction in terms of virtual electronic transitions between the energy states of the system. The response is essentially instantaneous. The second type of non-linearity occurs when the photon energy is in resonance with an interband or exciton transition producing a redistribution of carriers between the bands. The result can be sensitive non-linear optical effects with time constants set by the carrier dynamics and recombination times which are typically in the nanosecond or picosecond regime for II–VI semiconductors. The third type of non-linearity of interest is the effect of laser heating on the optical constants of semiconductors. Thermally induced optical non-linearities are larger than resonant electronically induced effects in most materials even on relatively short timescales. Typical recovery times are in the region of milliseconds although nanoseconds are possible for very small spot sizes and good heat sinking. In all three categories, the initial change in optical constants can be very rapid when short excitation pulses are employed. For intrinsically absorptive non-linearities, the recovery is set by a number of experimental conditions. Although these are very different types of phenomena, the formalism developed for non-resonant optical non-linearities is normally extended to describe resonant interactions.

14.2.1 Bound electrons

Modulation of the polarization of a material at optical frequencies demands an essentially instantaneous response (bound electron interaction). These non-linearities are characterized by susceptibilities defined by expanding the induced polarization in terms of the electric field:

$$P_i^{\omega} = \chi_{ij}^{(1)} E_j^{\omega} + \chi_{ijk}^{(2)} E_j^{\omega_1} E_k^{\omega_2} + \chi_{ijkl}^{(3)} E_j^{\omega_1} E_k^{\omega_2} E_l^{\omega_3} + \tag{14.1}$$

The real and imaginary parts of $\chi^{(1)}$ describe linear refraction and absorption. These are related to the refractive index, n, and absorption coefficient, α, by,

$$n^2(\omega) \approx 1 + 4\pi\, Re[\chi^{(1)}(\omega)] \tag{14.2}$$

and

$$\alpha(\omega) = \frac{4\pi\omega\, Im[\chi^{(1)}(\omega)]}{cn(\omega)} \tag{14.3}$$

Absorption and refraction may be linked causally via the Kramers–Kronig relation,

$$n(\omega) - 1 = \frac{c}{\pi}\int_0^{+\infty} \frac{\alpha(\omega')}{\omega'^2 - \omega^2}\, d\omega' \tag{14.4}$$

Non-linear phenomena are described by the higher-order terms in the expansion of equation (14.1). Probably the best known non-linear optical phenomenon is harmonic generation in which a new laser frequency can be generated after passing an intense beam of coherent light through a non-linear medium. For instance, second-harmonic generation (a three-wave mixing process) is described by $\chi^{(2)}$. The third term describes four-wave mixing phenomena such as third-harmonic generation and also accounts for non-linear refraction, two-photon absorption and absorption saturation through real and imaginary parts. Under non-resonant conditions, the non-linear refraction can be viewed in terms of virtual, non-energy conserving transitions whereby an excited population exists only as long as the optical field is present. During their brief existence in the excited state, electrons remain coherent with the light field and induce changes in the optical constants of the material. Non-linear refraction defined to third order is known as the optical Kerr effect and described by a parameter n_2.

$$n = n_0 + \Delta n = n_0 + n_2 I \tag{14.5}$$

where the intensity of the light, I, is related to the field amplitude, $\boldsymbol{E}_0$, by

$$I = \frac{cn_0 |\boldsymbol{E}_0|^2}{8\pi} \tag{14.6}$$

We can relate n_2 to the real part of a total degenerate $\chi^{(3)}$ via the relation,

$$n_2 = \frac{8\pi^2 Re[\chi^{(3)}]}{cn_0^2} \tag{14.7}$$

14.2.2 Resonant interactions

If the photon energy approaches a bandgap energy, optical non-linearities become resonantly enhanced (electrons spend longer periods in the excited states). Resonance also increases the probability of scattering by phonons, impurities or defects providing the energy required for electrons to remain in the excited state until recombination occurs. Coherence between the light and the free carriers is rapidly lost because of scattering. These 'real' carrier populations result in intensity-dependent changes in the first-order susceptibility, $\Delta\chi^{(1)}$. This situation is most conveniently described by transition probabilities combined with rate equations; however, since $I \propto \backslash \boldsymbol{E}_0 \backslash^2$, an effective third-order susceptibility, $\chi^{(3)}_{\text{eff}}$, is often quoted to enable comparison of the magnitudes of different types of optical nonlinearity, i.e.

$$P(\omega) = (\chi^{(1)} + \Delta\chi^{(1)})E(\omega) = (\chi^{(1)} + \chi^{(3)}_{\text{eff}} |E(t)|^2)E(\omega) \tag{14.8}$$

Care must be taken with the use of the effective susceptibility approach if the density of excess carriers is not proportional to the intensity. For instance, Kerr-type non-linear refraction cannot be assumed if the carrier lifetime is density dependent or the absorption is non-linear.

A major breakthrough for optical bistability was the realization that optical excitation of semiconductors under bandgap resonant conditions could provide very sensitive non-linear optical effects [7–10]. Thus, optically induced changes in absorption coefficient and refractive index via the generation of free carriers can provide the control of light by light at moderate optical powers. We may characterize these optical non-linearities as either absorptive or refractive. Refractive index changes, Δn, and absorption changes, $\Delta\alpha$, may be linked by modified Kramers–Kronig relations, i.e.

$$\Delta n(\omega) = \frac{c}{\pi}\int_0^{\infty} \frac{\Delta\alpha(\omega')}{\omega'^2 - \omega^2}\, d\omega' \tag{14.9}$$

Note that changes in refractive index at a given frequency are induced by changes in absorption at other frequencies.

Resonant optical non-linearities may also be distinguished as being local or non-local. In the former case, the material excitation is limited to the location of the optical field causing the excitation, i.e. any spreading of the excitation is small compared with the laser spot size or fringe spacing for two or more beams. Under non-local conditions, the excitation (electronic or thermal) may diffuse to adjacent regions; therefore, the experimental conditions together with material properties determine whether local or non-local conditions apply.

Considering resonant and non-resonant optical non-linearities together, a remarkably large range of non-linear optical coefficients are available in II–VI semiconductors. Measured values range from a non-resonant $\chi^{(3)} \sim 5 \times 10^{-12}$ esu in CdS [11] to $\chi^{(3)}_{\mathrm{eff}} > 1$ esu in narrowgap CdHgTe [12] under resonant conditions.

14.2.3 Free-carrier optical nonlinearities

Most schemes for optical bistability involve excitation of the semiconductor below the bandgap energy. Excess free carriers can be efficiently photogenerated by absorption into the band-tail states and subsequent scattering into the band [10]. The local excess carrier density, N, is given by the rate equation

$$\frac{\mathrm{d}N}{\mathrm{d}t} = \frac{\alpha I}{\hbar\omega} - \frac{N}{\tau} + D\nabla^2 N \tag{14.10}$$

where τ is the recombination time and D the ambipolar diffusion coefficient of the carriers. The generated free carriers can alter the optical constants of a semiconductor via several contributions.

The simplest contribution arises from the Drude–Lorentz classical model for a free-carrier plasma modified in semiconductors by the effective mass of the electrons, m_e^*. This predicts a negative refractive index change per generated free carrier per unit volume, n_p, given by

$$n_p = -\frac{2\pi e^2}{n_0 m_e^* \omega^2} \simeq -\frac{8\pi e^2}{3n_0}\frac{P^2}{E_g^3}\left(\frac{E_g}{\hbar\omega}\right)^2 \tag{14.11}$$

where

$$\Delta n = n_p N \tag{14.12}$$

The Kane interband momentum matrix element, P, is nearly constant for III–V and II–VI semiconductors.

A second contribution arises from the generated free-carrier population causing blocking of interband transitions in the spectral region above the bandgap energy. This results in a modification to the refractive index below the gap which can be calculated from the absorption change above the gap by the Kramers–Kronig integration (equation (14.9)). The refractive cross-section per generated carrier, n_b, due to this 'band filling' is given to a good approximation within the Boltzman limit by [13]

$$n_b = -\frac{4\sqrt{\pi}e^2}{3n_0}\frac{P^2}{kT(\hbar\omega)^2} J\left(\frac{\hbar\omega - E_g}{kT}\right) \tag{14.13}$$

$$J(a) = \int_0^\infty \frac{\sqrt{x}e^{-x}}{x - a} \tag{14.14}$$

where J(a) is a thermodynamic integral giving the resonance with the band gap, the $x^{1/2}$ factor reflecting the role of the density of conduction band states. Below the bandgap, the negative plasma and band-filling contributions add with the latter dominating at photon energies close to the gap. Combining both contributions with $\hbar\omega \sim E_g$ gives

$$n_2 \simeq -\frac{8\pi e^2}{3n_0}\frac{P^2}{E_g^4}\left[1 + \frac{1}{2\sqrt{\pi}}\frac{E_g}{kT} J\left(\frac{\hbar\omega - E_g}{kT}\right)\right]\alpha(\omega)\tau \tag{14.15}$$

We see that a strong bandgap dependence favours narrowgap materials. Degenerate four-wave mixing measurements by Ji *et al.* [14] using an excimer-pumped dye laser gave a refractive index cross section $n_e \sim 5 \times 10^{-20}$ cm^3 for bulk polycrystalline ZnSe consistent with the band-filling model in the region 470–478 nm close to the bandgap. The carrier recombination time was in the range 2–50 ns and density dependent under the conditions employed.

In widegap II–VI semiconductors many-body effects such as bandgap renormalization and screening of the Coulomb interactions are important. Bandgap renormalization red shifts the bandedge such as to give a positive contribution to the refractive index which tends to cancel the plasma and band filling effects. Interest in excitons results from large changes in refractive index associated with saturation of the exciton absorption feature.

Dagenais [15] observed saturation of a bound exciton in CdS at a temperature of 2 K (Fig. 14.1). A saturation power, of only 3.6 μW (58 W cm^{-2}) was deduced for this inhomogeneously broadened exciton absorption. This system decays radiatively in 500 ps, providing a sensitive and fast optical non-linearity which can be exploited for studies of optical switching and bistability. Saturation of free-exciton absorptions in molecular beam epitaxy (MBE) grown ZnSe thin films and ZnSe/ZnMnSe superlattices at 77 K were reported by Anderson *et al.* [16, 17]. The saturation intensity of ZnSe layers was 10.7 kW cm^{-2} and 1.3 kW cm^{-2} for a superlattice with 73 Å quantum wells. The saturation was interpreted as being primarily due to Coulomb screening of the exciton by generated free carriers with an additional contribution from phase space filling in the quantum wells. Phase space filling is the result of generated free carriers filling the states necessary for exciton formation. Peyghambarian *et al.* [18] have subsequently reported large excitonic optical non-linearities in MBE-grown thin films of ZnSe and modelled the results using the plasma theory of Banyai and Koch [19] which

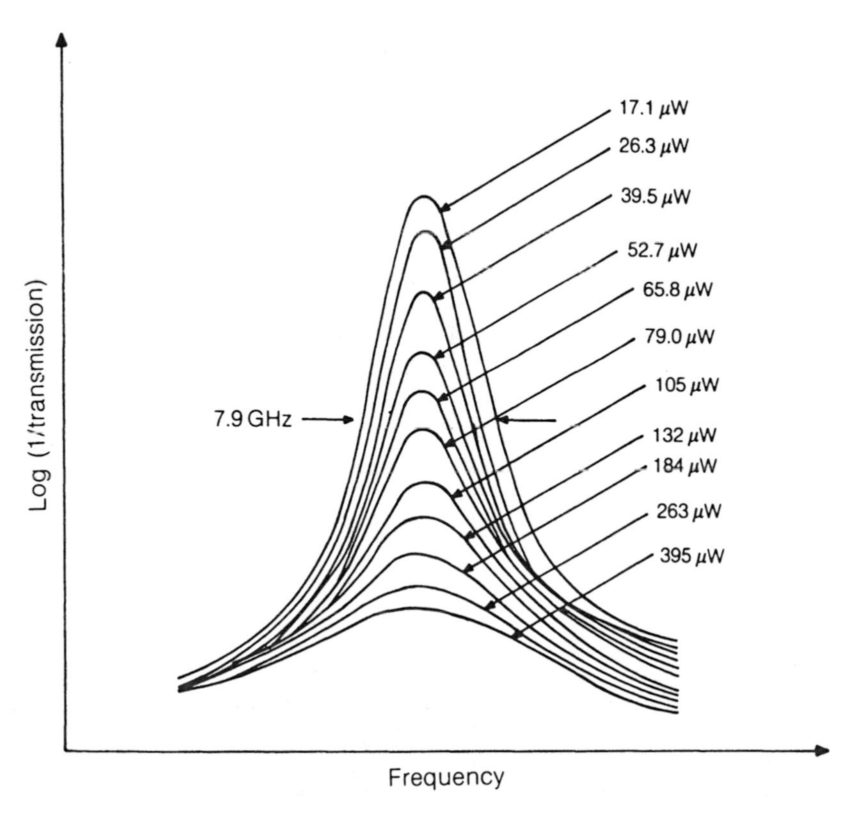

Fig. 14.1 Saturation of the I_2 bound exciton in a 20 μm thick CdS platelet at T = 2 K [15].

accounts for band filling along with many-body effects such exciton screening, bandgap renormalization and broadening of the tail states. The exciton is relatively strongly bound in ZnSe (binding energy ~ 18 meV) giving clearly resolved features at a temperature of 150 K (Fig. 14.2). Pump-probe measurements using wavelength-tunable 3 ns pulses around 435 nm were found to reduce the oscillator strength of the exciton at generating carrier densities above 10^{15} cm^{-3}. A Kramers–Kronig analysis of room-temperature results indicated that the maximum refractive index change associated with the exciton saturation should be $\Delta n = -0.017$ at 16 kW cm^{-2}.

The room-temperature saturation of well-resolved exciton absorption in room temperature $Cd_xZn_{1-x}Te/ZnTe$ ($x = 0.14$ and 0.25) quantum wells has been studied by Lee *et al.* [20]. The saturation intensity of the exciton was found to be 31.5 and 15 kW cm^{-2} for 50 and 100 Å wells which are

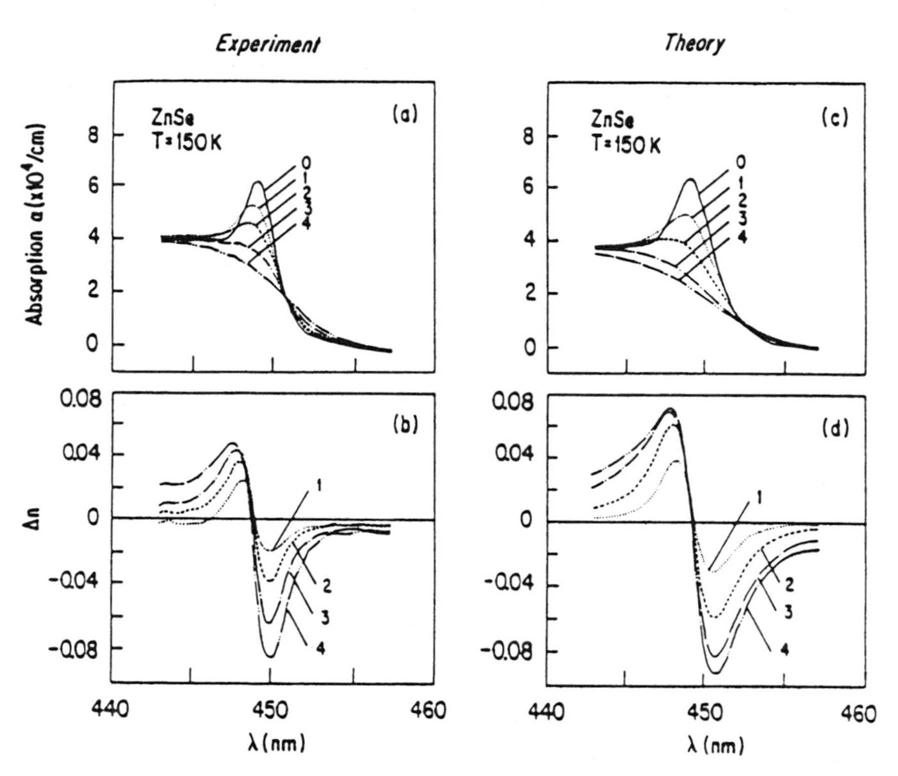

Fig. 14.2 Experimental and theoretical spectra of nonlinear absorption and refraction for a 0.55 μm ZnSe thin film at T = 150 K. (a) Experimental absorption for increasing pump powers: (0) no pump (1) 2.5 (2) 9 (3) 23 (4) 46.5 kW/cm^2. (b) Refractive index changes deduced from a Kramers-Kronig transformation of the nonlinear absorption (c) Calculated absorption spectra for different electron-hole pair densities. (0) 1×10^{15} (1) 5×10^{16}(2) 1×10^{17} (3) 2×10^{17} (4) 3×10^{17} cm^{-3}, (d) Calculated refractive index changes at these carrier densities [18].

significantly higher than for III–V multiple quantum wells (MQWs) because of the smaller exciton radius in this II–VI material. This makes room-temperature excitonic optical non-linearities less attractive in II–VIs but has advantages for hybrid bistable devices in which excitonic saturation should be avoided.

Jain and Lind [21] discovered that CdS_xSe_{1-x}-based colour filters consisting of semiconductor microcrystallites embedded in glass provide a source of readily available non-linear optical materials. Band-filling optical non-linearities in these filters have very fast response times because of rapid trapping of the excess carriers at the microcrystallite grain boundaries. Olbright *et al.* [22] and Yumoto *et al.* [23] measured values of n_2 on the order of 10^{-8} cm^2 kW^{-1} and response times on the order of 10 ps have been reported [24, 25]. A saturated index change of $\sim 1.5 \times 10^{-4}$ at intensities in excess of 1 GW cm^{-2} has been measured at a wavelength of 532 nm close to but below the bandgap [26]. Photo-induced darkening of colour filters under laser excitation is found to reduce the response time and the magnitude of the non-linearity [25, 26].

14.2.4 Two-photon-induced optical non-linearities

Two-photon absorption (TPA) is a non-linear process in which interband transitions are completed by the absorption of pairs of photons and can be related to the imaginary part of $\chi^{(3)}$. Interband TPA is actually forbidden at the zone centre of most common semiconductors because the wavefunctions of the band extrema at zero electron wavevector have opposite parity; however, transitions to states above the minimum energy gap become allowed because of mixing of the s- and p-like wavefunctions at finite wavevectors. The decrease in intensity, I, in the propagation direction, z, is given by

$$\frac{\mathrm{d}I}{\mathrm{d}z} = -\alpha I - \beta I^2 - \sigma N I \tag{14.16}$$

where β is the two-photon absorption coefficient. The third term describes absorption by generated carriers, where σ is the free-carrier absorption cross-section per carrier, and is usually significant. Two-photon absorption coefficients are now reliably established for most widegap II–VIs [27, 28] in good agreement with theory based on a $\boldsymbol{k}\cdot\boldsymbol{p}$ analysis [29, 30]. TPA is strongly bandgap dependent, β varying as E_g^{-3}. For example, narrowgap CdHgTe has $\beta \sim 10$ cm MW^{-1} at 10 μm [31] compared with $\beta \sim 5$ cm GW^{-1} for ZnSe at 0.5 μm [27]. TPA may cause non-linear refraction in several ways involving the generation of free carriers, biexciton excitation and virtual two-photon processes.

For photon energies in excess of half the bandgap energy, TPA may generate a free-carrier contribution to non-linear refraction in the same way as for single-photon transitions, i.e. Drude, band filling, bandgap renormalization and other many-body mechanisms. Ignoring dynamical effects and saturation of the non-linearity, we may expect the two-photon-induced free-carrier non-linear refraction to have an intensity dependence equivalent to a $\chi^{(5)}$ optical non-linearity. This contribution has been measured in large-gap II–VI semiconductors by self-focusing and degenerate four-wave mixing techniques. An example of the distortion in the spatial profile of 1.06 μm laser pulses after transmission through CdSe due to self-defocusing [32] is shown in Fig. 14.3. Relatively high intensities are needed to induce this effect in large-gap II–VIs since the bandgap dependence of both TPA and free-carrier non-linear refraction favours small gaps.

A second contribution to non-linear refraction arises as a direct consequence of the TPA. The transition of an electron to a virtual intermediate state by the first photon of the two-photon process induces an absorption change given by $\Delta\alpha = \beta I$. The predicted refractive index changes calculated from TPA theory and a Kramers–Kronig transformation (equation (14.9)) give good agreement with measured values of bound electronic non-linear refraction, n_2, in a very wide range of materials [33]. The bandgap dependence of the non-linear refraction is found to be consistent with $n_2 \sim E_g^{-4}$. This effect is not limited to photon energies larger than half the bandgap as the interaction involves only virtual interband transitions

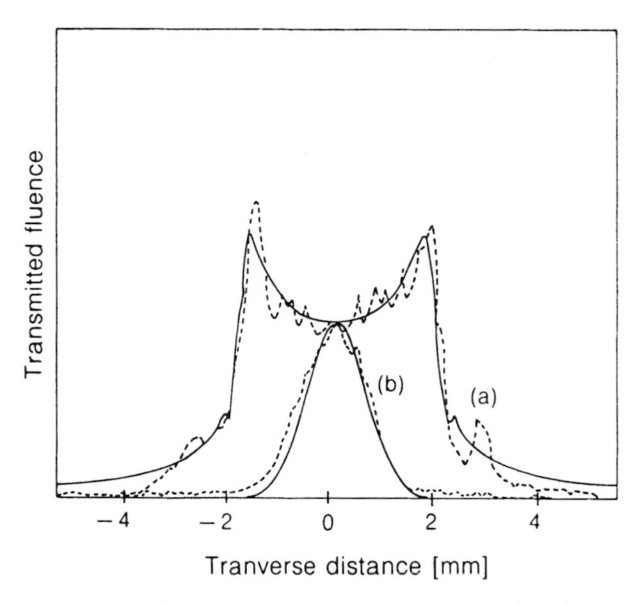

Fig. 14.3 Transverse profiles of a beam transmitted through polycrystalline CdSe at a distance of 0.5 m behind the sample (near field conditions) at (a) high irradiance (1GW/cm^2) and (b) low irradiance (0.3 GW/cm^3) [32]. The dashed line is experimental and the solid line is a theoretical fit. The pulse width used was 92 ps FWHM and the beam profiles are normalised to have the same on-axis fluence.

and is thus intrinsically very fast. The significance of these results for II–VI materials is that the bound electronic non-linearities can be accurately predicted in these materials. A full description requires inclusion of Raman and the a.c. Stark shift (i.e. virtual band blocking) for photon energies close to the bandgap [34].

Biexcitons also offer potential for large optical non-linearities via two-photon absorption. Excitonic molecules or biexcitons are created by two-photon excitation, $2\hbar\omega = E_{Bi} = 2E_{ex} - B$ where E_{ex} and E_{Bi} are the energies of the exciton and biexciton respectively and B is the binding energy of the two excitons in the molecule. Two-photon transitions to the biexciton state are always allowed if the exciton is dipole active; thus the biexciton is observed below the single-exciton energy. Since the biexciton transition has a large dipole moment resonantly enhanced by the single exciton providing an intermediate state, the biexciton is expected to have large non-linearities associated with it. Most studies of biexciton optical non-linearities have been in CuCl ($E_{Bi} = 3.186$ eV at low temperature). Biexcitons have been studied in CdS by two-photon absorption, luminescence-assisted two-photon spectroscopy, hyper Raman scattering [35] and laser-induced transient gratings [36].

14.2.5 Optothermal effects

Any absorption of light will invariably lead to heating. The temperature dependence of the bandgap energy causes absorption and hence refractive index changes which can be appreciable at energies close to the bandgap. Only in narrowgap semiconductors such as InSb and InAs do electronically induced optical non-linearities dominate over thermal effects under continuous-wave (CW) illumination. Thermal effects may be viewed as a nuisance if they compete with faster electronic effects; however, the most successful II–VI semiconductor optically bistable devices to date exploit the sensitive thermal non-linear optical effects in interference filters.

Heat is produced by the depletion of the excess energy of optically generated electrons into lattice vibrations and typically occurs on picosecond timescales. The time constant for the recovery is set by heat dissipation which depends on the laser spot size and heat sinking. The bandgap shifts with temperature for II–VI semiconductors are in the range from -3×10^{-4} to -7×10^{-4} eV K^{-1}. An estimate of the refractive index change with temperature for photon energies below the bandgap can be obtained from a Kramers–Kronig transformation of an absorption edge shifted to longer wavelengths [37]. Values of refractive index change with temperature, dn/dT, are in the order of 10^{-4} K^{-1} for most large-gap II–VIs with only a weak bandgap resonance. Optothermal effects have been of most interest for use in non-linear interference filters and are discussed in more detail later.

14.3 FUNDAMENTALS OF OPTICAL BISTABILITY

Optical bistability can be achieved by combining a non-linear response with some form of feedback. The optical non-linearity may be local or non-local, absorptive or refractive, and often described in terms of an effective third-order non-linear optical coefficient. Here we outline two types of bistabilities appropriate to the results discussed later, dispersive optical bistability employing a Fabry–Pérot etalon, and cavityless optical bistability due to increasing absorption.

14.3.1 Etalons

A convenient method of providing optical feedback is to insert the non-linear material between two mirrors to form a Fabry–Pérot cavity. For a semiconductor this may be as simple as polishing two opposite faces plane parallel, although coatings are usually required to enhance the reflectivity of the polished surfaces. Alternatively, etalons can be fabricated by multilayer thinfilm coating techniques which include the semiconductor as a spacer layer within a dielectric stack.

The transmission, T, and reflection, R, of a Fabry–Pérot etalon, with front and back mirror reflectivities, R_F, R_B, and a spacer of thickness, D, with linear absorption coefficient, α, are given by [38–40],

$$T = \frac{A}{1 + F\sin^2\phi} \qquad A = \frac{e^{-\alpha D}(1 - R_F)(1 - R_B)}{(1 - R_\alpha)^2} \tag{14.17}$$

$$R = \frac{B + F\sin^2\phi}{1 + F\sin^2\phi} \qquad B = \frac{R_F[1 - (R_\alpha/R_F)]^2}{(1 - R_\alpha)^2} \tag{14.18}$$

where

$$R_\alpha = (R_F R_B)^{1/2}\exp(-\alpha D) \tag{14.19}$$

is an effective mean reflectivity for the etalon, and

$$F = \frac{4R}{1 - R_\alpha} \tag{14.20}$$

is related to the cavity finesse, $f = \pi\sqrt{F}/2$.

The transmission of the etalon is thus described by an Airy function when plotted against the phase ϕ (Fig. 14.4). The transmission can be controlled by varying the intensity of the input light. Although optical bistability can be produced in etalons by either non-linear absorption or non-linear refraction, the latter dominates in most circumstances since changes in the etalon transmission occur only slowly with absorption whereas large swings in transmission can be produced by a fraction of a wavelength change in optical path length. Dispersive optical bistability may be understood by realizing that the phase change in the cavity is proportional

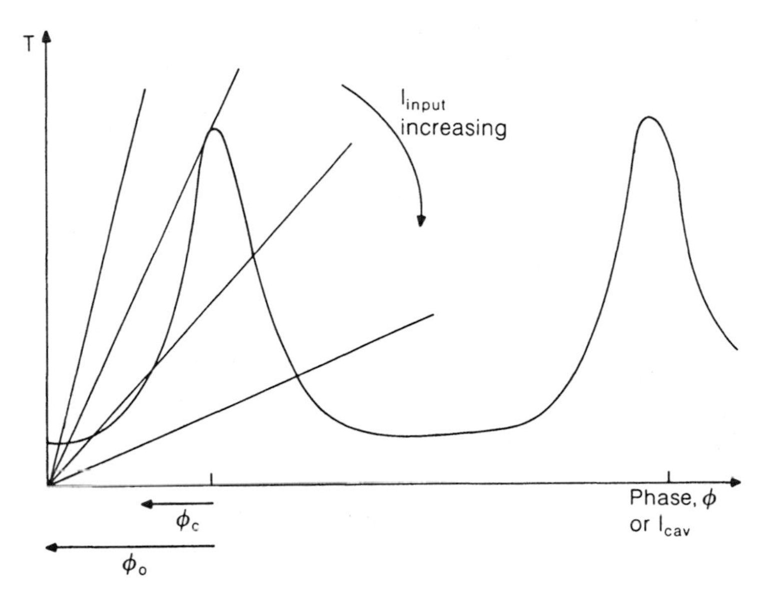

Fig. 14.4 Transmission of a Fabry-Perot etalon as a function of phase, ϕ.

to the cavity intensity, I_{cav}, rather than the input intensity, I_i. For a Kerr-like optical non-linearity, n_2, the phase can be divided into an initial detuning, ϕ_0 and an intensity-dependent part:

$$\phi = \frac{2\pi n D}{\lambda} = \phi_0 + \frac{2\pi D}{\lambda} n_2 I_{cav} \tag{14.21}$$

The cavity intensity is averaged within the etalon for convenience. For an etalon initially tuned off resonace such that an increase in cavity intensity tunes the etalon into resonance, an incremental increase in input power will increase the intensity within the cavity, which tunes the cavity into resonance allowing even more light into the cavity, which further tunes the cavity, etc. This positive feedback can lead to switching from low to high cavity transmission. A switch from high to low transmission will occur at lower input powers than for the switch-up because of the energy already stored within the cavity.

The conditions for bistability can be found graphically. A second equation describing the etalon comes from the proportionality between the transmitted power and the cavity intensity,

$$I_i = C \frac{I_{cav}}{T} \tag{14.22}$$

where

$$C = \frac{\alpha D \mathrm{e}^{-\alpha D}(1 - R_B)}{(1 - \mathrm{e}^{-\alpha D})(1 + R_B \mathrm{e}^{-\alpha D})} \tag{14.23}$$

Equations (14.17), (14.21) and (14.22) solved simultaneously to eliminate I_{cav} describe the non-linear Fabry–Pérot transmission as a function of *incident* intensity. The straight lines defined by equation (14.22) are plotted in Fig. 14.4 for several input intensities showing multiple intersections with the periodic function implying optical bistability (different transmission states for a single input level). Plotting the output as a function of input intensity produces an S-shaped curve (Fig. 14.5). The intermediate solution can be shown to be unstable, giving switching and an anticlockwise hysteresis loop. The shape of the loop depends on the cavity parameters, i.e. the finesse of the etalon, and the initial detuning from resonance. For some critical detuning, the periodic and straight lines in Fig. 14.4 touch tangentially which produces a vertical but non-hysteretic input–output response thus defining a limiting condition for bistability. This sets the minimum critical input intensity, I_c, necessary to achieve optical bistability.

The basis for optical memory and logic is thus achieved using a hold beam and one or more control or address beams. With the element in its opaque (logic '0') state, the hold beam biases the device close to its switch point. A small input signal can then flip the device through the transition to the upper state (logic '1'), allowing transmission of the higher power hold beam. Thus, optical signal gain can be achieved for delivery to further devices. The reflected signal gives the complementary levels and logic functions to transmission.

The switching speed is affected not only by the material parameters but

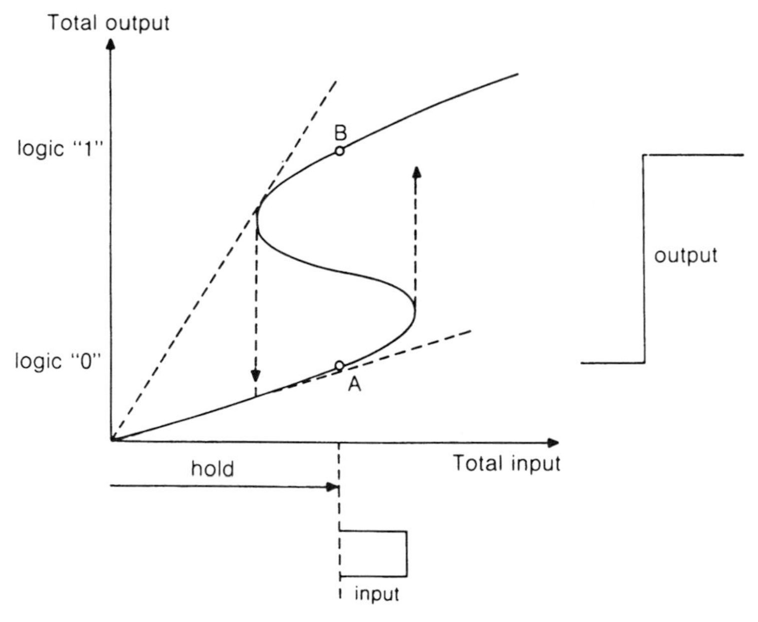

Fig. 14.5 Input-output characteristic of an optically bistable Fabry-Perot etalon.

also the bistable response. The switching times become very long when the device is biased very close to the switch point owing to a phenomenon referred to as critical slowing down. There is therefore a trade-off between switching pulse energy and switching speed. The advantage of critical slowing down can be the reduced susceptibility of the device to noise which would otherwise limit how critically the device can be set [41]. A complete characterization of optical bistability in etalons needs to include spatial effects, diffusion, diffraction, saturation of the non-linearity, the density dependence of the carrier recombination rate and non-linear absorption.

14.3.2 Increasing absorption bistability

Optical bistability can also appear under conditions in which the absorption increases as a function of laser excitation. In this case an etalon is not required as the feedback is either intrinsic to the absorption process or can be applied with electrical feedback. For instance, the bandedge of most semiconductors moves towards longer wavelengths with increasing temperature. Thus, laser heating by excitation into the band tail states shifts the bandedge such as to increase the absorption. Under appropriate conditions, this can become a runaway process whereby an incremental increase in intensity heats the sample thus increasing the absorption which in turn heats the sample more, etc. The characteristic is a clockwise hysteresis loop with a switch from high to low transmission on increasing the input intensity and a lower-power switch from low to high transmission when decreasing the input power. Figure 14.6 shows an example of increasing absorption bistability in a room-temperature CdTe sample [42] at three different wavelengths close to the bandedge. Wherrett *et al.* [43] have derived a general condition, $\alpha_0 L < 0.18$, necessary to achieve increasing absorption bistability for exponential band tails where α_0 is the initial absorption, and L is the length of the sample.

Bandgap renormalization at high carrier densities can also lead to increasing absorption optical bistability in semiconductors. An example is the shift to lower energy of an exciton absorption line with increasing carrier density. Figure 14.7 shows the calculated absorption of CdS at a temperature of 30 K for a photon energy below the bound exciton absorption as a function of carrier density [44]. Also shown is the absorption coefficient given by the expression for the carrier generation,

$$\alpha(N) = \frac{N}{I}\frac{\hbar\omega}{\tau} \tag{14.24}$$

For different values of intensity, I, one obtains one, two or three intersections, thus predicting bistable behaviour as a result of bandgap renormalization.

It should be noted that dynamic optical hysteresis may be observed in many systems, but bistability requires the demonstration that two stable states exist for one input power.

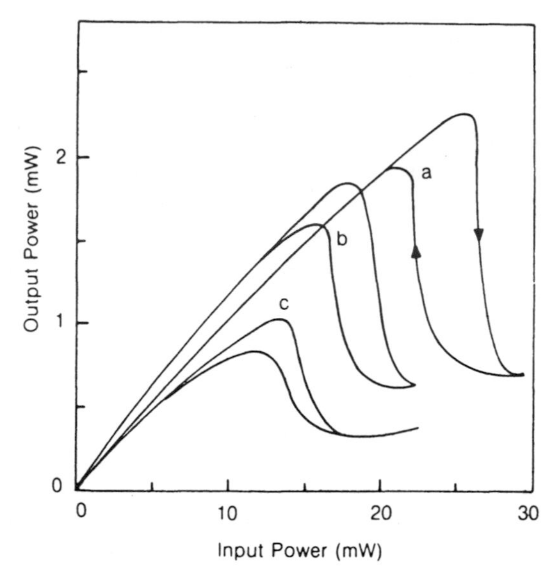

Fig. 14.6 Increasing absorption bistability in a 100 μm thick polycrystalline CdTe sample at room temperature at (a) 848.5 nm, (b) 847.5 nm, and (c) 844.5 nm [42].

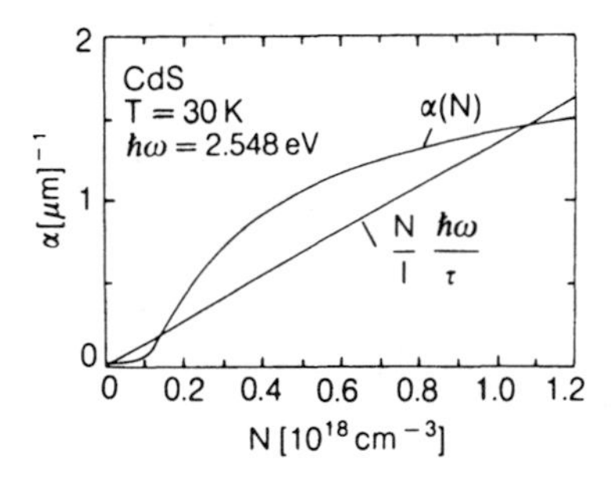

Fig. 14.7 Calculated density dependence of the absorption of CdS at T = 30 K and $\hbar\omega = 2.548$ eV Multiple intersections are shown with the line defined by eq. 14. 24 [44].

14.4 OPTICAL BISTABILITY IN BULK II–VI MATERIALS

Optical bistability has been achieved via a number of non-linear optical mechanisms in bulk II–VI semiconductors. Bound and free excitons with large oscillator strengths provide large optical non-linearities, making these materials particularly interesting for bistability studies.

Optical hysteresis in dye laser pulses transmitted by CdS was observed by Bohnert *et al.* [45] at a sample temperature of 1.8 K and by Rossmann and Henneberger [46, 47] at 77 K. This was interpreted as increasing absorption bistability although no distinct threshold switching was observed. In both cases, pulse intensities were $\sim 1\,\mathrm{MW\,cm^{-2}}$. Schmidt *et al.* [48] and Rossmann and Henneberger [49] modelled this effect in terms of a bandgap renormalized red shift of the bound exciton due to the generated electron–hole plasma, although an alternative interpretation based on a thermal bandedge shift has also been suggested [50, 51]. Later work by Fidorra *et al.* [52] repeated these measurements with improved temporal resolution in order to confirm that the induced absorption can be attributed to the formation of an electron–hole plasma. Dagenais and Sharfin [53] used thermally induced absorption shifts in CdS to obtain increased absorption bistability at a few milliwatts of CW power from a dye laser operating near 487 nm (Fig. 14.8). The 10 μm thick platelets produced very high contrast clockwise hysteresis loops when held at a temperature between 2 and 50 K with the laser wavelength detuned by up to $20\,\mathrm{cm^{-1}}$ from the bound exciton absorption. At temperatures above 50 K, the free exciton was considered to be more important. Dagenais and coworkers [54–77] also reported dispersive optical bistability in CdS platelets fabricated in the form of Fabry–Pérot

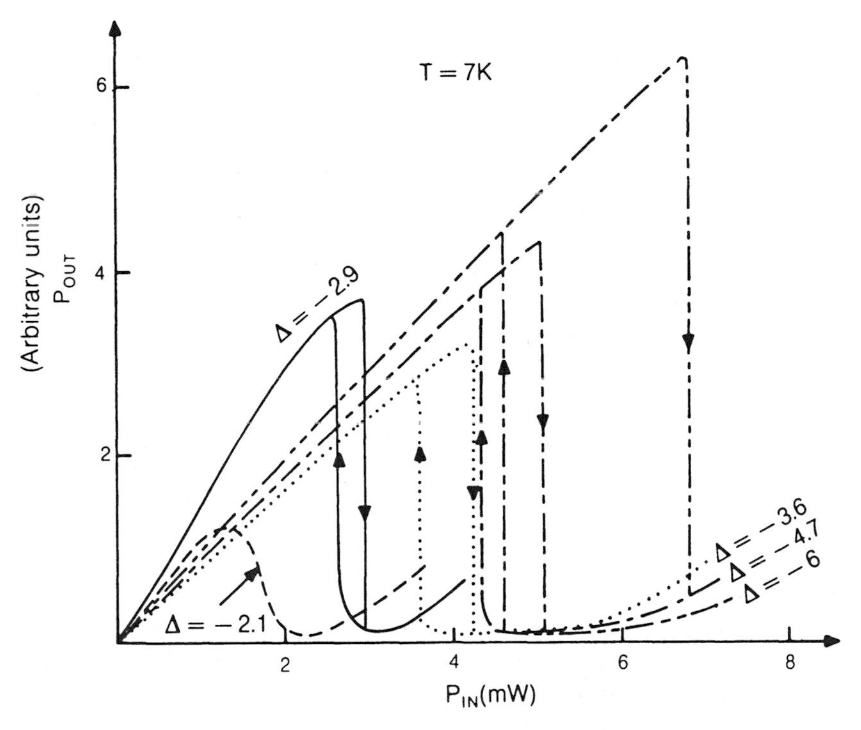

Fig. 14.8 Experimental observation of increasing absorption bistability in CdS at T = 7 K for different detunings, $\Delta = \omega - \omega_0$ below the bound exciton resonance, ω_0 [53].

etalons. Initial measurements [54] were made on uncoated, 25% natural reflectivity samples of thickness 5–20 μm held at a temperature of 2 K. The non-linearity is associated with the giant oscillator strength of the bound exciton which saturates at low optical powers (Fig. 14.1). Only transverse optical bistability (monitoring only the central spot of the transmitted beam) could be observed, but a refractive index change as large as 0.08 was deduced near the bound exciton resonance. Later experiments [55, 56] used coated samples of CdS with 90% reflectivities again at helium temperatures to achieve whole-beam-dispersive optical bistability with a switching energy of < 4 pJ and ON–OFF switching speed < 2 ns [56].

Optical bistability has been demonstrated in CdS platelets at room temperature by Wegener *et al.* [58] and Henneberger and coworkers [59, 60]. CdS samples 10 μm thick with 60% reflecting coatings produced optical bistability by absorption bleaching with 20 ns pulses of intensity in the order of 1 MW cm^{-2} [58]. The switching times were on sub-nanosecond timescales.

Increasing absorption bistability has been observed in bulk ZnSe samples owing to the thermally induced red shift of the bandgap energy. Taghizadeh *et al.* [61] used 2.1 mm thick polycrystalline ZnSe to observe clockwise hysteresis loops at around 100 mW input power with an argon laser operating at 476.5 nm and a 40 μm spot size. Switching times were 100 μs and 500 μs for switching to low and high transmission respectively. Strong self-focusing of the laser beam was also noted from distortions in the beam profile after transmission through the sample. For the sample and wavelength used, $\alpha_0 L = 1.5$ which contradicted the criterion $\alpha_0 L < 0.18$ for increasing absorption bistability [43]. In subsequent work, Kar and Wherrett [62] observed both absorptive and dispersive bistability in a 370 μm thick polycrystalline ZnSe etalon at 476.5 nm with spot sizes of 70 μm. The natural reflectivity of the polished ZnSe provided seven orders of refractive optical bistability before the onset of absorptive switching (Fig. 14.9). For these thick samples, the temperature rise for dispersive switching was estimated at only 2 K with the lowest order occurring at around 30 mW. Because the absorption edge in this material was not strictly exponential in shape, the criterion for absorptive switching was found to be modified to $\alpha_0 L < 0.45$. Increasing absorption optical bistability has also been reported in a 100 μm thick CdSe at 15 mW input power by Miller *et al.* [42]. Bistability occurred close to the bandedge at wavelengths greater than 844 nm with switching times of 10 μs observed for a 10 μm spot size (Fig. 14.6).

Kim *et al.* [63] have demonstrated increasing absorption bistability in II–VI waveguides. In this case, single-crystal epitaxial layers of ZnSe grown by metal–organic chemical vapour deposition (MOCVD) on GaAs substrates produced bistability at 15 mW at 488.8 nm with a 10 μs switching time due to laser heating. A lower limit for the switch-off time was estimated from these results to be around 400 ns for a strip-loaded channel waveguide.

Kinks in the input–output characteristics of optical bistability due to

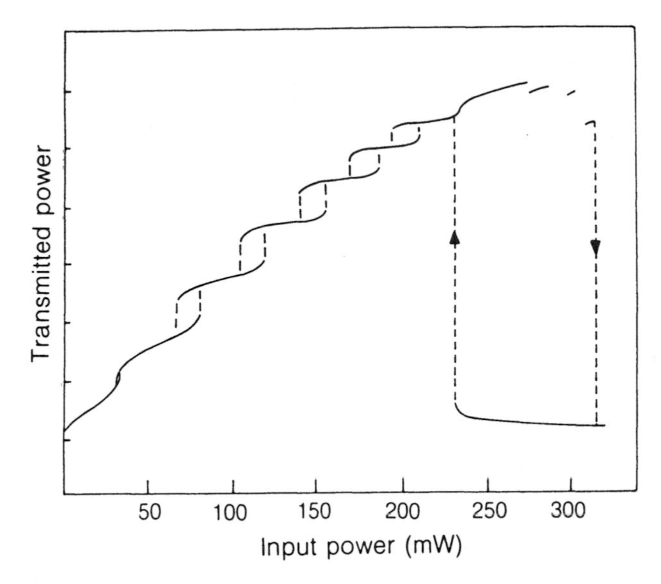

Fig. 14.9 Refractive optical bistability (six orders) and increasing absorptive bistability at high powers in room temperatute 370 μm thick polycrystalline ZnSe at 476 nm [62].

increasing absorption have been observed in CdS_xSe_{1-x} doped glasses by Gibbs *et al.* [64]. These were interpreted as being due to the discontinuity of excitation density which occurs between the front of the sample which is switched 'off' and the back which is still switched 'on'. The kink jumps discontinuously along the beam propagation direction causing a sawtooth temporal dependence of the output as the input light intensity is increased linearly at a rate comparable with the excitation lifetime (of thermal origin in this case).

Henneberger *et al.* [60] have proposed and demonstrated an alternative cavityless bistability scheme using a ZnSe prism. The polarized output of an argon ion laser at 514 nm was angled close to Brewster's angle at the rear face of the 30° prism (p-polarization) but less than the critical angle for total internal reflection. Thus the transmission of the prism is high at low input powers. By increasing the intensity of the laser, the thermo-optic effect in the ZnSe increases the refractive index and the Brewster condition is lost such that light is increasingly reflected back into the prism, providing feedback. The refractive index change also tunes the prism towards the critical angle for total internal reflection. A condition can be reached whereby the feedback regeneratively sweeps the prism beyond the critical condition for total internal reflection such that the transmission drops. Bistability was obtained in this way at around 200 mW input power with an unfocused beam.

Wegener *et al.* [65] have demonstrated a CdS self-electro-optic effect device (SEED) based on photothermal absorption changes in the presence of an electric field. This device is a variant on an electro-optic MQW SEED device which exploits the effect of electric fields on excitonic absorption via the quantum confined Stark effect [66]. In this case, the device operates with a CW argon laser as a room-temperature, increasing absorption bistable device, but a rise in photocurrent as the laser intensity is increased provides a second feedback mechanism via resistive heating, thus enhancing the bistability. Bistability could also be observed in the electrical current through the device.

14.5 INTERFERENCE FILTERS

Thermo-optic non-linear interference filters (NLIFs) based on II–VI semiconductors are attractive as prototype logic and memory devices for parallel all-optical digital computing schemes. They can be uniformly fabricated in large areas and have the convenience of room-temperature operation. The first report of passive optical bistability in a semiconductor in 1978 by Karpushko and Sinitsyn [1] in Minsk, USSR, employed ZnS as spacer layer between dielectric multilayer reflectors and used 100 mW of power from an argon ion laser operating at 514 nm focused to a 50 μm spot diameter. Although originally discussed in terms of an electronic non-linearity, the mechanism for the bistability has been confirmed as a thermally induced refractive index change by Smith *et al.* [67] at Heriot-Watt University, Edinburgh, and independently by Olbright *et al.* [68] at the Optical Sciences Centre in Tucson. Both the Edinburgh and Arizona groups have made extensive studies of optical bistability and relevant properties of NLIF devices. Most effort has involved ZnSe since it is a well-established thin film coating material and provides a bandgap energy compatible with the green output of the argon ion laser. ZnS and CdSe NLIFs have also been investigated in some detail. Research on digital optical circuits has benefited from the advantages of the availability of room temperature ZnSe and ZnS devices with visible light operation; however, the wavelength can be widely adjusted by the introduction of additional absorbing layers such that commercial laser diodes can now be used as sources. Most II–VI NLIF bistable devices have been fabricated by standard thermal evaporation coating techniques; however, ultrahigh vacuum molecular beam deposition has been adopted more recently to produce better-optimized devices with improved stability.

NLIFs have the general structure, $(m_1\text{H}'\text{L})(m\text{HH})(m_2\text{LH}')$ where H and H′ indicate quarter-wavelength ($\lambda/4$) optical thickness layers of high index (e.g. ~2.7 at 514 nm for ZnSe), m is an integer which determines the etalon spacer thickness in half optical wavelengths, and L is a $\lambda/4$ layer of low-index transparent material (e.g. ThF_4, $n = 1.55$). The high-low stacks act as high reflectors on either side of the spacer layer to form a Fabry–Pérot etalon. These etalons can be self-tuned when a laser beam causes a rise in temperature owing to absorption of some of the light in the spacer layer. The absorption edge of most semiconductors shifts to longer wavelengths with increasing temperature which leads to an increase in refractive index because of the positive dispersion in the region of the bandgap. The very broad band tail absorption, which seems to be characteristic of thermally evaporated II–VI semiconductors, can be utilized to achieve optical bistability as a function of increasing irradiance under appropriate conditions of resonator design and wavelength. Laser heating can alter the etalon phase via either a refractive index change or expansion–contraction, i.e.

$$\phi = \phi_0 + \frac{2\pi nD}{\lambda}\left(\frac{1}{n}\frac{\mathrm{d}n}{\mathrm{d}T} + \frac{1}{D}\frac{\mathrm{d}D}{\mathrm{d}T}\right)\Delta T \quad (14.25)$$

where D is the spacer thickness and ϕ_0 is the initial phase. The thermo-optic coefficient, $(1/n)\,\mathrm{d}n/\mathrm{d}T$, is at least an order of magnitude greater than the contribution from the linear expansion, $(1/D)\mathrm{d}D/\mathrm{d}T$, at wavelengths near the bandgap.

The absorption coefficient is a key parameter in the design of an optically bistable device. There is a trade-off between high absorption which will produce more heating and hence refractive index change, but will decrease the cavity finesse. The band tail absorption in thermally evaporated II–VIs is much less abrupt than in bulk material. For instance, at 514 nm the absorption coefficient of thin film ZnSe is $600\,\mathrm{cm}^{-1}$ [67] compared with about $5\,\mathrm{cm}^{-1}$ in bulk polycrystalline ZnSe.

Figure 14.10 shows optical bistability in a ZnSe NLIF at 514 nm as a

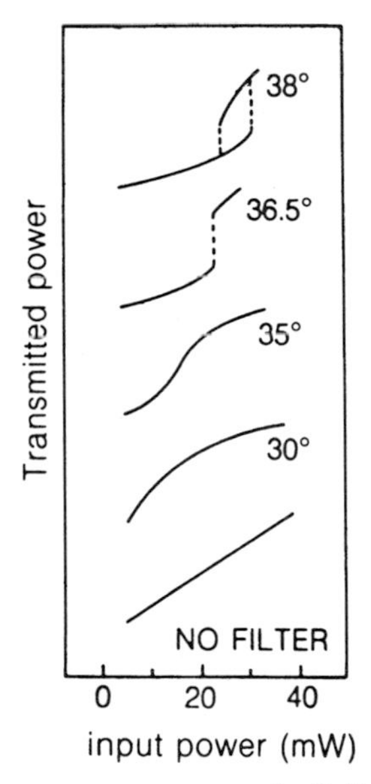

Fig. 14.10 A family of input-output responses of a ZnSe NLIF at different angles of incidence corresponding to different initial detunings of the etalon [67].

function of angle (i.e. different initial resonator tunings). In initial studies with thin spacer layers (typically eight quarter-wavelengths), the temperature rise needed for switching was in the order of 100 K. The response time of the switch depends on spot size and heat sinking. Smaller spot sizes reduce both the switching time and the switching power. Olbright *et al.* [68] measured switch-on and switch-off times in ZnSe and ZnS NLIFs in the range 10–50 μs for spot diameters less than 10 μm.

Janossy *et al.* [69] found that the critical switching power was linearly related to the spot size down to 50 μm in ZnSe NLIFs. This is consistent with a model which assumed a spacer much thinner than the spot size such that the heat diffusion is primarily into the substrate. Sahlen [70] confirmed this linearity between spot size and switching power down to a spot radius of 1.8 μm in a ZnS NLIF at 501.7 nm. At very small spot sizes, the switching power should increase again as a result of diffraction [71–73]. A number of authors [74–77] have made detailed theoretical studies of heat diffusion in thin film interference filters with the conclusion that the heat loss is primarily into the substrate but, in an array of unpixellated bistable devices, transverse heat flow imposes a critical separation of bistable elements below which cross-talk would become significant. For a thin ZnSe NLIF array, this distance was predicted to be $19\omega_0$ [74, 75] where ω_0 is the laser spot size radius. An absolute maximum switching speed was predicted at 2 ns from analysis of the transient heat diffusion for a diffraction-limited spot size of 0.21 μm, or 100 ns at 1.5 μm [77]. Janossy *et al.* [78] observed that the switch-on time is quadratic with spot size.

Wherrett *et al.* [79–82] have investigated in some detail the optimization of NLIFs for low-power switching in terms of the cavity parameters and spot size, ω_0. It was found that non-linearity within the reflecting stack layers does not contribute appreciable to the performance of the device and a plane wave cavity approach with the temperature rise spread evenly over the spacer layer is adequate. Optimization differs in thermo-optic compared with opto-electronic bistable devices because the total absorbed flux, $\alpha I D$, rather than the energy absorbed per unit volume gives the refractive index change. For spot sizes large compared with the filter thickness such that transverse heat diffusion is small, an effective non-linear refraction n_2^{T} can be defined and a critical intensity for bistability, I_c, derived:

$$n_2^{\mathrm{T}} \simeq \frac{\partial n}{\partial T}\frac{\alpha D \omega_0}{\kappa_s} \tag{14.26}$$

$$I_c = \frac{\lambda \alpha \kappa_s}{2\pi |\mathrm{d}n/\mathrm{d}T| \omega_0}\frac{f(R_F, R_B, \alpha D)}{\alpha D} \tag{14.27}$$

where κ_s is the substrate thermal conductivity. The critical intensity is the product of material parameters and a complicated function, $f/\alpha D$ of etalon parameters (front and back reflectivities, absorption coefficient and spacer thickness). Figure 14.11 shows the cavity function $f/\alpha D$ plotted against αD for ZnSe NLIFs with different numbers of multilayer HL periods, M, in the reflecting stacks [79]. For given mirror reflectivities, there is an optimum value of αD which minimizes $f/\alpha D$ at a value of 2.6. Using more layers in the stack to increase the reflectivity does not reduce the minimum cavity factor but allows flexibility in the choice of αD. The optimum positions correspond to the condition

$$\alpha D = 2 - R_F - R_B \tag{14.28}$$

For example, the optimum value of αD for a ZnSe NLIF with $M = 3$ is approximately 0.1, i.e. the optimum spacer thickness $D_{opt} = 1.7\,\mu m$ at 514 nm. This is thicker than the NLIFs produced by conventional thermal evaporation. Consideration of the material factor shows that, for optimized structures, one does not expect the choice of II–VI material or the operational frequency relative to the bandedge to be significant [80]. However, the correct choice of cavity parameters is crucial. Spot size and substrate thermal conductivities are the other adjustable parameters. Fabrication of a ZnSe NLIF with $M = 4$ mirror stacks has allowed optical bistability with a 633 nm He–Ne laser at 30 mW consistent with this model [83]. This demonstrates that bistability can be achieved at longer wavelengths with the advantage of a lower temperature switch; however, the larger irradiances at 633 nm due to the lower absorption, $\alpha \sim 100\,\mathrm{cm}^{-1}$, resulted in the devices being more unstable. With spot sizes greater than 17 μm, the linear dependence of critical switching power with spot size was again confirmed.

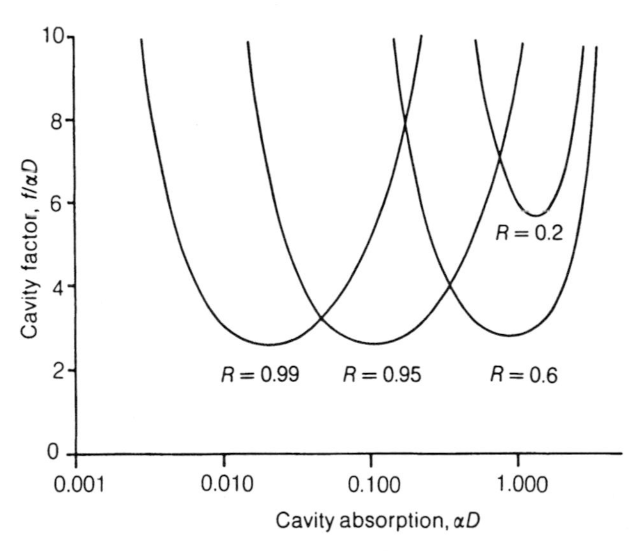

Fig. 14.11 Cavity factor $f/\alpha D$ giving switching irradiance, I_c, for optothermal NLIF bistable devices with different mirror reflectivities [79].

When tight focusing is used to minimize the switching power (or for thick films), the spot size may become smaller than or at least comparable with the thickness of the filter. In the limit of transverse heat conduction in the film [79],

$$n_2^{\mathrm{T}} \simeq \frac{\partial n}{\partial T} \frac{\alpha \omega_0^2}{\kappa_{\mathrm{f}}} \tag{14.29}$$

$$I_{\mathrm{c}} = \frac{\lambda \kappa_{\mathrm{f}}}{|\mathrm{d}n/\mathrm{d}T| \omega_0^2} f(R_{\mathrm{F}}, R_{\mathrm{B}}, \alpha D) \tag{14.30}$$

where κ_{f} is the thin film conductivity and, in this case, the minimum cavity factor f corresponds to the condition,

$$\alpha D = \frac{2 - R_{\mathrm{F}} - R_{\mathrm{B}}}{4} \tag{14.31}$$

The performance of optically bistable devices suitable for employing in 2D arrays demands tight tolerances on the material parameters such as refractive index, thicknesses, reflectivities of the mirror stacks, etc. Films prepared by conventional thermal evaporation are usually porous and contain water, making good stability difficult to achieve with sustained operation (i.e. the bistable hysteresis loop drifts and changes shape). Optimized structures also demand Fabry–Pérot etalons with spacer thicknesses which cannot be easily achieved by thermal evaporation. A significant improvement in the quality of polycrystalline ZnSe thin films and thus bistable devices has been achieved by employing molecular beam deposition (MBD) which has been developed by Lewis at the Royal Signals and Radar Establishment, UK, for advanced optical coatings [84]. This technique uses an ultrahigh vacuum (MBE) system to produce highly dense films free of porosity and with low contamination levels. Initial measurements were made on filters which were fabricated with MBD layers deposited between thermally evaporated multilayer stacks [83]. This exploited the advantage of the MBD technique to produce much thicker spacer layers with good mechanical stability and hence to better optimize the design. Optical bistability was achieved using a CW krypton ion laser at 647 nm in filters with thicknesses up to 6.6 μm. Complete multilayer stacks were subsequently developed using the MBD process employing barium fluoride ($n = 1.45$) as the low-index material. Figure 14.12 shows the transmission spectrum of a complete MBD-fabricated NLIF [85]. The filter consisted of a two-period stack comprising $\lambda/4$ thick layers of ZnSe and BaF_2 followed by a 6 μm thick ZnSe spacer and finally a matching two-period mirror. The etalons were exceptionally smooth with no evidence of surface texture visible by Nomarski interference microscopy. Transmission spectra were recorded at a number of sample temperatures in order to determine the dispersion in both the refractive index and the index change with temperature of the MBD

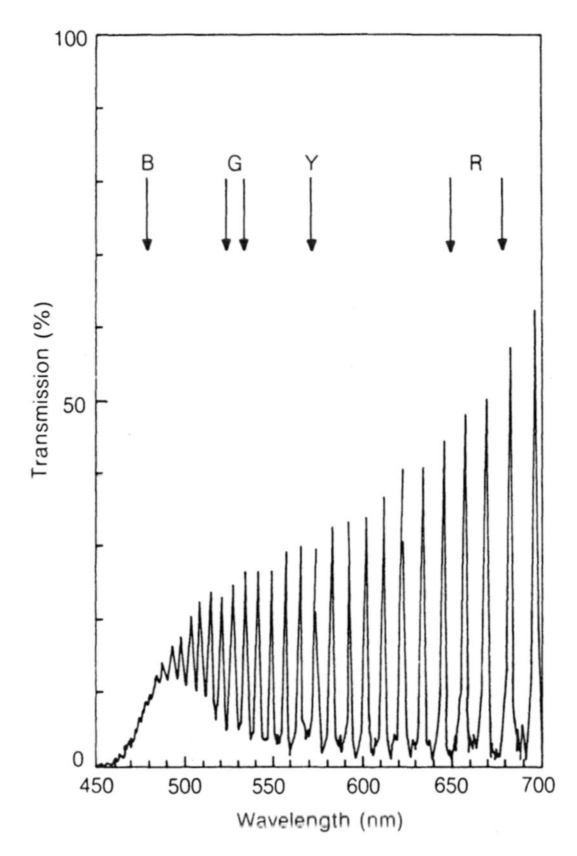

Fig. 14.12 Transmission spectrum of a molecular beam deposited $ZnSe/BaF_2$ filter with two period reflecting stacks centered on 640 nm [85].

material [86]. Because of the thickness of this filter (etalon order, $m \sim 120$), optical bistability could be achieved at five krypton laser output wavelengths between 676 and 521 nm, allowing a study of the wavelength dependence of thermo-optic bistability in a single device [85, 86]. The measured critical switching powers for optical bistability (Fig. 14.13) were in good agreement with both equation (14.27) and equation (14.30). In either case, the wavelength dependence of the critical switching power is given by $P_c \sim \lambda f/(dn/dT)$ [87]. The optimum wavelength of 568 nm for this particular filter gave a switching power of 10 mW using a 12 μm spot size. Increasing absorption bistability using the 514 nm output of an argon ion laser could be observed at an input power of 150 mW with a 100 μm spot size in this thick filter. Some irreversible changes in the optical constants of MBD-grown NLIFs are observed with sustained laser excitation [85] but much reduced compared with conventional thermally evaporated filters [84, 88]. This improvement is due partly to better

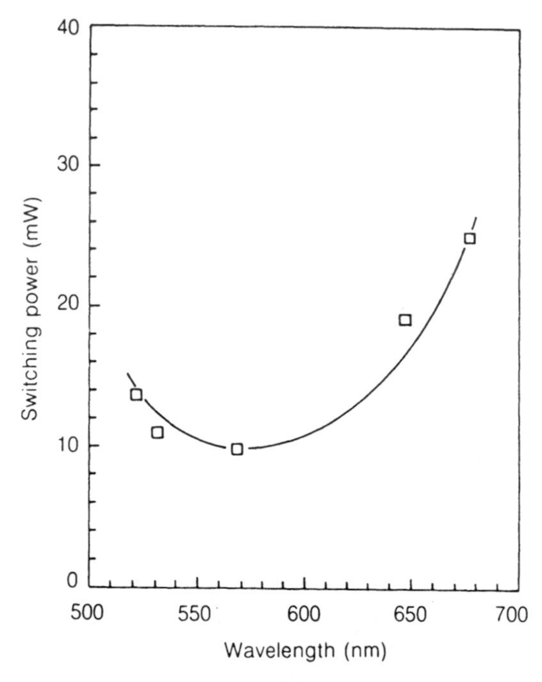

Fig. 14.13 Measured critical switching power for optical bistability as a function of wavelength compared to theory for filter shown in Fig. 14.12[86].

material quality and partly because the thicker layers allow better optimization of the devices.

The ultrafast switching dynamics of ZnSe NLIFs have been investigated by Bigot *et al.* [89–91] and Wherrett *et al.* [92]. In these experiments, carriers were generated by short optical pulses when the NLIF bistable device was biased with CW argon [89] or He–Ne [92] laser beams. Carrier recombination in these filters is very fast owing to trapping of the carriers at the small microcrystallite surfaces such that rapid switching can be induced by the heat generated by the relaxing carriers. Thermally induced refractive index changes were observed on a timescale less than 20 ps [92]. Opto-electronic bistability is therefore unlikely in NLIFs.

Sahlen [93, 94] has studied optical bistability in CdSe NLIFs with a bandgap energy around 710 nm. The reason for adopting this material was to exploit the band tail absorption in the wavelength region 800–850 nm for compatibility with commercial high-power semiconductor lasers. The evaporated structures consisted of ZnS/MgF_2 multilayer reflecting stacks with a spacer layer consisting of CdSe of thickness between 2 and 40 quarter-wavelengths. It was found that the thicker etalons produced lower critical switching power devices consistent with theory, and bistability could be achieved in the range 800–870 nm at input powers down to 1.4 mW with a spot radius of 4–5 μm. Some initial drift was observed in the operating point during the first hour of operation which then stabilized. Optical bistability was demonstrated at 3 mW using a commercial laser diode operating at 844 nm.

The usability of thermo-optic NLIFs in optical circuits has been significantly improved by adopting a bistable etalon with absorbed transmission (BEAT) structure [95]. In this case, a totally absorbing layer (e.g. aluminium) is introduced on one side of the filter (Fig. 14.14(a)). Exposure of this layer to an input beam can efficiently heat and tune the etalon. The hold beam and the reflected output enter and exit from the other side of the structure unaffected by the absorber. This device is therefore restricted to reflective operation, but this limitation is more than outweighed by the advantages of improved performance. Indeed reflective operation of bistable devices gives several advantages [39]. By removing the internal absorption and placing it outside the cavity, the BEAT device releases the restriction of operating close to the bandgap energy and can take advantage of the higher etalon finesse at longer wavelengths. While the absorption of the thin film ZnSe falls from 600 cm^{-1} at 514 nm to $< 100\,cm^{-1}$ at 830 nm, the optothermal coefficient decreases by less than a factor of 2. Walker [95] predicted a reduction in critical switching power by a factor of ~ 3.5 for symmetric BEAT devices with front and back mirror reflectivities equal compared with an otherwise optimised filter, and demonstrated a BEAT device with a critical switching power of 4 mW operated with an 834 nm dye laser and a 10 μm spot diameter. Because of the low cavity absorption, the contrast between upper and lower states can be large in a BEAT device, limited only by the minimum on-resonance reflectivity. Buller *et al.* [96] have demonstrated the advantage of using asymmetric mirror stacks for use with 830 nm laser diodes. By using off-axis address, a hard-limiting threshold response was obtained with a BEAT device [97] (Fig. 14.14(b)).

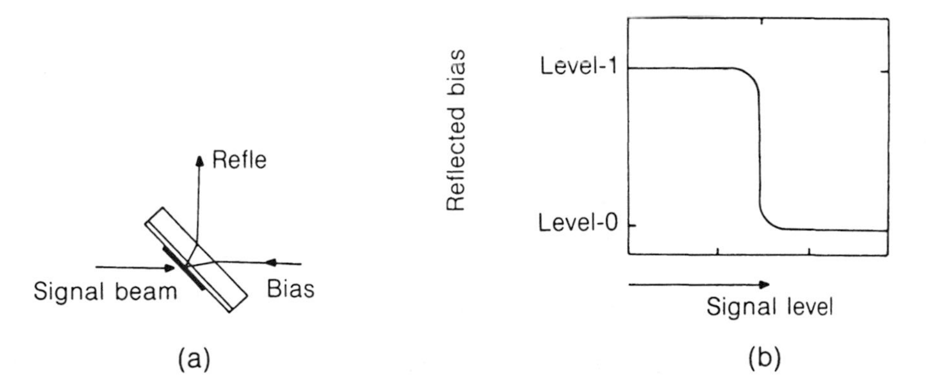

Fig. 14.14 (a) Beat device with fully absorbing layer placed on one side of the etalon, and (b) hard limiting characteristic using the configuration shown in (a) [97].

Abraham *et al.* [98] predicted through extensive theoretical studies of heat flow that switching powers could be significantly reduced to around 100 μW by pixellation of NLIFs. Pixellation also alleviates problems with cross-talk at high packing densities. A number of other theoretical studies have addressed this question [99–102]. Kar *et al.* [103] have reported the reduction of switch power by pixellation of ZnSe NLIFs using the 193 nm output of an ArF excimer laser. Using 10x10 arrays of 25 μm × 25 μm pixels of depth 20 μm, the critical switching power of BEAT–NLIF devices was measured at 2.4 mW compared with 4.2 mW in a uniform area of the same filter. It was anticipated that the switching power could be reduced to 100 μW with fast switching rates using smaller pixels.

A variant of the NLIF is an electrically addressable spatial high modulator (SLM) developed by the Heriot-Watt group [104]. In this case, the NLIF was incorporated on the inside of the glass face-plate of a vidicon tube prior to vacuum sealing. The device operates by heating of the NLIF by the electron beam causing tuning of the NLIF such that it can be read by an optical beam reflected from the front surface. The device can be operated in two regimes. In the optically linear case, the read beam is of insufficient irradiance to induce optical tuning of the interference filter and instead provides a purely passive interrogation of the SLM reflectivity. In the optically non-linear mode, higher optical irradiances are used in order to bias the NLIF into the strongly non-linear operating regime. If held close to an optically bistable switch point by the optical beam then the electron beam can be used to switch the device into the upper state. When the incident optical beam is configured as an array of spots, a binary image can be stored on the NLIF by the presence of the incident beams and read by monitoring the reflection of these beams.

14.6 OPTICAL CIRCUITS

Optically bistable elements have been considered for use in various forms of information processing from all-optical data routing switches in optical communications systems to general purpose optical computing [6]. Optically bistable devices are most suited to pipelined, iterative circuits using arrays of switching devices with a free-space interconnection scheme determining the algorithm for computation [105–107]. Interconnection can be provided by geometrical optics or holographic elements. Large-area II–VI NLIFs have allowed the first demonstration of a number of all-optical digital circuits which make use of the free-space interconnect properties of optics. NLIFs have been used to demonstrate parallel digital loop circuits [106], symbolic substitution [108–110], flip–flop [111] and full adder [111–113] circuits, a digital edge extractor [6, 111] and a cellular logic image processor [114].

There are a number of basic requirements which need to be satisfied for bistable elements to be usefully implemented in optical circuits [115]. A primary requirement is cascadability, i.e. the output of one device should be compatible with the input of the next device and the logic level restored at each output stage. A useful logic device should be capable of switching more than one subsequent device (up to ten in some architectures). This is known as the fan-out of the device. Cascadability requires optical gain between the input and output of the bistable device. High contrast between the bistable states and avoidance of multiple-beam effects such as interference are essential. The output of each device should be isolated from the input. To operate a large number of devices repetitively, uniformity and stability over a range of operating conditions are required. Particular care must be addressed to heat sinking. For large arrays, the switching energy must be minimized in order to keep the total thermal dissipation low and to minimize the requirements on the laser supplying optical power to the devices. Pixellation of the devices will probably be necessary in order to avoid cross-talk. The speed required of the devices will be governed by the application. For telecommunication purposes, a data rate of 1 Gbit s^{-1} requires recovery times of the devices to be less than a nanosecond. For parallel computing, a processing rate of 10^{10} gate switches cm^{-2} could be achieved by employing arrays of 100×100 devices with 1 μs cycle times. It is reasonable to assume that a 1 cm^2 plate could dissipate 10 W such that the power absorbed per device must be no more than 1 mW for this size of array.

A number of configurations of ZnSe NLIF bistable devices and circuits have been demonstrated by the group at Heriot-Watt University in Scotland. Basic to several of these is a configuration called 'lock and clock' [106] in which data can be passed continuously around an optical loop containing bistable plates to allow iterative computations (Fig. 14.15). Optical bistability provides temporary memory or optical time delay to pass data (or a complete image) from one plate to the next by the phased clocking of the holding beams on and off. Since the recovery times for optothermal devices are slow

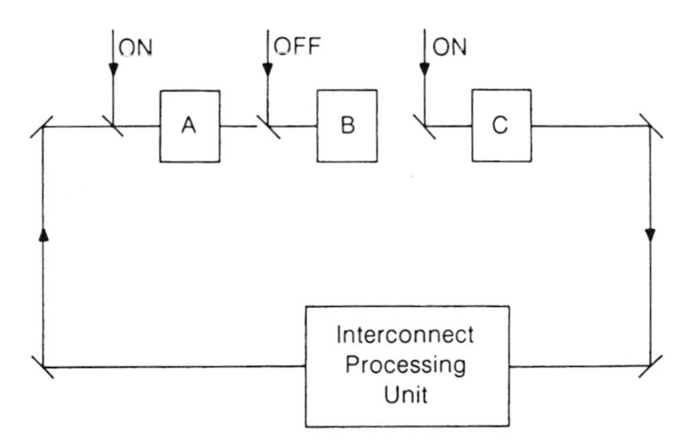

Fig. 14.15 Schematic of the lock and clock architecture [106].

compared with the time for light to complete a loop of any reasonable size, this architecture synchronizes the flow of images without the data contaminating itself. A minimum of three bistable plates in the loop are necessary whereby one plate acts as a shutter between data held on the plates on either side. An important aspect of circuits implemented in this way is the restoration of logic levels and the avoidance of error accumulation over multiple cycles. An implementation of the lock-and-clock architecture is shown in Fig. 14.16, which demonstrates a three-OR-gate circuit [106]. This circuit was demonstrated using a 514 nm argon ion laser with each bistable device operated in transmission mode and clocking achieved using computer-controlled acousto-optic modulators. A similar circuit was constructed with one of the bistable plates operated in reflection to give a NOR gate, thus producing signal inversion on each cycle.

Tooley *et al.* [111] have successfully constructed two other loop circuits; a four-channel flip–flop which employs the non-hysteretic response of ZnSe NLIFs, and a full adder which makes use of both the transmission and the reflection of a single device simultaneously to achieve this function with a single gate (compared with the five electronic gates needed to complete the same function electronically). In addition, the principle of digital image edge extraction has been demonstrated incorporating two NAND and one NOT gates in 2D arrays with holographic interconnection [6].

Considerable improvements in circuit operation have been achieved by adopting the BEAT principle. Craig *et al.* [116] have demonstrated an all-optical programmable logic gate with ZnSe NLIF–BEAT devices using an infrared dye laser to give the eight two-input logic functions, OFF, NOR, AND, XOR, NAND, XNOR, OR and ON, using two bistable gates [117]. Three ZnSe BEAT devices with five 833 nm laser diodes providing both power sources and input signals were employed to illustrate the operation of one channel of an optical–cellular logic image processing circuit (Fig. 14.17). The use of laser diodes in such circuits is a significant refinement over the use of bulky and expensive ion lasers.

The lock-and-clock data transfer principle has been demonstrated with 15×15 parallelism [114]. The ZnSe NLIF–BEAT structures for this work were deposited onto a $3\,\mu m$ thick layer of polyimide spun onto a sapphire substrate in order to control the thermal properties and operated with a CW Nd:YAG laser as the power source. Binary phase gratings (Dammann gratings) were used to generate two patterns of 15×15 focused spots for two NLIFs (Fig. 14.18). The reflected output of one, NLIF1 was imaged onto the second array, NLIF2. With NLIF1 'on' (low reflectivity), NLIF2 was 'off' (high reflectivity). Resetting NLIF1 by interrupting its input caused the array on NLIF2 to switch to 'on' (low reflectivity) and latch such that it remained in this state when the interconnect path was blocked. Arbitrary patterns could be imposed on NLIF1 and transfered to NLIF2 with $> 90\%$ accuracy.

Simultaneous bistable switching of 10 pixels in a ZnS NLIF with a potential

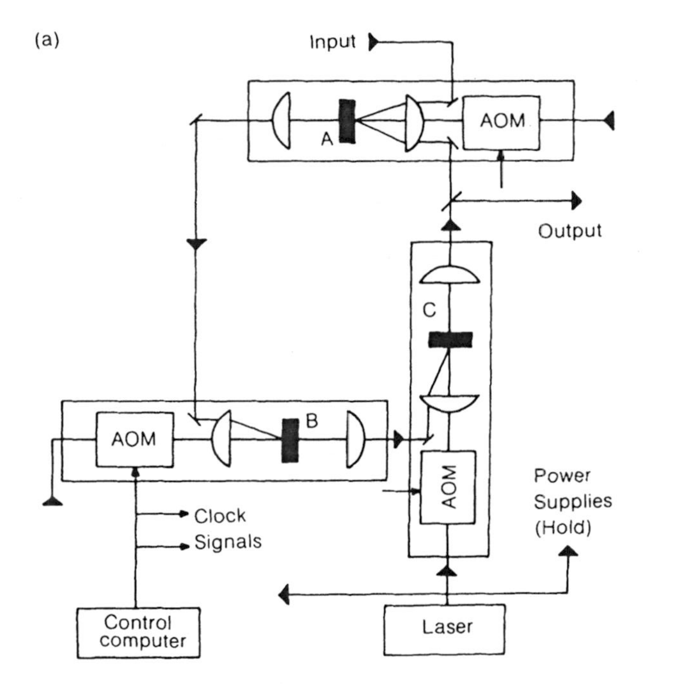

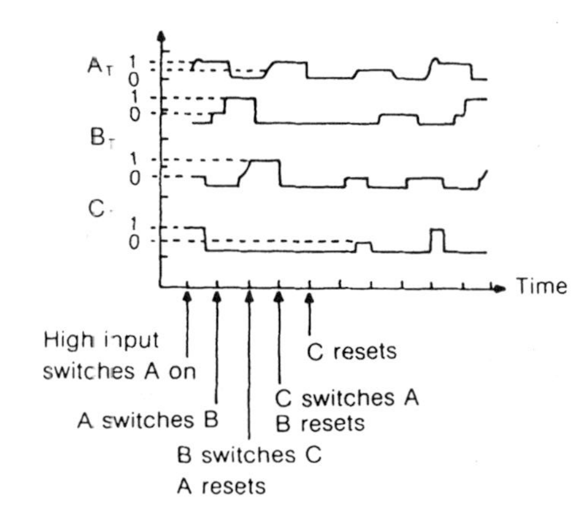

Fig. 14.16 (a) Schematic diagram of a three OR-gate loop circuit employing the lock and clock principle with ZnSe NLIFs and (b) transmission output waveforms from each gate [106].

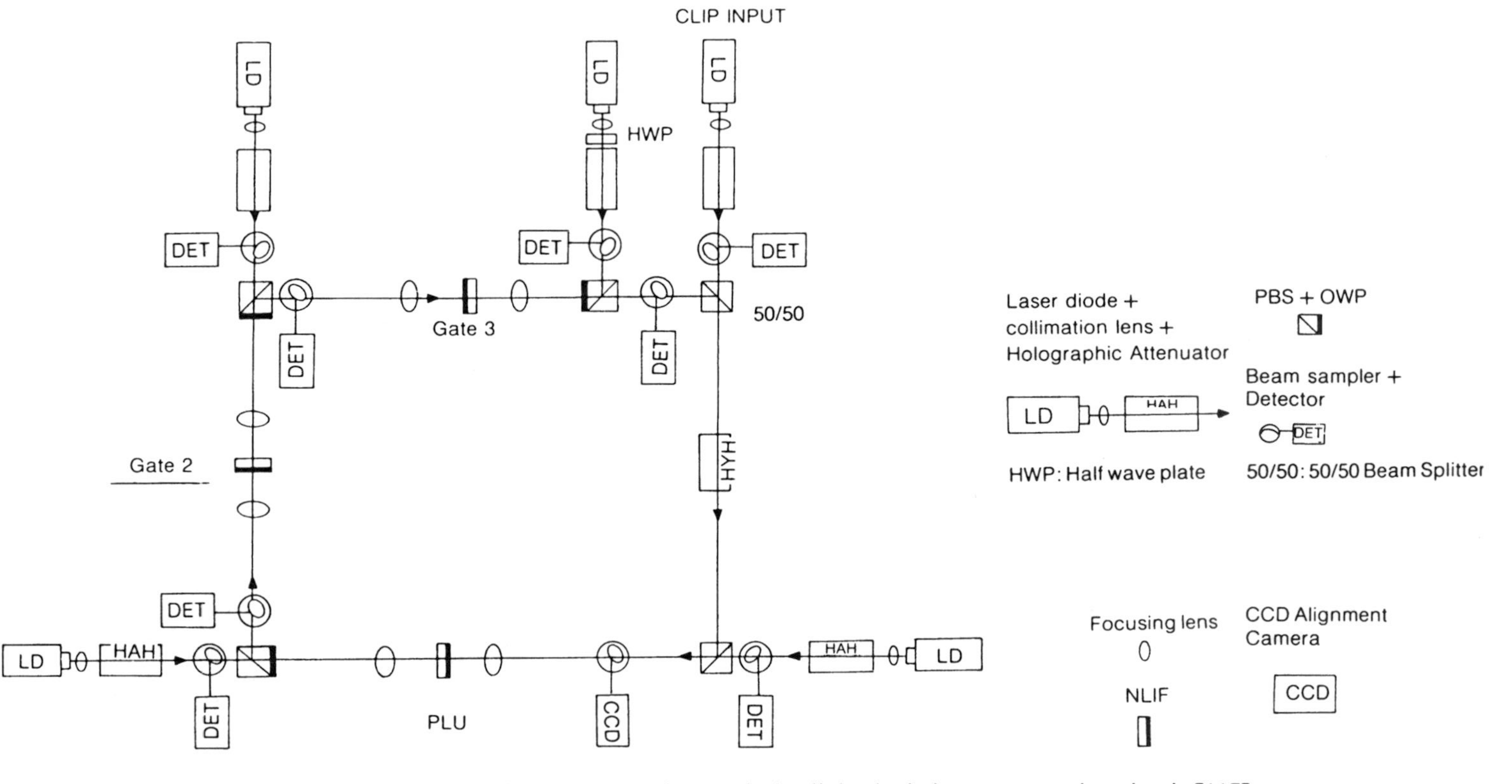

Fig. 14.17 Implementation of one channel of an optical cellular logic image processing circuit [117].

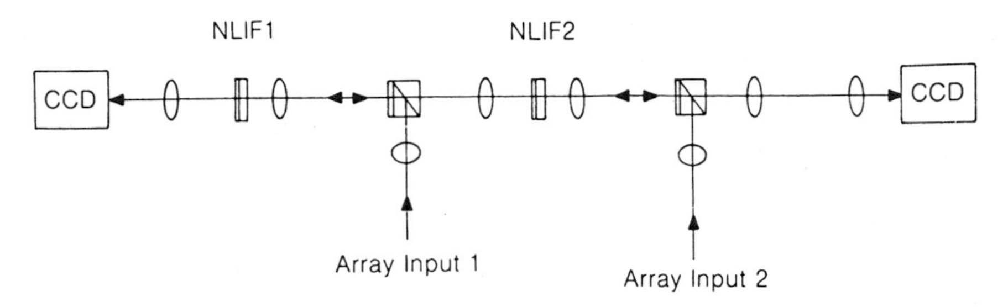

Fig. 14.18 Demonstation of lock and clock transfer of a 15 × 15 array between two ZnSe NLIFs /114/.

fan-out of 4 has been achieved by the group at the Optical Sciences Center in Tucson and used to demonstrate pattern recognition and symbolic substitution concepts [108–110]. For the pattern demonstration, a 2 × 3 array was produced using a fly's eye lens array. The array of beams was split into two and recombined on the NLIF with a relative spatial shift. The NLIF was configured as AND gates such as to provide pattern recognition of a masked pattern of spots. A second ZnS NLIF provided symbolic scription, thus demonstrating cascadability between the two arrays.

Lewis and West [118] have demonstrated the use of ZnSe NLIFs as an erasable memory for a human–machine interface. The output of an argon laser was divided into four holding beams on a ZnSe bistable plate. Some of the light was split off at a beam splitter and sent through an optical fibre to act as a light pen to switch the bistable devices individually. As the mechanism is optothermal, this could also be accomplished with a heated tip, thus avoiding the critical angular orientation of the light pen. Extending this concept to larger arrays could provide a scratch-pad for entering data or instructions into a computer.

14.7 CONCLUSIONS

A great deal of progress has been made in understanding optical non-linearities and bistability in widegap II–VI semiconductors over the last decade. Devices have been successfully developed for the demonstration of the first all-optical digital circuits while our knowledge of optical interactions in semiconductors has benefited greatly from research carried out to clarify the mechanisms responsible for optical bistability.

Speed is one of the potential advantages of exploiting optics in signal processing and routing. Unfortunately, ultrafast bound electron optical non-linearities in semiconductors are limited in their usefulness at photon energies greater than half the bandgap energy because of significant two-photon absorption at the high intensities required to induce sufficient phase shifts [34]. However, although ultrahigh speed switching is desirable, it is not necessarily an overriding requirement and indeed speed can generally be traded successfully for increased sensitivity when benefit is gained from the very high degree of parallel processing offered by optics. Although highly parallel systems have already been achieved in electronic array processors, optics offers capabilities far in excess of these if the potential of up to $10^6\,cm^{-2}$ switching elements can be combined with the complex interconnect capabilities achievable with either fixed holographic interconnects or reconfigurable interconnects in the future. In this case, it is essential that optical logic devices can be accessed at very low power levels, thus minimizing the overall system energy and making possible large arrays of devices. Resonant, free-carrier optical non-linearities offer greatly increased sensitivity; however, the strong bandgap dependence of free-carrier-induced optical non-linearities makes widegap II–VI materials much less attractive than lower-gap materials. Bistability can be achieved at low power by exploiting the saturation of bound excitons if the inconvenience of operating at low temperatures can be tolerated.

Widegap II–VI semiconductors have found greatest success in optically bistable devices which operate via optothermal non-linearities. Non-linear optical interference filters exhibiting optical bistability at a few milliwatts of input power have provided test-beds for exploring innovative ideas on parallel architectures for optical computing. These devices have allowed studies of the limitations and tolerances needed to achieve useful all-optical circuits for potential applications in data routing, image processing, radar array processing and all-optical digital optical computing. As II–VI semiconductors technology continues to advance in terms of doping, heterostructures and low-dimensional structures, we may expect to see hybrid optical switching devices being developed similar to III–V SEED-type devices [66].

REFERENCES

1. Karpushko, F.K. and Sinitsyn, G.V. (1978) *J. Appl. Spectrosc (USSR)*, **30**, 1323.
2. Mandel, P., Smith, S.D. and Wherrett, B.S. (eds), *"From Optical Bistability Towards Optical Computing: the EJOB Project*, (1987) North-Holland, Amsterdam.
3. Gibbs, H.M. (1985) *Optical Bistability: Controlling Light with Light*, Academic Press, New York.
4. Haug, H. (ed.) (1988) *Optical Nonlinearities and Instabilities in Semiconductors*, Academic Press, New York.
5. Miller, A., Staromlynska, J., Muirhead, I.T., Lewis, K.L., Craig, D. and Steward, G. (1988) *J. Cryst. Growth*, **86**, 858.
6. Wherrett, B.S., (1990) *Mol. Cryst. Liq. Cryst. A*, **182**, 163.
7. Woerdman, J.P. and Bolger, B. (1969) *Phys. Lett. A*, **30**, 164.
8. Miller, D.A.B., Mozolowski, M., Miller, A. and Smith, S.D. (1978) *Opt. Commun*, **27**, 133.
9. Gibbs, H.M., McCall, S.L., Venkatesan, T.N.C., Gossard, A.C., Passner, A. and Weigmann, W. (1979) *Appl. Phys. Lett.* **35**, 451.
10. Miller, A., Miller, D.A.B. and Smith, S.D. (1981) *Adv. Phys.*, **30**, 697.

11. Maker, D.D., and Terhune, R.W., (1965) *Phys. Rev. A*, **137**, 801.
12. Miller, A., Parry, G. and Daley, R. (1984) *IEEE J. Quantum Electron.*, **20**, 710.
13. Miller, D.A.B., Seaton, C.T., Prise, M.E. and Smith, S.D. (1981) *Phys. Rev. Lett.*, **47**, 197.
14. Ji, W., Milward, J.R., Kar, A.K., Wherrett, B.S. and Pidgeon, C.R. (1990) *J. Opt. Soc. Am. B.*, **7**, 868.
15. Dagenais, M. (1983) *Appl. Phys. Lett.*, **43**, 742.
16. Anderson, D.R., Kolodziejski, L.A., Gunshor, R.L., Dutta, S., Kaplan, A.E. and Nurmikko, A.V. (1986) *Appl. Phys. Lett.*, **48**, 1559.
17. Kolodziejski, L.A., Gunshor, R.L., Otsuka, N., Dutta, S., Becker, W.M. and Nurmikko, A.V. (1986) *IEEE J. Quantum Electron.*, **22**, 1666.
18. Peyghambarian, N., Park, S.H., Koch, S.W., Jeffery, A., Potts, J.E. and Cheng, H. (1988) *Appl. Phys. Lett.*, **52**, 182.
19. Banyai, L. and Koch, S.W., (1986) *Z. Phys. B*, **63**, 283.
20. Lee, D., Zucker, J.E., Johnson, A.M., Feldman, R.D. and Austin, R.F. (1990) *Appl. Phys. Lett.*, **57**, 1132.
21. Jain, R.K. and Lind, R.C. (1983) *J. Opt. Soc. Am.*, **73**, 646.
22. Olbright, G.R. and Peyghambarian, N. (1986) *Appl. Phys. Lett.*, **48**, 1184.
23. Yumoto, J., Fukushima, S. and Kubodera, K. (1987) *Opt. Lett.*, **12**, 832.
24. Cotter, D. (1986) *Digest International Quantum Electronics Conference, San Francisco*, 1986.
25. Mitsunaga, M., Shinojima, H. and Kubodera, K. (1988) *J. Opt. Soc. Am. B*, **5**, 1448.
26. Coutaz, J.-L. and Kull, M. (1991) *J. Opt. Soc. Am. B*, **8**, 95.
27. Van Stryland, E.W., Woodall, M.A., Vanherzeele, H. and Soileau, M.J. (1985) *Opt. Lett.*, **10**, 490.
28. Van Stryland, E.W., Vanherzeele, H., Woodall, M.A., Soileau, M.J., Smirl, A.L., Guha, S. and Boggess, T.F. (1985) *Opt. Eng.*, **24**, 613.
29. Pidgeon, C.R., Wherrett, B.S., Johnson, A.M., Dempsey, J. and Miller, A. (1979) *Phys. Rev. Lett.*, **42**, 1785.
30. Weiler, M. (1981) *Solid State Commun.*, **39**, 937.
31. Miller, A., Johnson, A., Dempsey, J., Smith, J. Pidgeon, C.R. and Holah, G.D. (1979) *J. Phys. C*, **12**, 4839.
32. Guha, S., Van Stryland, E.W. and Soileau, M.J. (1985) *Opt. Lett.* **10**, 285.
33. Sheik-Bahae, M., Hagan, D.J. and Van Stryland, E.W. (1990) *Phys. Rev. Lett.*, **65**, 96.
34. Sheik-Bahae, M., Hutchings, D.C., Hagan, D.J. and Van Stryland, E.W. (1991) *IEEE J. Quantum Electron.*, **27**, 1296.
35. Honerlage, B., Levy, R., Grun, J.B., Klingshirn, C. and Bohnert, K. (1985) *Phys. Rep.*, **124**, 161.
36. Kalt, H., Renner, R. and Klingshirn, C. (1986) *IEEE J. Quantum Electron.*, **22**, 1312.
37. Moss, T.S. (1979) *Phys. Status Solidi B*, **101**, 555.
38. Miller, D.A.B.(1981) *IEEE J. Quantum Electron.*, **17**, 306.
39. Wherrett, B.S., (1984) *IEEE J. Quantum Electron.*, **20**, 646.
40. Miller, A. and Parry, G. (1984) *Opt. Quantum Electron.*, **16**, 339.
41. Mathew, J.G.H., Taghizadeh, M.R., Abraham, E., Janossy, I. and S.D. Smith, In *Optical Bistability III* (eds H.M. Gibbs, P. Mandel, N. Peyghambarian and Smith, S.D.), Springer Proceedings in Physics & Springer Berlin, p. 57.
42. Miller, A., Craig, D. and Steward, G. In *Optical Bistability III* (eds H.M. Gibbs, P. Mandel, N. Peyghambarian and Smith S.D.), Springer Proceedings in Physics & Springer, Berlin p. 140.
43. Wherrett, B.S., Tooley, F.A.P. and Smith, S.D. (1984) *Opt. Commun.*, **52**, 301.

44. Haug, H., Koch, S.W. and Lindberg, M. (1986) *Phys. Scr.*, **13**, 178.
45. Bohnert, K., Kalt, H. and Klingshirn, C. (1983) *Appl. Phys. Lett.*, **43**, 1088.
46. Rossmann, H., Henneberger, F. and Voigt, J. (1983) *Phys. Status Solidi B*, **115**, K63.
47. Henneberger, F. and Rossmann, H. (1984) *Phys. Status Solidi B*, **121**, 685.
48. Schmidt, H.E., Haug, H. and Koch, S.W. (1984) *Appl. Phys. Lett.*, **44**, 787.
49. Rossmann, H. and Henneberger, F. (1985) *Phys. Status Solidi*, **131**, 185.
50. Dagenais, M. (1984) *Appl. Phys. Lett.*, **45**, 1267.
51. Nguyen, H.X. and Zimmermann, R. (1984) *Phys. Status Solidi B*, **124**, 191.
52. Fidorra, F., Wegener, M., Bigot, J.Y., Honerlage, B. and Klingshirn, C. (1986) *J. Lumin.* **35**, 43.
53. Dagenais, M. and Sharfin, W.F. (1984) *Appl. Phys. Lett.*, **45**, 210.
54. Dagenais, M. and Winful, H.G. (1984) *Appl. Phys. Lett.*, **44**, 574.
55. Dagenais, M. and Sharfin, W.F. (1985) *Appl. Phys. Lett.*, **46**, 230.
56. Dagenais, M. and Sharfin, W.F. (1986) *Opt. Eng.*, **25**, 219.
57. Dagenais, M., Surkis, A., Sharfin, W.F. and Winful, H.G. (1985) *IEEE J. Quantum Electron.*, **21**, 1458.
58. Wegener, M. Klingshirn, C., Koch, S.W. and Banyai, L. (1986) *Semicond. Sci. Technol.*, **1**, 366.
59. Henneberger, F., Rossmann, H. and Schulzgen, A. (1988) *Phys. Status Solidi B*, **145**, K83.
60. Henneberger, F., Puls, J., Rossmann, H., Kretzschmar, M., Spiegelberg, C. and Schulzgen, A. (1988) *J. Phys. (Paris), Suppl. 6*, **49**, C2-91.
61. Taghizadeh, M., Janossy, I. and Smith, S.D. (1985) *Appl. Phys. Lett.*, **46**, 331.
62. Kar, A.K. and Wherrett, B.S. (1986) *J. Opt. Soc. Am. B*, **3**, 345.
63. Kim, B.G., Garmire, E., Shibata, N. and Zembutsu, S. (1987) *Appl. Phys. Lett.*, **51**, 475.
64. Gibbs, H.M., Olbright, G.R., Peyghambarian, N., Schmidt, H.E., Koch, S.W. and Haug, H. (1985) *Phys. Rev. A*, **32**, 692.
65. Wegener, M., Witt, A., Klingshirn, C., Gnass, D., Iyechika, Y. and Jager, D. (1988) *J. Phys. (Paris), Suppl. 6*, **49**, C2-109.
66. Miller, D.A.B., Chemla, D.S., Damen, T.C., Gossard, A.C., Wiegmann, W., Wood, T.H. and Burns, C.A. (1984) *Appl. Phys. Lett.*, **45**, 13.
67. Smith, S.D., Mathew, J.G.H., Taghizadeh, M.R., Walker, A.C., Wherrett, B.S. and Hendry, A. (1984) *Opt. Commun.*, **51**, 357.
68. Olbright, G.R., Peyghambarian, N., Gibbs, H.M., Mcleod, H.A. and Van Milligan, F. (1984) *Appl. Phys. Lett.*, **45**, 1031.
69. Janossy, I., Taghizadeh, M.R., Mathew, J.G.H. and Smith, S.D. (1985) *IEEE J. Quantum Electron.*, **21**, 1447.
70. Sahlen O. (1986) *Opt. Commun.*, **59**, 238.
71. Firth, W.J. and Wright, E.M. (1982) *Opt. Commun.*, **40**, 233.
72. Firth, W.J., Abraham, E., Wright, E.M., Galbraith, I. and Wherrett, B.S. (1984) *Philos. Trans. R. Soc. London Sec. A*, **313**, 299.
73. Firth, W.J., Galbraith, I. and Wright, E.M. (1985) *J. Opt. Soc. Am. B*, **2**, 1005.
74. Abraham, E. and Ogilvy, I.J.M. (1987) *Appl. Phys. B*, **42**, 31.
75. Abraham, E. and Rae, C. (1987) *J. Opt. Soc. Am. B*, **4**, 490.
76. Halley, J.M. and Midwinter, J.E. (1986) *Opt. Quantum Electron.*, **18**, 57.
77. Halley, J.M. and Midwinter, J.E. (1987) *J. Appl. Phys.*, **62**, 4055.
78. Janossy, I., Taghizadeh, M.R., Mathew, J.G.H. and Smith, S.D. (1986) *IEEE J. Quantum Electron.*, **22**, 2224.
79. Wherrett, B.S., Hutchings, D. and Russell, D. (1986) *J. Opt. Soc. Am. B*, **3**, 351.
80. Wherrett, B.S., Kar A.K., Hutchings D., Russell, D. and Clement, H. (1986) *Opt. Acta*, **33**, 517.

81. Hutchings, D.C., Wherrett, B.S. and Frank, D. (1988) *J. Phys. (Paris)*, **49**, C2-119.
82. Hutchings, D.C., Wang, C.H. and Wherrett, B.S. (1991) *J. Opt. Soc. Am. B.*, **8**, 618.
83. Chow, Y.T., Wherrett, B.S., Van Stryland, Van, McGuckin, B.T., Hutchings, D., Mathew, J.G.H., Miller, A. and Lewis, K.L. (1986) *J. Opt. Soc. Am. B.*, **3**, 1535.
84. Lewis, K.L. and Savage, J.A. (1985) *Proceedings of the Symposium on Laser Induced Damage in Optical Materials*, National Bureau of Standards, Washington, DC, NBS Special Publication 688.
85. Miller, A., Staromlynska, J., Muirhead, I.T. and Lewis, K.L. (1988) *J. Mod. Opt.*, **35**, 529.
86. Miller, A., Muirhead, I.T., Lewis, K.L., Staromlynska, J. and Welford, K.R. (1988) *J. Phys. (Paris)*, **49**, C2-105.
87. Miller, A., Muirhead, I.T., Lewis, K.L., Staromlynska, J. and Welford, K.R. (1990) In *Nonlinear Optics and Optical Computing* (eds) (S. Martellucci and A.N. Chester), Plenum, New York, p. 51.
88. Cambell, R.J., Mathew, J.G.H., Smith, S.D. and Walker, A.C. (1989) *J. Mod. Opt.*, **36**, 323.
89. Bigot, J.Y., Daunois, A., Leonelli, R., Sence, M., Mathew, J.G.H., Smith, S.D. and Walker, A.C. (1986) *Appl. Phys. Lett.*, **49**, 844.
90. Bigot, J.Y., Daunois, A. and Grun, J.B. (1987) *Phys. Rev. A*, **35**, 3810.
91. Daunois, A., Bigot, J.Y., Leonelli, R. and Smith, S.D. (1987) *Opt. Commun.*, **62**, 360.
92. Wherrett, B.S., Darzi, A.K., Chow, Y.T., McGuckin, B.T. and Van Styland, E.W. (1990) *J. Opt. Soc. Am. B.*, **7**, 217.
93. Sahlen, O. (1988) *J. Opt. Soc. Am. B*, **5**, 82.
94. Sahlen, O. (1988) *J. Phys. (Paris)*, **49**, C2-127.
95. Walker, A.C. (1986) *Opt. Commun.*, **59**, 145.
96. Buller, G.S., Paton, C.R., Smith, S.D. and Walker, A.C. (1989) *Opt. Commun.*, **70**, 522.
97. Wherrett, B.S., Chow, Y.T., Rhoomy-Darzi, A.K. and Lloyd, A.D. (1989) *Phys. Scr.*, **25**, 247.
98. Abraham, E., Godslave, C. and Wherrett, B.S. (1988) *J. Phys. (Paris)*, **49**, C2-443.
99. Frank, D. and Wherrett, B.S. (1987) *Opt. Eng.*, **26**, 53.
100. Abraham, E., Godslave, C. and Wherrett, B.S. (1988) *J. Appl. Phys.*, **64**, 21.
101. Abraham, E., Kar, A.K., Suttie, M.R., Harris, R.M., Walker, A.C. and Smith, S.D. (1988) *J. Appl. Phys.*, **64**, 3393.
102. Walker, A.C., Taghizadeh, M.R., Mathew, J.G.H., Redmond, I., Cambell, R.J., Smith, S.D., Dempsey, J. and Lebreton, G. (1988) *Opt. Eng.*, **27**, 38.
103. Kar, A.K., Harris, R.M., Buller, G.S., Smith, S.D. and Walker, A.C. (1988) *J. Phys. (Paris)*, **49**, C2-443.
104. Walker, A.C., Smith, S.D., Cambell, R.J. and Mathew, J.G.H. (1988) *J. Phys. (Paris)*, **49**, C2-47.
105. Wherrett, B.S. (1988) *J. Phys. (Paris)*, **49**, C2-29.
106. Smith, S.D., Walker, A.C., Tooley, F.A.P. and Wherrett, B.S. (1987) *Nature* (London), **325**, 27.
107. Miller, (1986) A. *Nature (London)*, **323**, 13.
108. Jin, R., Wang, L., Sprague, R.W., Gibbs, H.M., Gigioli, G.C., Kulcke, H., Macleod, H.A., Peyghambarian, N., Olbright, G.R. and Warren, M. (1986) In *Optical Bistability III* (eds H.M. Gibbs, P. Mandel, N. Peyghambarian and S.D. Smith), Springer Proceedings in Physics 8, Springer, Berlin, p. 61.
109. Tsao, M.T., Wang, L., Jin, R., Sprague, R.W., Gigioli, G., Kulcke, H.M., Li, Y.D., Chou, H.M., Gibbs, H.M. and Peyghambarian, N. (1987) *Opt. Eng.*, **26**.
110. Peyghambarian, N. and Gibbs, H.M. (1988) In *Optical Nonlinearities and Instabilities in Semiconductors* (ed. H. Haug), Academic Press, New York, p. 295.

111. Tooley, F.A.P., Wherrett, B.S., Craft, N.C., Taghizadeh, M.R., Snowdon, J.F. and Smith, S.D. (1988) *J. Phys. (Paris)*, **49**, C2-459.
112. Tooley, F.A.P., Craft, N.C., Smith, S.D. and Wherrett, B.S. (1987) *Opt. Commun.*, **63**, 365.
113. Wherrett, B.S. (1985) *Opt. Commun.*, **56**, 87.
114. Wherrett, B.S., Craig, R.G.A., Snowdon, J.F., Buller, G.S., Tooley, F.A.P., Bowman, S., Pawley, G.S., Redmond, I.R., McKnight, D., Taghizadeh, M.R., Walker, A.C. and Smith, S.D. (1990) *Proceedings OE/LASE 90, Los Angeles*, Society of Photo-Optical Instrument Engineers.
115. Smith, S.D. (1984) *Philos. Trans. R. Soc. London*, Sec. A, **313**.
116. Craig, R.G.A., Buller, G.S., Tooley, F.A.P., Smith, S.D., Walker, A.C. and Wherrett, B.S. (1990) *Appl. Opt.*, **29**, 2148.
117. Wherrett, B.S. (1985) *Appl. Opt.*, **24**, 2867.
118. Lewis, M.F. and West, C.L. (1987) *Opt. Commun.*, **62**, 77.

15

Implanted widegap II–VI materials for electro-optic applications and electron-beam-pumped devices

M. Yamaguchi

15.1 IMPLANTED WIDEGAP II–VI MATERIALS FOR ELECTRO-OPTIC APPLICATIONS

Ion implantation into II–VI compound semiconductor materials is thought to be useful for p–n junction formation, especially for realizing minority-carrier injection light-emitting diodes (laser diodes). However, there are few studies on implantation into widegap II–VI materials [1, 2] compared with those into Si and III–V compound materials. In addition, in II–VI materials there are self-compensation problems resulting from lattice defects induced by ion implantation. In order to obtain high-performance widegap II–VI compound devices by ion implantation, it is necessary to understand defect structures and behaviours of defects and impurities. In this chapter, several approaches for ion implantation into widegap II–VI matcrials, cspecially ZnS, ZnSe CdS, are reviewed from the points of view of understanding defects and emission centres, and realizing p–n junction formation.

15.1.1 Implanted ion species in II–VI materials

Table 15.1 shows main implanted ion species in II–VI materials.

When group I impurities such as Li, Na, K and Ag substitute for group II atoms or group V impurities such as N, P, As and Sb substitute for group VI atoms, they will act as acceptor impurities in II–VI materials. In a similar way, groups III and VII impurities such as B, Al, Ga, In, F, Cl, Br and I will act as donor impurities in II–VI materials. These impurities are candidate

Table 15.1 Main implanted ion species into II–VI materials

Group	Ia	Ib	IIa	IIb	III	IV	V	VI	VII	VIII
type	p	SA			n		p		n	
	^{1}H									^{2}He
	^{3}Li				^{5}B		^{7}N		^{9}F	
	^{11}Na				^{13}Al		^{15}P	^{16}S	^{17}Cl	
	^{19}K	^{29}Cu	^{30}Zn		^{31}Ga		^{33}As	^{34}Se	^{35}Br	
		^{47}Ag	^{48}Cd		^{49}In		^{51}Sb	^{52}Te	^{53}I	

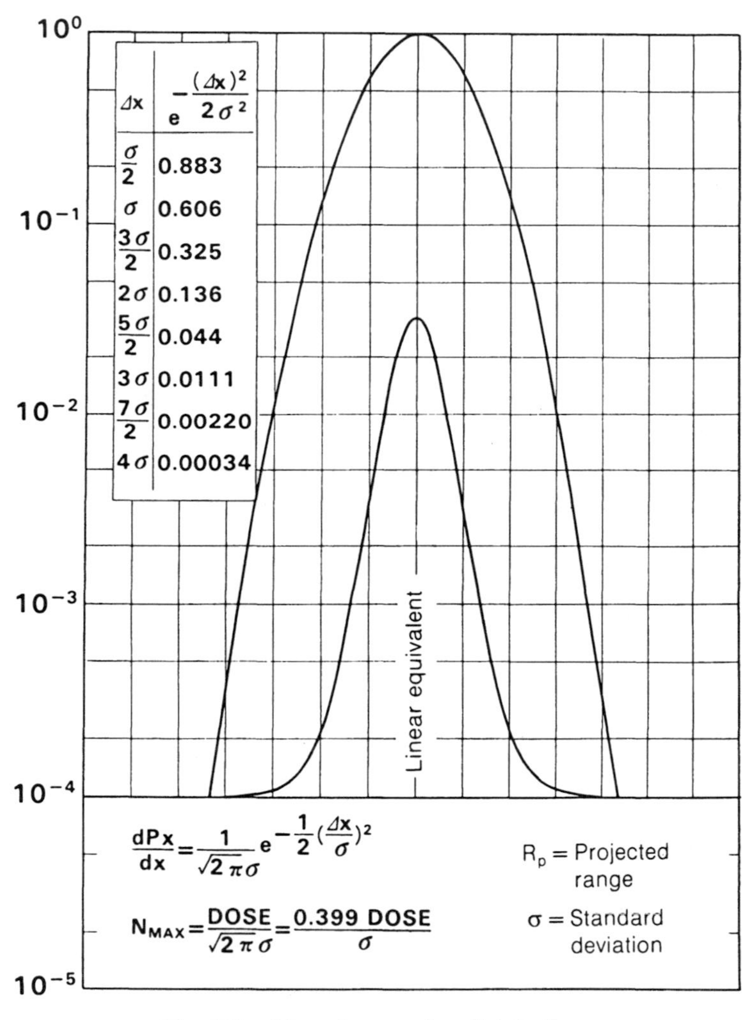

Fig. 15.1 Plot of a gaussian distribution.

elements for p–n junction formation. Moreover, light ions such as H and He are thought to be useful for insulating layer formation, that is, for Metal–insulator–semiconductor (MIS) structures. Recently, rare earth elements such as Nd, Yb and so forth have been implanted to study luminescence properties and to realize new luminescence devices.

In the choice of ion species for p–n junction formation, the following points should be considered:

1. shallow impurity level formation;
2. less defect introduction and defects that are easy to anneal out;
3. impurity profile.

The energy level of implanted impurities and defect introduction and

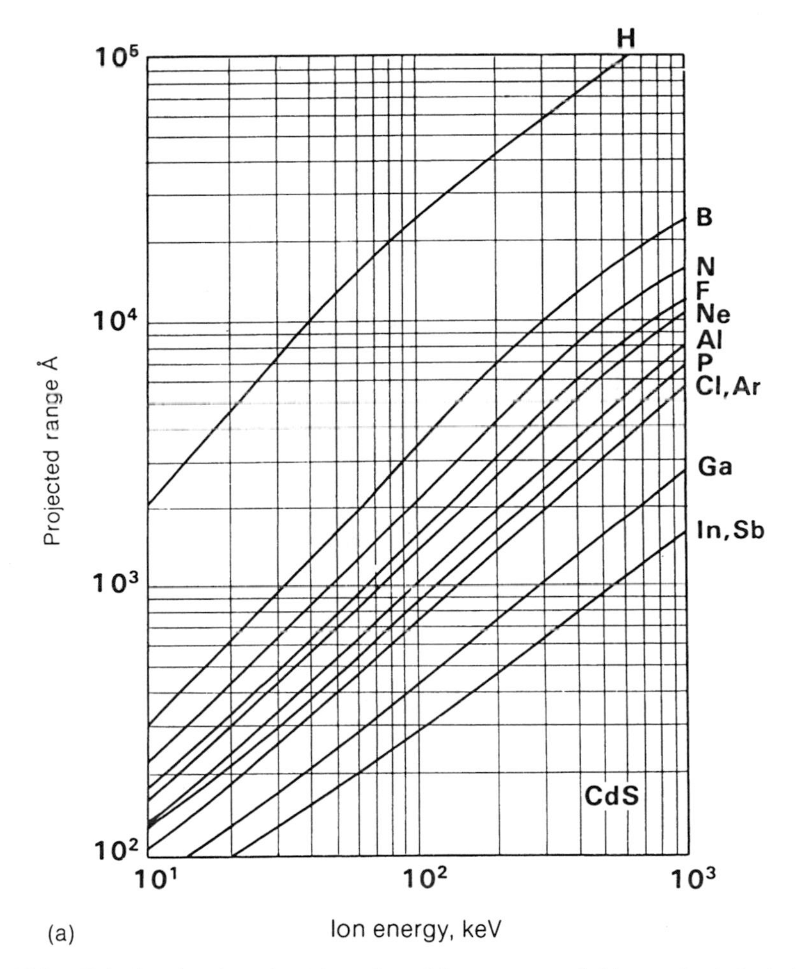

Fig. 15.2 Calculated values for (a) projected ion range and (b) standard deviation of main ion species in CdS as a function of ion energy [3].

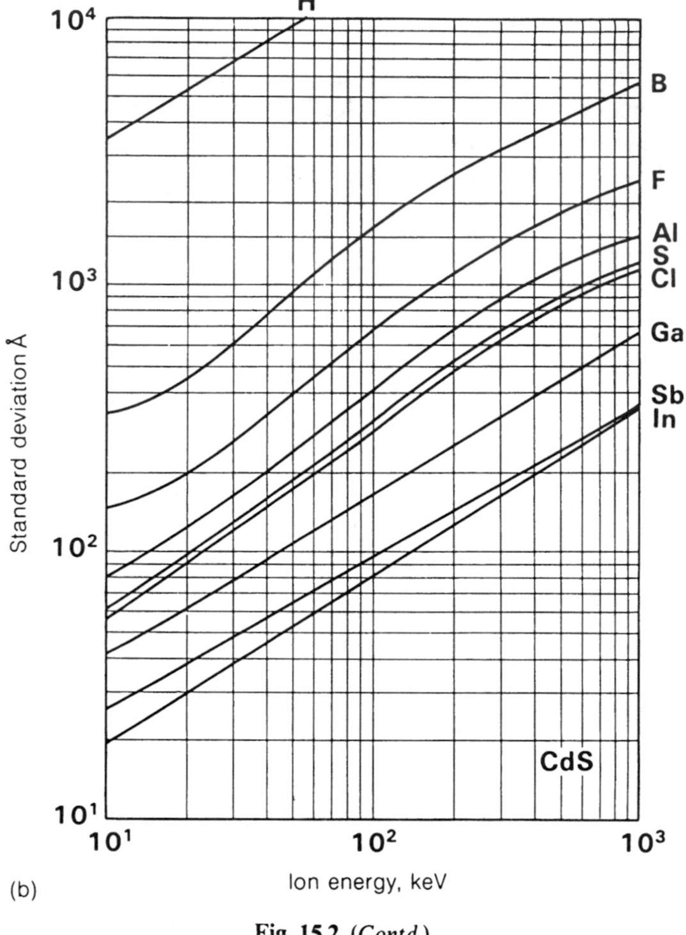

Fig. 15.2 (*Contd.*)

annealing behaviours are discussed in the following sections. As shown in Fig. 15.1, the depth profile of implanted ions $N(x)$ in II–VI materials is expected to be a gaussian distribution as a result of interaction between ions and target atoms and is expressed by the following equation:

$$N(x) = (N/\sqrt{2\pi}\sigma_p) \exp[-(x - R_p)^2/2\sigma_p^2] \qquad (15.1)$$

where x is the depth, R_p is the projected range of ions and σ_p is the standard deviation. Figure 15.2 shows calculated values for projected ion range and standard deviation of the main ion species in CdS as a function of ion energy [3].

15.1.2 Ion-induced defects and their annealing

When high-energy ions are implanted, lattice atoms are displaced by collisions of ions with II–VI semiconductor atoms. The maximum energy T_m transferred to lattice atoms by collision of implanted ions with lattice atoms is given by

$$T_m = 4M_1M_2E_0/(M_1 + M_2)^2 \tag{15.2}$$

where M_1 and M_2 are the masses of implanted ions and lattice atoms respectively and E_0 is the ion energy. When the transferred energy T is larger than the threshold displacement energy E_d of lattice atoms, vacancies and interstitials are induced by ion implantation. The number of displaced atoms N_d is expressed by

$$N_d = vn\phi \tag{15.3}$$

$$v = T/E_d \tag{15.4}$$

where n is the number of primary knock-on atoms per implanted ion, v is the average number of knock-on atoms by primary knock-on atom, ϕ is the number of implanted ions and T is the average energy of primary knock-on atoms. For example, in the case of ZnS, the threshold energies of Zn and S were reported to be 7.3 eV and 15.0 eV respectively [4].

Lattice defects in II–VI materials induced by ion implantation are thought to have similar depth profiles to that of implanted ions shown in Fig. 15.1.

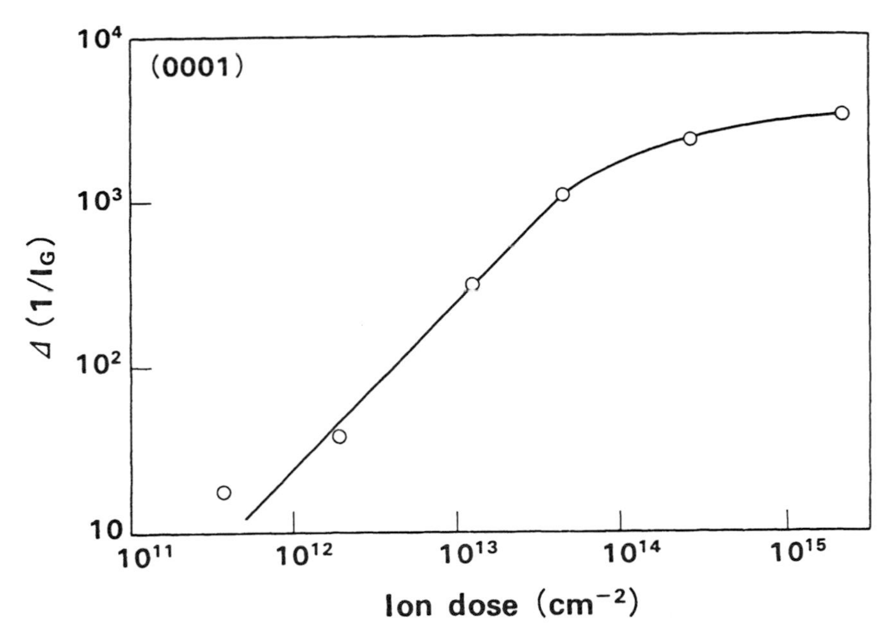

Fig. 15.3 Ion dose dependence of the green emission (near-band-edge emission at 5030 A at room temperature) intensity in 50 keV nitrogen-ion-implanted CdS [5].

It is necessary to understand and reduce the lattice defects in order to produce a high-quality implanted layer, because most lattice defects may act as trap centres, scattering centres and recombination centres, and thus reduce carrier concentration, carrier mobility and minority-carrier lifetime. Figure 15.3 shows the ion dose dependence of the green emission (near-band-edge emission at 5030 Å at room temperature) intensity in 50 keV nitrogen-ion-implanted CdS [5]. In this figure, changes of the inverse of the green emission intensity $\Delta(l/I_G)$ which indicates the concentration of the non-radiative recombination centres increase almost proportionally to the ion dose. These non-radiative recombination centres are thought to be due to lattice defects produced by ion implantation. In the worst case, lattice defects may change the semiconductor crystal into an amorphous state. The formation of an amorphous layer is dependent on ion dose, ion species, ion energy and substrate temperature. In general, ion implantation with ion doses less than the critical ion dose necessary to produce an amorphous layer is useful because, in high-ion-dose implantation, high-temperature annealing

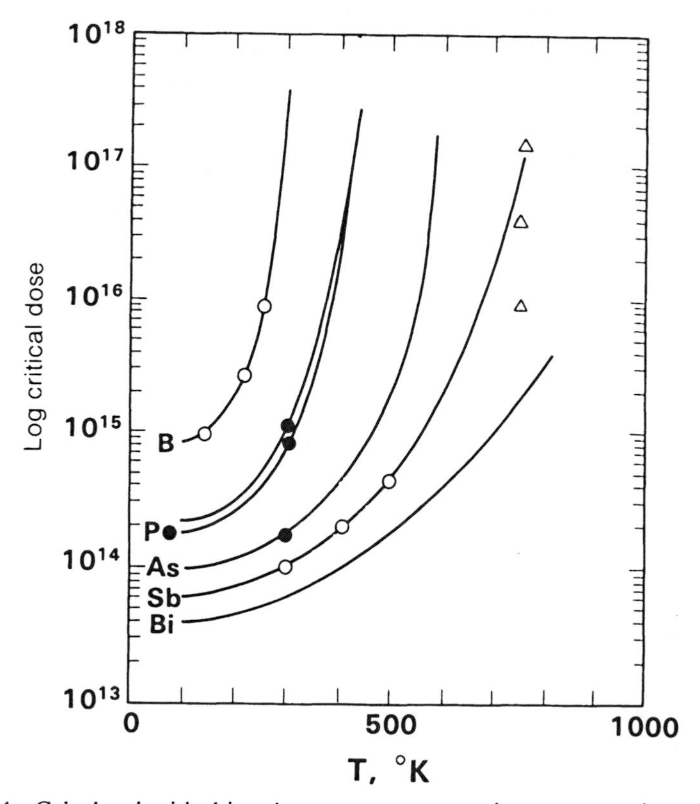

Fig. 15.4 Calculated critical ion dose necessary to produce an amorphous layer in Si as a result of ion bombardment [3].

is needed to recover the amorphous state. As an example, the critical ion dose is shown in Fig. 15.4 as a function of substrate temperature during implantation, for several ions into Si [3].

Ion implantation with an ion dose less than the critical ion dose is expected to produce point defects such as vacancies and interstitials, and complex defects such as divacancies, four-vacancies and vacancy–impurity complexes. Fundamental data on ion-induced defects in II–VI materials, for example the atomic structure of defects, are few. Therefore, physical aspects of lattice defects in radiation-induced defects in II–VI materials are discussed according to the results of electron-induced defects in ZnSe. Figure 15.5 shows a schematic diagram of the isochronal annealing properties of 1.5 MeV electron-irradiated ZnSe crystal at 20.4 K. According to the results of Watkins [6], the annealing stage of several radiation-induced defects are summarized as follows:

1. at 60–90 K, annealing of the Frenkel pair of Zn atoms;
2. at 90–130 K, annealing of the Frenkel pair of Se atoms;
3. at 420 K, annealing of Zn vacancy (V_{Zn}^-);
4. at 600–700 K, annealing of complex defects such as $V_{Zn}^-Cl^+$, V_{Zn}^-S and V_{Zn}^-Te.

From the results of electron irradiation of ZnSe, it is expected that high-temperature annealing at more than 700 K is needed to anneal out complex defects in II–VI materials induced by ion implantation. In fact, non-radiative recombination centres (ion-induced defects) in CdS induced by

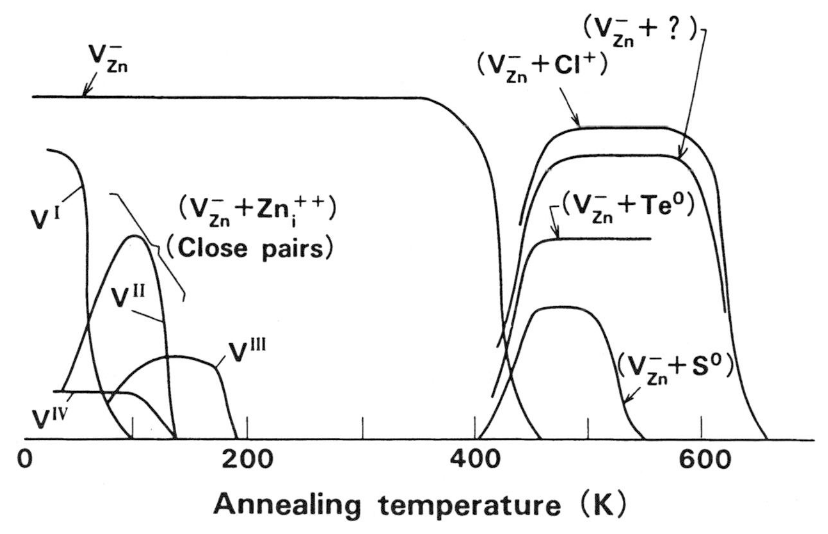

Fig. 15.5 Schematic diagram of the Zn vacancy annealing stage in ZnSe [6].

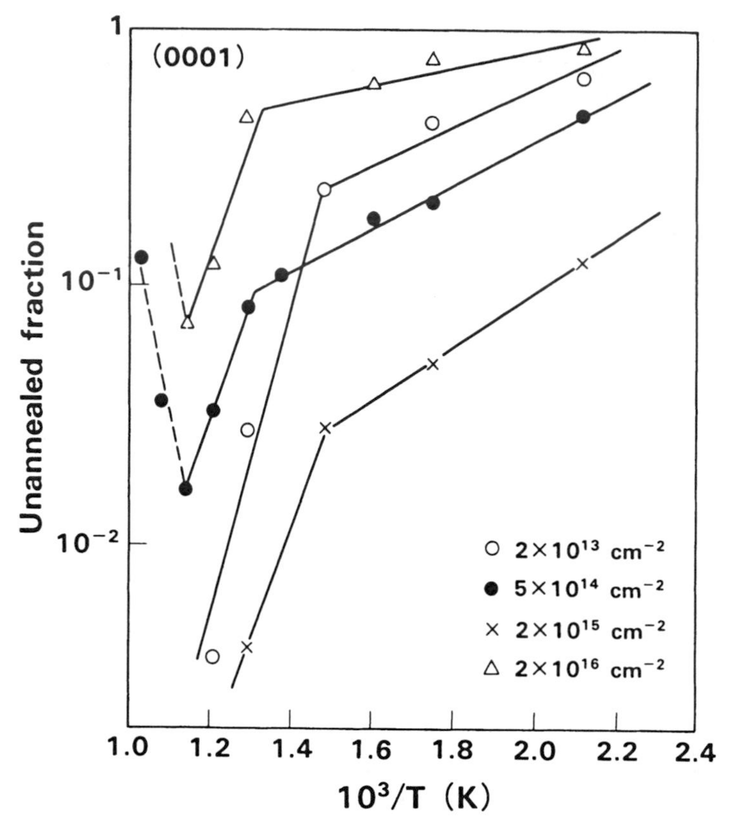

Fig. 15.6 Isochronal annealing curves (10 min) of N-implanted CdS with various doses [5].

nitrogen ion implantation anneal out at around 800 K as shown in Fig. 15.6 [5]. Figure 15.6 shows the fraction not annealed, $f = (1/I_a - 1/I_0)/(1/I - 1/I_0)$, where I_0, I and I_a represent the bandedge green emission intensity before and after ion implantation and after annealing respectively. Two recovery stages are observed; the first stage with an activation energy of 0.07–0.2 eV may be caused by recovery of mobile point defects, and the second stage of energy 0.95 eV may be caused by the recovery of complex defects. Hovever, the green emission intensity is decreased above 600 °C because of thermal decomposition of CdS.

Figure 15.7 shows the annealing temperature dependence of the absolute intensities of various photoluminescence peaks (bandedge, free-to-acceptor (FA) and self-activated (SA) emissions) in Na-implanted ZnSe, measured at 77 K [7]. It can be seen that the crystalline quality recovers completely at around 650 °C while the formation rate of acceptor levels effectively saturates at around 550 °C.

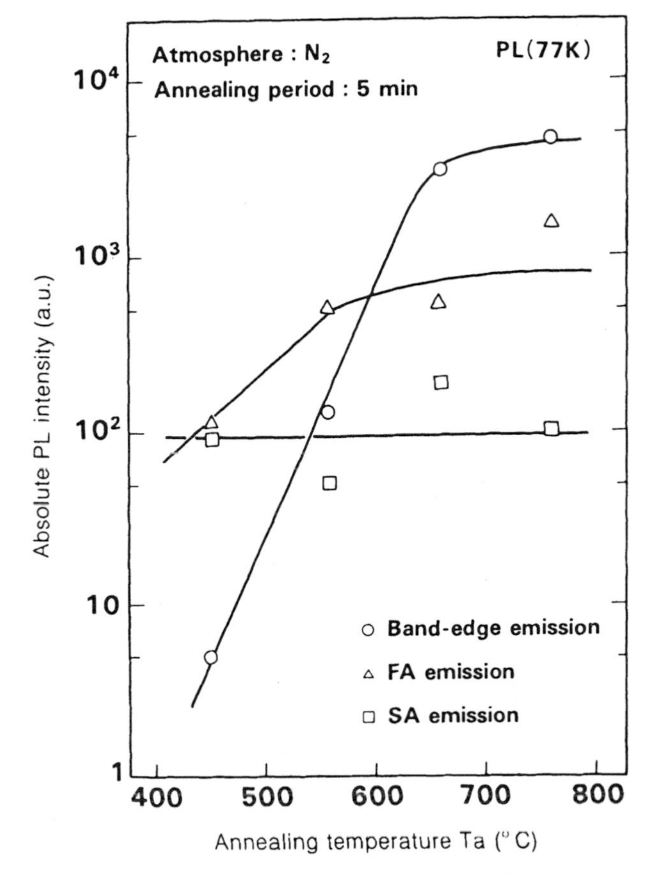

Fig. 15.7 Annealing temperature dependence of the absolute intensities of various photoluminescence peaks (bandedge, free-to-acceptor (FA) and self-activated (SA) emission) in Na-implanted ZnSe, measured at 77 K [7].

15.1.3 Luminescence centres due to implanted ions

(a) Defect centres

Table 15.2 shows energy levels and physical properties of vacancies in widegap II–VI materials [8]. Ion implantation into widegap II–VI materials is thought to produce luminescence centres related to deep levels that are due to complex defects as well as vacancies. Some deep centres act as recombination centres and thus reduce bandedge and near-bandedge emissions as shown in Figs 15.6 and 15.7. Therefore, annealing of ion-induced defects in II–VI materials is needed to remove recombination centres and to activate impurities doped by ion implantation or crystal growth.

Table15.2 Energy levels and physical properties of vacancies in widegap II–VI compound materials [8]

Material	Vacancy	Energy level (eV)	Emission wavelength (nm)	Annealing temperature T_a (K)	Migration energy (eV)
ZnS	V_{Zn}	$E_v + 1.1$	570	330	1.04
	V_S	$E_c - 0.17$	340	673	3
ZnSe	V_{Zn}	$E_v + 1.1$	720	373	1.26
	V_{Se}	$E_c - 0.7$	980	150	0.7
ZnTe	V_{Zn}	$E_v + 1.7$	720	340	0.78
	V_{Te}			280	0.73
CdS	V_{Cd}	$E_v + 0.8$		300	
	V_S				
CdTe	V_{Cd}	$E_c + 0.6$	1000	353	0.8
	V_{Te}	$E_c - 0.54$	1100	120	0.2

(b) Shallow acceptor centres

P-type impurity doping by ion implantation is one of the important approaches for realizing p–n junction formation in widegap II–VI compound materials. Until now, ion implantation of acceptor impurities such as Li [9, 10]. N [2, 11], Na [7], P [1] and so on into ZnSe and ZnS materials has been studied.

Table 15.3 shows shallow emission centres (acceptor levels) in ZnSe introduced by ion implantation with various ion species. Formation of shallow acceptor levels in ZnSe due to implantation with Li, N and Na and annealing between 400 and 500 °C have been observed at emission centres of bound excition (BE), FA and donor–acceptor (DA) pairs by photoluminescence studies as shown in Fig. 15.8 [7]. However, annealing

Table 15.3 Shallow emission centres in ZnSe introduced by ion implantation

Implanted ion	BE (eV)	FA (eV)	DA (eV)	T (K)	Implantation conditions	Annealing conditions	Reference
Li		2.695		77	50 keV 10^{14} cm^{-2}	757 °C	[10]
N	2.7803	2.702		77	260 keV, 7×10^{12} cm^{-2}	500 °C	[11]
Na	2.7917–2.7928		2.68	4.2	50 keV, 10^{14} cm^{-2}	451–557 °C	[7]

BE, bound exciton emission; FA, free-to-acceptor emission; DA, donor–acceptor pair emission.

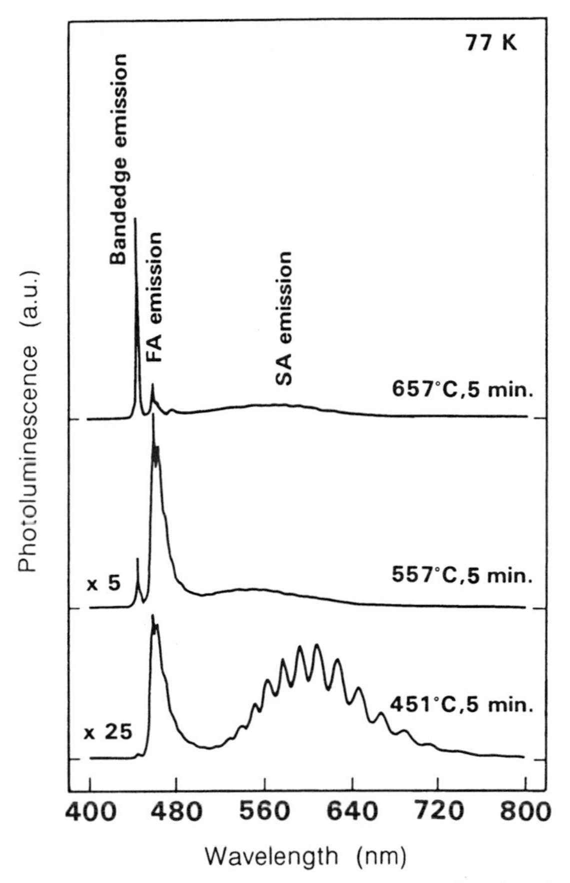

Fig. 15.8 Dependence on annealing temperature of photoluminescence of Na-implanted ZnSe measured at 77 K [7].

with a temperature higher than 550 °C causes saturation of the acceptor level formation rate and generates deep levels [7]. Acceptor binding energies of Li, N and Na in ZnSe, estimated from the measured energies of FA and DA emissions shown in Table 15.4, are 110 meV, 103 meV and 100 meV respectively.

(c) Shallow donor centres

Ion implantation of Al, Ga and In into ZnS and ZnSe has been studied [12, 13]. Formation of shallow donor centres has been observed using photoluminescence and resistivity measurements and the donor binding energy, estimated from the BE emission, for Al- and Ga-implanted ZnSe is 28 meV which accords with 26.3 meV (Al) and 27.9 meV (Ga) obtained by Merz *et al.* [14].

Table 15.4 Visible spectral lines and corresponding transitions of rare-earth-implanted ZnS light-emitting diodes

	Wavelength (nm)	Energy (cm^{-1})	Transition
ZnS:Nd^{3+}	540.0	18520	$^4G_{7/2}-^4I_{9/2}$
	588.0	17000	$^2G_{7/2}-^4I_{9/2}$
	603.0	16580	$^4G_{5/2}-^4I_{9/2}$
	663.0	15080	$^2H_{11/2}-^4I_{9/2}$
	680.0	14700	$^4F_{9/2}-^4I_{9/2}$
ZnS:Er^{3+}	384.5	26000	$^4G_{11/2}-^4I_{15/2}$
	410.5	24360	$^2H_{9/2}-^4I_{15/2}$
	532.0	18800	$^2H_{11/2}-^4I_{15/2}$
	550.0	18180	$^4S_{3/2}-^4I_{15/2}$
	663.5	15070	$^4F_{9/2}-^4I_{15/2}$
ZnS:Sm^{3+}	541.0	18480	$^4G_{7/2}-^6H_{7/2}$[a]
			$^4F_{3/2}-^6H_{5/2}$[a]
	568.0	17610	$^4G_{5/2}-^6H_{5/2}$
	603.0	16580	$^4G_{5/2}-^6H_{7/2}$
	649.0	15410	$^4G_{5/2}-^6H_{9/2}$
ZnS:Tm^{3+}	483.5	20680	$^1G_4-^3H_6$
	527.0	18980	$^1D_2-^3H_5$
	653.0	15310	$^1G_4-^3H_4$
	700.0	14290	$^3F_3-^3H_6$
	805.0	12420	$^1G_4-^3H_5$
ZnS:Tb^{3+}	438.0	22830	$^5D_3-^7F_4$
	491.0	20370	$^5D_4-^7F_6$
	547.0	18280	$^5D_4-^7F_5$
	587.0	17040	$^5D_4-^7F_4$
	621.0	16100	$^5D_4-^7F_3$
	658.0	15200	$^5D_4-^7F_2$

[a]Proposed transitions.

Low-resistivity ZnS crystal with a resistivity of less than 0.1 Ω cm has been obtained by In (indium)-implantation and annealing under Zn vapour pressure conditions [12].

(d) Rare earth impurity doping

Ion implantation of rare earth elements into widegap II–VI materials has the possibility of allowing different colours of electroluminescence emissions and selective doping into a substrate. Electroluminescence of various colours from rare-earth-implanted ZnS has been observed as shown in Fig. 15.9 [15].

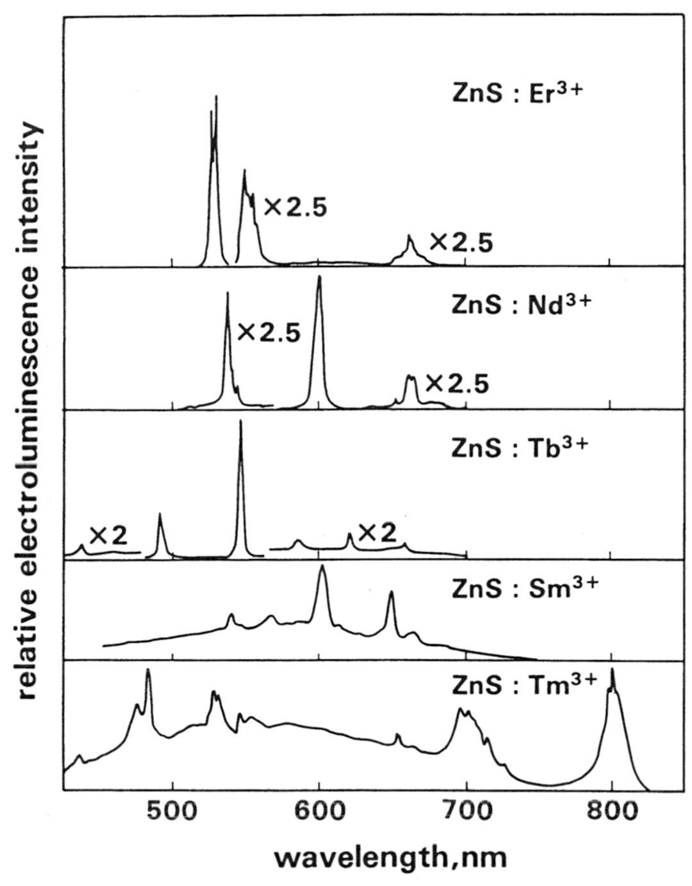

Fig. 15.9 Electroluminescence spectra of rare-earth-implanted ZnS diodes at 293 K [15].

Ion implantation with an ion energy of 18 keV and fluence range between 10^{15} and $9 \times 10^{15}\,\text{cm}^{-2}$ has been carried out and electroluminescence from reverse-biased Schottky ZnS diodes has been observed. The visible emission spectral lines and the corresponding transitions of the rare-earth-implanted ZnS light-emitting diodes are listed in Table 15.4. In particular in terms of both brightness and power efficiency, the best electroluminescence has been obtained in the ZnS:Nd^{3+} and ZnS:Er^{3+} diodes.

(e) Disordering by ion implantation

Selective disordering of superlattices is thought to be an attractive technique for fabricating three-dimensional optical waveguides and non-linear optical devices. Disordering of ZnSe/ZnS strained-layer superlattices (SLSs) with 10 periods of 140 Å ZnSe–140 Å ZnS, induced by Si ion implantation (100 keV,

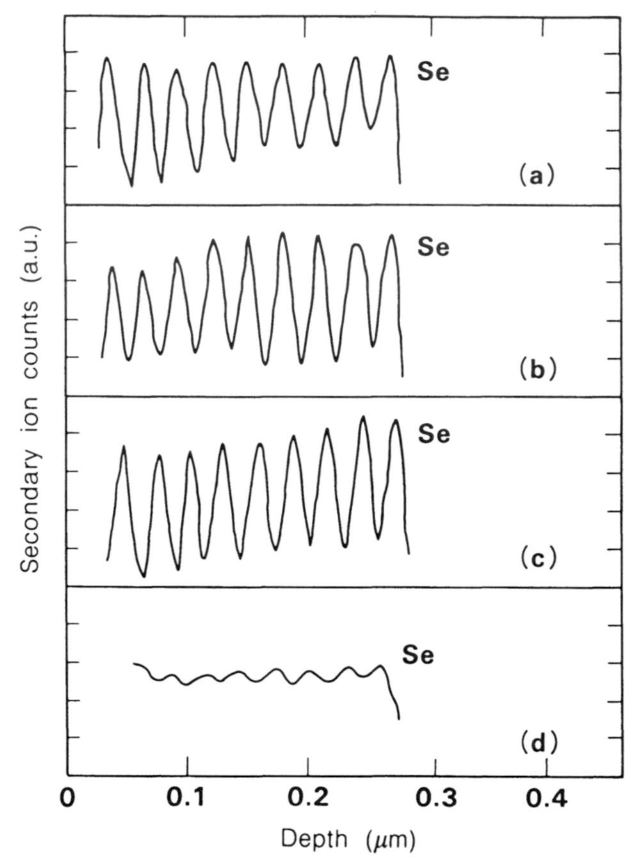

Fig. 15.10 Depth profiles of Se obtained by secondary ion mass spectrometry analysis (a) after metal–organic chemical vapour deposition growth. (b) after Si ion implantation (100 keV, $10^{16}\ cm^{-2}$), (c) after thermal annealing (450° C, 3 h) without ion implantation, and (d) after ion implantation and subsequent annealing (450 °C, 3 h). The superlattices are composed of ten periods of ZnSe (140 Å) and ZnS (140 Å) [16].

$10^{16}\ cm^{-2}$) and subsequent low-temperature thermal annealing at 450 °C for 3 h has been confirmed, as shown in Fig. 15.10 [16]. The disordering is mainly induced by the diffusion of defects generated by the ion implantation at the early stage of low-temperature annealing.

15.1.4. Junction formation and device application using ion implantation

(a) p–n or p–i–n junction formation

Widegap II–VI materials such as ZnS and ZnSe are promising materials for blue electroluminescence devices. The main problem has been how to inject

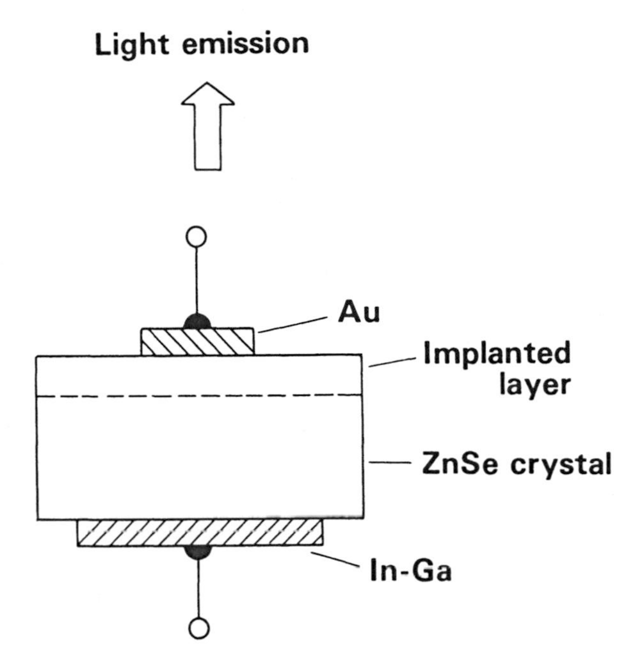

Fig. 15.11 Schematic diagram of ion-implanted ZnSe diode structure [2].

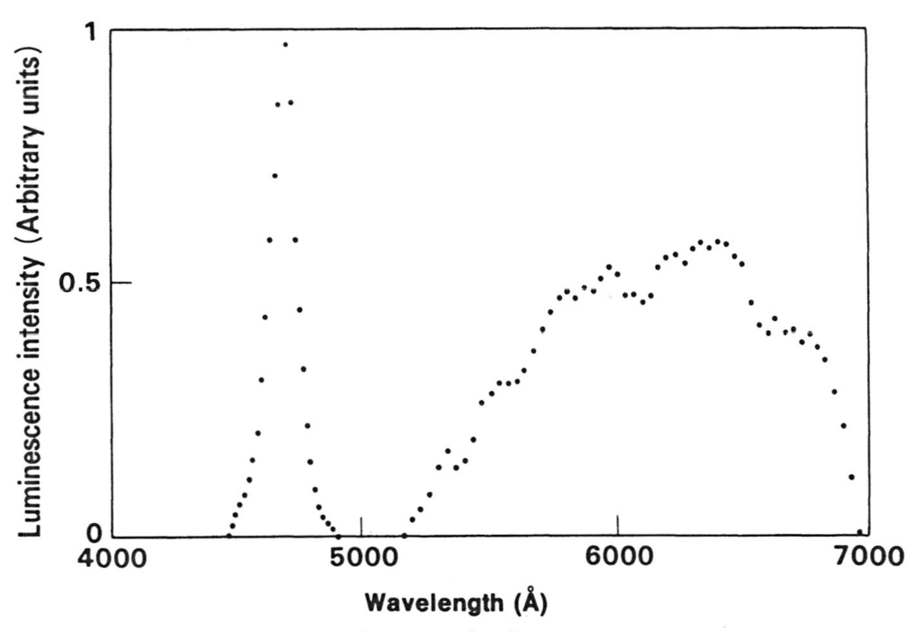

Fig. 15.12 Electroluminescence spectra from an N-ion-implanted ZnSe diode (annealed at 300 °C) in forward bias at room temperature [2].

the minority carrier into II–VI materials, and p–n junction formation due to ion implantation is expected to realize efficient minority carrier injection. P–n junction formation by ion implantation of dopants such as Li [9], N [2, 11] and P [1] ions into ZnSe has been carried out.

Blue electroluminescence has been observed from nitrogen-implanted ZnSe diodes in forward bias at room temperature by Yamaguchi *et al.* [2] Figure 15.11 shows the structure of the ion-implanted ZnSe diode. 50 keV nitrogen with an ion dose of $2 \times 10^{15}\,cm^{-2}$ was implanted into a ZnSe(1 1 0) substrate at room temperature. Annealing was performed at 300–400 °C for 10 min in an N_2 atmosphere. The nitrogen-implanted ZnSe diode is thought to have a p–i–n structure from *C–V* measurements. Figure 15.12 shows the electroluminescence spectrum at room temperature. The diode exhibits blue emission peak wavelength of 4670 Å and a half-wdith of 130 Å and the SA and Cu-R emissions at about 5950 and 6300 Å respectively. The blue emission is due to band-to-band and/or donor-to-free emissions. The orange and yellow bands are caused by the impurity copper and the defect centre produced by ion implantation. Figure 15.13. shows the *I–V* and *I–L* (luminescence intensity of the blue emission) characteristics for the diode.

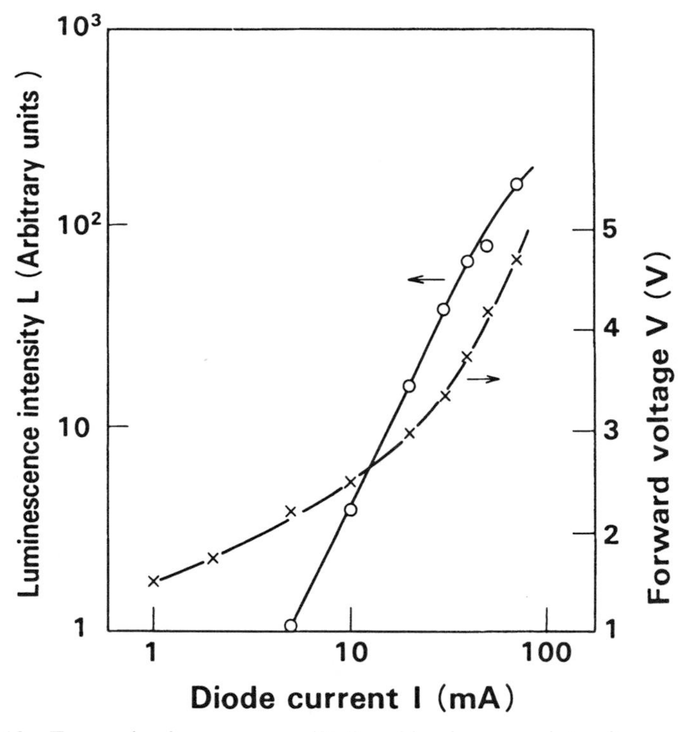

Fig. 15.13 Forward voltage–current (*V–I*) and luminescence intensity–current (*L–I*) curves for the N-implanted ZnSe diode (annealed at 300 °C) [2].

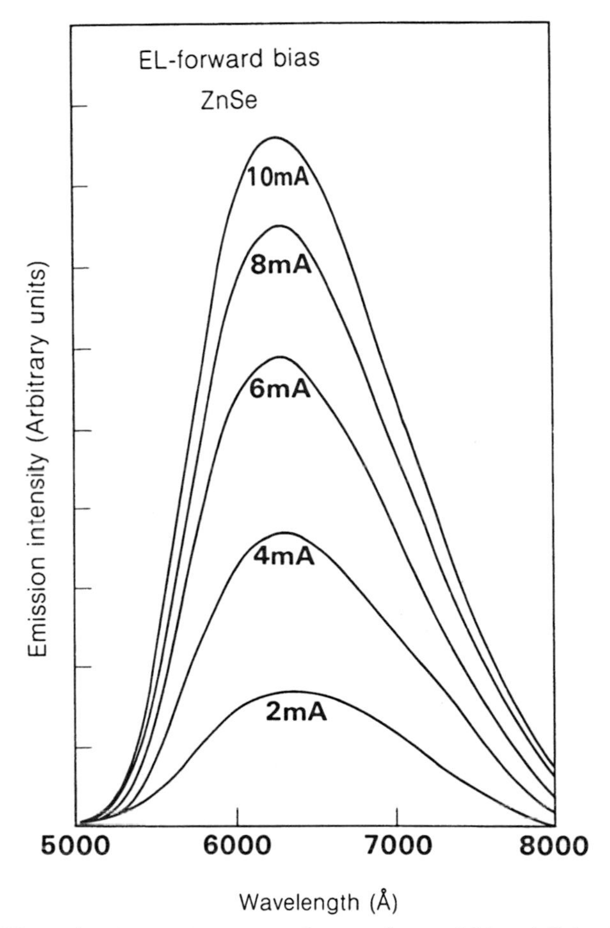

Fig. 15.14 Electroluminescence spectra from a forward-biased P-implanted ZnSe diode [1].

The electroluminescence intensity increases proportionally to the square of the forward current. The external quantum efficiency was about 10^{-3} % at an input current of 30 mA and a forward voltage of 4.7 V.

Injection electroluminescence from Li- or P-ion-implanted ZnSe diodes has also been observed by Park *et al.* [1, 9]. Figure 15.14 shows electroluminescence spectra from a forward-biased P-implanted ZnSe diode. The P ion implantation was carried out at room temperature at 400 keV. The total ion dose was about 10^{14} cm^{-2}. Implanted ZnSe samples were annealed at 450 °C for 5 min in an argon atmosphere to remove the damage induced by the room-temperature implantation. In this case, the diodes exhibit a red electroluminescence spectrum similar to the photoluminescence spectrum with a peak at 6300 Å giving an external quantum effficiency of 0.01%.

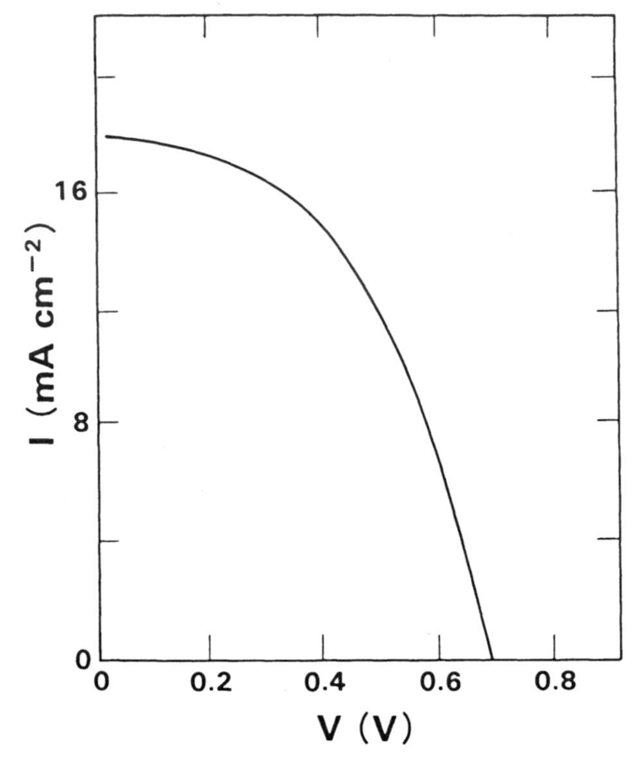

Fig. 15.15 Solar cell properties of the P-ion-implanted CdTe p–n junction diode [17].

II–VI compound materials such as CdS and CdTe are also useful materials for solar cells. P–n junction formation from CdTe by P ion implantation and its solar cell properties have been reported [17]. Phosphorus ions with an energy of 60 keV and a dose from 3×10^{15} to $3 \times 10^{16}\,cm^{-2}$ were implanted into n-CdTe with a carrier concentration from 2×10^{15} to $5 \times 10^{17}\,cm^{-3}$. Thermal annealing of ion-induced defects in CdTe was carried out in Cd vapour at temperatures of 753–943 K during a period of 0.5–3 h. The higher ion dose and annealing temperature resulted in an increase of open-circuit voltage V_{oc}. Figure 15.15 shows the solar cell properties of the CdTe p–n junction diode fabricated by P implantation under an incident illumination power of $85\,mW\,cm^{-2}$. A conversion efficiency of about 7% was obtained.

(b) Metal–insulator–semiconductor

The formation of MIS structures in ZnS and ZnSe by the implantation of light ions such as H^+ and He^+ was examined. N-implanted CdS also showed the MIS structure and high photoconductive gain owing to avalanche injection in the implanted–insulating layer [18].

15.1.5 Summary

In order to obtain high-performance opto-electronic devices such as highly efficient blue electroluminescence structures from ZnSe and ZnS by ion implantation, it is necessary to improve the photoluminescence properties of the ZnSe and ZnS crystals before implantation and to reduce ion-induced defects. Residual impurities such as Cu and Ag that act as deep centres should be reduced and the concentration of donor impurities should also be controlled in ZnSe and ZnS. For realizing p–n junction formation in ZnSe and ZnS by ion implantation, group I or V elements are needed to occupy a substitutional site on Zn, S or Se respectively. Substitutional group I or V atoms on Zn, S or Se sites act as acceptors and interstitial impurities act as donors. Most impurities can only create a low concentration of acceptor levels (10^{16}–10^{17} cm^{-3}). Figure 15.16 shows covalent radii of the group I and V atoms compared with those of Zn and Se. N or P impurities with a smaller covalent radius than that of Se, S or Zn may be easily substituted in Se, S or Zn respectively. On the other hand, the sizes of other impurities such as Li, Na, As or Sb are larger than those of Zn, Se or S. This means that incorporation of those impurities in ZnSe and ZnS requires a considerable elastic strain energy and may be more difficult to substitute compared with N and P. Ion species, implantation and annealing conditions should also be optimized.

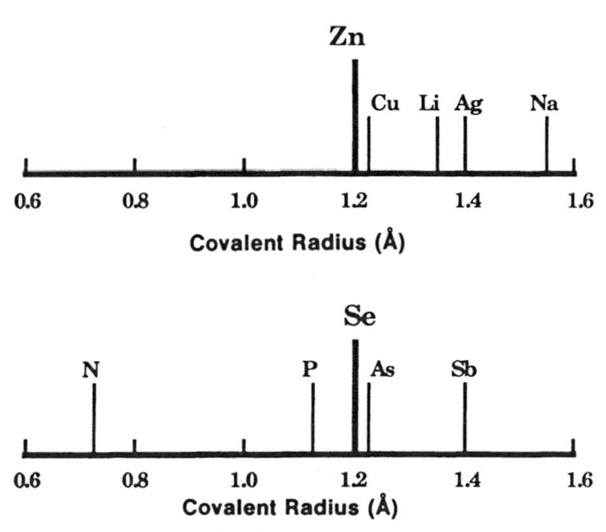

Fig. 15.16 Covalent radii of the group I and V atoms and of Zn and Se.

15.2 ELECTRON-BEAM-PUMPED DEVICES

A semiconductor laser pumped by an electron beam can be applied in optical scanning microscopy, photolithography and optical telemetry. Some research on electron beam pumped lasers has been carried out, mainly by researchers in the USSR. However, no such lasers are yet available.

The emission of ZnSe and ZnS at low temperatures and low excitation is due to processes related to excitons, impurities and defects. At medium excitation ($kW\,cm^{-2}$) at low temperatures, band emission starts dominating owing to higher-order processes coming from excitonic molecules and scattering. At higher excitation densities ($MW\,cm^{-2}$), emission is mostly in the form of strong DA pair bands. The variation in emission properties of ZnSe crystals with different impurities and properties has been studied [19] under high electron beam excitation at 300 K. As shown in Fig. 15.17, epitaxial ZnSe crystals showed a strong emission intensity. Samples of 10–30 μm thickness with silver mirrors showed laser emission with a lasing

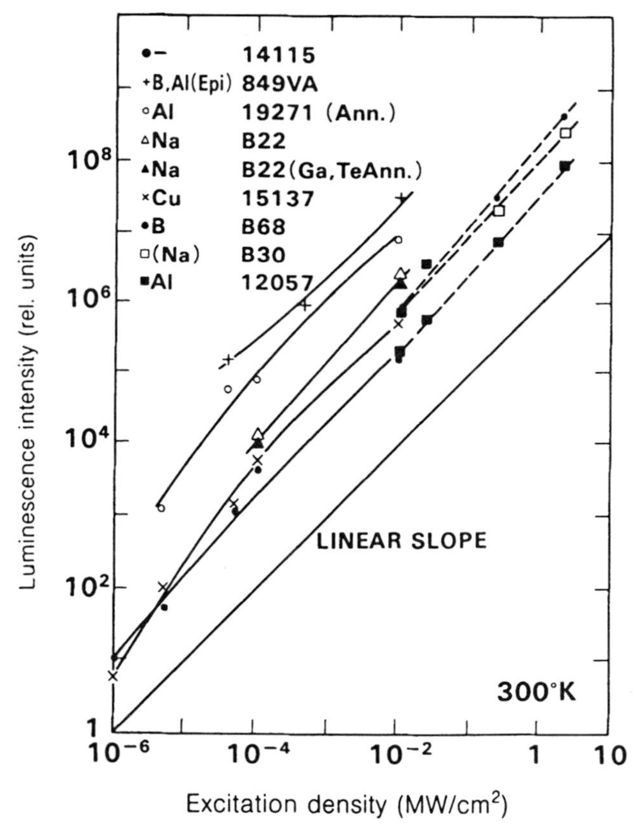

Fig. 15.17 Cathodoluminescence (CL) intensity variations for different ZnSe materials, measured at 300 K [19].

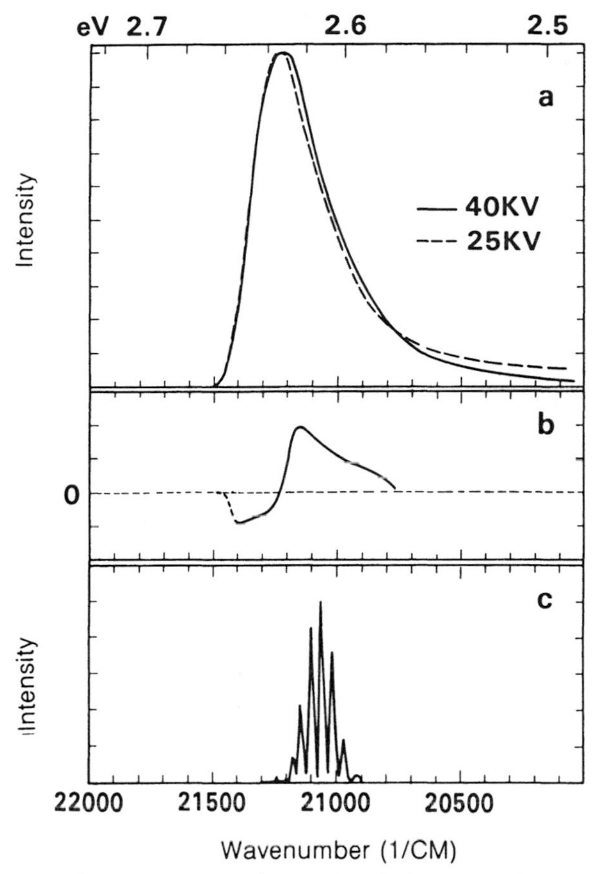

Fig. 15.18 (a) CL from ZnSe at 25 kV and 40 kV electron beam excitation, (b) difference between CL bands and (c) laser output spectrum from ZnSe excited by a 40 kV electron beam, at 300 K [19].

threshold of around $50\,\mathrm{A\,cm^{-2}}$ and emission energy of 2.615 eV, at 300 K with a 40 keV electron beam, as shown in Fig. 15.18(c).

A ZnS semiconductor laser pumped by an electron beam also showed lasing with a maximum power of 3 kW and wavelength of 345 nm in the pulsed regime at 300 K [20]. ZnS crystals annealed in liquid Zn showed a higher luminescence intensity than As-prepared ZnS.

REFERENCES

1. Park, Y.S. and Shin, B.K. (1974) *J. Appl. Phys.*, **45**, 1444.
2. Yamaguchi, M., Yamamoto, A. and Kondo, M. (1977) *J. Appl. Phys.*, **48**. 196.
3. Wilson, R.G. and Brewer, G.R. (1973) *Ion Beams*, New York, p. 367.
4. Bryant, F.J. and Hamid, S.A. (1969) *Phys. Rev. Lett.*, **23**, 304.

5. Yamaguchi, M. (1976) *Jpn. J. Appl. Phys.*, **15**, 1675.
6. Watkins, G.D. (1976) *Radiation Effects in Semiconductors, 1976, Inst. Phys. Conf. Ser.*, **31**, 95.
7. Yodo, T. and Yamashita, K. (1989) *Appl. Phys. Lett.*, **54**, 1778.
8. Taguchi, T. (1985) *Oyo Buturi*, **54**, 28.
9. Park, Y.S. and Chung, C.H. (1971) *Appl. Phys. Lett.*, **18**, 99.
10. Yodo, T. and Yamashita, K. (1988) *Appl. Phys. Lett.*, **53**, 2403.
11. Akimoto, K., Miyajima, T. and Mori, Y. (1989) *Jpn. J. Appl. Phys.* **28**, L528.
12. Bryant, F.J. (1980) *Extended Abstracts of the 2nd British Association for Crystal Growth II–VI Meeting, Lancaster, September 13, 1980.*
13. Rabago-Bernal, F., Heuryel, A., Triboulet, R., Legros, R. and Marfaing, Y. (1981) *Defects and Radiation Effects in Semiconductors 1980, Inst. Phys. Conf. Ser.*, **59**, 383.
14. Merz, J.L., Kukimoto, H., Nassau, K. and Shiever, J.W. (1972) *Phys. Rev. B*, **6**, 545.
15. Zhong, G.Z. and Bryant, F.J. (1982) *IEE Proc.*, **129** (Part 1), 85.
16. Saitoh, T., Yokogawa, T. and Narusawa, T. (1989) *Appl. Phys. Lett.*, **55**, 735.
17. Simashkevich, A.V., Gorchak, L.V., Gilan, E.V. and Sushkevich, K.D. (1989) *Phys. Status Solidi A* , **112**, 305.
18. Yamaguchi, M. (1976) *Jpn. J. Appl. Phys.*, **15**, 723.
19. Colak, S., Bhargava, R.N., Fitzpatrick, B.J. and Sicignano, A. (1984) *J. Lumin.*, **31–32**, 430.
20. Kozlovskii, V.I., Korostelin, Yu, V., Nasibov, A.S., Skasyrskii, Ya.K. and Shapkin, P.V. (1984) *Sov. J. Quantum Electron.*, **14**, 420.

Index

Printed in the United States
118224LV00002B/29/A

9 780412 391002